INTRODUCTION TO THE
THEORY OF NUMBERS

INTRODUCTION TO THE THEORY OF NUMBERS

HAROLD N. SHAPIRO
Courant Institute of Mathematical Sciences
New York University

A Wiley-Interscience Publication

JOHN WILEY & SONS

New York   Chichester   Brisbane   Toronto   Singapore

Library of Congress Cataloging in Publication Data:

Shapiro, Harold N.
    Introduction to the theory of numbers.

    (Pure and applied mathematics, ISSN 0079-8185)
    "A Wiley-Interscience publication."
    Includes bibliographies and indexes.
    1. Numbers, Theory of. I. Title. II. Series: Pure
    and applied mathematics (John Wiley & Sons))

QA241.S445   1982   512'.7   82-10929
ISBN 0-471-86737-3

Printed in the United States of America

10 9 8 7 6 5 4 3 2

To the memory of my parents

Julia and Will

PREFACE

As with much of science, the theory of numbers has experienced an extremely rapid growth since the turn of this century. Once a large collection of scattered fragments, it has evolved into a collection of rather extensive and sophisticated subdisciplines. This alone would indicate a need for some reappraisal of what constitutes an "introduction to the theory of numbers." In fact, it would seem appropriate to have many introductions to the different aspects of number theory integrated with some of the atmosphere of related subjects. It is the avowed objective of this text to begin such a development without sacrificing the very basic problem-oriented nature of the subject. Finally, the desire to carry this out is tempered by the necessity of avoiding an unpleasantly encyclopedic work, and still to reserve the prerogative to wander into an occasional narrow alley, strictly on the basis of taste and enthusiasm.

Chapters 1 through 5 contain the very basic material of number theory. The techniques presented are illustrated throughout by special problems, some of which are deeply rooted in the history of number theory.

The material of Chapters 6 through 10 carries the development forward at what might be called an "intermediate level." At this level, Chapters 7 and 8 present evolutions from the notion of congruence. Chapter 8 also brings together a variety of applications within the framework of a general philosophy of how to approach counting problems. Chapter 9 provides much of the beginnings of prime number theory, which are then further developed and applied in Chapter 10.

The choice of material herein presents only a small beginning. However, it does provide some view of the wide variety of techniques that relate to the theory of numbers. The exercises have been presented so that insofar as possible they relate to the section that precedes them. In addition, there are two "do-it-yourself" chapters that may be thought of as larger problem sequences. Their purpose is to enable the reader to simulate the experience of carrying out the successive steps of a small mathematical investigation, which, in turn, is based on previously presented material.

Wherever possible, a given main section is followed by sections with the same number and suffixes A, B, etc., in which specific applications are discussed (e.g., 5.5A, 5.5B). Also, historical and other notes, together with a related bibliography, have been placed at the end of each chapter.

The methodology of the book is completely elementary in that the theory of functions of a complex variable is not used. In its entirety, the content of this book is appropriate for a graduate course. However, selected chapters, sections, and exercises would constitute a feasible undergraduate course.

I am grateful to the many individuals who have provided assistance in the preparation of this volume. Mary Joan Collison of Mulloy College prepared the initial draft of all references, and carried out the bibliographical searches. John Brillhart gave much early encouragement and many suggestions. To him we owe the almost complete extermination of the semicolon. Gerson Sparer read and corrected the final version of the manuscript. The late Paul Brock taught from an early version of the text (at the University of Vermont) and shared his experience.

Needless to say, all errors and blunders contained herein are the sole responsibility of the author.

New York, New York   HAROLD N. SHAPIRO
November 1982

CONTENTS

INTRODUCTION TO THE
THEORY OF NUMBERS

1

DIVISIBILITY AND OTHER BEGINNINGS

§1.1 THE RING OF INTEGERS

The subject matter of number theory is bound together by its common focus on problems and properties that relate to the integers: $0, \pm 1, \pm 2, \ldots$. The integers may be subjected to the familiar operations of addition and multiplication, written as $a + b$ and ab respectively. Relative to these, the integers constitute a structure known as a ring.

The ring properties possessed by the integers are easily verified, and are the following:

(i) the sum and product of two integers is an integer;

(ii) multiplication and addition are commutative, that is,

$$ab = ba, \qquad a + b = b + a;$$

(iii) $a + 0 = a, \qquad a \cdot 1 = a;$

(vi) $a + (-a) = 0;$

(v) Distributive Law: $(a + b)c = ac + bc;$

(vi) Associative Laws:

$$a(bc) = (ab)c, \qquad (a + b) + c = a + (b + c).$$

The ring of integers will be denoted by Z. Because of the commutativity of multiplication, that is, $ab = ba$, rings such as Z are sometimes referred to as *commutative rings*. The ring Z possesses a further property:

Cancellation Law: $ab = ac$ and $a \neq 0$ imply $b = c$.

This additional property qualifies Z structurally to be called an *integral domain*.

Presuming, as we are, that we are working within the framework of the real numbers, the integers generate the rational numbers as follows.

1

Definition 1.1.1.   A *rational number* is a real number that equals the quotient of two integers. A real number that is not rational is called *irrational*.

Thus for example, $\frac{1}{3}$, .24, $\frac{9}{11}$ are rational, whereas, as we shall soon see, the real number $\sqrt{2}$ is irrational.

The rational numbers, which we denote by Q, may also be subjected to addition and multiplication and clearly form a ring with respect to these operations. However, in addition to the conditions (i)–(vi) listed above, Q satisfies a multiplicative analogue of (iv); namely, for $a \neq 0$

$$a \cdot \left(\frac{1}{a}\right) = 1.$$

As a result of this, the rational numbers constitute an example of a structure called a *field*. Clearly, the ring of integers Z is *not* a field.

Since Z is not a field, the simple equation

$$rx = s$$

with $r, s \in Z$, need not have a solution for x in Z. Considering this fact, one is naturally led to the following definition.

Definition 1.1.2.   An *integer r divides an integer s* if there exists an integer t such that

$$rt = s.$$

In this case we also say that s *is a multiple of r* and call r a *divisor of s*.

Utilizing this terminology we have another definition.

Definition 1.1.3.   An integer that is a multiple of 2 is called *even*; one which is not is called *odd*.

Thus the even integers are

$$0, \pm 2, \pm 4, \pm 6, \ldots$$

and the odd integers are

$$\pm 1, \pm 3, \pm 5, \ldots.$$

Hence generically, we may write *2n* for an even number, and *2n + 1* for an odd number.

EXERCISES

1.   Verify that the rational numbers satisfy the conditions (i)–(vi).

2.   Show that the rational numbers form an integral domain.

3. Prove that the product of two odd numbers is odd and the product of two even numbers is even.

4. Prove that if $m^2 = m \cdot m$ is even, then m must be even.

5. Prove that the sum of two even or two odd integers is even, whereas the sum of an even and an odd integer is odd.

6. Show that the square of an odd integer is one more than a multiple of 4.

7. A *polynomial* $f(x)$ with integer coefficients is an expression of the form $f(x) = a_n x^n + a_{n-1} x^{n-1} + \ldots + a_0$, $a_i \in Z$. If $a_n \neq 0$, $f(x)$ is said to be of *degree n*. Denoting the set of all such polynomials by $Z[x]$, show that $Z[x]$, is a ring with respect to a natural definition of addition and multiplication of polynomials. Show also that $Z[x]$ is an integral domain. Do the same for $Q[x]$ = the set of polynomials with rational coefficients.

8. Let $Q(x)$ denote the set of expressions of the form $f(x)/g(x)$, $g(x) \neq 0$, where $f(x)$, $g(x) \in Z[x]$. Show that $Q(x)$ is a field. Show that if we consider expressions $f(x)/g(x)$ where f and g are polynomials with rational coefficients (i.e., $f(x)$ and $g(x)$ are in $Q[x]$), we obtain the same set $Q(x)$. Note the inclusions

$$Q(x) \supset Q[x] \supset Z[x].$$

Utilizing the assertion of Ex. 4 above, we can now provide a proof of the following theorem.

Theorem 1.1.1. $\sqrt{2}$ is irrational.

Proof. Assume the theorem false, that is, that $\sqrt{2}$ is rational. Then $\sqrt{2}$ has a representation as a quotient of two integers

$$\sqrt{2} = \frac{r}{s},$$

where $r > 0$, $s > 0$. Further, of all such possible representations, we choose one for which s is a minimum. We then have

$$2s^2 = r^2,$$

so by Ex. 4 above, since r^2 is even, it follows that r is even. Thus we may write $r = 2u$, which implies

$$2s^2 = 4u^2 \quad \text{or} \quad s^2 = 2u^2.$$

Again, since s^2 is even, s must be even, and we write $s = 2v$. But then

$$\sqrt{2} = \frac{r}{s} = \frac{2u}{2v} = \frac{u}{v}$$

provides a representation of $\sqrt{2}$ as a quotient of two positive integers, with denominator $v < s$. This contradicts the choice of s as the minimum such denominator, and the theorem follows.

§1.2 THE DIVISION ALGORITHM

In connection with the notion of divisibility, the first important tool is a result called the *division algorithm*, which provides a canonical description of the process of dividing any integer by a positive integer.

DIVISION ALGORITHM. Given any two integers $a, b, b > 0$, there exists a unique representation of the form

$$(1.2.1) \qquad\qquad a = bq + r, \qquad 0 \le r < b,$$

where $q, r \in Z$. (q is called the quotient, r the remainder).

    Proof. For z any real number, let $[z]$ denote the largest integer $\le z$. Given $a, b \in Z, b > 0$, define q by

$$q = \left[\frac{a}{b} \right]$$

and set

$$r = a - bq.$$

Since

$$\frac{a}{b} - 1 < \left[\frac{a}{b} \right] = q \le \frac{a}{b}$$

$$a - b < bq \le a$$

or combining both inequalities

$$0 \le a - bq = r < b.$$

    This establishes the existence of the representation (1.2.1). For uniqueness, suppose we have another

$$(1.2.2) \qquad\qquad a = bq' + r', \qquad 0 \le r' < b.$$

Subtracting (1.2.2) and (1.2.1), we obtain

$$0 = b(q - q') + (r - r').$$

This implies that b divides $r - r'$, and since both r and r' lie between 0 and b, this yields $r = r'$, $q = q'$.

EXERCISES

1. Let $f(x), g(x) \in Q[x]$, so that

$$f(x) = a_n x^n + \cdots + a_0, \qquad a_n \neq 0$$
$$g(x) = b_m x^m + \cdots + b_0, \qquad b_m \neq 0$$

where the a_i and b_i are all rational numbers. If $n \geq m$, show that the degree of $h(x) = f(x) - (a_n/b_m)x^{n-m}g(x)$ is less than n.

2. The division algorithm has an analogue in $Q[x]$ that asserts: given two polynomials $f(x)$, $g(x)$, $g(x)$ not identically 0, there exist $q(x), r(x) \in Q[x]$ such that

(*) $$f(x) = g(x)q(x) + r(x),$$

and either $r(x) \equiv 0$, or

(**) $$0 \leq \text{degree of } r(x) < \text{degree of } g(x).$$

Prove this. One route is to proceed by the following steps:

a. If $f(x) \equiv 0$ or the degree of $f(x)$ is 0, [i.e., $f(x)$ is constant] prove the assertion.

b. Proceed by induction on the degree of $f(x)$ and assume results for all cases where $f(x)$ is of degree $< n$. Let $f(x)$ be of degree n.

Subcase (1). degree $g(x) >$ degree $f(x)$
Prove directly.

Subcase (2). degree $f(x) \geq g(x)$
Form polynomial $h(x)$ of Ex. 1 and use inductive hypothesis.

3. Prove the uniqueness of the representation (*) provided by Ex. 2.

4. Prove that the assertion of Ex. 2 can be carried out for $f(x), g(x), q(x), r(x)$ in $Z[x]$, provided we assume the coefficient of its highest power of x in $g(x)$ is 1 (i.e., $b_m = 1$).

§1.3 SUBSETS OF Z

Many of the problems of number theory are concerned with distinguishing, counting, or relating various subsets of Z. In fact, there have already been signs of this in the preceding sections. At various points, the positive integers $1, 2, 3, \ldots$, were singled out. Also noted were the familiar subsets consisting of the even integers and the odd integers. Other equally familiar subsets are

(i) *squares*.   an integer is a square if it equals

$$m \cdot m = m^2$$

(ii) *cubes*.   an integer is a cube if it equals

$$m \cdot m \cdot m = m^3,$$

and more generally,

(iii) *nth powers*.   an integer is an nth power if and only if it equals

$$\underbrace{m \cdot m \cdot \ldots \cdot m}_{n \text{ factors}} = m^n.$$

There are many other important subsets, the study of which is basic to the subject. Among these we have the following.

Definition 1.3.1.   An integer n, $n \neq \pm 1$, is called a *prime* if it is divisible only by ± 1, and $\pm n$. An integer other than ± 1 that is not a prime is called *composite*.

It follows that a positive integer greater than one is a prime if and only if its only positive divisors are itself and one. The first few positive primes are easily seen to be 2, 3, 5, 7, 11, 13, 17, ... Note at this point that we are turning our focus essentially on the *positive* integers. Those negative integers n less than -1 that are divisible only by $\pm n$ and ± 1 are easily seen to be of the form $-p$, where p is a positive prime.

Definition 1.3.2.   A positive integer n is called *squarefree* if 1 is the only square that divides n.

Clearly, all primes are squarefree. However, there are composite numbers that are squarefree, for example, 6, 10, 14, ...

A very simple type of subset of Z may be defined by considering all the multiples of a given integer s. We will denote this subset by M_s. The subset M_s has certain obvious properties.

(1.3.1) $m, n \in M_s$ implies $m - n \in M_s$;

(1.3.2) $t \in Z, \quad n \in M_s$ implies $tn \in M_s$.

From these we induce the following definition.

Definition 1.3.3.   A nonempty subset $\mathscr{T}$ of Z is called an *ideal* of Z if

(1.3.3) $$t_1, t_2 \in \mathscr{T} \qquad \text{implies } t_1 - t_2 \in \mathscr{T};$$

(1.3.4) $$m \in Z, \quad t \in \mathscr{T} \qquad \text{implies } mt \in \mathscr{T}.$$

Utilizing this terminology we have, in particular, that for any nonnegative integer, s, M_s is an ideal of Z. Clearly, M_0 consists only of the integer 0, and $M_1 = Z$. The natural question to ask next is: are there other ideals of Z besides the M_s? The answer is no, as we show in the following.

Theorem 1.3.1. Let $\mathscr{T}$ be an ideal of Z. Then there exists a nonnegative integer s such that $\mathscr{T} = M_s$.

Proof. If $\mathscr{T}$ contains only 0, we have that $\mathscr{T} = M_0$. Otherwise $\mathscr{T}$ contains a nonzero integer t. If $t < 0$, then $\mathscr{T}$ contains $(-1)t = -t > 0$, so that $\mathscr{T}$ contains positive integers. Let $s =$ the smallest positive integer in $\mathscr{T}$. This will be the s we are looking for, that is, $\mathscr{T} = M_s$. To demonstrate this, it suffices to show that s divides all integers in $\mathscr{T}$. Let $t \in \mathscr{T}$, $t \ne 0$, and apply the division algorithm to obtain

$$t = sq + r, \qquad 0 \le r < s.$$

Since $r = t - q \cdot s \in \mathscr{T}$, $r > 0$ would contradict the choice of s as smallest positive integer of $\mathscr{T}$. Thus $r = 0$, $t = sq$, and s divides t.

An ideal such as M_s, consisting of the multiples of a single element, is called a *principal ideal*. Thus the above theorem asserts that all ideals of Z are principal.

There is a somewhat different result, of the same genre as Theorem 1.3.1, that is frequently applicable in number theoretic arguments.

Theorem 1.3.2. Let $\mathscr{S}$ be a nonempty subset of the integers that is "closed under the division algorithm." That is, given integers a, b in $\mathscr{S}$, $a \ge b > 0$, the assertion $a = bq + r$ of the division algorithm can be made with $r \in \mathscr{S}$, $(0 \le r < b)$. Then, if $\mathscr{S}$ contains positive integers, they are all divisible by the smallest positive integer of $\mathscr{S}$.

The proof of the above is exactly the same as that of Theorem 1.3.1 and is left as an exercise. In practice, the fact that $\mathscr{S}$ is "closed under the division algorithm" will often follow from the following conditions.

(i) $m \ge n$, $m, n \in \mathscr{S}$ implies $m - n \in \mathscr{S}$
(ii) $m \in Z$, $m \ge 0$, $n \in \mathscr{S}$ implies $mn \in \mathscr{S}$.

These conditions are similar (but not identical) to those defining an ideal.

The subsets of the above discussion were motivated by considerations of exact divisibility by a given integer. Other subsets of importance are singled out by considering those integers that yield a given remainder when divided by a given positive integer. More precisely, for fixed integers k, l, $k > 0$, we have the set of integers

$$km + l, \qquad m = 1, 2, 3, \ldots .$$

Such a sequence of integers is referred to as an *arithmetic progression*. We note that the exact divisibility case corresponds to the remainder $l = 0$.

EXERCISES

1. Given two polynomials $f, g \in Q[x]$, g *divides* f (or f is a multiple of g) if and only if there exists an $h \in Q[x]$ such that $f = gh$. Extending the notions of this section to $Q[x]$, let M_f be the subset of $Q[x]$ consisting of the multiples of f. Show that M_f is an ideal of $Q[x]$.

2. Prove that every ideal of $Q[x]$ is of the form M_f for some $f \in Q[x]$.

3. Prove that if a nonempty subset $\mathcal{S}$ of Z satisfies (1.3.3), then it follows that it satisfies (1.3.4).

§1.4 GREATEST COMMON DIVISOR AND LEAST COMMON MULTIPLE

Two important functions of *pairs* of integers are known as the "greatest common divisor" (g.c.d.) and "least common multiple" (l.c.m.). These are defined as follows.

Definition 1.4.1. Let a, b be two integers, not both zero. The *greatest common divisor* $d = (a, b)$ of the two integers a, b, is an integer d such that

(1.4.1) 　　　　　d divides both a and b (i.e., it is a common divisor)

and

(1.4.2) 　　　　　d is the largest integer satisfying (1.4.1).

Definition 1.4.2. The *least common multiple* $e = \{a, b\}$ of two nonzero integers a, b is an integer $e > 0$ such that

(1.4.3) 　　a divides e and b divides e (i.e., e is a common multiple of a and b)

and

(1.4.4) 　　　　　e is the smallest positive integer satisfying (1.4.3).

The nature of the above definitions is such that both the g.c.d. and l.c.m. are easily seen to exist and to be uniquely defined. It is a consequence of the definition of g.c.d. that it is positive, whereas for l.c.m., the positivity is a part of the definition. There are, however, certain basic properties of g.c.d. and l.c.m. that are obscured by this definition. We proceed to derive these.

Theorem 1.4.1. If c is a common multiple of $a \neq 0$ and $b \neq 0$, then

$$e = \{a, b\} \quad \text{divides } c.$$

Proof. Consider the set $\mathcal{T}$ of all common multiples of a and b. It is easily seen that (1.3.3) and (1.3.4) are satisfied, (i.e., $\mathcal{T}$ is an ideal of Z). Hence from Theorem 1.3.1, $\mathcal{T}$ is a principal ideal M_s, where s is the smallest positive integer of $\mathcal{T}$. But this implies that $s = e$, so that e divides all common multiples of a and b.

Theorem 1.4.2. If a and b are not both 0, and c is a common divisor of a and b, then c divides $d = (a, b)$.

Proof. Without loss of generality, we may assume that $c > 0$. Consider $f = \{c, d\}$, so that the desired result "c divides d" is equivalent to $f = d$. Suppose that the theorem is false, in which case we must have $f > d$. Since a is a common multiple of c and d, by Theorem 1.4.1, a is a multiple of $f = \{c, d\}$. Similarly, b is a multiple of f. Thus f is a common divisor of both a and b which is greater than d and this contradicts the definition of d as the *greatest* common divisor.

With the above theorems as a base, one can proceed to develop nearly all the properties of g.c.d. and l.c.m. Inasmuch as most of these follow easily from results of Chapter 2, we proceed here only with those that play some immediate role. In the statement of these, and throughout the remainder of this section, all integers are assumed to be positive. We have

(1.4.5) $$(ca, cb) = c(a, b)$$

(1.4.6) $$\{ca, cb\} = c\{a, b\}$$

(1.4.7) $$(a, uv) = (a, (a, u)v).$$

Proof of (1.4.5). Since $c(a, b)$ divides both ca and cb, Theorem 1.4.2 yields that

(1.4.8) $$c(a, b) \quad \text{divides} \quad (ca, cb).$$

Note, in particular, that this in turn implies that c divides (ca, cb). Hence, since (ca, cb) divides ca, it follows that $(ca, cb)/c$ is an integer that divides a. Similarly, $(ca, cb)/c$ divides b, so that again by Theorem 1.4.2,

$$\frac{(ca,\ cb)}{c}\ \text{divides}\ (a,\ b)$$

or

(1.4.9) $\qquad\qquad\qquad\qquad (ca,\ cb)\quad\text{divides}\quad c(a,\ b).$

From (1.4.8) and (1.4.9) the desired result, (1.4.5), follows.

Proof of (1.4.6). $\{ca,\ cb\}/c$ is an integer that is a multiple of a and b, and hence by Theorem 1.4.1, it is a multiple of $\{a,\ b\}$. Therefore

(1.4.10) $\qquad\qquad\qquad \{ca,\ cb\}\quad\text{is a multiple of}\quad c\{a,\ b\}.$

Further, $c\{a,\ b\}$ is a multiple of ca and cb so that

(1.4.11) $\qquad\qquad\qquad c\{a,\ b\}\quad\text{is a multiple of}\quad \{ca,\ cb\}.$

From (1.4.10) and (1.4.11), we obtain (1.4.6).

Proof of (1.4.7). Since $(a,\ (a,\ u)v)$ divides both a and uv, we have that

(1.4.12) $\qquad\qquad\qquad (a,\ (a,\ u)v)\quad\text{divides}\ (a,\ uv).$

On the other hand,

$$(a,\ uv)\quad\text{divides}\ a\ \text{which, in turn, divides}\ av$$

and

$$(a,\ uv)\ \text{divides}\ uv,$$

so that

$$(a,\ uv)\quad\text{divides}\ (av,\ uv) = (a,\ u)v$$

[the last equality via (1.4.5)]. Thus since $(a,\ uv)$ divides $(a,\ u)v$ and a, we obtain

(1.4.13) $\qquad\qquad\qquad (a,\ uv)\quad\text{divides}\quad (a,\ (a,\ u)v).$

From (1.4.12) and (1.4.13), (1.4.7) follows.

The simple identity (1.4.7) has various consequences of interest.

Theorem 1.4.3. If a divides bc, and $(a,\ b) = 1$, then a divides c.

Proof.

$$a = (a,\ bc) \qquad \text{since } a \text{ divides } bc,$$
$$= (a,\ (a,\ b)c) \qquad \text{by (1.4.7),}$$
$$= (a,\ c) \qquad \text{since } (a,\ b) = 1,$$

which gives that a divides c.

Theorem 1.4.4. If a and b divide c and $(a,\ b) = 1$, then ab divides c.

Proof.

$$a = (a,\ c) = \left(a,\ \frac{c}{b} \cdot b\right)$$
$$= \left(a,\ (a,\ b)\frac{c}{b}\right) \qquad \text{by (1.4.7)}$$
$$= \left(a,\ \frac{c}{b}\right)$$

which yields that a divides c/b, or ab divides c. More directly, since b divides c, $c = b \cdot (c/b)$, and by Theorem 1.4.3 $(a,\ b) = 1$ and a divides c implies that a divides (c/b), or ab divides c.

Corollary 1. If $(a, b) = 1$, then $\{a, b\} = ab$.

Proof.   Since ab is a common multiple of a and b, Theorem 1.4.1 provides that $\{a, b\}$ divides ab. On the other hand, since $(a, b) = 1$, Theorem 1.4.4 asserts that ab divides $\{a, b\}$ and hence the corollary.

Corollary 2. If d divides ab, where $(a, b) = 1$, then

$$d = (d, a)(d, b).$$

Proof.   Since $(a, b) = 1$ implies $((d, a), (d, b)) = 1$, it follows from Theorem 1.4.4 that

$$(d, a)(d, b) \text{ divides } d.$$

Further, since d divides ab, using (1.4.7)

$$d = (d, ab) = (d, (d, a)b) = (d, (d, a)(d, b)),$$

so that

$$d \text{ divides } (d, a)(d, b)$$

and the corollary follows.

Theorem 1.4.5.

(1.4.14) $\{a, b\}(a, b) = ab.$

   Proof.   Letting $d = (a, b)$, (1.4.5) yields that

$$\left(\frac{a}{d}, \frac{b}{d}\right) = 1.$$

Then, by Corollary 1 of Theorem 1.4.4, we have

$$\left\{\frac{a}{d}, \frac{b}{d}\right\} = \frac{a}{d} \cdot \frac{b}{d}.$$

Multiplying both sides by d^2,

$$d^2\left\{\frac{a}{d}, \frac{b}{d}\right\} = ab$$

and applying (1.4.6),

$$d\{a, b\} = ab,$$

which is precisely (1.4.4).

   Among the many other properties of the greatest common divisor that stem from (1.4.7) is

(1.4.15) $(a, b)^2 = (a^2, b^2).$

   Proof of (1.4.15).   Using (1.4.7) repeatedly, we have

$$(a^2, b^2) = (a^2, (a^2, b)b) = (a^2, ((a, b)a, b)b)$$
$$= (a^2, ((a, b)^2, b)b),$$

So that if $(a, b) = 1$, then

$$(a^2, b^2) = (a^2, b) = ((a, b)^2, b) = 1.$$

Since $(a/(a, b), b/(a, b)) = 1$ we have

$$1 = \left(\frac{a^2}{(a, b)^2}, \frac{b^2}{(a, b)^2}\right) = \frac{(a^2, b^2)}{(a, b)^2},$$

which gives (1.4.15).

From this, in turn, we obtain the following simple theorem which is a useful source of certain applications.

Theorem 1.4.6. If $ab = s^2$ where $a > 0$, $b > 0$, and $(a, b) = 1$; then

$$a = (a, s)^2 \quad \text{and} \quad b = (b, s)^2.$$

Proof. Since $1 = (a, b)$, by (1.4.5),

$$a = a(a, b) = (a^2, ab) = (a^2, s^2).$$

Then from (1.4.15),

$$a = (a, s)^2.$$

Similarly,

$$b = b(a, b) = (ab, b^2) = (s^2, b^2) = (s, b)^2.$$

EXERCISES

1. Prove that (a, bc) divides (a, b) (a, c).

2. If $(a, b) = 1$, the pair of integers a, b are said to be *relatively prime*. The integers $a_1, \ldots, a_k$ are said to be *relatively prime by pairs* if every pair a_i, a_j, $i \neq j$ is relatively prime. Let $a_1, \ldots, a_k$ be relatively prime by pairs and such that a_i divides m, $i = 1, \ldots, k$. Prove that the product $a_1, \ldots, a_k$ divides m.

3. Prove that two distinct integers of the form $a^{2^m} + 1$, $a^{2^n} + 1$ are relatively prime if a is even and have g.c.d. 2 if a is odd.

4. For positive integers a, b such that $(a, b) = 1$, and r, s such that r divides s, determine

$$(a^r + b^r, a^s + b^s).$$

5. Show that for every integer $k \geq 1$,

$$(a, b)^k = (a^k, b^k).$$

6. Prove that if $ab = s^k$, k an integer ≥ 2, and $(a, b) = 1$, then $a = (a, s)^k$ and $b = (b, s)^k$.

7. Prove that if a^k divides b^k, then a divides b.

8. Prove that $(a, b) = (a + mb, b)$ for all $m \in Z$.

9. For k even, show that $3^{k+1} \pm 1$ is not divisible by 5. From this deduce that, in this case, $13^{k+1} \pm 1$ is not divisible by 5.

10. For k odd, show that $13^{k+1} - 1$ is divisible by 7.

11. Prove that the sum of two reduced fractions $(a/b) + (c/d)$ cannot be an integer unless $b = d$.

12. If $\alpha, \beta, \gamma, \delta$ are four integers such that

$$\Delta = \alpha\delta - \beta\gamma = \pm 1,$$

show that the g.c.d. of

$$m = \alpha a + \beta b$$

$$n = \gamma a + \delta b$$

is the same as that of a and b. What can be asserted if only $\Delta \neq 0$ is given?

13. If $(m, n) = (a, b)$, show that there exists a transformation such as described in Ex. 12, with $\Delta = 1$. (The material of §1.5 will help.)

14. Generalize the definitions of g.c.d. and l.c.m. to the greatest common divisor of n integers $a_1, \ldots, a_n$, denoted by $(a_1, \ldots, a_n)$ and the least common multiple of $a_1, \ldots, a_n$, denoted by $\{a_1, \ldots, a_n\}$. Prove the corresponding generalization of Theorem 1.4.1 and Theorem 1.4.2.

15. Prove each of the following (all integers assumed positive).

    (a) $(ca_1, \ldots, ca_n) = c(a_1, \ldots, a_n)$

    (b) $\{ca_1, \ldots, ca_n\} = c\{a_1, \ldots, a_n\}$

16. Prove that

    (a) $(a, b, c) = (a, (b,c)),$

    (b) $\{a, b, c\} = \{a, \{b,c\}\}.$

17. Let $a_1, \ldots, a_k$ be any positive integers. Show that

$$(\{a_1, \ldots, a_{k-1}\}, a_k) = \{(a_1, a_k), \ldots, (a_{k-1}, a_k)\}.$$

§1.5 AN ALTERNATE APPROACH TO GREATEST COMMON DIVISOR

As we've seen in Definition 1.4.1, the greatest common divisor may be introduced in an explicit way, so that the existence and uniqueness are obvious from the outset. The important property asserted in Theorem 1.4.1 must then be derived from the definition. The following is another approach to the development of the notion of greatest common divisor, one in which the essence of Theorem 1.4.1 is taken as the definition.

Definition of (a, b), an Alternate Approach.   The greatest common divisor e of two integers a, b, not both zero, is a positive integer such that

(1.5.1) e divides a, and e divides b;

(1.5.2)   any integer d that divides both a and b must, as a consequence, divide e. However, if this is taken as starting point, it is not clear that (a, b) exists. This then would be a theorem.

Theorem 1.5.1.   The greatest common divisor of two integers a, b, not both zero, exists and is unique.

Proof.   Consider the set $\mathcal{T} = \{ma + nb, m, n \in Z\}$; that is, $\mathcal{T}$ is the set of all linear combinations of a and b, with integer coefficients. It is easily verified that $\mathcal{T}$ is an ideal of Z, which contains nonzero integers. Thus $\mathcal{T}$ consists of the multiples of a single positive integer $e \in \mathcal{T}$. We claim that e is the sought for greatest common divisor.

Note first that since $e \in \mathcal{T}$ we have for some $m, n \in Z$

$$e = ma + nb.$$

Thus if d divides both a and b, d must divide e. Finally, since both a and b are in $\mathcal{T}$, e is a common divisor of a and b.

From the above proof we extract the following.

Corollary.   There exist integers x, y such that

(1.5.3) $(a, b) = xa + yb.$

The property of (a, b) provided by the above corollary is the source of alternate proofs of many of the results given in §1.4. As illustration, consider Theorem 1.4.3.

Since $(a, b) = 1$, we have from (1.5.3)

$$1 = xa + yb$$

for some x and y. Multiplying through by c we obtain

(1.5.4) $c = xac + ybc.$

Then, since a divides bc, it divides the right side of (1.5.4), and therefore a divides c, the desired conclusion.

In connection with (1.5.3), the question arises as to how to go about calculating the coefficients x, y in this representation; or what is closely related, how can one efficiently compute (a, b)? Both of these questions are answered by a procedure known as the *Euclidean Algorithm*. This method is based on the division algorithm, and is as follows: assuming $a \geq b$, we can write

$$a = bq_1 + a_1, \qquad 0 \leq a_1 < b,$$

and by Ex. 8 of §1.4,

$$(1.5.5) \qquad (a, b) = (a - bq_1, b) = (a_1, b).$$

Thus we see that (1.5.5) reduces the problem of computing the g.c.d. of a and b to that of computing the g.c.d. of the smaller (than a) number a_1 and b. Continuing this process, if

$$b = a_1 q_2 + a_2, \qquad 0 \le a_2 < a_1,$$

we have

$$(a, b) = (a_1, b) = (a_2, a_1).$$

In general, the ensuing steps appear as

$$(1.5.6) \qquad \begin{cases} a_1 = a_2 q_3 + a_3, & 0 \le a_3 < a_2 \\ \quad \cdot \ \cdot \ \cdot \ \cdot \ \cdot \\ a_{i-3} = a_{i-2} q_{i-1} + a_{i-1}, & 0 \le a_{i-1} < a_{i-2} \\ a_{i-2} = a_{i-1} q_i + a_i, & 0 \le a_i < a_{i-1}, \end{cases}$$

where

$$(1.5.7) \qquad (a, b) = (a_1, b) = (a_2, a_1) = (a_3, a_2) = \cdots = (a_i, a_{i-1}).$$

Since $a > a_1 > a_2 > \ldots > a_{i-1} > a_i \ge 0$, there must exist a first $i = i^*$ where $a_{i^*} = 0$, and the process stops. But then from (1.5.7),

$$(a, b) = (a_{i^*}, a_{i^*-1}) = (0, a_{i^*-1}) = a_{i^*-1},$$

so that the last nonzero remainder produced by this process is, in fact, the g.c.d. of a and b.

The same steps of the Euclidean Algorithm, given above, provide a means for computing the representation (1.5.3). Since $a_{i^*-1} = (a, b)$, the equation $a_{i^*-3} = a_{i^*-2} q_{i^*-1} + a_{i^*-1}$ yields

$$(1.5.8) \qquad (a, b) = a_{i^*-3} - a_{i^*-2} q_{i^*-1}.$$

From the next previous equation we have

$$a_{i^*-2} = a_{i^*-4} - a_{i^*-3} q_{i^*-2},$$

and inserting this in (1.5.8) yields a relation of the form

$$(a, b) = x_{i^*-3} a_{i^*-3} + y_{i^*-3} a_{i^*-4}.$$

Continuing to unravel the equations (1.5.6) ultimately yields an expression for (a, b) as a linear combination of a and b.

EXERCISES

1. Generalize the Euclidean Algorithm so as to provide a method for computing the greatest common divisor of any n integers $a_1, a_2, \ldots, a_n$ (not all zero).

2. Apply the algorithm developed as answer to Ex. 1 so as to provide a method for calculating integers x_i, $i = 1, \ldots, n$, such that

$$x_1 a_1 + x_2 a_2 + \cdots + x_n a_n = (a_1, \ldots, a_n).$$

3. Given any three integers a, b, c, consider the equation

$$ax + by = c.$$

Develop a criterion for whether this equation has integer solutions for x and y. Show that one can effectively determine *all* the integer solutions of this equation.

4. For $f(x)$, $g(x) \in Q[x]$ (not both identically 0), define a greatest common divisor of $f(x)$ and $g(x)$, denoted by $(f(x), g(x))$, to be a polynomial $d(x)$, such that
    (a) $d(x)$ divides both $f(x)$ and $g(x)$, and
    (b) if $h(x)$ divides both $f(x)$ and $g(x)$, then $h(x)$ divides $d(x)$.

    Prove the following:
    (a) Such a greatest common divisor $d(x) \in Q[x]$ exists.
    (b) $d(x)$ is unique apart from multiplication by a rational number; therefore, it is uniquely determined if we require the coefficient of the highest power of x in $d(x)$ to be 1.
    (c) There exist $a(x)$, $b(x) \in Q[x]$ such that

$$a(x)f(x) + b(x)g(x) = d(x).$$

5. If $f(x) \in Q[x]$, then for

$$f(x) = a_n x^n + a_{n-1} x^{n-1} + \cdots + a_o$$

    define

$$f'(x) = n a_n x^{n-1} + (n-1)a_{n-1} x^{n-2} + \cdots + a_1.$$

    Prove the assertion that $f(x)$ has no multiple roots {the complex number θ is a multiple root if $f(x) = (x - \theta)^2 g(x)$, where $g(x)$ is a polynomial not necessarily in $Q[x]$} is equivalent to $(f(x), f'(x)) = 1$.

6. The analogue of a prime in $Q[x]$ is a polynomial $f(x)$ such that $f(x)$ has no nonconstant divisors in $Q[x]$ of degree smaller than the degree of $f(x)$. Such an $f(x)$ is called *irreducible* (with respect to $Q[x]$). Prove that an irreducible polynomial has no multiple roots.

§1.6 PYTHAGOREAN TRIPLETS

It is interesting to pause at this point and make an application of the notion of greatest common divisor as developed in §1.5. The application relates to the determination of the so-called "Pythagorean Triplets," so familiar in elementary geometry. They are met there, for example, as the 3, 4, 5 and 5, 12, 13 right triangles. In numerical terms, they represents positive integer solutions x, y, and z in the equation

$$(1.6.1) \qquad\qquad x^2 + y^2 = z^2.$$

This type of equation, in which one seeks only integer solutions, is an instance of what are called diophantine equations.

Our objective will be to determine all of the positive integer solutions x, y, z to the equation (1.6.1). To begin, we note that if the triplet x, y, z is a solution, and t is any positive integer, then tx, ty, tz is also a solution. Conversely, a solution tx, ty, tz with the common factor t provides the smaller solution x, y, z. Thus it suffices to determine all solutions x, y, z in which there is no common factor of the three, other than one. Further, since (1.6.1) implies that

$$(x^2, y^2) = (x, y)^2 \text{ divides } z^2,$$

we see that (Ex. 7 of §1.4)

$$(x, y) \text{ divides } z.$$

Similarly,

$$(x, z) \text{ divides } y,$$

$$(y, z) \text{ divides } x,$$

for any solution x, y, z. Thus the assumption that the three integers x, y, z have no common factor is [for solutions of (1.6.1)] equivalent to

$$(1.6.2) \qquad\qquad (x, y) = (x, z) = (y, z) = 1.$$

Solutions x, y, z to (1.6.1) that satisfy (1.6.2) will be called *primitive*. The above discussion shows that it suffices to determine all *primitive* solutions in order to obtain all the solutions of (1.6.1).

If both x and y were even, 2 would divide $(x, y) = 1$. Since this is impossible, at least one of them is odd. We shall designate this one as x. Then x^2 is also odd, as is

$$(1.6.3) \qquad\qquad (z - y)(z + y) = z^2 - y^2 = x^2.$$

As a result, both $z - y$ and $z + y$ are odd. Then, if d is a common divisor of $z - y$ and $z + y$, it follows that

$$d \text{ divides } (z + y) + (z - y) = 2z$$

and

$$d \text{ divides } (z + y) - (z - y) = 2y,$$

whence

$$d \text{ divides } (2y, 2z) = 2(y, z) = 2.$$

However, since $z - y$ is odd, d is odd and as a positive odd divisor of two, we must have $d = 1$. Since 1 is the only positive common divisor of $z - y$ and $z + y$,

(1.6.4) $$(z + y, z - y) = 1.$$

Applying Theorem 1.4.6 with $a = z - y$, $b = z + y$, $s = x$, we conclude that $z + y$ and $z - y$ must be odd squares, that is,

(1.6.5)
$$\begin{cases} z + y = u^2, & z - y = v^2 \\ u, v \text{ odd}, & u > v > 0 \\ (u, v) = 1. \end{cases}$$

From (1.6.5) it follows that

(1.6.6) $$z = \frac{u^2 + v^2}{2}, \qquad y = \frac{u^2 - v^2}{2},$$

and consequently from (1.6.3)

(1.6.7) $$x = uv.$$

Conversely, if we start with x, y, z as given by (1.6.6) and (1.6.7), for odd integers u, v such that $u > v > 0$ and $(u, v) = 1$, then it is easily verified that x, y, z is a primitive solution. That we have a solution follows from

$$\left(\frac{u^2 - v^2}{2} \right) + u^2 v^2 = \left(\frac{u^2 + y^2}{2} \right)^2,$$

where the fact that u and v are odd insures that y and z are integers. Finally, that the solution is primitive will follow if we can show that $(x, y) = 1$. This, in turn, will be a consequence of $(u, v) = 1$. To see this, note first that

$$(u, u^2 - v^2) = (u, v^2) \text{ divides } (u^2, v^2) = (u, v)^2 = 1$$
$$(v, u^2 - v^2) = (v, u^2) \text{ divides } (v^2, u^2) = 1.$$

Then from this we obtain via two applications of (1.4.7),

$$(uv, u^2 - v^2) = (u^2 - v^2, (u, u^2 - v^2)(v, u^2 - v^2)) = 1.$$

Summarizing the above we have the following theorem.

Theorem 1.6.1. All positive primitive integer solutions to (1.6.1) are given by

$$x = uv, \qquad y = \frac{u^2 - v^2}{2}, \qquad z = \frac{u^2 + v^2}{2}$$

where u, v are odd integers such that $u > v > 0$, and $(u, v) = 1$ together with those solutions obtained by interchanging x and y.

EXERCISES

1. Show that for all primitive solutions of (1.6.1), z is odd.

2. From elementary geometry, we recall the primitive solution to (1.6.1) given by $x = 3$, $y = 4$, $z = 5$. Find other solutions in which x and y are consecutive integers.

3. Recalling the 5, 12, 13 right triangle, determine all solutions of (1.6.1) in which y and z are consecutive integers.

4. Show that the only solution of (1.6.1) in which x, y, z are three consecutive positive integers is $x = 3$, $y = 4$, $z = 5$ (or $x = 4$, $y = 3$, $z = 5$).

5. Show that there is no right triangle with integer sides whose perimeter equals a power of a prime.

6. Show that the only two non-negative consecutive integers whose product is a square is 0, 1.

7. For which integers n does the equation

$$x^2 - y^2 = n$$

have integer solutions x, y? What can be said about the number of such solutions?

8. Let p be a fixed prime, and n a fixed integer greater than 1. Prove that the equation

$$x^n - 1 = p^y$$

has at most a finite number of integer solutions for x, y.

9. Show that the only solution to the equation

$$x^u - 1 = v^y$$

in positive integers x, y, u, v, such that $u > 1$, $x > 2$, v prime, is $x = 3$, $u = 2$, $v = 2$, $y = 3$.

10. For $f(x)$, $g(x)$, $h(x) \; \varepsilon \; Q[x]$, determine all solutions of

$$(f(x))^2 + (g(x))^2 = (h(x))^2.$$

11.   If (a, b, c) is a solution of (1.6.1), show that 60 divides abc.

12.   Show that $(33, 56, 65)$, $(16, 63, 65)$ are the smallest sets of primitive positive solutions to (1.6.1) having equal largest elements. Are there infinitely many such pairs of solutions?

NOTES

§1.1.   The formal concepts of "ring" and "field" were introduced in the nineteenth century by Richard Dedekind and Nils Abel respectively [1.1].

§1.1.   Theorem 1.1.1 was discovered by the Pythagoreans, probably before 400 B.C. [1.2]. A scholium to Book X of Euclid's *Elements* indicates that the first of the Pythagoreans who made public investigations of this matter perished in a shipwreck. The related discovery of "incommensurable ratios" (quantities that could not be measured by a common unit, as e.g., $\sqrt{2}/2$), is attributed to Hippasus of Metapontum. The Pythagoreans were supposed to have been at sea at the time, and to have thrown Hippasus overboard. (This type of mathematical behavior has persisted to this day!)

It is not known how the irrationality of $\sqrt{2}$ was *first* proved. Plato mentions in his *Theatetus* [1.3] that his teacher, Theodorus of Cyrene, proved the irrationality of $\sqrt{3}, \sqrt{5}, \ldots, \sqrt{17}$. This is estimated to have occurred before 400 B.C. The fact that Theodorus did not proceed further than $\sqrt{17}$ suggests that he did not have a general method of proof.

The proof of Theorem 1.1.1 given in the text is essentially the one found in Aristotle's *Analytica Priora*. This proof was interpolated into Euclid's *Elements* as Proposition 117 of Book X.

§1.2.   The division algorithm was used by Euclid as the basis for his algorithm for finding the greatest common divisor (cf. §1.5). It is not proved or discussed independently by Euclid; he may have considered it a self-evident result.

§1.3.   The classification of integers as even and odd was made by the Pythagoreans. These concepts, as well as the definitions of squares, cubes, primes, and composites, appear in Book VII of Euclid's *Elements* [1.4].

§1.3.   Ideals are a relatively recent development. Ernst Kummer introduced a theory of "ideal numbers" in 1844. Richard Dedekind formulated the present concept of ideal in 1876, and gave it that name in honor of Kummer's contribution.

§1.4.   The identity (1.4.7) represents a formal "entrapment" of various basic facts concerning divisibility. Later, in §2.5, it is used as the basis of a proof of the Unique Factorization Theorem.

§1.5.   The Euclidean Algorithm is given by Euclid in Propositions 1 and 2 of Book VII of his *Elements* [1.5].

§1.6.   The Pythagoreans probably did not prove the "Pythagorean Theorem," but relied on the evidence of special cases. The proof given in Euclid's *Elements* (Book I, Proposition 47) was credited by Euclid himself to Proclus [1.6].

An alternate formulation of Theorem 1.6.1 asserts that all primitive positive integer solutions to (1.6.1) are given by $x = u^2 - v^2$, $y = 2uv$, $z = u^2 + v^2$, where u, v, are integers such that $u > v > 0$, $(u, v) = 1$, together with the solutions obtained by interchanging x and y.

REFERENCES

1.1.   Morris Kline, *Mathematical Thought from Ancient to Modern Times,* New York: Oxford University Press, 1972, pp. 821, 755.

1.2.   Sir Thomas Heath, *A History of Greek Mathematics,* 2 vols., London: Oxford University Press, 1921, p. 154.

1.3.   Ref. 1.2, pp. 90–91.

1.4.   *The Thirteen Books of Euclid's Elements,* 3 vols., 2nd edition, translation edited by Thomas Heath, Cabridge: Cambridge University Press, 1926. Vol. II, p. 278, Definitions 6, 7, 18, 19, 11, 13.

1.5.   Ref. 1.4, Vol. II, pp. 296–299.

1.6.   Ref. 1.1, p. 34.

2

THE UNIQUE
FACTORIZATION
THEOREM

§2.1 WHAT IS THE UNIQUE FACTORIZATION THEOREM?

We have already distinguished, among the integers, the subset consisting of those that are primes. The relationship between the positive integers, in general, and primes, in particular, emerges first in the fact that each integer $n > 1$, by its very existence, is a device for bringing a prime into existence.

Lemma 2.1.1. Every integer $n > 1$ is divisible by at least one prime.

Proof. Since $n > 1$, n has at least one divisor > 1, namely, n itself. Thus n must have a smallest divisor $q > 1$. This divisor q must be a prime. For if not, $q = q'q''$, $1 < q' < q$, and q' would be a divisor of n, which is smaller than q and greater than 1, a contradiction.

From this lemma one obtains the following theorem immediately.

Theorem 2.1.1. The set of primes is infinite.

Proof. Suppose $p_1, \ldots, p_k$ are a given set of primes. Form the integer

$$n = p_1 p_2 \cdots p_k + 1.$$

By Lemma 2.1.1 we have that n has at least one prime divisor p. This prime p must differ from all the p_i, $i = 1, \ldots, k$, since the p_i do not divide n (dividing n by p_i leaves a remainder 1). Thus we've shown that there always exist primes that are not included in any given finite set of primes. In other words, the set of primes is infinite.

In addition to being infinite in number, the primes are sufficiently numerous to provide a multiplicative representation of any integer.

Theorem 2.1.2. Every integer $n > 1$ has at least one representation as a product of primes.

Proof.   The integer $n = 2$ is a prime, and hence the theorem holds for $n = 2$. Proceeding by induction, assume the theorem for all integers n', $2 \leq n' < n$. If n is a prime, the theorem is immediate. Otherwise $n = n'n''$, $2 \leq n' < n$, and $2 \leq n'' < n$. By our inductive hypothesis, both n' and n'' have representations as products of primes, which implies the same for $n = n'n''$. This completes the induction.

In the representation of n as a product of primes, a given prime may occur more than once. That is, the primes that occur are not necessarily distinct, as in the example

(2.1.1) $$12 = 2 \cdot 2 \cdot 3.$$

Gathering together all the occurrences of each prime p and expressing their product as a power of p, a representation of $n > 1$ as a product of primes may be written as

(2.1.2) $$n = p_1^{\alpha_1} p_2^{\alpha_2} \cdots p_r^{\alpha_r},$$

where $1 < p_1 < p_2 < \ldots < p_r$ are *distinct* primes and the α_i are positive integers. In this manner, (2.1.1) would be rewritten as

$$12 = 2^2 \cdot 3.$$

For each integer $n > 1$, it follows from Theorem 2.1.2 that n has at least one representation of the form (2.1.2). The natural question is whether this representation is unique. The answer is in the affirmative, and is known as *The Unique Factorization Theorem* (UFT). In formal terms this may be stated as follows.

Theorem 2.1.3 (Unique Factorization Theorem).   Let

$$p_1^{\alpha_1} p_2^{\alpha_2} \cdots p_r^{\alpha_r} = q_1^{\beta_1} q_2^{\beta_2} \cdots q_s^{\beta_s}$$

where $1 < p_1 < \ldots < p_r$, $1 < q_1 < \ldots q_s$; the p_i and q_j are all primes and the integers α_i and β_j are positive. Then it follows that

$$\begin{cases} r = s \\ p_i = q_i, & i = 1, \ldots, r \\ \alpha_i = \beta_i, & i = 1, \ldots, r. \end{cases}$$

In the terms of the representations of an integer $n > 1$ as a product of primes (not necessarily distinct), the Unique Factorization Theorem is equivalent to the assertion that such representations are unique, apart from rearrangements of the order of the prime factors.

There are many proofs of the Unique Factorization Theorem that differ mainly in the manner in which they present (or obscure) the role of "addition" in the proof. Indeed, it should be noted that it is precisely the link between addition and multi-

plication that produces the validity of the Unique Factorization Theorem. In order
to highlight this point, we will exhibit a "nontrivial" subset $\mathcal{M} \subset Z$ which is closed
under multiplication, that is,

(2.1.3) $m_1, m_2 \in \mathcal{M}$ implies $m_1 m_2 \in \mathcal{M}$;

such that in the "arithmetic of $\mathcal{M}$" unique factorization fails. More precisely, if an
"$\mathcal{M}$-prime" is an integer of $\mathcal{M}$ that has no divisors in $\mathcal{M}$ other than possibly 1 and
itself, the representation of integers of $\mathcal{M}$ as a product of $\mathcal{M}$-primes (which always
exists) need not be "unique." It is perhaps amusing that a simple way to produce such
an example is to utilize the truth of the Unique Factorization Theorem for Z. This
is done by letting $\mathcal{M}$ be the subset of those positive integers of Z that are representable as a product of an even number of primes (not necessarily distinct). It is, of
course, the Unique Factorization Theorem that ensures that $\mathcal{M}$ is well defined.
Further, (2.1.3) clearly holds. What are the $\mathcal{M}$-primes? Again, from the Unique
Factorization Theorem it is easily seen that they are integers of the form p^2 (the
square of a prime) or pq (the product of two distinct primes). That unique factorization fails in $\mathcal{M}$ may then be seen from

(2.1.4) $p^2 q^2 = (pq)(pq) = (pq)^2,$

which provides equal but completely different products of $\mathcal{M}$-primes (the left side
is the product of two different $\mathcal{M}$-primes whereas the right side is the square of an
$\mathcal{M}$-prime).

EXERCISES

1. Verify that the $\mathcal{M}$-primes of the set $\mathcal{M}$ defined above are the squares of primes
    and products of two distinct primes.

2. Give other subsets of the integers for which the Unique Factorization Theorem
    is false, in the sense described above.

3. Show that:
    (a)   If $2^p - 1$ is a prime, then p is a prime (primes of the form $2^p - 1$ are
          called Mersenne primes).
    (b)   If $2^n + 1$ is a prime, then n is a power of 2 (primes of this form are called
          Fermat primes).
    It is not known whether there are infinitely many Mersenne primes or infinitely
    many Fermat primes.

4. Note that all primes > 2 are either of the form $4n + 1$ or $4n - 1$. Prove that
    there are infinitely many primes of the form $4n - 1$. Similarly, show that all
    primes > 3 are of the form $6n + 1$ or $6n - 1$.

5. Prove that for each integer $m \geq 1$, there exist m consecutive integers all of
    which are composite.

6. Let $f(x)$ be a polynomial of degree $n \geq 1$, with integer coefficients, such that $f(x) \to \infty$ as $x \to \infty$. Show that the prime divisors of the integers $f(m)$, $m = 1, 2, \ldots$, are infinite in number.

7. Under the same hypothesis as in Ex. 6, show that for infinitely many positive integers x, the value of $f(x)$ is composite.

8. Using the Unique Factorization Theorem, show that if

$$\prod_{i=1}^{r} p_i^{x_i} = 1$$

where the p_i are distinct primes and the x_i are rational numbers, it follows that all $x_i = 0$.

9. Using the Unique Factorization Theorem, prove that for distinct primes p_i, $i = 1, \ldots, r$, the quantities $\log (1 - 1/p_i)$ are linearly independent over the rationals, (i.e. $\Sigma \, r_i \log (1 - 1/p_i) = 0$ implies all $r_i = 0$).

10. Given integers a, b, and n, with $(a, b) = 1$, show that there is an integer x such that $(ax + b, n) = 1$.

§2.2   THE PRIME DIVISIBILITY LEMMA AS A ROUTE TO THE UFT

If, in Theorem 1.4.3, we take $a = p$, we obtain the following lemma.

Lemma 2.2.1.   If a prime p divides bc, and p does not divide b, then p divides c.

    That this follows from Theorem 1.4.3 by taking $a = p$ is immediate from the fact that because p is a prime, the condition $(p, b) = 1$ is equivalent to p not dividing b. It is Lemma 2.2.1 that we refer to as the Prime Divisibility Lemma. It may, of course, be derived directly by the methods of §1.4 or §1.5. In any event, it is the source of one type of proof of the Unique Factorization Theorem.

    In order to base a proof of the Unique Factorization Theorem on Lemma 2.2.1, one begins by setting up an induction. The theorem is clearly true for $n = 2$. We next assume it true for all n', $2 \leq n' < n$. Suppose then that we have two factorizations for n, as products of primes,

(2.2.1) $n = p_1 p_2 \cdots p_r = q_1 q_2 \cdots q_s.$

Without loss of generality, we may assume that these primes have been indexed so that

(2.2.2) $p_1 \leq p_2 \leq \cdots \leq p_r, \quad q_1 \leq q_2 \leq \cdots \leq q_s$

and

(2.2.3) $p_1 \leq q_1.$

Since p_1 divides $n = q_1 \ldots q_s$, there is a largest index i^* such that p_1 divides

$q_{i^*}q_{i^*+1} \ldots q_s$. If $i^* = s$, then p_1 divides q_s, and since they are both primes, $p_1 = q_s$. But then $p_1 = q_s \geq q_1 \geq p_1$, so that $p_1 = q_s$. If $i^* < s$, and p_1 does not divide q_{i^*}, then by Lemma 2.2.1, p_1 would divide $q_{i^*+1} \ldots q_s$, contradicting the definition of i^*. Thus p_1 divides q_{i^*}, implying $p_1 = q_{i^*} \geq q_1 \geq p_1$, so that in this case we have $p_1 = q_1$. Since $p_1 = q_1$ in all cases, (2.2.1) implies

$$n' = p_2 \cdots p_r = q_2 \cdots q_s$$

where $1 \leq n' < n$. If $n' = 1$, $r = s = 1$, and there is nothing more to prove; and if $n' \geq 2$, the inductive hypothesis provides $r = s$, and $p_i = q_i$, $i = 2, \ldots, r$, thereby completing the induction, and the proof of the Unique Factorization Theorem.

Apart from the mechanics of the induction, the principle of the above proof is quite simple. Given two representations of n, as in (2.2.1), Lemma 2.2.1 is used to show that the prime p must appear as a prime factor on the right. Then canceling it on both sides, one repeats this argument until one of the two sides of the equation is canceled out of primes. Then the other side must also equal one, so it is manifestly clear that the two representations given in (2.2.1) were the same.

§2.3 THE COPRIME DIVISIBILITY LEMMA
AS A ROUTE TO THE UFT

Theorem 1.4.4 with $a = p$, $b = q$ establishes the following lemma.

Lemma 2.3.1. If p and q are distinct primes both of which divide c, then pq divides c.

This follows from Theorem 1.4.4, as claimed, since for two distinct primes p, q, we have $(p, q) = 1$. We refer to Lemma 2.3.1 as the Coprime Divisibility Lemma, and will provide a route to the Unique Factorization Theorem which is based on it.

As in the previous section, we set up an induction on n, and consider two representations

$$(2.3.1) \qquad n = p_1 p_2 \cdots p_r = q_1 q_2 \cdots q_s$$

where

$$(2.3.2) \qquad p_1 \leq p_2 \leq \cdots \leq p_r, \quad q_1 \leq q_2 \leq \cdots \leq q_s.$$

If $p_1 = q_1$, we can cancel p_1 from both sides of (2.3.1) and conclude the proof by means of the inductive hypothesis, as in §2.2. On the other hand, if $p_1 \neq q_1$, we follow another route, provided by Lemma 2.3.1, which leads to a contradiction. This lemma allows us to assert that $p_1 q_1$ divides n, and thus there must be a representation of n as a product of primes in which both p_1 and q_1 appear. That is,

$$(2.3.3) \qquad n = p_1 q_1 r_1 \cdots r_l$$

where the r_i are primes. Then since

$$n = p_1 p_2 \cdots p_r = p_1 q_1 r_1 \cdots r_l$$

or

$$p_2 \cdots p_r = q_1 r_1 \cdots r_l$$

so that via the inductive hypothesis, q_1 must equal one of the p_i, $i \geq 2$. Therefore, $q_1 \geq p_1$. Similarly,

$$q_1 q_2 \cdots q_s = p_1 q_1 r_1 \cdots r_l$$

or

$$q_2 \cdots q_s = p_1 r_1 \cdots r_l$$

and the induction provides that p_1 is one of the q_i, $i \geq 2$, and $p_1 \geq q_1$. Finally then, $p_1 = q_1$ and we have a contradiction.

§2.4 PURELY INDUCTIVE PROOFS OF THE UFT

There are various proofs of the Unique Factorization Theorem in which it is arranged that the induction carry the entire burden of the proof. That is, in some way, the role of such machinery as the "divisibility lemmas" is played by some more primitive device. Roughly speaking, we label such proofs as "purely inductive."

For example, the proof of §2.3 may be reorganized so that the use of Lemma 2.3.1 is replaced by the Division Algorithm itself. We start with the basic induction hypothesis, to the effect that the Unique Factorization Theorem is true for all n', $2 \leq n' < n$. Since we make this assumption, for such $n' < n$, if p and q are two distinct primes dividing n', it follows from the unique factorization of n' that pq divides n'. Consider then two representations of n as a product of primes

(2.4.1) $$n = p_1 p_2 \cdots p_r = q_1 q_2 \cdots q_s$$

where

(2.4.2) $$p_1 \leq p_2 \leq \cdots \leq p_r, \quad q_1 \leq q_2 \leq \cdots \leq q_s.$$

If either r or s equals 1, or $p_1 = q_1$, the result follows easily. Thus we may assume that $r > 1$, $s > 1$, and $p_1 \neq q_1$. Then $n \geq p_1^2$ implies $p_1 \leq \sqrt{n}$; $n \geq q_1^2$ gives $q_1 \leq \sqrt{n}$; hence

(2.4.3) $$p_1 q_1 \leq n.$$

Dividing n by $p_1 q_1$, the division algorithm provides

(2.4.4) $$n = p_1 q_1 t + n', \quad 0 \le n' < p_1 q_1.$$

From (2.4.3) and (2.4.4) we get that $n' < n$ so that since p_1 and q_1 divide n, they each divide n' and hence $p_1 q_1$ divides n'. Finally, from (2.4.4) we see that $p_1 q_1$ divides n. From this point, the proof concludes as in §2.3.

We note in the above that the elimination of Lemma 2.3.1 from the argument is somewhat superficial. Actually, Lemma 2.3.1 is being inductively proved in the background, as part of the main induction.

A less transparent and more elegant proof of the Unique Factorization Theorem (somewhat in the spirit of §2.3), in which the Divisibility Lemma does not appear explicitly, proceeds as follows.

The basic induction is set up and one assumes two representations satisfying (2.4.1) and (2.4.2). If $p_1 = q_1$, they may be canceled in (2.4.1) and the induction completed. Thus we may assume $p_1 > q_1$, and consider

$$m = n - (q_1 p_2 \cdots p_s) < n.$$

From (2.4.1),

$$m = (p_1 - q_1)\left(\frac{n}{p_1}\right) = q_1\left(\frac{n}{q_1} - \frac{n}{p_1}\right) > 1.$$

Since q_1 divides m and does not divide $p_1 - q_1$ (if it did, q_1 would divide p_1), the inductive hypothesis applied to $m < n$ allows us to conclude that q_1 divides n/p_1. Then, again by the inductive hypothesis, for some i,

$$q_1 = p_i \ge p_1 > q_1,$$

a contradiction.

§2.5   A PURELY NONINDUCTIVE PROOF OF THE UFT

Finally, the question may be raised as to whether the inductive element may be removed completely and a proof given of the Unique Factorization Theorem that rests completely on "greatest common divisor machinery." Indeed, such a proof can be constructed that stems from (1.4.7).

Since (1.4.7) asserts that

(2.5.1) $$(a, uv) = (a, (a, u)v),$$

applying this to itself gives

(2.5.2) $(a, uv) = (a, (a, u)\,(a, v))$

and this extends to (this is where the inductive element is hidden)

(2.5.3) $$\left(a, \prod_{i=1}^{r} u_i\right) = \left(a, \prod_{i=1}^{r} (a, u_i)\right).$$

Taking $a = \prod_{j=1}^{s} v_j$ in (2.5.3) and again using (2.5.3) upon itself, we obtain

(2.5.4) $$\left(a, \prod_{i=1}^{r} u_i\right) = \left(a, \prod_{i=1}^{r} \left(\prod_{j=1}^{s} (v_j, u_i),\ u_i\right)\right).$$

This is used first in order to establish the following lemma.

Lemma 2.5.1.   If p and q are primes, α and β nonnegative integers, then

(2.5.5) $$(p^\alpha, q^\beta) = \begin{cases} p^{\min(\alpha, \beta)} & \text{if } p = q \\ \\ 1 & \text{if } p \neq q \end{cases}.$$

Proof.   If $p = q$,

$$(p^\alpha, q^\beta) = p^{\min(\alpha, \beta)}(p^{\alpha - \min(\alpha, \beta)}, p^{\beta - \min(\alpha, \beta)})$$

$$= p^{\min(\alpha, \beta)}.$$

If $p \neq q$, taking $r = \beta$, $s = \alpha$, $v_j = p$, $u_i = q$ in (2.5.4), we obtain (note $a = p^\alpha$)

$$(p^\alpha, q^\beta) = \left(a, \prod_{i=1}^{r} \left(\prod_{j=1}^{s} (p, q),\ q\right)\right) = 1,$$

which completes the proof of the lemma.

Next assume that we have two factorizations of the integer n in the canonical form (2.1.2)

$$n = p_1^{\alpha_1} \cdots p_r^{\alpha_r} = q_1^{\beta_1} \cdots q_s^{\beta_s}$$

where the p_1 are distinct primes and the q_j are also distinct primes. Then we apply (2.5.4) with

$$u_i = p_i^{\alpha_i}, \quad i = 1, \ldots, r,$$

$$v_j = q_j^{\beta_j}, \quad j = 1, \ldots, s,$$

$$a = n \;;$$

so that

(2.5.6) $$n = \left(n, \prod_{i=1}^{r} \left(\prod_{j=1}^{s} (q_j^{\beta_j}, p_i^{\alpha_i}), p_i^{\alpha_i} \right) \right).$$

Using (2.5.5), this yields

$$n = \left(n, \prod_{\substack{i=1 \\ p_i = \text{some } q_j}}^{r} p_i^{\min(\alpha_i, \beta_j)} \right)$$

and since $n = \prod_{i=1}^{r} p_i^{\alpha_i}$, we see that

(i) for each i, p_i equals some $q_j = q_{j(i)}$, and
(ii) $\min(\alpha_i, \beta_{j(i)}) = \alpha_i$ [i.e., $\beta_{j(i)} \geq \alpha_i$].

Since $n = \prod_{j=1}^{s} q_j^{\beta_j}$, this implies that $\beta_{j(i)} = \alpha_i$, $r = s$, and the q_j are simply a permutation of the p_i. This again is the Unique Factorization Theorem.

EXERCISES

1. For each of the proofs of the Unique Factorization Theorem provided above, locate the places in the proof where the operation of addition of integers plays a role.

2. Prove the analogue of the Unique Factorization Theorem for polynomials of $Q[x]$ that asserts that each polynomial $f(x) \in Q[x]$ has a factorization

$$f(x) = c \prod_{i=1}^{r} f_i^{\alpha_i}(x)$$

where the $f_i(x)$ are distinct nonconstant irreducible polynomials of $Q[x]$ and c is a rational number. Further, if $f(x) = d \prod_{i=1}^{s} g_i^{\beta_i}(x)$ is another such factorization, we must have $r = s$, and to each i there is a $j = j(i)$ such that $\alpha_i = \beta_{j(i)}$ and $f_i(x) = e_i g_{j(i)}(x)$, where the e_i are rational numbers.

3. Show that the assertion of Lemma 2.5.1 evaluating (p^α, q^β) in the case $p \neq q$ also follows from the observation that (p^α, q^β) divides (p^γ, q^γ) for $\gamma = \max(\alpha, \beta)$.

4. Prove that every positive *rational* number r has a unique representation in the form

$$r = p_1^{\alpha_1} p_2^{\alpha_2} \cdots p_t^{\alpha_t}$$

where the p_i are distinct positive primes and the α_i are integers (positive or negative).

§2.6 A NEW VIEW OF G.C.D. AND L.C.M.

Given two positive integers a, b, they each have a canonical representation in the form

(2.6.1)
$$a = \prod_{i=1}^{r} p_i^{\alpha_i}, \quad b = \prod_{i=1}^{r} p_i^{\beta_i}.$$

Note that we may assume a common set of prime factors p_i if we allow the α_i and β_i to assume the value 0. Thus $\alpha_i \geq 0$, $\beta_i \geq 0$.

In terms of the representations (2.6.1), we have

(2.6.2)
$$(a, b) = \prod_{i=1}^{r} p_i^{\min(\alpha_i, \beta_i)}$$

(2.6.3)
$$\{a, b\} = \prod_{i=1}^{r} p_i^{\max(\alpha_i, \beta_i)}.$$

That these relations hold is easily verified from the definitions and will be left as an exercise.

Utilizing (2.6.2) and (2.6.3), all of our previously formulated results concerning greatest common divisors and least common multiples are easily verified. For example, the assertion

$$\{a, b\} (a, b) = ab$$

reduces to

$$\min(\alpha, \beta) + \max(\alpha, \beta) = \alpha + \beta.$$

The basic identity

$$(a, uv) = \left(a, (a, u)v \right)$$

reduces similarly to

$$\min(\alpha, \beta + \gamma) = \min(\alpha, \min(\alpha, \beta) + \gamma)$$

for $\alpha \geq 0$, $\beta \geq 0$, $\gamma \geq 0$.

EXERCISES

1. Prove (2.6.2) and (2.6.3).

2. Prove that

$$\{a, b, c\} = \frac{abc\,(a,b,c)}{(a,b)\,(b,c)\,(a,c)}$$

and generalize this to the least common multiple of more than three integers. Do the same for

$$(a, b, c) = \frac{abc\,\{a,b,c\}}{\{a,b\}\,\{a,c\}\,\{b,c\}}.$$

§2.7 CHARACTERIZING SUBSETS OF THE INTEGERS

Various subsets of the integers may be very simply characterized in terms of the canonical representation of integers provided by the Unique Factorization Theorem. An instance of this has already been met in the treatment of the subset $\mathcal{M}$ introduced in §2.1.

If we have

(2.7.1)
$$n = \prod_{i=1}^{r} p_i^{\alpha_i},$$

where the p_i are distinct primes and $\alpha_i > 0$, $i = 1, \ldots, r$, we say n is *squarefree* if and only if $\alpha_i = 1$, for $i = 1, \ldots, r$, or $n = 1$. It is easily seen that this is in agreement with Definition 1.3.2. The notion of squarefree integers may now be extended naturally to *cubefree* integers n which are such that in (2.7.1) $\alpha_i \leq 2$, $i = 1, \ldots, r$; and, in general, we introduce *kth-powerfree* integers as those such that

$$\alpha_i \leq k - 1, \quad i = 1, \ldots, r.$$

It is easily seen that a positive integer n is kth-powerfree if and only if the only kth power that divides it is 1.

As a kind of antithesis to squarefree integers, one has the notion of *squareful* integers. An integer n is *squareful* if and only if in its canonical representation (2.7.1) $\alpha_i \geq 2$, $i = 1, \ldots, r$.

Also, returning once again to integers that are "powers," more precisely *kth powers*, for $m = n^k$ we have via (2.7.1),

$$m = \prod_{i=1}^{r} p_i^{\alpha_i k}.$$

Thus we conclude that an integer m is a *kth power* if and only if in its canonical representation $m = \Pi_{i=1}^{s} q_i^{\beta_i}$, each β_i, $i = 1, \ldots, s$, is divisible by k.

This type of characterization can be quite critical in carrying out a proof. To illustrate this, we now extend Theorem 1.1.1.

Theorem 2.7.1. If the integer $A > 0$ is not a kth power, then $\sqrt[k]{A}$ is irrational.

Proof. Assume the assertion false, that is, $\sqrt[k]{A}$ is rational. Then we would have a representation

(2.7.2)
$$\sqrt[k]{A} = \frac{r}{s},$$

where r and s are positive integers, and we may assume that s is the smallest > 0 that appears in any such representation. Then, raising both sides of (2.7.2) to the kth power,

(2.7.3)
$$r^k = A \cdot s^k.$$

Using the Unique Factorization Theorem we may write

(2.7.4)
$$\begin{cases} A = p_1^{\alpha_1} p_2^{\alpha_2} \cdots p_r^{\alpha_r} \\ r = p_1^{\beta_1} p_2^{\beta_2} \cdots p_r^{\beta_r} \\ s = p_1^{\gamma_1} p_2^{\gamma_2} \cdots p_r^{\gamma_r}. \end{cases}$$

Note that in (2.74) we have used the same primes p_i, $i = 1, \ldots, r$, in the representations of A, r, and s. This may be arranged as before by allowing the exponents α_i, β_i, γ_i, to take on the value 0. For example,

$$10 = 2^1 \cdot 3^0 \cdot 5^1, \qquad 15 = 2^0 \cdot 3^1 \cdot 5^1, \qquad 27 = 2^0 \cdot 3^3 \cdot 5^0.$$

Clearly, the Unique Factorization Theorem provides that once a big enough set of primes p_i is fixed, the exponents $\alpha_i \geq 0$, $\beta_i \geq 0$, $\gamma_i \geq 0$, are uniquely determined.

Inserting (2.7.4) in (2.7.3), we obtain

(2.7.5)
$$\prod_{i=1}^{r} p_i^{k\beta_i} = \prod_{i=1}^{r} p_i^{k\gamma_i + \alpha_i}.$$

By the Unique Factorization Theorem, (2.7.5) implies that the exponents of each p_i on the two sides of (2.7.5) must be equal, that is

$$k\beta_i = k\gamma_i + \alpha_i, \qquad i = 1, \ldots, r.$$

Thus

$$(2.7.6) \qquad \alpha_i = k(\beta_i - \gamma_i), \qquad i = 1, \ldots, r.$$

Since (2.7.6) asserts that each α_i is divisible by k, it would follow that A is a kth power, which is a contradiction.

EXERCISES

1. Show that every positive integer n has a unique representation in the form $n = s^2 q$, where s and q are positive integers and q is squarefree.

2. Show that a positive squareful integer has a unique representation in the form

$$n = s^2 q^3$$

where q is squarefree, $s > 0$.

3. Let A, B be positive integers such that at least one of them is not a square. Prove that $\sqrt{A} + \sqrt{B}$ is irrational.

4. Show that the equation $x^y = y^x$ has as solutions in positive integers only the possiblities $y = x$; $x = 2$, $y = 4$; and $x = 4$, $y = 2$.

5. Show that if m, n, x, y are positive integers such that

$$(m!)^x = (n!)^y,$$

then $x = y$ and $m = n$. (This exercise is an easy consequence of results which are given later. It seems to be quite difficult to give a completely primitive solution.) $(n! = n(n - 1) \cdots 1)$

§2.8 SOME IMPLICATIONS FOR THE PRIMES

The relationship between the integers and the primes established by the Unique Factorization Theorem is the basic starting point for obtaining more detailed information about the primes. Letting $\pi(x)$ denote the number of primes $\leq x$, Theorem 2.1.1 asserts that $\pi(x) \to \infty$ as $x \to \infty$. This may now be sharpened.

Theorem 2.8.1. There exists a constant $c > 0$ such that for all integers $N \geq 3$.

$$(2.8.1) \qquad \pi(N) > c \frac{\log N}{\log \log N}$$

(in particular $c = \frac{1}{3}$ would suffice for $N \geq e^2$).

Proof. Each integer n, $2 \leq n \leq N$ has a unique representation of the form

$$(2.8.2) \qquad n = p_1^{\alpha_1} p_2^{\alpha_2} \cdots p_m^{\alpha_m}, \qquad \alpha_i \geq 0,$$

where the p_i, $i = 1, \ldots, m$, are distinct primes. Since each such $p_i \le n \le N$, it follows that

(2.8.3) $$m \le \pi(N).$$

Further, since $N \ge n \ge 2^{\alpha_i}$, each $\alpha_i \le \log N / \log 2$ so that for each prime there are at most

$$\left(1 + \frac{\log N}{\log 2}\right)$$

(including $\alpha_i = 0$) possible exponents. This together with (2.8.3) implies that the number of possibilities for n is at most

$$\left(1 + \frac{\log N}{\log 2}\right)^{\pi(N)}.$$

Thus we obtain

(2.8.4) $$N \le \left(1 + \frac{\log N}{\log 2}\right)^{\pi(N)}.$$

This implies (2.8.1); in fact, $c = \frac{1}{3}$ will suffice for $N \ge e^2$. For then

$$1 + \frac{\log N}{\log 2} = \frac{\log 2N}{\log 2} \le \frac{2}{\log 2} \log N < 4 \log N \le (\log N)^3$$

so that (2.8.4) yields

$$N \le (\log N)^{3\pi(N)}$$

and the result follows.

The above argument is obviously a crude one and serves mainly as a first illustration of how the factorization of integers provides information about the primes. In Theorem 2.8.1, the count of the primes $\pi(N)$ was considered directly. However, the factorization of integers into primes may be used to focus on other expressions that are related to the primes. An illustration of this is provided by the following theorem.

Theorem 2.8.2. For every integer $N \ge 3$

(2.8.5) $$\prod_{p \le N} \left(1 - \frac{1}{p}\right)^{-1} > \log N$$

(the product on the left is taken over all primes $p \le N$).

Proof.   We have

$$\prod_{p\leq N}\left(1 - \frac{1}{p}\right)^{-1} = \prod_{p\leq N} (1 + \frac{1}{p} + \frac{1}{p^2} + \ldots + \frac{1}{p^\alpha} + \ldots) > \sum_{n \leq N} \frac{1}{n},$$

where the last inequality follows from the fact that each integer $\leq N$ has a represent-ation as a product of primes $\leq N$, and different integers have different represent-ations. Then (2.8.5) follows since

$$\sum_{n=1}^{N} \frac{1}{n} = \frac{1}{N} + \sum_{n=1}^{N-1} \left(\frac{1}{n} - \log (1 + \frac{1}{n})\right) + \log N,$$

and from $\log (1 + 1/n) < 1/n$, it follows that this last expression is $> \log N$.

As will result from the inequalities discussed in a later chaper, the effectiveness of the above arguments differs considerably. The result (2.8.1) is very weak and quite far from the true picture. In fact, a theorem known as the *Prime Number Theorem* asserts that $x/\log x$ is an approximation to $\pi(x)$, with a percentage error that tends to zero as $x \rightarrow \infty$. On the other hand, (2.8.5) is fairly close to the truth in that the order of magnitude, $\log N$, is correct apart from a constant factor. In any event, it should be noted that, in both of the above illustrations, the uniqueness part of the Unique Factorization Theorem is not used.

EXERCISES

1.   Prove that the sum of reciprocals of the primes diverges.

2.   Prove that there exists a constant $c > 0$, such that for all real $x > 1$,

$$\prod_{p\leq x} (1 - \frac{1}{p}) \leq \frac{c}{\log x}.$$

3.   Deduce from the assertion of Ex. 2 that for some positive constant c_1 and all real $x \geq 3$,

$$\sum_{p\leq x} \frac{1}{p} > c_1 \log \log x.$$

4.   For real numbers $s > 1$, show that the series $\sum_{n=1}^{\infty} 1/n^s$ converges so that for $s > 1$, we may use it to define the Riemann Zeta function as

$$\zeta(s) = \sum_{n=1}^{\infty} \frac{1}{n^s}.$$

By means of the Unique Factorization Theorem show that for any fixed positive integer N

$$\zeta(s) \geq \prod_{p \leq N} \left(1 - \frac{1}{p^s}\right)^{-1} \geq \sum_{n \leq N} \frac{1}{n^s}.$$

Then letting $N \to \infty$, derive the Euler factorization

$$\zeta(s) = \prod_{p} \left(1 - \frac{1}{p^s}\right)^{-1}.$$

§2.9   VALUATIONS: ANOTHER CONSEQUENCE OF THE UFT

For a given prime p, we define $\nu_p(n)$ to be the exponent of the highest power of p that divides the integer n, $[\nu_p(0) = \infty]$. For a rational number $r = m/n$, define $\nu_p(r) = \nu_p(m) - \nu_p(n)$. Because of the Unique Factorization Theorem, it is clear that the definition of $\nu_p(r)$ is independent of the particular representation of r as the ratio of two integers. Moreover, in general, for any two rational numbers r, s,

$$(2.9.1) \qquad\qquad \nu_p(rs) = \nu_p(r) + \nu_p(s).$$

Fixing any real constant c, $0 < c < 1$, we can introduce a kind of "p-absolute value" over the rationals that corresponds to the prime p. Namely, let

$$(2.9.2) \qquad\qquad |r|_p = c^{\nu_p(r)}.$$

The basic properties of this "p-absolute value" are given by the following theorem.

Theorem 2.9.1.   For any fixed prime p, we have

$$(2.9.3) \qquad |r|_p \geq 0 \text{ for all } r, \text{ and } |r|_p = 0 \text{ if and only if } r = 0,$$

$$(2.9.4) \qquad\qquad |rs|_p = |r|_p |s|_p$$

$$(2.9.5) \qquad\qquad |r + s|_p \leq \max\left(|r|_p, |s|_p\right).$$

Proof.   The assertion (2.9.3) is immediate from (2.9.2), and (2.9.4) is a consequence of (2.9.1) and (2.9.2).

Next, suppose that $|r|_p \geq |s|_p$, so that if $r = 0$, (2.9.5) is clearly true. Assuming $r \neq 0$, using (2.9.4) we obtain

$$(2.9.6) \qquad\qquad |r + s|_p = |r|_p \left|1 + \frac{s}{r}\right|_p.$$

Here

(2.9.7)
$$\left|\frac{s}{r}\right|_p = \frac{|s|_p}{|r|_p} \le 1$$

and (2.9.5) will follow from (2.9.6) if we can show that this implies

(2.9.8)
$$\left|1 + \frac{s}{r}\right|_p \le 1.$$

But (2.9.7) implies that $s/r = p^\alpha m/n$, where $\alpha \ge 0$, and m, n are integers such that $v_p(m) = v_p(n) = 0$. Hence

$$1 + \frac{s}{r} = 1 + p^\alpha \frac{m}{n} = \frac{n + p^\alpha m}{n}$$

Since $v_p(n) = 0$ and $v_p(n + p^\alpha m) \ge 0$, it follows that

$$v_p\left(1 + \frac{s}{r}\right) = v_p(n + p^\alpha m) - v_p(n) \ge 0,$$

which yields (2.9.8) and completes the proof of the theorem.

Note that (2.9.5) implies the usual triangle inequality

(2.9.9)
$$|r + s|_p \le |r|_p + |s|_p$$

and suggests the following.

Definition 2.9.1. A real valued function $V(r)$ defined over the rationals is called a *valuation* if

(2.9.10) $V(r) \ge 0$, and $V(r) = 0$ if and only if $r = 0$,

(2.9.11) $V(rs) = V(r)V(s)$,

(2.9.12) $V(r + s) \le V(r) + V(s)$.

Thus the "p-absolute values" $|r|_p$ defined above are valuations. In addition, ordinary absolute value

(2.9.13)
$$|r| = \begin{cases} r & \text{if } r \ge 0, \\ -r & \text{if } r < 0, \end{cases}$$

is a valuation of the rationals. Further, it can be shown that for any fixed real number η, $0 < \eta \le 1$, $V(r) = |r|^\eta$ is also a valuation. (See Ex. 10 of §2.9.)

The special valuation defined by taking $V(r) = 1$ for all $r \neq 0$, and $V(0) = 0$ is called the *trivial valuation*. (Any other one is a *nontrivial* valuation.)

From (2.9.2), we see that $|n|_p \leq 1$ for *all* integers n. In fact, this property distinguishes the "p-absolute values" from the ordinary $|r|$.

Definition 2.9.2. A valuation such that $V(n) \leq 1$ for all integers n is called *nonarchimedean*. Otherwise the valuation is called *archimedean*.

Note that (2.9.11), together with the fact that $V(\pm 1) > 0$, implies $V(1) = V(-1) = 1$. Thus $V(r) = V(-r)$, so that whether a valuation is archimedean is determined by the values of $V(n)$ for n a positive integer.

The $|r|_p$ are all nonarchimedean and the $V(r) = |r|^\eta, 0 < \eta \leq 1$ are archimedean. In fact, these include all the possible nontrivial valuations of the rationals.

Theorem 2.9.2. All the nontrivial valuations of the rationals are either of the form (2.9.2) for some prime p or given by a power of the ordinary absolute value.

Proof. Consider any two integers m, n, both bigger than 1. Then for a given valuation $V(r)$, let

$$(2.9.14) \qquad M = M_m = \max_{0 \leq i < m} V(i)$$

(note that $M \geq 1$). We will show that

$$(2.9.15) \qquad V(n) \leq \frac{\log (mn)}{\log m} \{\max [1, V(m)]\}^{\log n / \log m} M$$

holds for $n \geq 1$, $m > 1$. This is clear for $n = 1$ and any $m > 1$, since $1 = V(1) \leq M = M_m$. Proceeding by induction, we assume (2.9.15) for any n', $1 \leq n' < n$, and all $m > 1$. Thus by the division algorithm

$$n = mq + i, \qquad 0 \leq i < m, \qquad 0 \leq q < n.$$

If $q = 0$, then

$$V(n) = V(i) \leq M \leq \frac{\log mn}{\log m} \max [1, V(m)]^{\log n / \log m} M.$$

Otherwise $q \geq 1$, and

$$V(n) \leq V(mq) + V(i) = V(m)V(q) + V(i)$$

$$\leq V(m) \frac{\log qm}{\log m} \{\max [1, V(m)]\}^{\log q / \log m} M + M$$

$$\leq \frac{\log n}{\log m} \{\max [1, V(m)]\}^{\log n / \log m} M + M$$

$$\leq \frac{\log nm}{\log m} \{\max [1, V(m)]\}^{\log n/\log m} M,$$

which completes the induction and establishes (2.9.15).

Replacing n by n' in (2.9.15) gives

$$(V(n))^t = V(n^t) \leq (1 + t\frac{\log n}{\log m}) \{\max [1, V(m)]\}^{t \log n/\log m} M.$$

Taking the tth root of both sides and letting $t \to \infty$ yields

$$(2.9.16) \qquad V(n) \leq \{\max [1, V(m)]\}^{\log n/\log m}, \qquad (n \geq 1, m > 1).$$

We now have two cases.

CASE I. $V(m) \leq 1$ for some integer $m > 1$

Thus from (2.9.16) it follows that $V(n) \leq 1$ for *all* integers n. This implies that the valuation is nonarchimedean, and hence that (see Ex. 4 of 2.9)

$$(2.9.17) \qquad\qquad\qquad V(r + s) \leq \max [V(r), V(s)].$$

Further, since $V(r)$ is nontrivial, there is some integer n such that $V(n) \neq 1$, and therefore $V(n) < 1$. Letting $\mathcal{S}$ denote the nonempty subset of integers n for which $V(n) < 1$, we note that $\mathcal{S}$ is an ideal. Namely, if $n \in \mathcal{S}$, and m any integer, $V(m) \leq 1$ so that $V(mn) = V(m)V(n) < 1$ (i.e. $mn \in \mathcal{S}$). Also, by (2.9.17) for $m, n \in \mathcal{S}$, $V(m - n) \leq \max [V(m), V(n)] < 1$ (i.e., $m - n \in \mathcal{S}$). Therefore $\mathcal{S}$ consists of the multiples of a positive integer p (which is the smallest positive one in $\mathcal{S}$). Here p is a prime, for if $p = ab$, $1 > V(p) = V(a)V(b)$ implies $V(a)$ or $V(b)$ less than 1. Thus if $(n, p) = 1$, n is not in $\mathcal{S}$, and $V(n) = 1$.

There is an argument due to J. Collison that establishes the above without requiring (2.9.17). Namely, if $V(n) \leq 1$ for all integers n, and there exists some integer $j \neq 0$ such that $V(j) < 1$, it follows that for some prime p, $V(p) = c < 1$. Consider an integer n such that $(n, p) = 1$. Then for each positive integer k, there exist integers a_k, b_k such that

$$a_k n^k + b_k p = 1.$$

If $V(n) < 1$, it follows from

$$1 = V(1) = V(a_k n^k + b_k p) \leq V(n^k) + V(p),$$

by letting k tend to infinity, that $1 \leq V(p) = c < 1$. Thus $(n, p) = 1$ implies $V(n) = 1$, as desired.

For any given integer m, we can write $m = p^{v_p(m)}s$, $(s, p) = 1$, so that

$$V(m) = (V(p))^{\nu_p(m)}.$$

This implies that the valuation is of the form (2.9.2) with $c = V(p)$.

CASE II. $V(m) > 1$ for all integers $m > 1$
  Equation (2.9.16) yields

$$V(n)^{1/\log n} \leq V(m)^{1/\log m}$$

for all integers m, n larger than 1. Since the roles of m and n are symmetric, this implies

$$V(n)^{1/\log n} = V(m)^{1/\log m}$$

for all integers m, n larger than 1. That is, there is a constant $c > 0$ such that

$$V(n) = c^{\log n} = n^{\log c}$$

for all integers $n > 1$. This, in turn, implies that $V(r) = |r|^{\log c}$ for all rational r, which completes the proof of the theorem.
  In the above case we must, in fact, have that $0 < \log c \leq 1$ (see Ex. 10 of §2.9). If we normalize the p-valuation of choosing $c = 1/p$, we have

$$(2.9.18) \qquad\qquad |r|_p = \frac{1}{p^{\nu_p(r)}}.$$

The set of these is then related to the ordinary absolute value by the Product Formula.

Theorem 2.9.3 (The Product Formula). For the nonarchimedean valuations normalized as in (2.9.18), we have for all rational $r \neq 0$,

$$(2.19.19) \qquad\qquad |r| \prod_p |r|_p = 1,$$

where the product is taken over all primes p.

Proof. For any rational $r \neq 0$, $\nu_p(r) = 0$ and hence $|r|_p = 1$ for all but a finite number of primes. Thus the product in (2.9.19) only contains a finite number of terms with $|r|_p \neq 1$. Then by the UFT

$$\prod_p |r|_p = \prod_p \frac{1}{p^{\nu_p(r)}} = \frac{1}{|r|},$$

which yields (2.9.19).

EXERCISES

1. Verify that $\nu_p(r) = \nu_p(m) - \nu_p(n)$ is independent of the representation $r = m/n$. Also prove (2.9.1).

2. Prove that if a valuation of the rationals has the property that $V(r) \leq 1$ implies $V(1 + r) \leq 1$, then for all r, s, (2.9.17) holds.

3. Prove that if a valuation $V(r)$ satisfies (2.9.17), it is nonarchimedean.

4. Show that if $V(r)$ is nonarchimedean, then (2.9.17) holds. Give a direct proof not using Theorem 2.9.2.
   Hint. Consider $V((1 + r)^k)$ and use binomial theorem and triangle inequality.

5. Give a direct proof that if a valuation $V(r)$ is bounded on the integers, then it is nonarchimedean.

6. Show that a valuation $V(r)$ always satisfies

$$V(r + s) \leq 2 \max (V(r), V(s)).$$

7. If a function $V(r)$ over the rationals satisfies (2.9.10), (2.9.11), and

$$V(r + s) \leq c \max (V(r), V(s)),$$

   for a fixed constant c, $1 \leq c \leq 2$, then for $m = 2^k$,

$$V(r_1 + \cdots + r_m) \leq c^k \max V(r_i).$$

8. Using the result of Ex. 7, show that under the same hypothesis it follows that $V(t) \leq t$ for all positive integers t.

9. Prove that the hypothesis of Ex. 7 implies (2.9.12); that is, $V(r)$ is a valuation.
   Hint. Using Ex. 7 and Ex. 8, derive

$$V((r + s)^n) \leq c^k(V(r) + V(s))^n$$

   for $n = 2^k - 1$.

10. For ordinary absolute value $|r|$ show that $V(r) = |r|^\eta$ is a valuation if and only if $0 \leq \eta \leq 1$.

NOTES

§2.1. Lemma 2.1.1 was given explicitly by Euclid in Book VII, Proposition 31, of his *Elements*. He states there, "Any composite number is measured (divisible) by some prime number." He proves this by factoring the composite which he calls A, and repeating the process if only composite factors are obtained. Euclid observes then [2.1] that a prime factor must eventually be obtained, because "if it is not found, an infinite series of (positive) numbers will measure the number A, each of which is less than the other, which is impossible in numbers. Therefore some prime number will be found which will measure the one before it, which will also measure A."

Theorem 2.1.1 appears in Euclid's *Elements* as Proposition 20 of Book IX. Euclid proves, "Prime numbers are more than any assigned multitude of prime numbers," by exactly the same method as given in the text. This proof can be simplified somewhat by using the recursion $a_n = a_1 a_2 \ldots a_{n-1} + 1$ to define the infinite sequence of positive integers a_n, (for any given initial $a_1 > 0$). Then it is easily seen that *every pair* of integers in this sequence is relatively prime, and the prime dividing each, provided by Lemma 2.1.1, yields Theorem 2.1.1 [2.2].

With regard to Theorem 2.1.3, it is widely believed that the Unique Factorization Theorem was not stated before 1801, when Gauss presented it in his *Disquisitiones Arithmeticae* (Section 16) [2.3]. This view is expressed by Hardy and Wright [2.4]. They in turn attribute their information to remarks made by S. Bochner. In his article, "Mathematical Reflections" [2.5], Bochner recalls this incident and states, "the nearest that Euclid himself came to this theorem was his proposition (VII,24): 'if two numbers be prime to any number, their product will be prime to the same.'" Bochner also quotes from his book, *The Role of Mathematics in the Rise of Science* (1966, p. 16): "There is no substance to assertions that the fundamental theorem had been consciously known to mathematicians before Gauss, but that they had neglected to make the fact known. We think that the seventeenth and even the eighteenth century was not yet ready for the peculiar kind of mathematical abstraction which the fundamental theorem involves . . ."

This view appears to be based on an oversight. Indeed, Euclid came quite close to a statement of the Unique Factorization Theorem in Book IX of his *Elements*. There, Proposition 14 states: "If a number be the least that is measured by prime numbers, it will not be measured by any other prime number except those originally measuring it." In Euclid's terminology "measured by" corresponds to "divisible by." There was no system for expressing exponents (with the exception that squares and cubes were explicitly defined and discussed). Also, it was rare for a concretely specified number to occur with more than three distinct prime divisors. Thus, it is not surprising that Euclid's proof of Proposition 14 begins by assuming that A is the smallest number measured by the primes B, Γ, and Δ. Then A cannot have any other prime divisor, for if another prime E measured A, where E is different from B, Γ, and Δ, A would have another factor Z so that (in modern notation) $A = EZ$. Then each of the primes B, Γ, Δ, measures EZ, which forces each of these primes to measure Z. This follows from Proposition 30 of Book VII of *Elements*, which states: "If two numbers by multiplying one another make some number, and any prime number measure the product, it will also measure one of the original numbers." But Z is smaller then A, which was assumed to be the least number measured by the primes B, Γ, Δ. This contradiction shows that A cannot have any other prime factors.

However, it is true that Euclid does not have the explicit statement that every integer is expressible as a product of primes. But in the terminology of Euclid's time, the closest one could get to a statement concerning a product was "E by multiplying Z has made A ("E ipsum Z multiplicans ipsum A fecit")." No word existed for "product," yet this was obviously Euclid's intent conceptually when he let A be "the smallest number measured by the primes B, Γ, Δ," in his proof of the Unique Factorization Theorem.

Thus we see that Euclid was notation and vocabulary limited. He had no notation for exponents and couldn't express a number with an arbitrary number of factors, but he was *not* notion limited. He saw all the difficulties that had to be dealt with in proving The Unique Factorization Theorem and developed appropriate means for coping with them. His Proposition 14 is essentially the Unique Factorization Theorem.

It is interesting to note that a completely different situation arises in the work of Euler and Legendre. Euler [2.6] does not state the Unique Factorization Theorem at all, but he does give a method for "resolving any number into its simple factors." The method is simply that of dividing by the different primes until all factors are found. He then states: "When therefore, we have represented any number, assumed at pleasure, by its simple factors, it will be very easy to show all the numbers by which it is divisible." With this, Euler is clearly *assuming* that any factorization thus obtained is essentially unique (in the sense of the UFT).

Legendre [2.7] presents the same information, but in a more detailed form. He states: "Un nombre quelconque N, s'il n'est pas premier, peut-être représenté par le produit de plusieurs nombres premiers α, β, γ, etc., élevés chacun à une puissance quelconque, de sorte qu'on peut toujours supposer $N = \alpha^m \beta^n \gamma^p$ etc. . . . . ." Legendre gives the division method for finding the representation, and then states: "Un nombre N étant réduit à la forme $\alpha^m \beta^n \gamma^p$ etc., tout diviseurs de ce nombre sera aussi de la forme $\alpha^\mu \beta^\nu \gamma^\pi$ etc., où les exposants μ, ν, π, etc., ne pourront surpasser m, n, p, etc." Of course, this is the Unique Factorization Theorem. No proof is given. Legendre seems to consider it selfevident.

Thus, both Euler and Legendre, without Euclid's linguistic or notational handicaps, give complete statements of the Unique Factorization Theorem. But, although they prove the existence of the representation as a product of primes, neither of them explicitly treats the uniqueness.

The exact statement of the Unique Factorization Theorem as given in the text appears in the work of Gauss. In his *Disquisitiones Arithmeticae* [2.8], he states, "Any composite number can be resolved into prime factors in only one way." The proof that Gauss gives is discussed in a note on §2.2.

§2.1. Exercise 3. Mersenne and Fermat primes have an interesting history. Numbers of the form $M_p = 2^p - 1$ are called Mersenne numbers after Marin Mersenne, a Minimite friar of the seventeenth century. He served as a "clearing house for mathematical information," in that he promoted a correspondence among the most prominent mathematicians of his day [2.9]. Among others, he corresponded with Descartes, Fermat, and Pascal. Interest in determining whether numbers of specific forms gave primes was particularly intense at that time. Mersenne produced the first 12 Mersenne primes (one of which was actually a composite) in the preface to his book *Cogitata-Physico-Mathematica* in 1644. The values of p that he correctly gave as yielding Mersenne primes are: (1), 2, 3, 5, 7, 13, 17, 19, 31, 127, and 257. He mistakenly included 67, and below 257 he overlooked 61, 89, and 107 [2.10].

Fermat primes are primes of the form $F_n = 2^{2^n} + 1$. These are the occasion of one of the two conjectural errors known to have been made by Pierre de Fermat (1601–1665), considered to be the "founding father of number theory" [2.11]. From the fact that F_n is prime for $n = 0, 1, 2, 3, 4$, Fermat conjectured that F_n was prime for all values of n. After searching 14 years for a proof of this conjecture, Fermat began to doubt its truth. In a letter to Pascal, admitting this growing doubt, he entreated Pascal to find a proof [2.12]. Nearly 100 years later, in 1739, Euler showed that $2^{2^5} + 1$ has the factor 641 [2.13].

§2.2. As described in the notes on §2.1, Euclid stated and proved the Prime Divisibility Lemma as Proposition 30 of Book VII of his *Elements*. He then used it to prove the Unique Factorization Theorem (Book IX, Proposition 14) in the special case where the number A has three distinct prime factors. His proof can easily be generalized.

In his *Théorie des Nombres* (based on the *Essai* of 1798), Legendre, who makes no explicit mention of the uniqueness of factorization, does state the contrapositive of the Prime Divisibility Lemma. Chronologically, Legendre's *Essai* predates Gauss' *Disquisitiones Arithmeticae* (1801). However, Gauss indicates that he had already finished the major portion of his book when the *Essai* appeared. (It is surprising that Gauss cared!)

For his proof of the Unique Factorization Theorem, Gauss uses the contrapositive of the Prime Divisibility Lemma:

"Si nullus numerorum a, b, c, d etc., per numerorum primum p dividi potest, etiam productum $abcd$ etc., per p dividi non poterit." Thus, if a number is expressed as a product of prime powers, it follows that the same primes must appear in every such factorization. Gauss then shows that if a number has two different factorizations in which the same prime appears with unequal exponents, dividing out the smaller power of this prime from both factorizations produces a contradiction of the lemma.

§2.3. The Coprime Divisibility Lemma is usually not stated as such. It is also a hidden implicit element of the purely inductive proofs given in §2.4.

§2.4. The second proof given in this section is similar to that given by F. A. Lindemann in 1933 [2.14]. The same proof appears in Davenport's *The Higher Arithmetic* [2.15] with the comment, "I am not aware of having seen this proof elsewhere, but it is unlikely that it is new!"

§2.6. Another proof of the Unique Factorization Theorem can be constructed that rests directly on the identity $(a, b)\{a, b\} = ab$. This proof mimics that of a theorem in group theory, called the Jordan–Holder Theorem, which actually implies the Unique Factorization Theorem. Namely, for a given integer n, one considers chains

(I) $$a_0 = n \to a_1 \to a_2 \to \cdots a_r = 1,$$

where each a_i is divisible by a_{i+1}, and $a_i/a_{i+1} = p_i$, a prime. The claim is that for all such chains out of a given integer n, the length r is the same, and the set of "factors," the primes p_i, is the same apart from rearrangement. The proof is by induction on n. Assuming the result for all smaller positive integers, consider another chain out of n,

(II) $$b_0 = n \to b_1 \to b_2 \to \cdots \to b_s = 1.$$

If $a_1 = b_1$, we're through. Thus we have $a_1 \neq b_1$, and note that $\{a_1, b_1\} = n$. We then form the chains

(I') $$n \to a_1 \to (a_1, b_1) \to c_3 \to \cdots c_t = 1,$$

(II') $$n \to b_1 \to (a_1, b_1) \to c_3 \to \cdots c_t = 1,$$

where the c_i represent any admissible chain out of (a_1, b_1). It is then easily seen that both of these are chains of the type under consideration. (This follows, since $(n/a_1) = \{a_1, b_1\}/a_1 = b_1/(a_1, b_1)$, and $(n/b_1) = \{a_1, b_1\}/b_1 = a_1/(a_1, b_1)$.) But then (I') and (II') have the same prime factors in some order. By our inductive hypothesis, this is true for (I) and (I') as well as (II) and (II'). The desired result then follows for (I) and (II).

§2.7. Exercise 5. This problem is easily solved using a result called Bertrand's Postulate, which is proved in Chapter 9. It should be possible to give a simple direct proof, but the author knows of no such proof.

§2.8. Theorem 2.8.2 is credited to Euler, who proved in 1737 that both $\Sigma (1/p)$ and $\Pi (1 - 1/p)^{-1}$ diverge. The latter is given as Theorem 7 [2.16]. Euler then used this to prove that $\Sigma(1/p)$ diverges (Theorem 19 of the same work).

§2.8. Exercise 4. The product expression given for $\zeta(s)$ is known as an "Euler factorization," and was given by Euler in Chapter 15 of his *Introductio Analysin Infinitorum*. It is one of many similar expressions that he formulated involving infinite sums and products. In particular, Euler derived the values of $\zeta(s)$ for s an even positive integer [2.17].

Nevertheless, the Riemann Zeta Function bears the name of Riemann because he extended the definition to complex s, and established many of the basic properties of $\zeta(s)$.

Georg Riemann (1826–1866), a student of Gauss and Weber, began his studies in theology but quickly turned to mathematics. He remained at Göttingen during his relatively short life, where he succeeded Dirichlet as professor of mathematics in 1859.

§2.9. The concept of valuation was introduced by Josef Kurschak in 1913. In his paper [2.18], Kurschak identifies Kurt Hensel's work as the inspiration for this concept, namely, Hensel's investigations on p-adic numbers in 1907. Hensel had invented an analogue of the absolute value that he denoted by $|A| = p^a$, where p^a is the highest power of the prime p that divides A [2.19]. Kurschak adapted this idea, replacing p by $1/p$, and later by $1/e$. This provided a model whose properties he pursued, not only over the rationals, but for algebraic number fields.

In 1918, Ostrowski continued the development of valuations [2.20]. He distinguished the archimedean and nonarchimedean cases, deriving the inequality (2.9.5) in the latter case. Focusing on the field of rationals, he proved Theorem 2.9.2. Ostrowski's work was later much simplified by E. Artin and it is substantially Artin's presentation that is given here. Artin also proved Theorem 2.9.3 in a more general form [2.21].

REFERENCES

2.1. Ref. 1.4, Vol. I, p. 332.

2.2. Ref. 1.4, Vol. I, pp. 412–413.

2.3. C. F. Gauss, *Disquisitiones Arithmeticae, Werke*, I, Gottingen: Koniglichen Gesellschaft Der Wissenschaften, 1863, pp. 1–466 (original: 1801).

2.4. G. H. Hardy and E. M. Wright, *An Introduction To The Theory of Numbers*, 3rd edition, New York: Oxford University Press, 1960, p. 10.

2.5. S. Bochner, Mathematical Reflections, *American Mathematical Monthly*, **81**, 1974, 827–852.

2.6. L. Euler, *Elements of Algebra*, London, 1797 (original: 1770), pp. 17, 18, 27.

2.7. A.-M. Legendre, *Théorie des Nombres*, 2 vols. (4th edition, 2nd edition, 1830), Paris: Librairie Scientifique et Technique A. Blanchard, 1955, p. 5.

2.8. Ref. 2.3, p. 15.

2.9. C. B. Boyer, *A History of Mathematics*, New York: Wiley 1968, p. 367.

2.10. F. Cajori, *A History of Mathematics*, New York: Macmillan, 1919, p. 167.

2.11. O. Ore, *Number Theory and Its History*, New York: McGraw-Hill, 1948, p. 54.

2.12. R. Noguès, *Théorèmes De Fermat*, Paris: Albert Blanchard, 1966, pp. 8–9.

2.13. Ref. 2.11, p. 74.

2.14. F. A. Lindemann, The Unique Factorization Of A Positive Integer, *Quarterly Journal of Mathematics*, **4**, 1933, 319–320.

2.15. H. Davenport, *The Higher Arithmetic*, 2nd Edition, London: Hutchinson and Co., 1962 (1st edition 1952), pp. 19–20.

2.16. L. Euler, Variae Observationes Circa Series Infinitas, *Opera Omnia*, Leipzig: B. G. Teubner, 1924, I, 14, pp. 216–244 (original: 1748).

2.17. L. Euler, *Introductio In Analysin Infinitorum*, Ref. 2.16, I, p. 8.

2.18. J. Kurschak, Uber Limesbildung und Allgemeine Korpertheorie, *Crelle's Journal*, **142**, 1913, 211–253.

2.19. K. Hensel, *Theorie Der Algebraischen Zahlen*, Leipzig: B. G. Teubner, 1908.

2.20. A. Ostrowski, Uber Einige Losungen Der Funktionalgleichung $\varphi(x)\varphi(y) = \varphi(xy)$, *Acta Mathematica*, **41**, 1918, 271–284.

2.21. E. Artin, *Algebraic Numbers and Algebraic Functions*, New York: Gordon and Breach, 1967.

3

ARITHMETIC
FUNCTIONS

§3.1 THE FUNDAMENTAL ARITHMETIC FUNCTIONS

In the course of the investigation of properties of the integers, various special functions and types of functions play a central role. One such class of functions are called *arithmetic functions* and are defined as follows.

Definition. 3.1.1. An *arithmetic function* $f(n)$ is a complex valued function defined for all integers $n \geq 1$.

In many cases of interest, the values of $f(n)$ are also integers. Among such arithmetic functions, those that occur most frequently include

(3.1.1) $\qquad\qquad d(n) =$ number of positive divisors of n

(3.1.2) $\qquad\qquad \sigma(n) =$ sum of the positive divisors of n

(3.1.3) $\qquad \phi(n) =$ number of positive integers $\leq n$ that are relatively prime to n

(3.1.4) $\qquad\qquad \nu(n) =$ number of distinct prime factors of n

(3.1.5) $\qquad\qquad \Omega(n) =$ number of prime factors of n

(3.1.6) $\qquad \mu(n) = \begin{cases} 0 & \text{if } n \text{ is not squarefree} \\ (-1)^{\nu(n)} & \text{if } n \text{ is squarefree} \end{cases}, \qquad \mu(1) = 1$

Note that the nature of the definitions of $d(n)$, $\sigma(n)$, and $\phi(n)$ is such that it is immediate that they are well defined. As for the functions $\nu(n)$, $\Omega(n)$, $\mu(n)$, that they are well defined is a consequence of the Unique Factorization Theorem.

In counting the number of divisors $d(n)$, the numbers 1 and n must be included. Thus, for example,

$$d(1) = 1, \quad d(2) = 2, \quad d(3) = 2, \quad d(4) = 3.$$

Similarly, $\sigma(n)$ includes both 1 and n in summing up all the divisors of n, so that

$$\sigma(1) = 1, \qquad \sigma(2) = 3, \qquad \sigma(3) = 4, \qquad \sigma(4) = 7,$$
$$\sigma(5) = 6, \qquad \sigma(6) = 12.$$

The functions $\nu(n)$ and $\Omega(n)$ differ in that $\Omega(n)$ counts the total number of prime factors in the representation of n as a product of primes, including multiplicities, whereas $\nu(n)$ counts only the distinct primes. Thus

$$\nu(1) = 0 \qquad \nu(2) = 1, \qquad \nu(3) = 1, \qquad \nu(4) = 1,$$
$$\nu(5) = 1, \qquad \nu(27) = 1;$$

$$\Omega(1) = 0, \qquad \Omega(2) = 1, \qquad \Omega(3) = 1, \qquad \Omega(4) = 2,$$
$$\Omega(5) = 1, \qquad \Omega(27) = 3.$$

The function $\phi(n)$ is known as the *Euler function,* with typical values

$$\phi(1) = 1, \qquad \phi(2) = 1, \qquad \phi(3) = 2, \qquad \phi(4) = 2,$$
$$\phi(5) = 4, \qquad \phi(6) = 2, \qquad \phi(7) = 6.$$

The function $\mu(n)$ is known as the *Möbius function.* Its early values are

$$\mu(1) = 1, \qquad \mu(2) = -1, \qquad \mu(3) = -1, \qquad \mu(4) = 0,$$
$$\mu(5) = -1, \qquad \mu(6) = 1.$$

This chapter will be devoted to the consideration of the basic properties of each of the above functions. In addition, we shall consider various illustrative problems of number theory that relate to these functions. Further, we will develop certain more general properties of arithmetic functions from the example provided by the special cases. (All integers considered are nonnegative, unless otherwise specified.)

§3.2 THE NUMBER OF DIVISORS OF n

A given integer $n > 1$ has a canonical representation of the form

$$(3.2.1) \qquad\qquad n = p_1^{\alpha_1} p_2^{\alpha_2} \ldots p_r^{\alpha_r},$$

where the p_i are distinct primes and $\alpha_i > 0, i = 1, \ldots, r$. Any divisor m of n must have a representation

$$(3.2.2) \qquad\qquad m = p_1^{\beta_1} p_2^{\beta_2} \ldots p_r^{\beta_r},$$

where $0 \leq \beta_i \leq \alpha_i, i = 1, \ldots, r$, and any expression of the form (3.2.2) such that $0 \leq \beta_i \leq \alpha_i, i = 1, \ldots, r$ is a divisor of n. Further, by the Unique Factorization Theorem, each divisor of n occurs *exactly* once in these representations. Hence, the number of divisors of n equals the number of possible ways of choosing integers $\beta_1, \ldots, \beta_r$ that satisfy

$$0 \le \beta_i \le \alpha_i, \quad i = 1, \ldots, r.$$

Thus we obtain

$$(3.2.3) \qquad d(n) = (\alpha_1 + 1)(\alpha_2 + 1) \ldots (\alpha_r + 1) = \prod_{i=1}^{r} (\alpha_i + 1),$$

which provides a formula for $d(n)$.

Noting that $d(p_i^{\alpha_i}) = \alpha_i + 1$, (3.2.3) implies that

$$(3.2.4) \qquad\qquad\qquad d(n) = \prod_{i=1}^{r} d(p_i^{\alpha_i}).$$

EXERCISES

1. Prove that $d(n)$ is odd if and only if n is a square.

2. Prove that if $(m, n) = 1$, then $d(mn) = d(m)d(n)$.

3. Prove that if n is squarefree, then $d(n) = 2^{v(n)}$.

4. Prove that the number of solutions for nonnegative integers x, y, in

$$d(2^x 3^y) = m$$

   equals $d(m)$.

5. Show that for infinitely many integers n, $d(n) < d(n + 1)$, and also that $d(n) > d(n + 1)$ for infintely many n.

6. For $p_1, \ldots, p_r$ a fixed set of primes and m a fixed positive integer, show that the number of nonnegative integer solutions $x_1, \ldots, x_r$, of

$$d(p_1^{x_1} p_2^{x_2} \cdots p_r^{x_r}) = m,$$

   is finite.

7. Prove that the product of the positive divisors of a positive integer n equals $n^{(1/2)d(n)}$.

8. Prove that for a fixed positive integer k, the equation

$$n = kd(n)$$

   has at most a finite number of solutions for the integer n.

9. Prove that for any fixed $\epsilon > 0$

$$\lim_{n \to \infty} \frac{d(n)}{n^\epsilon} = 0.$$

§3.3 THE SUM OF THE DIVISORS OF n

We now introduce the notation $\sum_{d/n}$ to denote a summation taken over all the divisors d of n. With this notation we may write

$$(3.3.1) \qquad\qquad\qquad d(n) = \sum_{d/n} 1.$$

Similarly, for the sum of the divisors,

$$(3.3.2) \qquad\qquad\qquad \sigma(n) = \sum_{d/n} d.$$

Since, as d runs over the divisors of n so does (n/d), this formalism provides

$$\sigma(n) = \sum_{d/n} \frac{n}{d} = n \sum_{d/n} \frac{1}{d}$$

or

$$(3.3.3) \qquad\qquad\qquad \frac{\sigma(n)}{n} = \sum_{d/n} \frac{1}{d}.$$

Thus, $\sigma(n)/n$ equals the sum of the reciprocals of the divisors of n.

From Corollary 2, Theorem 1.4.4, we obtain that if d divides ab where $(a, b) = 1$, then $d = d_1 d_2$ where d_1/a and d_2/b. Further, from the Unique Factorization Theorem, it follows that such a representation is unique; that is, if

$$d_1 d_2 = d_1' d_2', \qquad d_1'/a, \qquad d_2'/b,$$

then $d_1 = d_1'$, $d_2 = d_2'$. Thus we can conclude that as d_1 runs over the divisors of a, and d_2 runs independently over the divisors of b, $d_1 d_2$ covers all of the divisors of ab exactly once [provided of course that $(a, b) = 1$]. This observation is the basis of the proof of the following theorem.

Theorem 3.3.1. If $(a, b) = 1$, then

$$(3.3.4) \qquad\qquad\qquad \sigma(ab) = \sigma(a)\sigma(b).$$

    Proof. We have

$$\sigma(ab) = \sum_{d|ab} d = \sum_{\substack{d_1|a \\ d_2|b}} d_1 d_2$$

$$\sigma(ab) = \left(\sum_{d_1|a} d_1\right)\left(\sum_{d_2|b} d_2\right) = \sigma(a)\sigma(b).$$

Corollary. For $n = \prod_{i=1}^{r} p_i^{\alpha_i}$ we have

(3.3.5)
$$\sigma(n) = \prod_{i=1}^{r} \frac{p_i^{\alpha_i+1} - 1}{p_i - 1}.$$

Proof. From the definition,

$$\sigma(p_i^{\alpha_i}) = 1 + p_i + p_i^2 + \cdots + p_i^{\alpha_i}$$

$$= \frac{p_i^{\alpha_i+1} - 1}{p_i - 1}.$$

Then, repeated application of Theorem 3.3.1 provides that

$$\sigma(n) = \prod_{i=1}^{r} \sigma(p_i^{\alpha_i}) = \prod_{i=1}^{r} \frac{p_i^{\alpha_i+1} - 1}{p_i - 1}.$$

One of the oldest notions in mathematics relates to the function $\sigma(n)$. This is the notion of *perfect number*.

Definition 3.3.1. A positive integer n is said to be *perfect* if and only if

(3.3.6)
$$\sigma(n) = 2n.$$

No odd perfect numbers are known and it is suspected that none exist. As for even perfect numbers, they can be determined, in a certain sense.

Theorem 3.3.2. An even integer n is perfect if and only if

(3.3.7)
$$n = 2^{q-1}(2^q - 1),$$

where q and $2^q - 1$ are primes.

Proof. If n is of the form (3.3.7), since $(2^{q-1}, 2^q - 1) = 1$, we have

$$\sigma(n) = \sigma(2^{q-1})\sigma(2^q - 1)$$

$$(2^q - 1)2^q = 2n,$$

where we've used the assumption that $2^q - 1$ is a prime.

Conversely, if n is even and perfect, we have $n = 2^{\alpha-1}s$, s odd, $\alpha > 1$, and $\sigma(n) = 2n$. This, in turn, implies

$$2^{\alpha}s = 2n$$

$$= \sigma(n) = \sigma(2^{\alpha-1}s) = \sigma(2^{\alpha-1})\sigma(s)$$

or

(3.3.8) $$2^{\alpha}s = (2^{\alpha} - 1)\sigma(s).$$

In (3.3.8), solving for $\sigma(s)$ yields

(3.3.9) $$\sigma(s) = s + \frac{s}{2^{\alpha} - 1}.$$

From (3.3.9) we see that $(s/(2^{\alpha} - 1))$ is an integer, say t, and, of course, $t \neq s$ would be a divisor of s. Thus from (3.3.9),

$$\sigma(s) = s + t;$$

that is $\sigma(s)$ equals the sum of two divisors of s, s and t. However, by definition, $\sigma(s)$ equals the sum of *all* of the divisors of s. From this it follows that s and t are *all* the divisors of s; so $t = 1$ and s is a prime. Since

$$1 = t = \frac{s}{2^{\alpha} - 1},$$

we must have $s = 2^{\alpha} - 1$. From Exercise 3(a), of §2.1, it follows that α must also be a prime. Hence $s = 2^{\alpha-1}(2^{\alpha} - 1)$ where α and $2^{\alpha} - 1$ are primes as required.

There is an element of incompleteness in Theorem 3.3.2 due to the fact that it is not known whether there are infinitely many Mersenne primes $2^q - 1$.

EXERCISES

1. Show that if m divides n, then

$$\frac{\sigma(n)}{n} \geq \frac{\sigma(m)}{m}$$

with equality holding only if $m = n$. From this deduce that no proper divisors of a perfect number can be perfect.

2. Prove that for all positive integers n

$$\sigma(n^2) \leq \sigma^2(n)$$

with equality holding only for $n = 1$. More generally, prove that

$$\sigma(ab) \leq \sigma(a)\sigma(b)$$

3. Consider the equation

$$(m - 1)\sigma(n) = mn$$

and show that the solutions in integers m and n, $m > 2$, are precisely

$$n = m - 1 = \text{a prime.}$$

4. Prove that no perfect number is a power of a prime.

5. Show that if m and n are two prime powers (not necessarily of the same prime) such that

$$\frac{\sigma(m)}{m} = \frac{\sigma(n)}{n}$$

then it follows that $m = n$.

6. Show that $n = 2^{15} \cdot 3^7 \cdot 5 \cdot 7 \cdot 11 \cdot 17 \cdot 41 \cdot 43 \cdot 257$ satisfies the equation $\sigma(n) = 5n$. Find another solution.

7. Show that an odd perfect number must have the form

$$n = p^{4m+1}s^2$$

where p is a prime of the form $4k + 1$, and $(p, s) = 1$.

8. Show that all numbers in the closed interval $[0, 1]$ are limit points of the sequence $n/\sigma(n)$.

9. Prove that for every fixed positive integer n, the equation $\sigma(x) = n$ has at most n solutions for x.

10. Prove that the number of integers $n \leq x$ such that n is perfect is $\leq \sqrt{x}$.

§3.3A   ODD PERFECT NUMBERS

The question as to whether there exist odd perfect numbers is as yet unsolved. (Not even a wrong proof has ever been published.) The general suspicion is that odd perfect numbers do not exist. There is, in fact, some meager evidence to support this. Exercise 4 of §3.3 provides that there can be no odd perfect number that is a power of a prime; this suggests the consideration of odd perfect numbers with a given number of distinct prime factors. In this connection, it is known that an odd perfect number must have at least seven distinct prime factors, although here we prove only that it must have four.

Theorem 3.3A.1.   An odd perfect number must have at least four distinct prime factors.

Proof.   If $n = \prod_{i=1}^{M} q_i^{\alpha_i}$ is the unique factorization of n, where the q_i are distinct primes $[M = \nu(n)]$, then

$$\frac{\sigma(n)}{n} = \prod_{i=1}^{M} \frac{q_i - 1/q_i^{\alpha_i}}{q_i - 1},$$

so that

(3.3A.1)
$$\frac{\sigma(n)}{n} < \prod_{i=1}^{M} \frac{q_i}{q_i - 1}.$$

Thus if n is perfect, (3.3A.1) yields the *necessary* condition

(3.3A.2)
$$2 < \prod_{i=1}^{M} \frac{q_i}{q_i - 1}.$$

With this simple inequality we may now quickly dispose of the case $M = 2$ of the theorem. For suppose we have an odd perfect number of the form $n = q_1^{\alpha_1} q_2^{\alpha_2}$ and $q_1 < q_2$. Then, from (3.3A.2) we have

$$2 < \frac{q_1}{q_1 - 1} \cdot \frac{q_2}{q_2 - 1} \leq \frac{3}{2} \cdot \frac{5}{4} = \frac{15}{8},$$

since $q_1 \geq 3$ and $q_2 \geq 5$. This, of course, is a contradiction.

If we attempt the same argument for $M = 3$, with $n = q_1^{\alpha_1} q_2^{\alpha_2} q_3^{\alpha_3}$, $q_1 < q_2 < q_3$ all primes, then (3.3A.2) yields

(3.3A.3)
$$2 < \frac{q_1}{q_1 - 1} \cdot \frac{q_2}{q_2 - 1} \cdot \frac{q_3}{q_3 - 1} \leq \frac{3}{2} \cdot \frac{5}{4} \cdot \frac{7}{6} = \frac{105}{48}.$$

But since 105/48 is larger than 2, there is no contradiction. However, the argument may be modified so as to produce some information. *Suppose* that $q_1 \geq 5$. Then $q_2 \geq 7$, $q_3 \geq 11$, and

$$\frac{q_1}{q_1 - 1} \cdot \frac{q_2}{q_2 - 1} \cdot \frac{q_3}{q_3 - 1} \leq \frac{5}{4} \cdot \frac{7}{6} \cdot \frac{11}{10} = \frac{77}{48} < 2,$$

which contradicts (3.3A.3). Thus we conclude that $q_1 = 3$. Similarly, *suppose* that $q_2 \geq 7$, so that $q_3 \geq 11$. We then have

$$\frac{q_1}{q_1 - 1} \cdot \frac{q_2}{q_2 - 1} \cdot \frac{q_3}{q_3 - 1} \leq \frac{3}{2} \cdot \frac{7}{6} \cdot \frac{11}{10} = \frac{231}{120} < 2.$$

This again contradicts (3.3A.3), so we conclude that $q_2 = 5$.

Thus, at this point, we've ascertained that an odd perfect number with three distinct prime factors would have to be of the form

$$(3.3A.4) \qquad\qquad n = 3^\alpha \cdot 5^\beta \cdot p^\gamma,$$

where p is a prime ≥ 7. From (3.3A.2) we have

$$2 < \frac{3}{2} \cdot \frac{5}{4} \cdot \frac{p}{p - 1} \qquad \text{or} \qquad \frac{16}{15} < \frac{p}{p - 1}.$$

this last inequality implies $p < 16$, so in (3.3A.4) the only possibilities for p are 7, 11, and 13.

From the above argument we see that in order to complete the proof of the theorem it must be shown there can be no perfect number that is of any of the following three forms:

$$(3.3A.5) \qquad\qquad n = 3^\alpha 5^\beta 7^\gamma$$

$$(3.3A.6) \qquad\qquad n = 3^\alpha 5^\beta 11^\gamma$$

$$(3.3A.7) \qquad\qquad n = 3^\alpha 5^\beta 13^\gamma,$$

where α, β, γ are positive integers.

Each of these cases will be treated separately. However, a useful tool on several occasions is a lower bound inequality analogous to (3.3A.3). More precisely, for $n = \Pi_{i=1}^{M} q_i^{\alpha_i}$

$$(3.3A.8) \qquad \frac{\sigma(n)}{n} = \prod_{i=1}^{M} \left(1 + \frac{1}{q_i} + \frac{1}{q_i^2} + \cdots + \frac{1}{q_i^{\alpha_i}} \right)$$

so if $\sigma(n)/n = 2$, and $0 \leq \delta_i \leq \alpha_i$, $i = 1, \ldots, M$, we must have

$$(3.3A.9) \qquad 2 \geq \prod_{i=1}^{M} \left(1 + \frac{1}{q_i} + \cdots + \frac{1}{q_i^{\delta_i}} \right).$$

CASE OF (3.3A.5). $n = 3^\alpha 5^\beta 7^\gamma$

Since $\sigma(n) = 2n$, and n is odd, 4 cannot divide $\sigma(n)$. Consequently, 4 cannot divide $\sigma(3^\alpha)$. Thus $\alpha = 1$, $\sigma(3^\alpha) = \sigma(3) = 4$ is impossible, so that $\alpha \geq 2$. Similarly, for $\gamma = 1$, $\sigma(7^\gamma) = \sigma(7) = 8$ is untenable, so that $\gamma \geq 2$. Hence we may take $\delta_1 = 2$, $\delta_2 = 1$, $\delta_3 = 2$ in (3.3A.9), which yields

$$2 \geq \left(1 + \frac{1}{3} + \frac{1}{3^2}\right)\left(1 + \frac{1}{5}\right)\left(1 + \frac{1}{7} + \frac{1}{7^2}\right) = \frac{494}{245} > 2.$$

From this contradiction we conclude that no integer of the form (3.3A.5) can be perfect.

CASE OF (3.3A.6). $n = 3^\alpha 5^\beta 11^\gamma$

For $\alpha = 1, 2, 3$, $\sigma(3^\alpha) = 4, 13, 40$. As noted above, $\alpha = 1, 3$ are then not possible since 4 does not divide $\sigma(n)$. In this case, 13 also does not divide $\sigma(n) = 2n$, and we conclude that $\alpha \geq 4$. For $\gamma = 1$, $\sigma(11) = 12$, which is impossible, so that $\gamma \geq 2$.

We now have two subcases according as $\beta \geq 2$ or $\beta = 1$.

Subcase I. $\beta \geq 2$
Taking $\delta_1 = 4$, $\delta_2 = 2$, $\delta_3 = 2$ in (3.3A.9) yields

$$2 \geq \left(1 + \frac{1}{3} + \frac{1}{3^2} + \frac{1}{3^3} + \frac{1}{3^4}\right)\left(1 + \frac{1}{5} + \frac{1}{5^2}\right)\left(1 + \frac{1}{11} + \frac{1}{11^2}\right) = \frac{4123}{2025} > 2,$$

a contradiction.

Subcase II. $\beta = 1$
Here we make upper estimates that take advantage of the fact that $\beta = 1$. Namely,

$$2 = \frac{\sigma(n)}{n} < \frac{3}{2} \cdot \frac{\sigma(5^\beta)}{5^\beta} \cdot \frac{11}{10} = \frac{3}{2} \cdot \frac{6}{5} \cdot \frac{11}{10} = \frac{99}{50} < 2,$$

a contradiction.

CASE OF (3.3A.7). $n = 3^\alpha 5^\beta 13^\gamma$
As in the above, $\alpha \neq 1, 3$ and $\beta \geq 1$. Further, for $\gamma = 1$, $\sigma(13^\gamma) = 14$, which is impossible since 7 does not divide n in this case. Thus $\gamma \geq 2$.

We consider three cases, as follows.

Subcase I. $\alpha \geq 4$ and $\beta \geq 2$
Then, taking $\delta_1 = 4$, $\delta_2 = 2$, $\delta_3 = 2$ in (3.3A.9) yields

$$2 \geq \left(1 + \frac{1}{3} + \frac{1}{3^2} + \frac{1}{3^3} + \frac{1}{3^4}\right)\left(1 + \frac{1}{5} + \frac{1}{5^2}\right)\left(1 + \frac{1}{13} + \frac{1}{13^2}\right) = \frac{228{,}811}{114{,}075} > 2,$$

a contradiction.

Subcase II. $\beta = 1$
Using an upper estimate, we have

$$2 = \frac{\sigma(n)}{n} < \frac{3}{2} \cdot \frac{\sigma(5^\beta)}{5^\beta} \cdot \frac{13}{12} = \frac{3}{2} \cdot \frac{6}{5} \cdot \frac{13}{12} = \frac{39}{20} < 2,$$

a contradiction.

Subcase III. $\alpha = 2$

Again, estimating from above,

$$2 = \frac{\sigma(n)}{n} < \frac{\sigma(3^\alpha)}{3^\alpha} \cdot \frac{5}{4} \cdot \frac{13}{12} = \frac{13}{9} \cdot \frac{5}{4} \cdot \frac{13}{12} = \frac{845}{432} < 2,$$

a contradiction.

The above proof of Theorem 3.3A.1 is purely numerical in nature. There are, however, other approaches that are more arithmetic in nature. This can be illustrated by the following treatment of Case (3.3A.7).

Applying Exercise 7 of §3.3, $n = p^{4k+1}s^2$, where p is a prime of the form $4t + 1$. Since $n = 3^\alpha 5^\beta 13^\gamma$, we have two subcases according as $p = 5$ or 13.

Subcase A. $p = 5$, β is odd, α and γ even.
Then since $\sigma(n)/n = 2$,

$$(3.3A.10) \qquad \left(\frac{3^{\alpha+1} - 1}{2}\right)\left(\frac{5^{\beta+1} - 1}{4}\right)\left(\frac{13^{\gamma+1} - 1}{12}\right) = 2 \cdot 3^\alpha 5^\beta 13^\gamma.$$

We have that 5 does not divide $(5^{\beta+1} - 1)/4$, and by Exercise 9 of §1.4, the fact that α and γ are even precludes the possibility of 5 dividing either $(3^{\alpha+1} - 1)/2$ or $(13^{\gamma+1} - 1)/12$. Thus in this subcase, 5 cannot divide the left side of (3.3A.10) which is a contradiction.

Subcase B. $p = 13$, γ odd, α and β even.
Again, the equation (3.3A.10) holds. But by Exercise 10 of §1.4, γ odd implies that 7 divides $(13^{\gamma+1} - 1)/12$, which is absurd since 7 does not appear as a factor on the right side of (3.3A.10).

EXERCISES

1. Prove that there exists at most a finite number of odd perfect numbers with a given number of distinct prime factors.

2. For a real number $\alpha > 1$, define a *primitive α-abundant* number to be an integer n such that

$$\sigma(n) \geq \alpha n,$$

and no proper divisor of n has this property. For each irrational $\alpha > 1$, prove that the number of primitive α-abundant integers, with a fixed number of distinct prime factors, is finite.

3. For α rational, a *necessary* condition that there exist infinitely many primitive α-abundant numbers with k distinct prime factors is that α have a representation of the form

$$\alpha = \frac{\sigma(a)}{a} \prod_{p/b} \frac{p}{p-1}$$

where $(a, b) = 1$, $b > 1$, and $v(ab) < k$.

4. For every positive prime q, show that there exists at most a finite number of primitive $q^2/(q^2 - 1)$-abundant numbers with a fixed number of distinct prime factors.

5. Show that if there is, at most, a finite number of solutions to the equation $\sigma(n) = 3n$, then there is, at most, a finite number of odd perfect numbers.

6. For polynomials $f(x) \in Q[x]$, define $\sigma[f(x)] = $ the sum of those polynomials, with leading coefficient 1, that divide $f(x)$. Call $f(x)$ pseudoperfect if for some fixed rational number r

$$\sigma[f(x)] = rf(x).$$

Show that a pseudoperfect polynomial must be a constant (i.e., of degree zero).

§3.4   MULTIPLICATIVE ARITHMETIC FUNCTIONS

Both the functions $d(n)$ and $\sigma(n)$ reveal a common structural property as seen in Exercise 2, of §3.2 and Theorem 3.3.1. The property may be abstracted as folows.

Definition 3.4.1. An arithmetic function $f(n)$ is called *multiplicative* if and only if for $(m, n) = 1$ we have

(3.4.1) $f(mn) = f(m)f(n).$

Having noted that $d(n)$ and $\sigma(n)$ are multiplicative, it is natural to look for some common source of these occurrences. This is found in the following theorem.

Theorem 3.4.1. If $f(n)$ is multiplicative, then

(3.4.2) $g(n) = \sum_{d/n} f(d)$

is also multiplicative.

Proof. For $(m, n) = 1$, we have that as d_1 and d_2 run independently over the divisors of m and n respectively, the product $d_1 d_2$ covers each divisor of mn once and only once. Thus

$$g(mn) = \sum_{d/mn} f(d) = \sum_{\substack{d_1/m \\ d_2/n}} f(d_1 d_2)$$

$$= \sum_{\substack{d_1/m \\ d_2/n}} f(d_1)f(d_2) = \left(\sum_{d_1/m} f(d_1)\right)\left(\sum_{d_2/n} f(d_2)\right)$$

$$= g(m)g(n).$$

We now see that since 1 and n are multiplicative arithmetic functions, the theorem quickly reaffirms that

$$d(n) = \sum_{c/n} 1$$

and

$$\sigma(n) = \sum_{c/n} c$$

are multiplicative.

Multiplicativity can be useful in proving identities in which both sides of the identity are clearly multiplicative. In such cases, it suffices to verify the identity for prime powers. For example, consider the assertion

(3.4.3)
$$\sum_{c/n} d^3(c) = \left(\sum_{c/n} d(c)\right)^2,$$

in which, from Theorem 3.4.1, it is clear that both sides are multiplicative. Thus, in order to establish (3.4.3), it suffices to verify it for $n = p^\alpha$, a power of a prime. In this case,

$$\sum_{c/p^\alpha} d^3(c) = d^3(1) + d^3(p) + \cdots + d^3(p^\alpha)$$

$$= 1^3 + 2^3 + \cdots + (\alpha + 1)^3$$

and

$$\left(\sum_{c/p^\alpha} d(c)\right)^2 = (d(1) + d(p) + \cdots + d(p^\alpha))^2$$

$$= (1 + 2 + \cdots + (\alpha + 1))^2$$

$$= \left(\frac{(\alpha + 1)(\alpha + 2)}{2}\right)^2.$$

Consequently, the completion of the proof of (3.4.3) reduces to verifying the identity

$$(3.4.4) \qquad 1^3 + 2^3 + \cdots + (\alpha + 1)^3 = \left(\frac{(\alpha + 1)(\alpha + 2)}{2} \right)^2.$$

This is clearly true for $\alpha = 0$ and we may proceed by induction on α, *assuming*

$$1^3 + 2^3 + \cdots + \alpha^3 = \left(\frac{\alpha(\alpha + 1)}{2} \right)^2.$$

Then

$$1^3 + 2^3 + \cdots + \alpha^3 + (\alpha + 1)^3 = \left(\frac{\alpha(\alpha + 1)}{2} \right)^2 + (\alpha + 1)^3$$

and in order to complete the induction, we need only establish that

$$(3.4.5) \qquad \left(\frac{\alpha(\alpha + 1)}{2} \right)^2 + (\alpha + 1)^3 = \left(\frac{(\alpha + 1)(\alpha + 2)}{2} \right)^2.$$

EXERCISES

1. For any real number ρ, let $\sigma_\rho(n) =$ the sum of the ρth powers of the divisors of n. Show that $\sigma_\rho(n)$ is multiplicative, and that for $n = \prod_{i=1}^r p_i^{\alpha_i}$ (the canonical factorization of n)

$$\sigma_\rho(n) = \prod_{i=1}^r \frac{p_i^{\rho(\alpha_i + 1)} - 1}{p_i - 1}.$$

2. Let $d_0(n) \equiv 1$, and inductively define for each integer $k \geq 1$

$$d_k(n) = \sum_{c/n} d_{k-1}(c).$$

   Prove that all the $d_k(n)$ are multiplicative. Interpret $d_k(n)$ arithmetically in terms of the divisors of n.

3. Prove that $f(n)$ is multiplicative if and only if for $n = p^\alpha s$, $(s, p) = 1$, p a prime, $f(p^\alpha s) = f(p^\alpha)f(s)$.

4. Prove that if $f(n)$ is multiplicative, and $n = \prod_{i=1}^r p_i^{\alpha_i}$ is the unique factorization of n, then

$$\sum_{d/n} f(d) = \prod_{i=1}^r \left(\sum_{j=0}^{\alpha_i} f(p_i^j) \right).$$

5. Prove that if $f(n)$ and $h(n)$ are multiplicative arithmetic functions, then

$$g(n) = \sum_{d/n} f(d)h\left(\frac{n}{d}\right)$$

is also multiplicative. This is a generalization of Theorem 3.4.1, in that taking $h(n) \equiv 1$ reduces the above to the case of the theorem.

6. A multiplicative function $P(n)$ is called *completely multiplicative* if for *all* pairs of positive integers, m, n,

$$P(mn) = P(m)P(n).$$

Show that the assertion that $P(n)$ is not identically 0 is equivalent to $P(1) = 1$. Also prove that $P(n)$ is completely determined by the values of $P(p)$ for p a prime.

7. Given integers $n_1, \ldots, n_r$, consider the vectors $(d_1, \ldots, d_r)$ such that d_i divides n_i, $i = 1, \ldots, r$. The number of such vectors is clearly $d(n_1)d(n_2) \cdots d(n_r)$. Show that the number of these vectors such that the components d_i are relatively prime by pairs equals $d(n_1 n_2 \cdots n_r)$.

§3.5 THE MÖBIUS FUNCTION

From the definition of the function $\mu(n)$, for an integer n with canonical representation

$$n = p_1^{\alpha_i} p_2^{\alpha_2} \cdots p_r^{\alpha_r},$$

we have $\mu(1) = 1$ and

$$(3.5.1) \qquad \mu(n) = \begin{cases} 0 & \text{if } \max_i \alpha_i > 1 \\ (-1)^r & \text{if all } \alpha_i = 1. \end{cases}$$

Then, it follows readily that $\mu(n)$ is multiplicative; that is, for $(m, n) = 1$,

$$(3.5.2) \qquad \mu(mn) = \mu(m)\mu(n).$$

The basic property of $\mu(n)$ that is the source of its deep role in the theory of numbers is the equation

$$(3.5.3) \qquad \sum_{d/n} \mu(d) = \begin{cases} 1 & \text{if } n = 1 \\ 0 & \text{if } n > 1. \end{cases}$$

Taking $n = 1$, we have

$$\sum_{d/1} \mu(d) = \mu(1) = 1.$$

For $n > 1$, we note that by Theorem 3.4.1, since $\mu(n)$ is multiplicative,

$$\Delta(n) = \sum_{d/n} \mu(d)$$

is also multiplicative. Since, for a prime power p^α we have

$$\Delta(p^\alpha) = \sum_{d/p^\alpha} \mu(d) = \mu(1) + \mu(p) = 0,$$

the proof of (3.5.3) is completed.

As we shall see in a later chapter, the property (3.5.3) of the μ-function, makes it very relevant for various counting processes. At this point, however, we consider rather its role in the formulation of inversion formulas, which also stems from (3.5.3).

Theorem 3.5.1. If $f(x)$ and $g(x)$ are any complex-valued functions of the real variable x, such that

$$(3.5.4) \qquad f(w) = \sum_{n \leq w} g\left(\frac{w}{n}\right) \qquad \text{for all } 1 \leq w \leq x,$$

then

$$(3.5.5) \qquad g(w) = \sum_{n \leq w} f\left(\frac{w}{n}\right) \mu(n) \qquad \text{for all } 1 \leq w \leq x.$$

Conversely, (3.5.5) implies (3.5.4).

Note that w ranges over all real numbers in the interval $1 \leq w \leq x$. Thus the above theorem can be viewed as asserting that the continuous system of equations (3.5.4) in the continuous set of unknowns $g(w)$, $1 \leq w \leq x$, has the solution given by (3.5.5) and this solution is unique.

Proof. Assuming (3.5.4), we have

$$\sum_{n \leq w} f\left(\frac{w}{n}\right) \mu(n) = \sum_{n \leq w} \mu(n) \sum_{m \leq w/n} g\left(\frac{w}{mn}\right)$$

$$= \sum_{mn \leq w} \mu(n) g\left(\frac{w}{mn}\right)$$

and introducing a new index of summation $c = mn$, this becomes

$$= \sum_{c \leq w} g\left(\frac{w}{c}\right) \sum_{n/c} \mu(n)$$

$$= g(w).$$

Similarly, if we assume (3.5.5), we have

$$\sum_{n \leq w} g\left(\frac{w}{n}\right) = \sum_{n \leq w} \sum_{m \leq w/n} \mu(m) f\left(\frac{w}{mn}\right)$$

$$= \sum_{mn \leq w} \mu(m) f\left(\frac{w}{mn}\right)$$

$$= \sum_{c \leq w} f\left(\frac{w}{c}\right) \sum_{m/c} \mu(m)$$

$$= f(w).$$

This completes the proof of the theorem.

Given an arithmetic function $f(n)$, we may extend it to a function $\tilde{f}(x)$ of a real variable x, by defining

$$(3.5.6) \qquad \tilde{f}(x) = \begin{cases} f(x) & \text{if } x \text{ is a positive integer} \\ 0 & \text{otherwise.} \end{cases}$$

If $g(n)$ is a second arithmetic function and $\tilde{g}(x)$ is similarly associated with it, the equations

$$(3.5.7) \qquad \tilde{f}(w) = \sum_{n \leq w} \tilde{g}\left(\frac{w}{n}\right), \qquad w \leq N$$

are equivalent to the equations

$$(3.5.8) \qquad f(w) = \sum_{n/w} g\left(\frac{w}{n}\right) \qquad \text{for all integers } w \leq N.$$

To see this, note first that for w *not* an integer, the (w/n) in (3.5.7) are not integers and for this w, the equation (3.5.7) reduces to $0 = 0$. In the remaining equations of (3.5.7) where w is an integer $\leq N$, $\tilde{g}(w/n) = 0$ unless n divides w, in which case it equals $g(w/n)$. Thus we see that for an integer $w \leq N$, the equation (3.5.7) becomes the corresponding one of (3.5.8). Similarly, the equations

(3.5.9) $$\tilde{g}(w) = \sum_{n \leq w} \tilde{f}\left(\frac{w}{n}\right)\mu(n) \qquad \text{for all } w \leq N$$

become

(3.5.10) $$g(w) = \sum_{n/w} f\left(\frac{w}{n}\right)\mu(n) \qquad \text{for all integers } w \leq N.$$

Thus Theorem 3.5.1 applied to $\tilde{f}$ and $\tilde{g}$ translates into the Möbius Inversion Formula.

Theorem 3.5.2.   (Möbius Inversion Formula).   If $f(n)$ and $g(n)$ are two arithmetic functions, then equations (3.5.8) and (3.5.10) are equivalent.

The above theorem can also be proved directly in an entirely similar fashion to the proof of Theorem 3.5.1. However, it is of interest to deduce it as a special case of that theorem, as we carried out above. There is also a refinement of Theorem 3.5.2, which can be proved in either fashion, to the effect that the equations (3.5.9) for all integers w *that divide* a fixed integer N, are equivalent to the equations (3.5.10) for this same range of w.

As a consequence of Theorem 3.5.2, we can establish the converse of Theorem 3.4.1.

Corollary.   For an arithmetic function $f(n)$,

$$g(n) = \sum_{d/n} f(d)$$

is multiplicative if and only if $f(n)$ is multiplicative.

Proof.   Theorem 3.4.1 takes care of the "if" part of the assertion. Next, assume that $g(n)$ is multiplicative. From Theorem 3.5.2

(3.5.11) $$f(n) = \sum_{d/n} \mu(d)g\left(\frac{n}{d}\right)$$

and since $\mu(n)$ and $g(n)$ are multiplicative, Exercise 5 of §3.4 provides that $f(n)$ is multiplicative.

The above theorems provide means for obtaining new identities from old ones, "by inversion." For example:

(i)   From

$$d(n) = \sum_{c/n} 1$$

we obtain via Theorem 3.5.2,

(3.5.12)
$$\sum_{c/n} \mu(c)d\left(\frac{n}{c}\right) = 1.$$

(ii)   From

$$\sigma(n) = \sum_{d/n} d$$

we obtain via Theorem 3.5.2,

(3.5.13)
$$\sum_{d/n} \mu(d)\sigma\left(\frac{n}{d}\right) = n.$$

(iii)   Taking $f(x) \equiv 1$ in (3.5.4), we have

$$f(w) = \sum_{n \leq w} 1 = [w].$$

Then, by Theorem 3.5.1, we have

(3.5.14)
$$\sum_{n \leq w} \mu(n)\left[\frac{w}{n}\right] = 1.$$

The inversion formulas presented above are all in what might be called "standard form." These can, in turn, be transformed into many other forms which appear to be (but are not) essentially different. Of course, in each instance the "new" version is easily proved directly. A useful class of such formulas is obtained by considering a function $\lambda(x, y)$ of two real variables $x \geq 1$, $y \geq 1$ which is such that for each fixed x, $\lambda(x, y)$ is a 1–1 map of the interval $1 \leq y \leq x$ into some initial segment of itself. More precisely, there is some function $\xi(x)$, satisfying $x \geq \xi(x) \geq 1$, such that $\eta_x(y) = \lambda(x, y)$ is a 1–1 map of $1 \leq y \leq x$ onto $1 \leq \eta_x(y) \leq \xi(x)$. For a fixed pair x, y, $1 \leq y \leq x$, and any positive integer n, $n \leq \lambda(x, y)$,

$$1 \leq \frac{\lambda(x, y)}{n} \leq \lambda(x, y) \leq \xi(x),$$

so that there exists a unique $y_n^* = y_n^*(x, y, n)$ such that $1 \leq y_n^* \leq x$, and

$$\frac{\lambda(x, y)}{n} = \lambda(x, y_n^*)$$

We then have

Theorem 3.5.3. For any fixed real number $x \geq 1$, the set of equations

$$(3.5.15) \quad f(x, y) = \sum_{n \leq \lambda(x,y)} g(x, y_n^*) \quad \text{for all } y \text{ such that } 1 \leq y \leq x$$

is equivalent to the set of equations

$$(3.5.16) \quad g(x, y) = \sum_{n \leq \lambda(x,y)} \mu(n) f(x, y_n^*) \quad \text{for all } y \text{ such that } 1 \leq y \leq x.$$

Proof. For fixed x, let $\lambda^{-1}(x, t)$ denote the function defined for $1 \leq t \leq \xi(x)$ such that $\lambda^{-1}(x, \lambda(x, y)) = y$ for $1 \leq y \leq x$. This exists and is unique because of our assumptions concerning $\lambda(x, y)$. Then define

$$\hat{f}(x, y) = f(x, \lambda^{-1}(x, y))$$
$$(3.5.17) \qquad \hat{g}(x, y) = g(x, \lambda^{-1}(x, y)).$$

For fixed x, Theorem 3.5.1 asserts that

$$(3.5.18) \qquad \hat{f}(x, y) = \sum_{n \leq y} \hat{g}\left(x, \frac{y}{n}\right) \quad \text{for all } y \leq \xi(x)$$

is equivalent to

$$(3.5.19) \qquad \hat{g}(x, y) = \sum_{n \leq y} \hat{f}\left(x, \frac{y}{n}\right) \mu(n) \quad \text{for all } y \leq \xi(x).$$

Replacing y by $\lambda(x, y)$, (3.5.18) is equivalent to

$$\hat{f}(x, \lambda(x, y)) = \sum_{n \leq \lambda(x,y)} \hat{g}\left(x, \frac{\lambda(x, y)}{n}\right)$$
$$= \sum_{n \leq \lambda(x,y)} \hat{g}(x, \lambda(x, y_n))$$

and noting that $\lambda^{-1}(x, \lambda(x, y)) = y$, by (3.5.17) this becomes

$$(3.5.20) \qquad f(x, y) = \sum_{n \leq \lambda(x, y)} g(x, y_n^*) \quad \text{for } 1 \leq y \leq x.$$

Similarly, (3.5.19) transforms into

$$(3.5.21) \qquad g(x, y) = \sum_{n \le \lambda(x, y)} f(x, y_n^*)\mu(n) \qquad \text{for } 1 \le y \le x$$

and the theorem is proved.

For the case $\lambda(x, y) = y$, $y_n^* = y/n$ and the result of this theorem is identical to that of Theorem 3.5.1. An interesting case is provided by

$$\lambda(x, y) = \frac{x}{y},$$

which is a 1–1 map of $1 \le y \le x$ into itself. Further, in this case, the solution of

$$\frac{\lambda(x, y)}{n} = \lambda(x, y_n^*) \qquad \text{for } n \le \lambda(x, y)$$

is given by

$$y_n^* = ny.$$

The theorem then becomes the following corollary.

Corollary 1. For each fixed $x \ge 1$, the set of equations

$$(3.5.22) \qquad f(x, y) = \sum_{n \le x/y} g(x, ny) \qquad \text{for all } y \le x$$

is equivalent to the set of equations

$$(3.5.23) \qquad g(x, y) = \sum_{n \le x/y} f(x, ny)\mu(n) \qquad \text{for all } y \le x.$$

As a last illustration, for any fixed integer $k \ge 1$ let

$$\lambda(x, y) = y^{1/k}.$$

Then for each fixed $x \ge 1$, $\eta_x(y) = \lambda(x, y)$ is a 1–1 map of $1 \le y \le x$ onto $1 \le \eta_x(y) \le x^{1/k}$, that is, $\xi(x) = x^{1/k}$. Further, the equation

$$\frac{\lambda(x, y)}{n} = \lambda(x, y_n^*), \qquad n \le \lambda(x, y)$$

has the solution

$$y_n^* = \frac{y}{n^k}.$$

The theorem then becomes a second corollary.

Corollary 2. For any fixed integer $k \geq 1$, the set of equations

$$(3.5.24) \qquad f(w) = \sum_{n \leq w^{1/k}} g\left(\frac{w}{n^k}\right) \qquad \text{for all } w \leq x$$

is equivalent to the set of equations

$$(3.5.25) \qquad g(w) = \sum_{n \leq w^{1/k}} f\left(\frac{w}{n^k}\right) \mu(n) \qquad \text{for all } w \leq x.$$

EXERCISES

1. Prove that

$$(d(n))^2 = \sum_{c/n} d\left(\frac{n^2}{c^2}\right).$$

2. Let $\sigma_p(n)$ be the arithmetic function introduced in Ex. 1 of §3.4. Prove that

$$(\sigma_p(n))^2 = \sum_{d/n} d^p \sigma_p\left(\frac{n^2}{d^2}\right).$$

3. Prove that for all real numbers $w \geq 1$

$$\left| \sum_{n \leq w} \frac{\mu(n)}{n} \right| \leq 1.$$

4. Let $F(u_1, \ldots, u_n), G(u_1, \ldots, u_n)$ be complex valued functions of n positive integer valued variables $u_1, \ldots, u_n$, (i.e., arithmetic functions of n variables). Show that the set of equations

   a. $\qquad G(u_1, \ldots, u_n) = \sum_{\substack{d_i/u_i \\ i=1,\ldots,n}} F(d_1, \ldots, d_n) \qquad \text{for all } (u_1, \ldots, u_n)$

   is equivalent to the set of equations

   b. $\qquad F(u_1, \ldots, u_n) = \sum_{\substack{d_i/u_i \\ i=1,\ldots,n}} \mu(d_1) \cdots \mu(d_n) G\left(\frac{u_1}{d_1}, \ldots, \frac{u_n}{d_n}\right)$

   $$\text{for all } (u_1, \ldots, u_n).$$

5. Let $F(u_1, \ldots, u_n), G(u_1, \ldots, u_n)$ be arithmetic functions of the n variables $u_1, \ldots, u_n$. Show that the set of equations

a.
$$G(u_1, \ldots, u_n) = \sum_{d/(u_1, \ldots, u_n)} f\left(\frac{u_1}{d}, \frac{u_2}{d}, \ldots, \frac{u_n}{d}\right)$$

for all integers $u_1, \ldots, u_n$ is equivalent to the set of equations

b.
$$F(u_1, \ldots, u_n) = \sum_{d/(u_1, \ldots, u_n)} \mu(d)G\left(\frac{u_1}{d}, \ldots, \frac{u_n}{d}\right)$$

for all integers $u_1, \ldots, u_n$.

6. Let $P(n)$ be a completely multiplicative function such that $P(1) = 1$. (See Ex. 6 of §3.4.) Show that the set of equations

a.
$$g(x) = \sum_{n \le x} P(n)f\left(\frac{x}{n}\right) \qquad \text{for all } x \text{ such that } 1 \le x \le W.$$

is equivalent to the set of equations

b.
$$f(x) = \sum_{n \le x} P(n)\mu(n)g\left(\frac{x}{n}\right) \qquad \text{for all } x \text{ such that } 1 \le x \le W.$$

(This is a generalization of Theorem 3.5.1, and reduces to it upon taking $P(n) \equiv 1$.)

7. Let $P(n)$ be completely multiplicative, $P(1) = 1$, and $f(n)$ an arithmetic function. Show that the set of equations

a.
$$g(n) = \sum_{c/n} P(c)f\left(\frac{n}{c}\right) \qquad \text{for all } n \text{ such that } n/N$$

is equivalent to the set of equations

b.
$$f(n) = \sum_{c/n} P(c)\mu(c)g\left(\frac{n}{c}\right) \qquad \text{for all } n \text{ such that } n/N.$$

8. Let $P(n)$ be a completely multiplicative arithmetic function, $P(1) = 1$, and set

$$F(n) = \sum_{d/n} P(d).$$

Prove that

$$\sum_{d/(u,v)} P(d)F\left(\frac{uv}{d^2}\right) = F(u)F(v).$$

Use this to give a solution to Ex. 2 above.

9. State and prove a generalization of the assertion of Ex. 5 so as to include a completely multiplicative function $P(n)$ as a factor inside the summations.

10. Deduce from Ex. 8 and Ex. 9 that for a completely multiplicative function $P(n)$, $P(1) = 1$, for

$$F(n) = \sum_{d/n} P(d)$$

we have

$$F(uv) = \sum_{d/(u,v)} \mu(d)P(d)F\left(\frac{u}{d}\right)F\left(\frac{v}{d}\right).$$

11. Formulate and prove the analogue of Theorem 3.5.3 for arithmetic functions, in which the summations are taken over the divisors of an integer.

12. The inversion formulas discussed in this section have analogous formulations for products. As illustration of this, prove that the equations

$$f(n) = \prod_{d/n} g(d) \qquad \text{for all } n/N,$$

are equivalent to

$$g(n) = \prod_{d/n} \left(f\left(\frac{n}{d}\right)\right)^{\mu(d)} \qquad \text{for all } n/N.$$

§3.6   ADDITIVE ARITHMETIC FUNCTIONS

Analogous to the notion of multiplicative arithmetic function, we have that of *additive arithmetic function*.

Definition 3.6.1.   An arithmetic function $f(n)$ is called *additive* if and only if for $(m, n) = 1$

$$(3.6.1) \qquad\qquad f(mn) = f(m) + f(n).$$

If $n = \prod_{i=1}^{r} p_i^{\alpha_i}$ is the unique factorization of n, for an additive function $f(n)$ we have

$$(3.6.2) \qquad\qquad f(n) = \sum_{i=1}^{r} f(p_i^{\alpha_i}).$$

Similarly, by analogy with the notion of the completely multiplicative function given in Ex. 6 of §3.4 we have the completely additive function.

Definition 3.6.2.   An additive arithmetic function is called *completely additive* if and only if (3.6.1) holds for *all* pairs of integers, m, n.

Still a third subclass of the additive arithmetic functions is singled out as strongly additive.

Definition 3.6.3.   An additive arithmetic function is called *strongly additive* if for all primes p

(3.6.3) $$f(p^\alpha) = f(p), \qquad \alpha \geq 1.$$

For $n = \prod_{i=1}^{r} p_i^{\alpha_i}$, the function $\nu(n)$ (= the number of distinct prime factors of n) may be evaluated as $\nu(n) = r$. It is clear that $\nu(n)$ is strongly additive. Similarly, since $\Omega(n) = \sum_{i=1}^{r} \alpha_i$, it follows that $\Omega(n)$ is completely additive.

Additive and multiplicative functions are simply related in that for nonzero constant c, in the equation

(3.6.4) $$g(n) = c^{f(n)},$$

$f(n)$ is additive if and only if $g(n)$ is multiplicative. Further, when $f(n)$ is strongly or completely additive, $g(n)$ is correspondingly called strongly or completely multiplicative.

EXERCISES

1. Prove that

$$\sum_{\substack{d/n \\ (d,n/d)=1}} 1 = \sum_{d/n} |\mu(d)| = 2^{\nu(n)}.$$

2. Prove that

$$\sum_{\substack{d/n \\ \nu(d)\,\text{even}}} |\mu(d)| = 2^{\nu(n)-1}.$$

3. Prove that for any complex number c

$$c^{\nu(n)} = \sum_{\substack{d/n \\ (d,n/d)=1}} (c-1)^{\nu(d)}.$$

4. Prove that

$$\pi(x) = \sum_{mn \leq x} \mu(m)\nu(n).$$

5. Prove that

$$\sum_{n \leq x} d(n) = \sum_{m \leq x} 2^{\nu(m)} \left[\sqrt{\frac{x}{m}} \right].$$

§3.7 THE EULER FUNCTION

The Euler function $\phi(n)$ is defined as the number of positive integers $\leq n$ that are relatively prime to n. A formula for $\phi(n)$ in terms of the canonical factorization of n is provided by the following theorem.

Theorem 3.7.1.   If $n = \prod_{i=1}^{r} p_i^{\alpha_i}$ is the unique factorization of n, then

(3.7.1) $$\phi(n) = \prod_{i=1}^{r} p_i^{\alpha_i-1}(p_i - 1).$$

This may be derived in many different ways, and two proofs will be discussed here. The first will depend on the following simple lemma.

Lemma 3.7.1.   The number of positive integers $\leq tm$ that are relatively prime to m equals $t\phi(m)$.

Proof.   The positive integers $\leq tm$ are contained in the subintervals

$$I_i : [im < n \leq (i + 1)m], \qquad i = 0, 1, \ldots, t - 1$$

If $(n, m) = 1$ and $n \in I_i$, then from Ex. 8 of §1.4,

$$(n - im, m) = (n, m) = 1$$

and

$$0 < n - im \leq m,$$

or

$$n' = n - im \in I_0.$$

Since these steps are reversible, the correspondence $n \leftrightarrow n'$ shows that the number of integers of I_i that are relatively prime to m equals the number of these that are in I_0, namely, $\phi(m)$. Thus since each I_i contains $\phi(m)$ integers that are relatively prime to m, the I_i, $i = 0, \ldots, t - 1$ taken together contain $t\phi(m)$ integers that are relatively prime to m.

First Proof of Theorem 3.7.1.   This proof consists of two main steps.

 (i)   If p is a prime such that p divides m, then

(3.7.2) $$\phi(pm) = p\phi(m).$$

Since p divides m, the condition $(n, pm) = 1$ is equivalent to $(n, m) = 1$. Thus from Lemma 3.7.1, the number of these (i.e., the number of integers $\leq$ pm) that are relatively prime to m equals $p\phi(m)$.

 (ii)   If p is a prime that does *not* divide m, then

(3.7.3) $$\phi(pm) = (p - 1)\phi(m).$$

We note first that the following set identity holds

(3.7.4) $\{1 \le n \le pm, (n, pm) = 1\} = \{1 \le n \le pm, (n, m) = 1\}$
$$- \{1 \le n \le pm, (n, m) = 1, p/n\}.$$

From Lemma 3.7.1, the number of integers in the first set on the right of (3.7.4) equals $p\phi(m)$. As for the second set on the right of (3.7.4), since p/n, or $n = pl$, then $n \le pm$ is equivalent to $l \le m$. Further, the condition $(pl, m) = 1$ is equivalent to $(l, m) = 1$ (since p does not divide m). Thus the second set on the right of (3.7.4) has as many elements as integers l, $1 \le l \le m$, such that $(l, m) = 1$, that is, $\phi(m)$. Hence, finally, the count in (3.7.4) yields

$$\phi(pm) = p\phi(m) - \phi(m) = (p - 1)\phi(m),$$

which is precisely (3.7.3).

From (3.7.2) and (3.7.3), it follows that for p a prime $(p, s) = 1$, we have

$$\phi(p^{\alpha}s) = p^{\alpha-1}\phi(ps)$$
(3.7.5) $$= p^{\alpha-1}(p - 1)\phi(s).$$

Taking $s = 1$ in (3.7.5) yields

(3.7.6) $$\phi(p^{\alpha}) = p^{\alpha-1}(p - 1)$$

so that (3.7.5) becomes

(3.7.7) $$\phi(p^{\alpha}s) = \phi(p^{\alpha})\phi(s).$$

From Ex. 3 of §3.4, it follows that $\phi(n)$ is multiplicative and this together with (3.7.6) yields (3.7.1).

As one might expect, from the very nature of the definition of $\phi(n)$, the proof of (3.7.1) given above really amounts to detailed counting. In the above proof, the counting is directed at the underlying multiplicativity and it is this that makes it work. We are thus led to ask whether one can carry out the counting in such a way that both the multiplicativity *and* (3.7.1) are natural by-products. That this can be done with the aid of the Möbius function is the substance of the second proof.

Second Proof of Theorem 3.7.1.   From (3.5.3) we have

(3.7.8) $$c(n) = \sum_{d/(n,m)} \mu(d) = \begin{cases} 1 & \text{if } (n, m) = 1 \\ 0 & \text{if } (n, m) > 1. \end{cases}$$

Hence

(3.7.9) $$\phi(m) = \sum_{n=1}^{m} c(n)$$

and, inserting (3.7.8) in (3.7.9), we obtain

$$(3.7.10) \qquad \phi(m) = \sum_{n=1}^{m} \sum_{d/(n,m)} \mu(d).$$

In (3.7.10), the first summation is over n and the inner one over d. Interchanging the two sum yields

$$\phi(m) = \sum_{d/m} \mu(d) \sum_{\substack{1 \le n \le m \\ n \text{ divisible by } d}} 1$$

$$= \sum_{d/m} \mu(d) \frac{m}{d}$$

$$(3.7.11) \qquad \phi(m) = m \sum_{d/m} \frac{\mu(d)}{d}.$$

Then since $\mu(d)$ and d are multiplicative, so is $\mu(d)/d$. Thus by Theorem 3.4.1, so is $\sum_{d/n} (\mu(d)/d$. Finally then, since n is multiplicative, it follows that $m \sum_{d/m} (\mu(d)/d = \phi(m)$ is also multiplicative. Also, from (3.7.11), for p a prime

$$\phi(p^\alpha) = p^\alpha \left(1 - \frac{1}{p}\right) = p^{\alpha-1}(p - 1),$$

which again completes a proof of (3.6.1).

Noting that (3.7.11), which asserts

$$\phi(m) = \sum_{d/m} \mu(d) \frac{m}{d},$$

is of the form (3.5.10) with $f(n) = n$, we can apply the Möbius Inversion, Theorem 3.5.2, to conclude that

$$(3.7.12) \qquad n = \sum_{d/n} \phi(d).$$

The function $\phi(n)$ appears in many identities and occurs naturally in many contexts. We note here a somewhat odd but amusing old chestnut known as Smith's Determinant. This determinant

$$(3.7.13) \qquad S_n = \begin{vmatrix} (1,1) & (1,2) & \cdots & (1,n) \\ & & (i,j) & \\ (n,1) & (n,2) & \cdots & (n,n) \end{vmatrix},$$

which corresponds to the integer n, is the determinant of the $n \times n$ array in which (i, j) appears in the ith row, jth column. The somewhat unexpected result is that

(3.7.14) $$S_n = \phi(1)\phi(2) \cdots \phi(n).$$

To prove this we start by noting that (3.7.12) gives

(3.7.15) $$(i, j) = \sum_{d/(i,j)} \phi(d).$$

Introducing the notation

$$\delta_{uv} = \begin{cases} 1 & \text{if } v \text{ divides } u \\ 0 & \text{if } v \text{ does not divide } u, \end{cases}$$

we may rewrite (3.7.15) as

(3.7.16) $$(i, j) = \sum_{d=1}^{n} \delta_{id}\delta_{jd}\phi(d).$$

Then, by the multiplication theorem for determinants, (3.7.16) implies that

$$\begin{aligned} \mathrm{Det}((i, j)) &= \mathrm{Det}(\delta_{id}) \, \mathrm{Det}(\delta_{jd}\phi(d)) \\ &= \phi(1) \cdots \phi(n)[\det(\delta_{id})]^2 \\ &= \phi(1) \cdots \phi(n), \end{aligned}$$

since $\det(\delta_{id}) = 1$.

EXERCISES

1. A multiplicative arithmetic function is called *strongly multiplicative* if and only if for every prime p,

   $$f(p^\alpha) = f(p), \qquad \text{for all } \alpha \geq 1.$$

   Show that $\phi(n)/n$ is strongly multiplicative.

2. Show that for $m > 2$, $\phi(m)$ is always even.

3. Prove that for $n > 1$

   $$\sum_{\substack{m \leq n \\ (m,n)=1}} m = \frac{1}{2}n\phi(n)$$

4. Prove that

$$\sum_{d=1}^{n} \phi(d)\left[\frac{n}{d}\right] = \frac{n(n+1)}{2}$$

5. Show for $d = (m, n)$ that

$$\phi(mn) = \phi(m)\phi(n)\frac{d}{\phi(d)}.$$

6. Show that

$$\frac{1}{\phi(n)} = \frac{1}{n}\sum_{d/n}\frac{\mu^2(d)}{\phi(d)}.$$

7. Prove that all solutions for x of the equation

$$\phi(x) = 4n - 2, \qquad n > 1,$$

are of the form p^α and $2p^\alpha$, where p is a prime of the form $4s - 1$.

8. Show for a fixed integer n, that the equation $\phi(x) = n$ has, at most, a finite number of solutions for x.

9. Show that for each integer $l \geq 1$, the equation

$$\phi(x) = \phi(x + l)$$

has at least one solution for x.

10. Prove that for any fixed integer $c > 1$, the equation $\phi(x) = x - c$ has, at most, a finite number of solutions.

11. Prove that the only solutions to the inequality

$$\phi(x) \geq x - \sqrt{x}$$

are $x = p$ and p^2 where p is a prime. If we require $\phi(x) > x - \sqrt{x}$, the only solutions are primes.

12. (Unsolved) Prove that if for a fixed integer m,

$$\phi(x) = m$$

has one integer solution x, then it must have a second solution.

13. Determine all solutions of

$$k\phi(n) = n.$$

14. For a given arithmetic function $f(n)$, let

$$F(m) = \sum_{d/m} f(d),$$

and for a given integer $n \geq 1$, let $D_n(f)$ denote the determinant of the $n \times n$ array $a_{ij} = F((i, j))$. Show that

$$D_n(f) = f(1)f(2) \cdots f(n).$$

15. Show that the number of pairs of positive integers a, b, $a \leq n$, $b \leq n$, such that $(a, b, n) = 1$ equals

$$n^2 \prod_{p/n} \left(1 - \frac{1}{p^2}\right).$$

16. An arithmetic function $f(n)$ (not identically zero) is called *over-multiplicative* if their exists an arithmetic function $g(d)$, such that for all pairs of positive integers m, n,

$$f(mn) = f(m)f(n)g((m, n)).$$

Prove that if we required that $f(1) = 1$ and $f(n) \neq 0$ for all n, then there exists a completely multiplicative arithmetic function $A(n)$ such that for $n = \prod_{i=1}^{r} p_i^{\alpha_i}$,

$$f(n) = \prod_{i=1}^{r} [A(p_i)]^{\alpha_i - 1} f(p_i)$$

$$= A(n) \prod_{i=1}^{r} \frac{f(p_i)}{A(p_i)}.$$

In other words,

$$g(d) = \prod_{p/d} \frac{A(p)}{f(p)}$$

(this product taken over the distinct primes dividing d), and hence $g(d)$ must be strongly multiplicative. Conclude then that, in this case, an arithmetic function is over-multiplicative if and only if it is the product of a completely multiplicative one and a strongly multiplicative one.

17. For every integer $n > 2$, $\phi(n)$ is even and $< n$. Thus if we define

$$\phi^{(2)}(n) = \phi(\phi(n)) \qquad \text{and in general} \qquad \phi^{(k)}(n) = \phi(\phi^{(k-1)}(n)),$$

for each $n > 2$, there is a unique integer k such that $\phi^{(k)}(n) = 2$. Define the function $C(n) = k$ for $n > 2$ and let $C(1) = C(2) = 0$. Prove that $C(n)$ is an additive function. Even more generally, prove that for all positive integers m, n

$$C(mn) = C(m) + C(n) + \epsilon(m, n),$$

where

$$\epsilon(m, n) = \begin{cases} 1 & \text{if both } m, n \text{ even} \\ 0 & \text{otherwise.} \end{cases}$$

(Note that this is a kind of over-additivity.)

18. Show that the product of the positive integers $\leq n$ and relatively prime to n equals.

$$n^{\phi(n)} \prod_{d/n} \left(\frac{d!}{d^d} \right)^{\mu(n/d)}.$$

§3.7A A PROPERTY OF THE NUMBER 30

If we consider the $\phi(30) = 8$ integers that are ≤ 30 and relatively prime to 30, we find these to be

$$1, 7, 11, 13, 17, 19, 23, 29.$$

Thus, apart from 1, all of these integers are primes. It is an interesting and somewhat startling fact that 30 is the largest integer with this property. The following theorem states this formally.

Theorem 3.7A.1. The number 30 is the largest integer such that all the integers less than it that are relatively prime to it are prime (apart from 1).

It is even more surprising that the proof of this can be carried out in a way that reduces, essentially, to the verification of a small number of simple cases. One could almost say that it is a proof by "computing the theorem to be true."

The idea of the proof is to consider $F_n =$ the number of positive integers $\leq \sqrt{n}$, and greater than 1, that are relatively prime to n. Then, if for a given n, one can show that $F_n > 0$, there is an integer s, $1 < s \leq \sqrt{n}$ such that $(s, n) = 1$. It follows that $s^2 \leq n$, $(s^2, n) = 1$, so that n would not have the property of the theorem.

As in (3.7.8), since

$$c(n) = \begin{cases} 1 & \text{if } (n, m) = 1 \\ 0 & \text{if } (n, m) > 1 \end{cases}$$

we have

$$F_n = \left(\sum_{m=1}^{\sqrt{n}} c(m) \right) - 1$$

$$= \sum_{m=1}^{\sqrt{n}} \sum_{d/(m,n)} \mu(d) - 1$$

$$= \sum_{d/n} \mu(d) \left[\frac{\sqrt{n}}{d} \right] - 1$$

so that

(3.7A.1)
$$F_n = \sum_{d/n} \mu(d) \left[\frac{\sqrt{n}}{d} \right] - 1.$$

Noting that

$$\frac{\sqrt{n}}{d} - 1 < \left[\frac{\sqrt{n}}{d} \right] \le \frac{\sqrt{n}}{d}$$

we have

$$F_n = \sum_{\substack{d/n \\ \mu(d)=+1}} \left[\frac{\sqrt{n}}{d} \right] - \sum_{\substack{d/n \\ \mu(d)=-1}} \left[\frac{\sqrt{n}}{d} \right] - 1$$

and hence

$$F_n > \sum_{\substack{d/n \\ \mu(d)=+1}} \left(\frac{\sqrt{n}}{d} - 1 \right) - \sum_{\substack{d/n \\ \mu(d)=-1}} \left(\frac{\sqrt{n}}{d} \right) - 1$$

or

(3.7A.2)
$$F_n > \sum_{d/n} \sqrt{n} \frac{\mu(d)}{d} - \sum_{\substack{d/n \\ \mu(d)=+1}} 1 - 1.$$

From Ex. 2 of §3.6,

$$\sum_{\substack{d/n \\ \mu(d)=+1}} 1 = 2^{\nu(n)-1}$$

so that (3.7A.2) yields

(3.7A.3)
$$F_n > \sqrt{n} \sum_{d/n} \frac{\mu(d)}{d} - (2^{\nu(n)-1} + 1)$$

or

$$F_n > \frac{\phi(n)}{\sqrt{n}} - (2^{\nu(n)-1} + 1).$$

Since $F_n > 0$ implies that n cannot have the property of the theorem, this is certainly the case whenever

(3.7A.4)
$$\frac{\phi(n)}{\sqrt{n}} \geq 2^{\nu(n)-1} + 1.$$

Setting $n = \prod_{i=1}^{r} p_i^{\alpha_i}$, (3.7A.4) becomes

(3.7A.5)
$$\prod_{i=1}^{r} p_i^{(1/2)(\alpha_i - 1)} \left(\frac{p_i - 1}{\sqrt{p_i}} \right) \geq 2^{r-1} + 1.$$

Dividing by 2^{r-1}, this is equivalent to

(3.7A.6)
$$2 \prod_{i=1}^{r} p_i^{(1/2)(\alpha_i - 1)} \left(\frac{p_i - 1}{2\sqrt{p_i}} \right) \geq 1 + \frac{1}{2^{r-1}}.$$

Assuming that $p_1 < p_2 < \ldots < p_r$, we now work numerically with (3.7A.6). First, we observe that if n is not divisible by 2, it is relatively prime to 2^2; if it is not divisible by 3, it is relatively prime to 3^2; if not divisible by 5, it is relatively prime to 5^2. Thus for $n > 25$, we need only consider integers n that are divisible by $2 \cdot 3 \cdot 5$. Then $r \geq 3$ and $p_1 = 2$, $p_2 = 3$, $p_3 = 5$. If $r > 3$, (3.7A.6) provides that if

$$2 \left(\frac{1}{2\sqrt{2}} \right) \left(\frac{2}{2\sqrt{3}} \right) \left(\frac{4}{2\sqrt{5}} \right) \left(\frac{p_r - 1}{2\sqrt{p_r}} \right) \geq 1 + \frac{1}{8}$$

[note that for $p \geq 7$, $(p - 1)/2\sqrt{p} \geq 1$], then n does not have the property of the theorem. But this last numerical inequality reduces to $p_r + 1/p_r \geq 39.9 \ldots$. Thus

$$7 \leq p_r \leq 39.$$

If $r > 3$, $n \geq 30.7 > 13^2 > 11^2 > 7^2$, so we must have that $p_4 = 7$, $p_5 = 11$, $p_6 = 13$. Hence n is divisible by $2 \cdot 3 \cdot 5 \cdot 7 \cdot 11 \cdot 13 > (41)^2$ and this implies that n is divisible by 41, which is a contradiction. Thus we must have $r = 3$ and

$$n = 2^\alpha 3^\beta 5^\gamma, \qquad \alpha > 0, \beta > 0, \gamma > 0.$$

If max $(\alpha, \beta, \gamma) > 1$, then $n \geq 60 > 7^2$, so that n could not have the property of the theorem. Thus, finally, $\alpha = \beta = \gamma = 1$ and we must have $n = 30$ as the only possibility > 25, which completes the proof of the theorem.

It is also of interest to note that the above theorem translates into a property of the sequence of primes as follows.

Theorem 3.7A.2. Let $p_1 = 2$, $p_2 = 3$, and, in general, p_i denote the ith prime. Then for $i \geq 4$,

(3.7A.7) $$p_1 p_2 \cdots p_i \geq p_{i+1}^2.$$

Proof. Since for $i \geq 4$,

$$n_i = p_1 \cdots p_i \geq 210 > 30,$$

there must be an integer s, $1 < s \leq n_i$ such that $(s, n_i) = 1$ and s is not a prime. Let s be the smallest such positive integer > 1. Then since every proper divisor of s is > 1, smaller than s, and relatively prime to n_i, every such proper divisor must be a prime. Thus $s = p^2$ or pq, p and q distinct primes. In the latter case, if $q > p$, $s = pq > p^2$, and p^2 should be a smaller composite integer that is relatively prime to n_i. Hence $s = p^2$, p a prime. Since p does not divide n_i, $p \geq p_{i+1}$, so that

$$n_i \geq s = p^2 \geq p_{i+1}^2,$$

which is precisely the assertion (3.7A.7).

EXERCISES

1. Show that Theorem 3.7A.2 implies Theorem 3.7A.1.

2. Let $\mathcal{W}_k$ be the set of integers n such that all the integers $1 < s \leq n$ such that $(s, n) = 1$ satisfy $\Omega(n) < k$. Prove for all integers $k > 2$ that the sets $\mathcal{W}_k$ are each finite. Theorem 3.7A.1 is essentially this result for $k = 2$.

3. Deduce from Ex. 2 above that for each fixed integer $k \geq 2$, there exists an $i^* = i^*(k)$ such that for all $i \geq i^*$

$$p_1 p_2 \cdots p_i \geq p_{i+1}^k.$$

§3.8 AVERAGES OF ARITHMETIC FUNCTIONS

In general, there are many ways to approach the study of the behavior of specific arithmetic functions and several of these will be met in later discussions. At this point, we consider the notion of mean value or average order of an arithmetic function. For an arithmetic function $f(n)$, we form the average of its values for all positive integers $n \leq x$,

(3.8.1)
$$M(f, x) = \frac{1}{x} \sum_{n \leq x} f(n).$$

Any statement that can be made about the behavior of $M(f, x)$ as $x \to \infty$ is then viewed as an assertion concerning "the average order of $f(n)$." On occasion, one can obtain more precise information so that it is possible to give good estimates for $xM(f, x)$ as $x \to \infty$.

We illustrate these notions with the Euler function, as follows.

Theorem 3.8.1.   We have

(3.8.2)
$$\sum_{n \leq x} \phi(n) = c_1 x^2 + R(x)$$

where for $x \geq 2$,

(3.8.3)
$$|R(x)| \leq x \log(x + 1) + \tfrac{9}{8}x.$$

The constant c_1 is given by

$$c_1 = \frac{1}{2} \sum_{n=1}^{\infty} \frac{\mu(n)}{n^2} > 0,$$

and, in fact, it can be shown that $c_1 = 3/\pi^2$.

Proof of Theorem 3.8.1.   We note first that

$$\sum_{n \leq x} \phi(n) = \sum_{n \leq x} n \sum_{d/n} \frac{\mu(d)}{d}$$

$$= \sum_{d \leq x} \frac{\mu(d)}{d} \sum_{\substack{n \leq x \\ n \, \text{divisible} \\ \text{by} \, d}} n$$

$$= \sum_{d \leq x} \mu(d) \sum_{t \leq x/d} t.$$

Using the fact that for an integer N

$$\sum_{t \leq N} t = \tfrac{1}{2}N(N + 1),$$

this last equation becomes

$$(3.8.4) \qquad \sum_{n \le x} \phi(n) = \frac{1}{2} \sum_{d \le x} \mu(d) \left[\frac{x}{d} \right] \left(\left[\frac{x}{d} \right] + 1 \right).$$

Since

$$\left[\frac{x}{d} \right] = \frac{x}{d} - \theta_d, \qquad 0 \le \theta_d < 1, \qquad \theta_d = \theta_d(d, x),$$

we have

$$\left[\frac{x}{d} \right] \left(\left[\frac{x}{d} \right] + 1 \right) = \left(\frac{x}{d} - \theta_d \right) \left(\frac{x}{d} - \theta_d + 1 \right)$$

$$= \left(\frac{x}{d} \right)^2 + E_d$$

where

$$|E_d| = \left| \frac{x}{d} (2\theta_d - 1) + \theta_d (1 - \theta_d) \right|$$

$$\le \frac{x}{d} + \frac{1}{4}.$$

Inserting this in (3.8.4) we obtain

$$\sum_{n \le x} \phi(n) = \frac{1}{2} \sum_{d \le x} \mu(d) \left(\frac{x^2}{d^2} + E_d \right)$$

$$= \frac{1}{2} \sum_{d \le x} \frac{\mu(d)}{d^2} \cdot x^2 + \frac{1}{2} \sum_{d \le x} \mu(d) E_d$$

$$= \left(\frac{1}{2} \sum_{d=1}^{\infty} \frac{\mu(d)}{d^2} \right) x^2 + \frac{1}{2} \sum_{d \le x} \mu(d) E_d - \left(\frac{1}{2} \sum_{d > x} \frac{\mu(d)}{d^2} \right) x^2$$

$$= c_1 x^2 + R(x)$$

where

$$(3.8.5) \qquad R(x) = \frac{1}{2} \sum_{d \le x} \mu(d) E_d - \left(\frac{1}{2} \sum_{d > x} \frac{\mu(d)}{d^2} \right) x^2.$$

Utilizing the fact that $|\mu(d)| \leq 1$, and $|E_d| \leq x/d + 1/4$, we obtain from (3.8.5) that

(3.8.6)
$$\frac{1}{2} \sum_{d \leq x} \left(\frac{x}{d} + \frac{1}{4} \right) \leq \frac{x}{2} \sum_{d \leq x} \frac{1}{d} + \frac{x}{8}$$

$$\leq x \sum_{d \leq x} \log\left(1 + \frac{1}{d} \right) + \frac{x}{8}$$

$$\leq x \log(x + 1) + \frac{x}{8},$$

and

$$\frac{1}{2} x^2 \left(\sum_{d > x} \frac{1}{d^2} \right) \leq \frac{x^2}{2} \sum_{d > x} \frac{1}{d(d - 1)} = \frac{1}{2} x^2 \sum_{d > x} \left(\frac{1}{d - 1} - \frac{1}{d} \right)$$

$$\leq \frac{1}{2} x^2 \cdot \frac{1}{[x]} \leq \left(\frac{1}{2} \right) \frac{x^2}{x - 1}$$

$$\leq x \quad \text{for } x \geq 2.$$

Combining the above, we obtain from (3.8.6) that for $x \geq 2$

$$|R(x)| \leq x \log(x + 1) + \tfrac{9}{8} x,$$

as required by (3.8.3). To complete the proof of the theorem it remains only to show that $c_1 > 0$. This follows from

$$c_1 = \frac{1}{2} \sum_{d=1}^{\infty} \frac{\mu(d)}{d^2}$$

$$> \frac{1}{2} \left(1 - \sum_{d=2}^{\infty} \frac{1}{d^2} \right) \geq \frac{1}{2} \left(1 - \sum_{d=2}^{\infty} \frac{1}{d(d - 1)} \right) = 0.$$

To translate the result of the above theorem into information about the average of $\phi(n)$, we divide both sides of (3.8.2) by x and obtain

(3.8.7)
$$M(\phi, x) = c_1 x + \frac{R(x)}{x}$$

where

$$\left| \frac{R(x)}{x} \right| \leq \log(x + 1) + \frac{9}{8}.$$

As x tends to infinity, this estimate for $R(x)/x$ is small with respect to the term c_1x, so that $M(\phi, x)$ is approximately c_1x. More precisely

$$\lim_{x \to \infty} \frac{M(\phi, x)}{c_1 x} = 1,$$

so again, roughly speaking, we say that "the average order of $\phi(n)$ is approximately $c_1 n$."

A similar analysis can be developed for the divisor function. In order to facilitate this, various estimates are gathered together in a preliminary lemma.

Lemma 3.8.1. We have for $x \geq 3$, that

(3.8.8) $$\sum_{d \leq x} \frac{1}{d} = \log x + c_2 + K(x)$$

where

(3.8.9) $$|K(x)| \leq \frac{4}{x}.$$

The constant

(3.8.10) $$c_2 = \sum_{d=1}^{\infty} \left(\frac{1}{d} - \log\left(1 + \frac{1}{d}\right) \right)$$

is known as Euler's constant, and satisfies the inequality $C_2 0$

(3.8.11) $$0 < c_2 < 2.$$

Proof. Assume first that x is an integer. Then

(3.8.12) $$\sum_{d \leq x} \frac{1}{d} = \sum_{1 \leq d \leq x} \left(\frac{1}{d} - \log \frac{d+1}{d} \right) + \log(x + 1)$$

$$= \sum_{d=1}^{\infty} \left(\frac{1}{d} - \log\left(1 + \frac{1}{d}\right) \right) - \sum_{d > x} \left(\frac{1}{d} - \log\left(1 + \frac{1}{d}\right) \right) + \log(x + 1).$$

Since

$$0 \leq \frac{1}{d} - \log\left(1 + \frac{1}{d}\right) \leq \frac{1}{d^2},$$

the infinite series converges to the constant c_2 as given in (3.8.10); and the inequality $c_2 > 0$ is immediate. Further,

$$\left| \sum_{d>x} \left(\frac{1}{d} - \log\left(1 + \frac{1}{d}\right) \right) \right| \le \sum_{d>x} \frac{1}{d^2}$$

$$\le \sum_{d>x} \frac{1}{d(d-1)} = \frac{1}{x}$$

and

$$\left| \log(x+1) - \log x \right| \le \frac{1}{x}$$

so that (3.8.12) assumes the form

$$\sum_{d \le x} \frac{1}{d} = \log x + c_2 + K(x)$$

with $|K(x)| < 2/x$. Finally if x is not an integer,

$$\sum_{d \le x} \frac{1}{d} = \sum_{d \le [x]} \frac{1}{d} = \log[x] + c_2 + K([x])$$

so that

$$K(x) = K([x]) + \log \frac{[x]}{x}.$$

Since $|K([x])| < 2/[x]$, this yields

$$|K(x)| < \frac{2}{[x]} + \frac{1}{[x]} \le \frac{4}{x} \qquad \text{for } x \ge 3,$$

which establishes (3.8.9). Also, $c_2 < 2$ follows by noting that

$$\sum_{d=2}^{\infty} \left(\frac{1}{d} - \log\left(1 + \frac{1}{d}\right) \right) < 1$$

Returning to the divisor function, we have the following theorem.

Theorem 3.8.2.   For all $x \ge 1$,

(3.8.13) $$\sum_{n \le x} d(n) = x \log x + R_1(x)$$

where

(3.8.14)
$$|R_1(x)| \leq 3x + 4.$$

Proof. From the definition of the divisor function

$$\sum_{n \leq x} d(n) = \sum_{n \leq x} \sum_{d/n} 1 = \sum_{d \leq x} \sum_{\substack{n \leq x \\ (n \text{ divisible} \\ \text{by } d)}} 1$$

$$= \sum_{d \leq x} \left[\frac{x}{d}\right]$$

$$= \sum_{d \leq x} \left(\frac{x}{d} - \theta_d\right).$$

Hence

(3.8.15)
$$\sum_{n \leq x} d(n) = x \sum_{d \leq x} \frac{1}{d} - \sum_{d \leq x} \theta_d.$$

Introducing (3.8.8) into the right side of (3.8.15), we get

$$\sum_{n \leq x} d(n) = x(\log x + c_2 + K(x)) - \sum_{d \leq x} \theta_d$$

so that for the $R_1(x)$ in (3.8.13),

(3.8.16)
$$R_1(x) = c_2 x + x K(x) - \sum_{d \leq x} \theta_d.$$

Then using (3.8.9), (3.8.11), and $|\theta_d| < 1$, we have for $x \geq 3$ that

$$|R_1(x)| < 2x + 4 + x = 3x + 4$$

as required by (3.8.14).

Since for $1 \leq x < 3$, $|\sum_{n \leq x} d(n)| \leq 3$, and $|x \log x| < 3x$, (3.8.14) holds also in this range, the proof of the theorem is completed.

Dividing both sides of (3.8.13) by x we obtain

(3.8.17)
$$M(d(n), x) = \log x + \frac{R_1(x)}{x}$$

where

$$\left|\frac{R_1(x)}{x}\right| \leq 3 + \frac{4}{x} \leq 7, \quad \text{for } x \geq 1.$$

Thus we see that as $x \to \infty$

$$\lim_{x \to \infty} \frac{M(d(n), x)}{\log x} = 1,$$

that is, "the average order of $d(n)$ is log x."

A close examination of the above argument reveals that the order of magnitude of the estimate for $R(x)$, obtained in (3.8.9), was forced by the order of magnitude of the estimate made for the term

$$\sum_{d \leq x} \theta_d,$$

which appears in (3.8.16). The term $c_2 x$ appearing in (3.8.16) is then "wasted" in being absorbed into an inequality. This inefficiency can be removed and Theorem 3.8.2 improved.

Theorem 3.8.3.   For all $x \geq 1$,

(3.8.18) $$\sum_{n \leq x} d(n) = x \log x + c_3 x + R_2(x)$$

where $c_3 = 2c_2 - 1$, and

(3.8.19) $$|R_2(x)| < 18\sqrt{x}.$$

Proof.   The proof begins exactly as with the earlier result, that is,

$$\sum_{n \leq x} d(n) = \sum_{n \leq x} \sum_{d/n} 1$$

and writing $n = de$, this becomes

(3.8.20) $$\sum_{n \leq x} d(n) = \sum_{de \leq x} 1,$$

where the sum on the right is a double sum over the indices d and e. The pairs of indices $[d, e]$ counted in this sum are precisely the points $[d, e]$ with *positive* integer

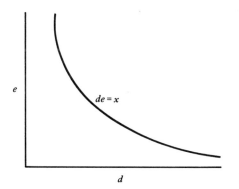

FIGURE 3.8.1

coordinates lying under or on the hyperbola $de = x$. Such points (with integer coordinates) are called *lattice points*; in fact, the sum on the right of (3.8.20) equals the number of lattice points in the first quadrant and under or on the hyperbola $de = x$. (See Fig. 3.8.1)

The device applied to the sum on the right of (3.8.20) counts the lattice points in the region

$$0 < d \leq \sqrt{x}, \qquad 0 < e \leq x/d,$$

plus the equal number in the symmetrical region

$$0 < e \leq \sqrt{x}, \qquad 0 < d \leq x/e.$$

Then noting that the number of lattice points in the square (See Fig. 3.8.2)

$$0 < d \leq \sqrt{x}, \qquad 0 < e \leq \sqrt{x}$$

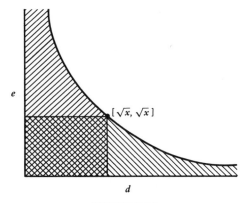

FIGURE 3.8.2

has been counted twice, this number is subtracted once. In detail, this may be written as

$$(3.8.21) \qquad \sum_{de \leq x} 1 = 2 \sum_{d \leq \sqrt{x}} \sum_{e \leq x/d} 1 - \left(\sum_{d \leq \sqrt{x}} 1 \right)^2.$$

The first term on the right of (3.8.21) equals

$$2 \sum_{d \leq \sqrt{x}} \left[\frac{x}{d} \right] = 2 \sum_{d \leq \sqrt{x}} \frac{x}{d} - 2 \sum_{d \leq \sqrt{x}} \theta_d.$$

From Lemma 3.8.1, for $x \geq 9$

$$2 \sum_{d \leq \sqrt{x}} \frac{x}{d} = 2x(\log \sqrt{x} + c_2 + K(\sqrt{x})),$$

where

$$|K(\sqrt{x})| \leq \frac{4}{\sqrt{x}} \qquad \text{and} \qquad \left| 2 \sum_{d \leq \sqrt{x}} \theta_d \right| \leq 2\sqrt{x}.$$

Hence for $x \geq 9$,

$$(3.8.22) \qquad 2 \sum_{d \leq \sqrt{x}} \left[\frac{x}{d} \right] = x \log x + 2c_2 x + K_1(x)$$

where

$$(3.8.23) \qquad |K_1(x)| \leq 10\sqrt{x}.$$

For the second term on the right of (3.8.21), we have

$$\left(\sum_{d \leq \sqrt{x}} 1 \right)^2 = [\sqrt{x}]^2 = (\sqrt{x} + \theta)^2, \qquad 0 \leq \theta < 1,$$

so that

$$(3.8.24) \qquad \left(\sum_{d \leq \sqrt{x}} 1 \right)^2 = x + K_2(x),$$

where

(3.8.25) $|K_2(x)| \le 2\sqrt{x} + 1.$

Combining (3.8.24) and (3.8.22) into (3.8.21) yields

$$\sum_{n \le x} d(n) = x \log x + (2c_2 - 1)x + K_1(x) - K_2(x).$$

Hence $R_2(x) = K_1(x) - K_2(x)$, and by (3.8.25) and (3.8.23), for $x \ge 9$,

$$|R_2(x)| \le |K_1(x)| + |K_2(x)| \le 12\sqrt{x} + 1 < 18\sqrt{x}$$

as required by (3.8.19). For $1 \le x < 9$, (3.8.19) is easily verified numerically. To see this, note that from (3.8.11)

$$-1 < c_3 < 3$$

so that from (3.8.18), for $x \ge 1$,

$$\sum_{n \le x} d(n) + x > R_2(x) > -x \log x - 3x.$$

For $2 \le x < 9$, this in turn implies

$$18\sqrt{x} > 20 + x > R_2(x) > -18\sqrt{x};$$

for $1 \le x < 2$, it implies

$$18\sqrt{x} \ge 3 > R_2(x) > -18\sqrt{x}.$$

Hence (3.8.19) is verified in all cases, and the proof of the theorem is complete.

The above theorem provides a sharper evaluation of $\sum_{n \le x} d(n)$ in that the terms through the order of x are included and the error term $R_2(x)$ has a much smaller order (less than $18\sqrt{x}$). The problem of investigating the behavior of $R_2(x)$ is known as the "Dirichlet Divisor Problem." In this connection, it can be proved, for example, that $|R_2(x)| < cx^{1/3}$ but that $|R_2(x)| < cx^{1/4}$ is not true.

It is now considered somewhat old fashioned to write out the estimates in the explicit form given above. More in vogue, and certainly for many arguments more facile, is the use of a scheme of notation which is as follows:

(i) $f(x) = 0(g(x))$ [read "$f(x)$ is large 0 of $g(x)$"]

as $x \to \infty$, denotes the assertion that there exists a constant K such that for all sufficiently large x

(3.8.26) $|f(x)| < K|g(x)|.$

If the expression $0(g(x))$ stands alone, it represents a function of x, say $f(x)$, that would satisfy (3.8.26). Thus for example $0(1)$ denotes a quantity that is bounded for x large.

(ii) $f(x) = o(g(x))$ [read "$f(x)$ is little o of $g(x)$"]

as $x \to \infty$, denotes the assertion that

(3.8.27) $$\lim_{x \to \infty} \frac{f(x)}{g(x)} = 0.$$

If the expression $o(g(x))$ stands alone, it denotes a function of x, say $f(x)$, that would satisfy (3.8.27). Thus, for example, $o(1)$ represents a quantity that tends to 0 as $x \to \infty$.

(iii) $f(x) \sim g(x)$ [read "$f(x)$ is asymptotic to $g(x)$"]

as $x \to \infty$ denotes the assertion that

(3.8.28) $$\lim_{x \to \infty} \frac{f(x)}{g(x)} = 1.$$

It is easily seen that (3.8.28) is equivalent to

$$f(x) = g(x) + o(g(x)) \qquad \text{or} \qquad g(x) = f(x) + o(f(x)).$$

There are various simple and useful rules for manipulating this symbolism. Among these we have

(1)   If $f(x) = 0(g(x))$, then

$$0(f(x)) + 0(g(x)) = 0(g(x)).$$

More generally, if $f_i(x) = 0(g(x))$, $i = 1, \ldots, r$, then

$$0(f_1(x)) + \cdots + 0(f_r(x)) = 0(g(x)).$$

(2)   Somewhat in the same spirit as (1), but for a sum of variable length, we have that if $g(x) > 0$ and $f(x) = 0(g(x))$, then

$$\sum_{n \leq x} f(n) = 0\left(\sum_{n \leq x} g(n)\right).$$

(3) $f(x) = o(g(x))$ implies $f(x) = 0(g(x))$.

These are easily verified. Other rules will emerge from usage of the notation.
In terms of the notation introduced above, (3.8.2) implies

(3.8.29)
$$\sum_{n \le x} \phi(n) = c_1 x^2 + 0(x \log x)$$

and (3.8.18) yields

(3.8.30)
$$\sum_{n \le x} d(n) = x \log x + c_3 x + 0(\sqrt{x}).$$

On many occasions a given average order result, of the type given above, may be transformed so as to yield other average order results. In the following, we provide two illustrations of such a process involving two different ideas.

Theorem 3.8.4. For the divisor function $d(n)$,

(3.8.31)
$$\sum_{n \le x} \frac{d(n)}{n} = \frac{1}{2} \log^2 x + (c_3 + 1) \log x + c_4 + 0\left(\frac{1}{\sqrt{x}}\right).$$

For the proof we need the following preliminary lemma.

Lemma 3.8.2. For every $\alpha > 1$, we have as $x \to \infty$

(3.8.32)
$$\sum_{n > x} \frac{1}{n^\alpha} = 0\left(\frac{1}{x^{\alpha-1}}\right).$$

Also, for $x \ge 1$

(3.8.33)
$$\sum_{n \le x} \frac{\log n}{n} = \frac{1}{2} \log^2 x + c_5 + 0\left(\frac{\log x}{x}\right).$$

Proof. To establish (3.8.32) we note that

$$\sum_{n > x} \frac{1}{n^\alpha} = \sum_{n > x} \int_{n-1}^n \frac{du}{n^\alpha} \le \sum_{n > x} \int_{n-1}^n \frac{du}{u^\alpha}$$

so that

$$\sum_{n > x} \frac{1}{n^\alpha} \le \int_{x-1}^\infty \frac{du}{u^\alpha} = \frac{1}{(\alpha - 1)(x - 1)^{\alpha-1}} = 0\left(\frac{1}{x^{\alpha-1}}\right).$$

As for (3.8.33), we have

$$\sum_{n \le x} \frac{\log n}{n} = \sum_{2 \le n \le x} \left(\frac{\log n}{n} - \int_{n-1}^{n} \frac{\log u}{u} du \right) + \sum_{2 \le n \le x} \int_{n-1}^{n} \frac{\log u}{u} du$$

$$= \sum_{2 \le n \le x} \int_{n-1}^{n} \left(\frac{\log n}{n} - \frac{\log u}{u} \right) du + \int_{1}^{[x]} \frac{\log u}{u} du.$$

Then, since

(3.8.34) $$\int_{1}^{x} \frac{\log u}{u} du = \frac{1}{2} \log^2 x$$

and

(3.8.35) $$\int_{n-1}^{n} \left(\frac{\log n}{n} - \frac{\log u}{u} \right) du = 0 \left(\frac{\log n}{n^2} \right),$$

the above yields

(3.8.36) $$\sum_{n \le x} \frac{\log n}{n} = \frac{1}{2} \log^2 x + \sum_{n=2}^{\infty} \int_{n-1}^{n} \left(\frac{\log n}{n} - \frac{\log u}{u} \right) du$$

$$- \sum_{n > x} \int_{n-1}^{n} \left(\frac{\log n}{n} - \frac{\log u}{u} \right) du.$$

By (3.8.35), the infinite series on the right of (3.8.36) converges. Denote its value by c_5. As for the last term

$$\left| \sum_{n > x} \int_{n-1}^{n} \left(\frac{\log n}{n} - \frac{\log u}{u} \right) du \right| \le \sum_{n > x} \int_{n-1}^{n} \left(\frac{\log (n - 1)}{n - 1} - \frac{\log n}{n} \right) du$$

$$\le \sum_{n > x} \left(\frac{\log (n - 1)}{n - 1} - \frac{\log n}{n} \right) = \frac{\log [x]}{[x]} = 0 \left(\frac{\log x}{x} \right),$$

so that (3.8.36) produces (3.8.33).

Proof of Theorem 3.8.4.   The derivation of (3.8.31) will be based on (3.8.30), and will be carried out by a procedure called "summation by parts." Setting

$$D(x) = \sum_{n \le x} d(n), \qquad D(0) = 0,$$

we have

$$(3.8.37) \qquad \sum_{n \le x} \frac{d(n)}{n} = \sum_{n \le x} \frac{1}{n}(D(n) - D(n-1)).$$

Rearranging the terms in the sum on the right of (3.8.37), we obtain

$$(3.8.38) \qquad \sum_{n \le x} \frac{d(n)}{n} = \frac{D([x])}{[x]} + \sum_{n \le x-1} D(n)\left(\frac{1}{n} - \frac{1}{n+1}\right)$$

(this is the essence of summation by parts). Inserting (3.8.30) in (3.8.38), we obtain

$$(3.8.39) \qquad \sum_{n \le x} \frac{d(n)}{n} = \left\{\log[x] + c_3 + 0\left(\frac{1}{\sqrt{x}}\right)\right\} + \sum_{n \le x-1} \frac{n \log n + c_3 n + R_2(n)}{n(n+1)}$$

[recall that $R_2(n) = 0(\sqrt{n})$]. The sum on the right of (3.8.39) equals

$$(3.8.40) \qquad \sum_{n \le x-1} \frac{\log n}{n+1} + c_3 \sum_{n \le x-1} \frac{1}{n+1} + \sum_{n \le x-1} \frac{R_2(n)}{n(n+1)}.$$

Consider each of these terms separately. First,

$$\sum_{n \le x-1} \frac{\log n}{n+1} = \sum_{1 \le n \le x-1} \frac{\log(n+1)}{n+1} - \sum_{n \le x-1} \frac{1}{n+1} \log\left(1 + \frac{1}{n}\right)$$

$$= \sum_{n \le x} \frac{\log n}{n} - \sum_{n=1}^{\infty} \frac{1}{n+1} \log\left(1 + \frac{1}{n}\right) + \sum_{n \ge x} \frac{1}{n+1} \log\left(1 + \frac{1}{n}\right).$$

Using (3.8.33) and the fact that $\log(1 + 1/n)$ equals $0(1/n)$, this

$$= \frac{1}{2} \log^2 x + c_5 - c_6 + 0\left(\frac{\log x}{x}\right) + 0\left(\sum_{n \ge x} \frac{1}{n^2}\right),$$

where $c_6 = \sum_{n=1}^{\infty} 1/(n+1) \log(1 + 1/n)$. Then applying (3.8.32) to the last sum, with $\alpha = 2$, we get

$$(3.8.41) \qquad \sum_{n \le x-1} \frac{\log n}{n+1} = \frac{1}{2} \log^2 x + c_7 + 0\left(\frac{\log x}{x}\right).$$

Next, from Lemma 3.8.1

$$c_3 \sum_{n \le x-1} \frac{1}{n+1} = c_3 \left(\log x + \left(c_2 - 1 \right) + 0\left(\frac{1}{x}\right) \right)$$

$$(3.8.42) \qquad\qquad = c_3 \log x + c_3(c_2 - 1) + 0\left(\frac{1}{x}\right).$$

The last term of (3.8.40) is

$$\sum_{n \le x-1} \frac{R_2(n)}{n(n+1)} = \sum_{n=1}^{\infty} \frac{R_2(n)}{n(n+1)} - \sum_{n > x-1} \frac{R_2(n)}{n(n+1)}$$

$$= c_8 + 0\left(\sum_{n > x-1} \frac{1}{n^{3/2}} \right),$$

so that using (3.8.32) with $\alpha = 3/2$ yields

$$(3.8.43) \qquad\qquad \sum_{n \le x-1} \frac{R_2(n)}{n(n+1)} = c_8 + 0\left(\frac{1}{\sqrt{x}}\right).$$

Inserting (3.8.41), (3.8.42), and (3.8.43) in (3.8.40) yields

$$(3.8.44) \qquad\qquad \frac{1}{2} \log^2 x + c_3 \log x + c_9 + 0\left(\frac{1}{\sqrt{x}}\right).$$

Returning then to (3.8.38) and recalling that (3.8.44) is an evaluation of the sum on the right, we obtain

$$\sum_{n \le x} \frac{d(n)}{n} = \frac{1}{2} \log^2 x + (c_3 + 1) \log x + c_4 + 0\left(\frac{1}{\sqrt{x}}\right)$$

as required by (3.8.31).

Turning to another type of transformation of one "average evaluation" into another, we will show that (3.8.30) can be used in this way so as to produce an estimation of

$$\sum_{n \le x} 2^{\nu(n)}.$$

This is achieved by means of an identity, which is provided by the following lemma.

Lemma 3.8.3. For $D(x) = \sum_{n \le x} d(n)$, we have

$$(3.8.45) \qquad \sum_{n \le x} 2^{\nu(n)} = \sum_{n \le \sqrt{x}} \mu(n) D\left(\frac{x}{n^2}\right).$$

Proof. As we have seen, (3.8.20) asserts that

$$(3.8.46) \qquad D(x) = \sum_{de \le x} 1.$$

Letting $\tau = (d, e)$ and writing $d = \tau d'$, $e = \tau e'$, where $(d', e') = 1$, (3.8.46) becomes

$$D(x) = \sum_{\substack{\tau^2 d' e' \le x \\ (d', e')=1}} 1$$

or

$$(3.8.47) \qquad D(x) = \sum_{\tau \le \sqrt{x}} \left(\sum_{m \le x/\tau^2} \sum_{\substack{d' e' = m \\ (d', e')=1}} 1 \right).$$

From Ex. 1 of §3.6,

$$\sum_{\substack{d' e' = m \\ (d', e')=1}} 1 = 2^{\nu(m)},$$

so that (3.8.47) becomes

$$(3.8.48) \qquad D(x) = \sum_{\tau \le \sqrt{x}} \left(\sum_{m \le x/\tau^2} 2^{\nu(m)} \right).$$

Then, by Corollary 2 of Theorem 3.5.3, applied with $k = 2$, (3.8.45) is an immediate consequence of (3.8.48).

The identity (3.8.45) may also be proved directly using (3.5.3) instead of the related machinery utilized above. Such a proof proceeds as follows:

$$\sum_{n \le x} 2^{\nu(n)} = \sum_{n \le x} \sum_{\substack{d/n \\ (d, n/d)=1}} 1 \qquad \text{Ex. 1 of §3.6}$$

$$= \sum_{\substack{dd' \le x \\ (d, d')=1}} 1 \qquad \text{setting } n = dd'$$

$$= \sum_{dd' \leq x} \sum_{\delta/(d,d')} \mu(\delta) \qquad \text{by (3.5.3)}$$

$$= \sum_{\delta \leq \sqrt{x}} \mu(\delta) \sum_{tt' \leq x/\delta^2} 1 \qquad \text{setting } d = \delta t, \ d' = \delta t'$$

$$= \sum_{\delta \leq \sqrt{x}} \mu(\delta) D\left(\frac{x}{\delta^2}\right) \qquad \text{by (3.8.20).}$$

From this lemma we now deduce the following theorem.

Theorem 3.8.5.

$$(3.8.49) \qquad \sum_{n \leq x} 2^{\nu(n)} = 2c_1 x \log x + c_{10} x + 0(\sqrt{x} \log x)$$

(it can be shown that $2c_1 = 6/\pi^2$).

   Proof.   Inserting (3.8.18) in (3.8.45) yields

$$\sum_{n \leq x} 2^{\nu(n)} = \sum_{n \leq \sqrt{x}} \mu(n)\left(\frac{x}{n^2} \log \frac{x}{n^2} + c_3 \frac{x}{n^2} + 0\left(\frac{\sqrt{x}}{n}\right)\right)$$

$$= x \log x\left(\sum_{n \leq \sqrt{x}} \frac{\mu(n)}{n^2}\right) - 2x \sum_{n \leq \sqrt{x}} \frac{\mu(n)\log n}{n^2}$$

$$+ c_3 x \sum_{n \leq \sqrt{x}} \frac{\mu(n)}{n^2} + 0(\sqrt{x} \log x).$$

Since

$$\sum_{n \leq \sqrt{x}} \frac{\mu(n)}{n^2} = 2c_1 + 0\left(\frac{1}{\sqrt{x}}\right)$$

once one verifies that

$$(3.8.50) \qquad \sum_{n \leq \sqrt{x}} \frac{\mu(n)\log n}{n^2} = c_{11} + 0\left(\frac{\log x}{\sqrt{x}}\right),$$

the estimate asserted in (3.8.49) follows. To establish this, note that

$$\left| \sum_{n > \sqrt{x}} \frac{\mu(n)\log n}{n^2} \right| \leq \sum_{n > \sqrt{x}} \frac{\log n}{n^2} = 0\left(\frac{\log \sqrt{x}}{x^{1/4}}\right) \sum_{n > \sqrt{x}} \frac{1}{n^{3/2}},$$

so that by applying (3.8.32) with $\alpha = 3/2$, this in turn is

$$0\left(\frac{\log \sqrt{x}}{x^{1/4}} \cdot \frac{1}{x^{1/4}}\right) = 0\left(\frac{\log x}{\sqrt{x}}\right).$$

EXERCISES

1. Using (3.8.29) estimate $\sum_{n \leq x} \phi(n)/n$ by summation by parts. Compare this with what you get by estimating $\sum_{n \leq x} \phi(n)/n$ directly.

2. Using (3.8.49) estimate $\sum_{n \leq x} 2^{\nu(n)}/n$.

3. Prove that

$$\sum_{n \leq x} \sigma(n) = c_{12}x^2 + 0(x \log x).$$

4. Prove that for fixed integer $l > 0$

$$\sum_{n \leq x} \frac{\phi(n + l) - \phi(n)}{n} \sim c_{13}l \log x$$

(where c_{13} is a constant independent of l and x).

5. Use (3.8.30) to estimate $\sum_{n \leq x} d(An)$, where A is a fixed positive integer. [*Hint.* Use. Ex. 10 of §3.5 with $P(n) \equiv 1$].

NOTES

§3.1.   The function $d(n)$ is also represented by the notation $\tau(n)$ in some texts. For example, Dickson, Kubilius, and Nagell all use the latter. Hardy and Wright, as well as Ramanujan, are among those who use $d(n)$.

The function $\sigma(n)$ has been a source of interest since the days of Euclid because of its connection with perfect numbers (discussed in §3.3). Various different notations, other than $\sigma(n)$, have been used over the past two centuries. For example, Euler employed $\int m$ in 1750, a practice continued by J. J. Sylvester in 1888 [3.1, 3.2]. In 1902 Cunningham used $s(n)$ [3.3], whereas in 1927 Landau chose the notation $S(n)$ [3.4]. However, $\sigma(n)$ is now the widely accepted notation.

The function $\varphi(n)$ was introduced by Euler in 1760 as a component of his generalization of a theorem of Fermat (discussed in Chapter 5) [3.5]. He did not employ any symbol for the function until 1780, when he used the notation πn [3.6]. The φ notation was introduced by Gauss in Article 38 of his *Disquisitiones Arithmeticae* [2.3].

The function $\nu(n)$ is also represented by the notation $\omega(n)$ by some authors. Interest in the ν function was stimulated in this century by the work of Hardy and Ramanujan (1920) [3.7].

The function $\mu(n)$ occurred implicitly in Euler's work as early as 1748, in connection with investigations concerning series. Euler never denoted the function by a symbol or studied it as a separate entity, but he did give the rule for determining the sign to be associated with each integer in his *Introductio in Analysin Infinitorum*. The first to systematically investigate the properties of this function was Augustus F. Möbius, a professor of astronomy at the University of Leipzig. Möbius' work appeared in 1832 [3.8], but even here the μ symbol was not used. The present notation was introduced by F. Mertens in

1874 [3.9]. In his paper, Mertens referred to Euler's work but did not mention Möbius. It is interesting to note that L. E. Dickson refers to the Möbius function as Mertens' function in the index of his *History* [3.10].

The arithmetic functions $d(n)$, $\sigma(n)$, $\phi(n)$, $\mu(n)$, are known to be algebraically independent [3.65].

§3.2. Formulas for $d(n)$ have existed since the Renaissance. In 1537 Cardan stated that a product P of k distinct prime factors has $1 + 2 + \ldots + 2^{k-1}$ divisors (excluding P). M. Stifel proved this rule in 1544. In 1732 Newton gave a method in his text *Arithmetica Universalis* that was tantamount to proving that $d(a^m b^n \ldots) = (m + 1)(n + 1) \ldots$ if a, b, $\ldots$ are distinct primes. This rule was proved by E. Waring in 1770. All of these formulas were stated verbally, without the use of a function symbol [3.10].

§3.3. (Theorem 3.3.1). The multiplicativity of $\sigma(n)$, as well as the formula for $\sigma(p^a)$, was known to Descartes in the seventeenth century. However, the first published proof of Theorem 3.3.1 and its corollary are credited to Euler [3.1].

§3.3. (Theorem 3.3.2). Interest in perfect numbers dates back to the Pythagoreans (and before). However, the first significant result is found in Euclid's *Elements*. There, Proposition 36 of Book IX provides the "if" part of Theorem 3.3.2. Centuries later, Euler proved the converse, that is, every even perfect number has the form (3.3.7) [3.11]. This was published posthumously. Prior to Euler, the converse had been stated by others, including Descartes. The proof given in the text is similar to that given by L. E. Dickson in 1911 [3.12].

In view of Theorem 3.3.2, every Mersenne prime gives rise to a perfect number. This has stimulated a continuing search for Mersenne primes. The twenty-third such prime is

$$2^{11,213} - 1,$$

and was discovered by D. Gillis in 1963. More recently, Bryant Tuckerman verified that

$$2^{19,937} - 1$$

is the twenty-fourth Mersenne prime [3.13].

§3.3A. The methods used in the proof of Theorem 3.3A.1 are attributed to J. J. Sylvester. He showed that an odd perfect number must have at least five distinct prime factors [3.14]. He also conjectured that it must have at least six and later claimed to have a proof, which he never published. This was later proved independently by Gradstein (1925), Kuhnel (1949), and Webber (1951) [3.15, 3.16, 3.17].

Recently (1972), N. Robbins and C. Pomerance independently proved that an odd perfect number must have at least seven distinct prime factors [3.18, 3.19]. In 1975, P. Hagis announced that an "extensive computer based study" showed that an odd perfect number must have at least eight distinct prime factors [3.20, 3.21].

Kuhnel's work implied that the smallest odd perfect number must be at least $(2.2) 10^{12}$. In 1957, H. Kanold raised this lower bound to 10^{20} [3.22]. B. Tuckerman further improved this to 10^{36} in 1967 [3.23], and P. Hagis to 10^{50} in 1973 [3.24]. In 1977, M. Buxton and S. Elmore announced that they had established the lower bound 10^{400} [3.25, 3.26].

Another approach to the consideration of odd perfect numbers is to limit the exponents of the distinct primes rather than their number. In this connection, W. McDaniel and P. Hagis (1975) showed that for M squarefree, p a prime not dividing M, $p^a M^{2b}$ cannot be an odd perfect number for $b = 5, 12, 17, 24,$ or 62 [3.27]. Older results of this type include the case $b = 1$ of Steuerwald, 1937 [3.28], and that of $b = 2$ by Kanold, 1941 [3.29]. The case $b = 3$ was provided by Hagis and McDaniel, 1972 [3.30], and for $b = 3k + 1$, $k \geq 0$, by McDaniel, 1970 [3.31].

In 1973, Hagis and McDaniel proved that an odd perfect number must have a prime divisor greater than 100,110 [3.32]. Using this result, Pomerance showed that the second largest prime divisor must be at least 139 [3.33].

Estimates for $V(x) =$ the number of perfect numbers less than or equal to x has evolved roughly as follows:

(i) $V(x) < x^{1/2}$ Hornfeck, 1955 [3.34]

(ii) $\overline{\lim} \dfrac{V(x)}{x^{1/2}} \leq \dfrac{1}{2 \cdot 5^{1/2}}$ Hornfeck, 1956 [3.35]

(iii) $V(x) < c\dfrac{x^{1/4} \log x}{\log \log x}$ Kanold, 1956 [3.36]

(iv) $V(x) < \exp\left(c\dfrac{\log x \log \log \log x}{\log \log x}\right)$ Hornfeck and Wirsing, 1957 [3.37]

(v) $V(x) < \exp\left(c\dfrac{\log x}{\log \log x}\right)$ Wirsing, 1959 [3.38]

§3.3A. Exercise 1. This assertion was first proved by Dickson in 1913 [3.39]. It was rediscovered by Gradstein in 1925 [3.15].

§3.3A. Exercise 3. The condition given here is also known to be sufficient [3.40].

§3.4. The multiplicativity of $\sigma(n)$ was proved by Euler in 1750. Prior to this, it had been stated by Descartes in 1638 (approximately) [3.1, 3.10].

The identity (3.4.3) is due to Liouville (1857), who derived it directly from the fact that $1^3 + 2^3 + \ldots + n^3 = (1 + 2 + \ldots + n)^2$ [3.41].

§3.5. As already stated in the notes on §3.1, Möbius (1832) was chiefly responsible for focusing on the properties of the μ-function [3.8]. However, Mertens (1874) seems to have been the first to explicitly state the property (3.5.3) [3.9].

Möbius was seeking the solution of problems involving inversion of series. For example, he deduced from

$$f(x) = \sum_{s=1}^{\infty} s^n F(x^s)$$

that

$$F(x) = \sum_{s=1}^{\infty} s^n f(x^s)\mu(s).$$

On the other hand, Mertens was motivated by the consideration of the average order of the Euler φ function.

Theorem 3.5.2 was proved simultaneously by Dedekind and Liouville in 1857 (without explicit mention of the μ-function) [3.42, 3.43]. The formulation given in the text was provided by Laguerre in 1872 [3.44].

The formula (3.5.14) was first given by Meissel in 1850 [3.45].

Generalizing the investigations of Möbius, in 1884 J. W. Glaisher [3.46] proved that if $P(n)$ is completely multiplicative, $P(1) = 1$, then

$$F(x) = \sum_{n} P(n)f(x^n)$$

implies

$$f(x) = \sum_{n} P(n)F(x^n)\mu(n).$$

§3.7. Formula (3.7.1) was proved by Euler in 1760. He derived it first for prime powers and then obtained the general case as a consequence of the multiplicativity of the φ-function [3.47].

Without the explicit use of the Möbius function, (3.7.11) was given by both Dedekind and Liouville in 1857 as consequences of their inversion formulas.

The identity (3.7.12) was proved by Gauss in Article 39 of his *Disquisitiones*.

Smith's Determinant, (3.7.13), was introduced and evaluated, (3.7.14), by H. J. S. Smith in 1876 [3.48].

§3.7A. The inequality (3.7A.7) is credited to Bonse [3.49]. The original source of Theorem 3.7A.1 was as a problem posed by de Rocquigny in 1899 [3.50]. Solutions were given by Maillet and others in 1900 [3.51].

It is also possible to derive Theorem 3.7A.1 as a consequence of Theorem 3.7A.2 [3.52].

§3.8. Dirichlet first proved a version of Theorem 3.8.1 with an error term $R(x) = O(x^\delta)$, for some δ, $1 < \delta < 2$ [3.53]. In fact, δ was chosen so as to satisfy $\sum_{n=2}^{\infty} (1/n^\delta) = 1$. This was improved to $R(x) = O(x \log x)$ by F. Mertens [3.9]. Many years later, in 1956, this was improved by Walfisz to $R(x) = O(x \log^{2/3} x(\log \log x)^{4/3})$ [3.54].

The essence of Lemma 3.8.1 appears explicitly in Dirichlet's work, in particular, in connection with the proof of Theorem 3.8.3 [3.53]. Euler's constant was calculated by Euler in 1755 [3.54]. It is sometimes called the "Mascheronian" constant.

Theorem 3.8.3 was first proved by Dirichlet [3.53, 3.56]. In 1903, Voronoi improved the result to $R_2(x) = O(x^{1/3} \log x)$ [3.57]. An incomplete proof of this had been given previously in 1886 by Pfeiffer [3.58]. Later, in 1912, Landau showed that Pfeiffer's method could be used to obtain a correct proof [3.59]. In 1915, Hardy and Landau independently proved that $R_2(x) = O(x^{1/4})$ was not true [3.60, 3.61]. In 1922, Van der Corput showed that $R_2(x) = (x^\theta)$ for some $\theta < (33/100)$, and in 1928, improved this to $R_2(x) = O(x^{27/82}(\log x)^{11/41})$ [3.62, 3.63]. In 1953, this was further improved by Reichert to $R_2(x) = O(x^{15/46} \log^{30/23} x)$ [3.64].

Theorem 3.8.5 was first obtained by Dirichlet with the weaker error term $O(x^{\delta/2+\epsilon})$, where δ is the same constant described above in connection with the divisor function [3.53].

REFERENCES

3.1.   L. Euler, De Numeris Amicabilus, *Opera Omnia*, Leipzig: B. G. Teubner, 1915, I, 2, pp. 86–162 (original: 1750).

3.2.   J. J. Sylvester, On The Divisors of the Sum of a Geometrical Series Whose First Term is Unity and Common Ratio Any Positive or Negative Integer, *Mathematical Papers*, New York: Cambridge University Press, 1912 (original: 1888).

3.3.   E. Cunningham, *Proceedings of the London Mathematical Society*, **35**, 1902.

3.4.   E. Landau, Vorlesungen Uber Zahlentheorie, Leipzig: S. Hirzel, 1927.

3.5.   L. Euler, Theoremata Arithmetica Nova Methodo Demonstrata, *Opera Omnia*, 1915, I, 2, pp. 531–555 (original: 1760).

3.6.   L. Euler, Speculationes Circa Quasdam Insignes Proprietates Numerorum, *Opera Omnia*, 1941, I, 4, pp. 105–115 (original: 1780).

3.7.   G. H. Hardy and S. Ramanujan, The Normal Number of Prime Factors of a Number n, *Quarterly Journal of Mathematics*, **48**, 1920, 76–93.

3.8.   A. F. Möbius, Uber eine besondere Art von Umkehrung der Reihen, *Crelle's Journal*, **9**, 1832, 105–123.

3.9.   F. Mertens, Ueber einige asymptotische Gesetze der Zahlentheorie, *Crelle's Journal*, **77**, 1874, 289–338.

3.10. L. E. Dickson, *History of the Theory of Numbers*, 3 vols, New York: Chelsea, 1919.

3.11. L. Euler, Tractatus de numerorum doctrina, *Opera Omnia*, 1944, I, 5 (original: 1749).

3.12. L. E. Dickson, Notes on The Theory of Numbers, *American Mathematical Monthly*, **18**, 1911, 109.

3.13. B. Tuckerman, *Notices of the American Mathematical Society*, **18**, 1971, 608.

3.14. J. J. Sylvester, Sur l'impossibilité de l'existence d'un nombre parfait impair qui ne contient pas au moins 5 diviseurs premiers distincts, Mathematical Papers, Cambridge University Press, 1912, pp. 611–614.

3.15. I. S. Gradstein, On Odd Perfect Numbers, *Matematicheskii Sbornik*, **32**, 1925, 476–510.

3.16. U. Kuhnel, Verscharfung der notwendigen Bedingungen für die Existenz von ungeraden vollkommenen Zahlen, *Mathematische Zeitschrift*, **52**, 1949, 202–211.

3.17. G. C. Webber, Nonexistence of Odd Perfect Numbers of The Form $3^{2\beta}p^\alpha s_1^{2\beta_1} s_2^{2\beta_2} s_3^{2\beta_3}$, *Duke Mathematics Journal*, **18**, 1951, 741–749.

3.18. N. Robbins, The nonexistence of odd perfect numbers with less than seven distinct prime factors, *Notices of the American Mathematical Society,* **19,** 1972, A-52.

3.19. C. Pomerance, Odd perfect numbers are divisible by at least seven distinct primes, *Acta Arithmetica,* **25,** 1973/74, 265–300.

3.20. P. Hagis, Jr., Every odd perfect number has at least 8 prime factors, *Notices of the American Mathematical Society,* **22,** 1975, A-60.

3.21. M. Kishore, Odd perfect numbers not divisible by 3 are divisible by at least 10 distinct primes, *Mathematics of Computation,* **31**(137), 1977, 274–279.

3.22. H. Kanold, Uber mehrfach vollkommene Zahlen II, *Journal für die reine und angewandte Mathematik,* **197,** 1957, 82–97.

3.23. B. Tuckerman, Odd perfect numbers: a search procedure and a new lower bound of 10^{36}, *Mathematics of Computation,* **27,** 1973, 943–949.

3.24. P. Hagis, A lower bound for the set of odd perfect numbers, *Mathematics of Computation,* **27,** 1973, 951–954.

3.25. M. Buxton and S. Elmore, An extension of lower bounds for odd perfect numbers, *Notices of the American Mathematical Society,* **23,** 1976, A-55.

3.26. B. Stubblefield, Greater lower bounds for odd perfect numbers, *Notices of the American Mathematical Society,* **20,** 1973, A-515.

3.27. W. McDaniel and P. Hagis, Some results concerning the non-existence of odd perfect numbers of the form $p^{\alpha}M^{2\beta}$, *Fibonacci Quarterly,* **13,** 1975, 25–28.

3.28. R. Steuerwald, *Verscharfung einen notwendigen Bedingung für die Existenz einen ungeraden volkommenen Zahl,* Bayerische Akad. der Wiss. Math.-Naturwiss. Klasse, Sitzungsberichte, 1937, pp. 69–72.

3.29. H. Kanold, Untersuchungen uber ungerade volkommene Zahlen, *Journal für die reine und angewandte Mathematik,* **183,** 1941, 98–109.

3.30. P. Hagis and W. McDaniel, A New Result Concerning the Structure of Odd Perfect Numbers, *Proceedings of the American Mathematical Society,* **32,** 1972, 13–15.

3.31. W. McDaniel, The Non-existence of Odd Perfect Numbers of a Certain Form, *Archiv der Mathematik,* **21,** 1970, 52–53.

3.32. P. Hagis and W. McDaniel, On the largest prime divisor of an odd perfect number II, *Mathematics of Computation,* **27,** 1973, 955–957.

3.33. C. Pomerance, The second largest prime factor of an odd perfect number, *Mathematics of Computation,* **29,** 1975, 914–921.

3.34. B. Hornfeck, Zur Dichte der Menge der volkommenen Zahlen, *Archiv der Mathematik,* **6,** 1955, 442–443.

3.35. B. Hornfeck, Bemerkung zu meiner Note uber vollkommene Zahlen, *Archiv der Mathematik,* **7,** 1956, 273.

3.36. H. Kanold, Uber die Verteilung der vollkommenen Zahlen und allgemeinerer Zahlenmengen, *Mathematische Annalen,* **132,** 1957, 442–450.

3.37. B. Hornfeck and E. Wirsing, Uber die Haufigkeit vollkommene Zahlen, *Mathematische Annalen,* **133,** 1957, 431–438.

3.38. E. Wirsing, Bemerkung zu der Arbeit uber vollkommene Zahlen, *Mathematische Annalen,* **137,** 1959, 316–318.

3.39. L. E. Dickson, Finiteness of the odd perfect and primitive abundant numbers with n distinct prime factors, *American Journal of Mathematics,* **35,** 1913, 413–422.

3.40. H. N. Shapiro, On Primitive Abundant Numbers, *Communications on Pure and Applied Mathematics,* **21,** 1968, 111–118.

3.41. J. Liouville, Généralisation d'un Théorème de l'Arithmétique Indienne, *Journal de Mathématiques Pures et Appliqués,* **2**(2), 1857, 393–396.

3.42.   R. Dedekind, Abriss einer Theorie der hoheren Congruenzen in bezug auf einen reelen Primzahl-Modulus, *Journal für die reine und angewandte Mathematik*, **54**, 1857, 1–26.

3.43.   J. Liouville, Sur l'expression $\varphi(n)$ qui marque combien la suite $1, 2, \ldots, n$ contient de nombres premiers à n, Journal de Mathématiques Pures et Appliqués, **2**(2), 1857, 110–112.

3.44.   E. N. Laguerre, *Bulletin Soc. Math. de France*, 1, 1872–1873, 77–81.

3.45.   E. Meissel, Observationes quaedam in theoria numerorum, *Crelle's Journal*, **48**, 1854, 301–316 (original: 1850).

3.46.   J. W. Glaisher, *Dublin Philosophical Magazine*, **18**, 1884, 518–546.

3.47.   L. Euler, Theoremata Arithmetica Novo Methodo Demonstrata, *Opera Omnia*, I, 2, pp. 531–555.

3.48.   H. J. S. Smith, On the Value of a Certain Arithmetical Determinant, *Collected Mathematical Papers*, New York: Chelsea Publishing Co., reprint 1965 (original: 1876).

3.49.   H. Bonse, Uber eine bekannte Eigenschaft der Zahl 30 und ihre Verallgemeinerung, *Archiv der Mathematik und Physik*, **3**(12), 1907, 292–295.

3.50.   G. de Rocquigny, *Intermediaire des Mathématiciens*, **6**, 1899, 243, Question No. 1656.

3.51.   E. Maillet, *Intermediaire des Mathématiciens*, **7**, 1900, 254.

3.52.   J. Uspensky and M. Heaslet, *Elementary Number Theory*, New York: McGraw-Hill, 1939.

3.53.   G. L. Dirichlet, *Uber die Bestimmung der mittleren Werthe in der Zahlentheorie, Werke*, G. Riemer, 1897, II, pp. 49–66 (original 1849).

3.54.   A. Walfisz, *Weylsche Exponentialsummen in der neueren Zahlentheorie*, Leipzig: B. G. Teubner, 1963, pp. 114–145.

3.55.   L. Euler, Institutiones Calculi Differentialis, *Opera Omnia*, 1913, I, 10 (original: 1755).

3.56.   G. L. Dirichlet, Sur l'usuage des séries infinies dans la théorie des nombres, *Crelle's Journal*, **18**, 1838, 259–274.

3.57.   G. Voronoi, Sur un problème du calcul des fonctions asymptotiques, *Journal für die reine und angewandte Mathematik*, **126**, 1903, 241–282.

3.58.   E. Pfeiffer, Uber die Periodicität in der Teilbarkeit der Zahlen und über die Verteilung der Klassen positiver quadratischer Formen auf ihre Determinanten, Jahresbericht der Pfeiffer'schen Lehr- und Erziehungs-Anstalt zu Jena, 1886, pp. 1–21.

3.59.   E. Landau, Die Bedeutung der Pfeiffer'schen Methode für die analytische Zahlentheorie, *Wiener Sitzungsberichte*, **121**, 1912, 2a, 2195–2332.

3.60.   G. H. Hardy, The Average Order of the Arithmetical Functions $P(x)$ and $\Delta(x)$, *Proceedings of the London Mathematical Society*, **15**(2), 1916, 192–213.

3.61.   E. Landau, Uber die Gitterpunkte in einem Kreise, Akademie der Wissenschaften, *Göttingen Nachrichten*, **5**, 1915, 161–171.

3.62.   J. G. Van der Corput, Verscharfung der Abschatzung beim Teilerproblem, *Mathematische Annalen*, **87**, 1922, 39–65.

3.63.   J. G. Van der Corput, Zum Teilerproblem, *Mathematische Annalen*, **98**, 1928, 697–716.

3.64.   H. Richert, Verscharfung der Abschatzung beim Diricletschen Teilerproblem, *Mathematische Zeitschrift*, **58**, 1953, 204–218.

3.65.   R. Bellman and H. N. Shapiro, The Algebraic Independence of Arithmetic Functions, *Duke Mathematical Journal*, **15**, 1948, 229–235.

4

THE RING OF
ARITHMETIC FUNCTIONS
A Do-It-Yourself Chapter

§4.1 RINGS OF FUNCTIONS AND CONVOLUTIONS

There are many number theoretic investigations that are based on identities between various special functions. These identities result from manipulations of the type introduced in the previous chapter, and they are evidence of a more formal structure surrounding the arithmetic functions. The setting for this structure is that of rings of functions and operations thereon, a setting which often provides simplicity and elegance to proofs. It is, however, just another way of viewing things and has not as yet produced essentially new insights.

The classes of functions that are distinguished are denoted by $\mathscr{S}$ and $\mathscr{A}$, and are defined as follows.

Definition 4.1.1. The set of $\mathscr{S}$ consists of all complex valued functions $f(x)$, defined for x real, such that $f(x) = 0$ for $x < 1$.

Definition 4.1.2. The set $\mathscr{A}$ of *arithmetic functions* consists of those functions of $\mathscr{S}$ that vanish for all nonintegral x.

Clearly $\mathscr{A} \subset \mathscr{S}$.

The functions of $\mathscr{S}$ form a ring in the sense of §1.1, [the properties (i)—(vi) hold] under the *natural* definition of addition and multiplication of functions; that is, for f, g in $\mathscr{S}$

$$(4.1.1) \qquad (f + g)(x) = f(x) + g(x),$$

$$(4.1.2) \qquad (fg)(x) = f(x)g(x).$$

With respect to these operations, $\mathscr{A}$ is a subring of $\mathscr{S}$.

We next introduce a third binary operation on $\mathscr{S}$.

Definition 4.1.3. For f, $g \in \mathcal{S}$ we define the *convolution* $f * g$ in $\mathcal{S}$ by

$$(4.1.3) \qquad (f * g)(x) = \sum_{1 \leq n \leq x} f\left(\frac{x}{n}\right) g(n).$$

The "action" of this definition on functions of $\mathcal{A}$ is given by the following lemma.

Lemma 4.1.1. If $f \in \mathcal{A}$, $g \in \mathcal{S}$, then $f * g \in \mathcal{A}$ and for integral n,

$$(4.1.4) \qquad (f * g)(n) = \sum_{d \mid n} f\left(\frac{n}{d}\right) g(d).$$

Proof. . . . .

It is clear that, in general, we do not have "commutativity" for the convolution operation; that is, $g * f$ is not necessarily the same as $f * g$, for all f, $g \in \mathcal{S}$. However, the following lemma is easily seen.

Lemma 4.1.2. If f, $g \in \mathcal{A}$ then $f * g = g * f$.

Proof. . . . .

For the general case, we do have a kind of "inner-commutativity" as follows.

Lemma 4.1.3. For f, g, $h \in \mathcal{S}$ we have

$$(4.1.5) \qquad (f * g) * h = (f * h) * g.$$

Proof. . . . .

Consider the function

$$e(x) = \begin{cases} 1 & \text{for } x = 1 \\ 0 & \text{otherwise.} \end{cases}$$

It is easily seen that $e \in \mathcal{A}$, and for $f \in \mathcal{S}$

$$(4.1.6) \qquad f * e = f$$

and

$$(4.1.7) \qquad e * f = \begin{cases} f(x) & \text{for } x \text{ an integer} \\ 0 & \text{otherwise.} \end{cases}$$

Thus e serves as a *right identity under convolution* for all of $\mathcal{S}$, but is a *left identity* only in $\mathcal{A}$.

The relation (4.1.7) suggests that for each $f \in \mathcal{S}$ we define an image $f_0 \in \mathcal{A}$ by

(4.1.8) $$f_0 = e * f.$$

Here $f \in \mathcal{A}$ is equivalent to $f = f_0$. For $g, h \in \mathcal{A}$, taking $f = e$ in (4.1.5) yields $g * h = h * g$, another verification of the commutativity in $\mathcal{A}$.

We now turn to an analysis of the associative law for convolutions. Again, this does not hold in all generality in $\mathcal{S}$. The reason for this apparent "asymmetry" will emerge later. At this point, we content ourselves with a "middle associativity."

Lemma 4.1.4. For $f, g, h \in \mathcal{S}$ we have

(4.1.9) $$(f * g) * h = f * (g_0 * h).$$

Proof. . . . .

Corollary. The associative law holds in $\mathcal{A}$. That is, for $f, g, h \in \mathcal{A}$,

(4.1.10) $$(f * g) * h = f * (g * h).$$

If we consider the operation of addition $(+)$ together with the operation of convolution $(*)$, $\mathcal{S}$ fails to constitute a ring in the sense of §1.1. For although the distributive laws

$$(f + g) * h = f * h + g * h$$
$$h * (f + g) = h * f + h * g$$

hold, the principal deficiencies are that the convolution operation provides neither commutativity nor associativity. However, the subset $\mathcal{A}$ of arithmetic functions does satisfy all of these properties, and therefore constitutes a ring with respect to the operations of addition and convolution. In this ring, the role of the "1" or "identity for convolution" is played by the arithmetic function $e(x)$.

§4.2 INVERSES AND UNITS

Since the ring structure of $\mathcal{A}$ with respect to addition and convolution is analogous to that of Z with respect to addition and multiplication, it is natural to pose certain questions. First of these is whether $\mathcal{A}$ is an integral domain; that is, if $g \neq 0$ (the 0 function is the one which is identically 0), does $f * g = h * g$ imply $f = h$; or (what is the same thing) does $f * g = 0$ imply $f = 0$? The answer is provided by the following theorem.

Theorem 4.2.1. $\mathcal{A}$ is an integral domain.

Proof. . . . .

The property of the integers ± 1 inherent in $1 \cdot 1 = 1$, and $(-1)(-1) = 1$ abstracts as follows.

Definition 4.2.1. An element $f \in \mathcal{S}$ is called a *left unit* in $\mathcal{S}$ if there exists a $g \in \mathcal{S}$ such that

$$(4.2.1) \qquad\qquad\qquad f * g = e.$$

It is called a *right unit* if there exists a $g \in \mathcal{S}$ such that

$$(4.2.2) \qquad\qquad\qquad g * f = e.$$

As a companion terminology, if (4.2.1) holds, g is called a *right inverse* for f, and f a *left inverse* for g.

The investigation of these concepts may be initiated with the following theorem.

Theorem 4.2.2. A necessary and sufficient condition for $f \in \mathcal{S}$ to have a left inverse (i.e., to be a right unit) is that

$$(4.2.3) \qquad\qquad\qquad f(1) \neq 0.$$

Proof. . . . .

Corollary 1. If $f(1) \neq 0$, the left inverse of f is in $\mathcal{A}$.

Proof. . . . .

Corollary 2. If $f(1) \neq 0$, and $f \in \mathcal{A}$, then f has a unique two-sided inverse in $\mathcal{A}$.

Proof. . . . .

The study of right inverses for functions $f \in \mathcal{S}$, in general, leads to nothing new, since they can exist only for functions in $\mathcal{A}$. The following theorem states this more precisely.

Theorem 4.2.3. If $f \in \mathcal{S}$ has a right inverse, then $f \in \mathcal{A}$.

Proof. . . . .

Many methods in number theory hinge on the explicit presentation of the inverses of given functions in $\mathcal{A}$. As an important illustration, consider the function $1(x)$ which is identically 1 for all $x \geq 1$. From Corollary 2 of Theorem 4.2.2 above, we know that 1_0 has a unique inverse in $\mathcal{A}$. In fact, this inverse is the Möbius function $\mu(n)$; therefore, from (3.5.3), we may write

$$(4.2.4) \qquad\qquad\qquad 1_0 * \mu = \mu * 1_0 = e.$$

In order to produce other explicit inverses, we are led to consider functions h in

$\mathcal{A}$ that are such that multiplication by it distributes with convolution in $\mathcal{A}$. That is, for $f, g \in \mathcal{A}$

(4.2.5) $h(f * g) = (hf) * (hg).$

We require, in addition, that $h(1) = 1$, and denote the subset of these functions of $\mathcal{A}$ [satisfying (4.2.5)] by $\mathcal{M}$. The usefulness of such functions in producing explicit inverses is given by the following theorem.

Theorem 4.2.4.   For $h \in \mathcal{M}$, the inverse of h is μh.

   Proof.   . . . .

The functions of $\mathcal{M}$ are easily characterized.

Theorem 4.2.5.   A function h of $\mathcal{A}$ is in $\mathcal{M}$ if and only if it is completely multiplicative and not identically zero.

   Proof.   . . . .

§4.3   INVERSION FORMULAS

The standard Möbius inversion formula, Theorem 3.5.1, asserts:

   (i)   if $f \in \mathcal{S}$ and $g = f * 1 \in \mathcal{S}$, then $f = g * \mu$; and
   (ii) if $g \in \mathcal{S}$ and $f = g * \mu \in \mathcal{S}$, then $g = f * 1$.

 This is clearly a special case of the following.

Theorem 4.3.1.   If $\alpha \in \mathcal{A}$ has an inverse $\alpha^{\langle -1 \rangle} \in \mathcal{A}$, then
   (i) $f \in \mathcal{S}$ and $g = f * \alpha$ implies $f = g * \alpha^{\langle -1 \rangle} \in \mathcal{A}$; and
   (ii) $g \in \mathcal{S}$ and $f = g * \alpha^{\langle -1 \rangle}$ implies $g = f * \alpha$.

   Proof.   . . . .

   Corollary.   If $h \in \mathcal{M}, f \in \mathcal{S}$, then $g = f * h$ implies $f = g * \mu h$.

   Proof.   . . . .

§4.4   THE NATURAL VALUATION IN $\mathcal{A}$

 There is a natural way to introduce metric considerations into the set $\mathcal{A}$ which are useful for certain types of manipulations with arithmetic functions. For this we require the following definition.

Definition 4.4.1. For $f \in \mathcal{A}, f \neq 0$, let $\mathcal{O}(f)$ equal the smallest positive integer k such that $f(k) \neq 0$. This quantity $\mathcal{O}(f)$ is called the *order of* f.

For $f \neq 0$, the integer $\mathcal{O}(f)$ exists and is positive. $\mathcal{O}(f) = 1$ is equivalent to $f(1) \neq 0$, so that from Theorem 4.2.2, the elements of $\mathcal{A}$ of order 1 are precisely the units.

In terms of the order, we introduce the "size" of an $f \in \mathcal{A}$ via the following.

Definition 4.4.2. For $f \in \mathcal{A}, f \neq 0$ set

$$(4.4.1) \qquad\qquad \|f\| = \frac{1}{\mathcal{O}(f)},$$

and for $f = 0$, let $\|f\| = 0$.

The following theorem contains the basic properties of $\|f\|$.

Theorem 4.4.1. For functions of $f, g \in \mathcal{A}$ we have

$$(4.4.2) \qquad 0 \leq \|f\| \leq 1, \qquad \text{where } \|f\| = 0 \text{ if and only if } f = 0,$$

$$(4.4.3) \qquad\qquad \|f * g\| = \|f\| \, \|g\|,$$

$$(4.4.4) \qquad\qquad \|f + g\| \leq \text{Max}(\|f\|, \|g\|).$$

> **Proof.**   . . . .

The above theorem asserts essentially that $\|f\|$ defines what is called a *non-archimedian* valuation over $\mathcal{A}$. The assertion (4.4.4) is, of course, a sharpening of the usual triangle inequality

$$\|f + g\| \leq \|f\| + \|g\|.$$

We can proceed now to do simple analysis in $\mathcal{A}$, with respect to this valuation.

Definition 4.4.3. A sequence $\{f_n\}$ of functions of $\mathcal{A}$ is said to converge to a function $f \in \mathcal{A}$, if and only if $\|f_n - f\|$ tends to 0 as $n \to \infty$. In this case, we write

$$(4.4.5) \qquad\qquad \|\lim_{n \to \infty}\| f_n = f.$$

We say that f_n converges to f in the valuation.

The import of the above definition is really that the "convergence" of the sequence f_n to f is equivalent to a rather strong version of pointwise convergence.

Theorem 4.4.2. Given a sequence $\{f_n\}$ of functions of $\mathcal{A}$, (4.4.5) is equivalent to the statement that for each positive integer m, there exists an $n_0 = n_0(m)$ such that for all $n > n_0$,

$$f_n(m) = f(m).$$

Proof. . . . .

We may also consider infinite series $\sum_{n=1}^{\infty} f_n$ of functions of $\mathcal{A}$. In the usual way, convergence of such series to a function $f \in \mathcal{A}$, is defined as follows.

Definition 4.4.4. We write for $f_n, f \in \mathcal{A}$,

(4.4.6)
$$\sum_{n=1}^{\infty} f_n = f$$

If and only if

(4.4.7)
$$\|\lim_{N \to \infty}\| \sum_{n=1}^{N} f_n = f.$$

We say then that the infinite series on the left of (4.4.6) converges to f in the valuation.

The following is a general property of nonarchimedean valuations (complete rings), but appears here as a consequence of Theorem 4.4.2.

Theorem 4.4.3. The series $\sum_{n=1}^{\infty} f_n$ converges in the valuation if and only if

$$\|\lim_{n \to \infty}\| f_n = 0.$$

Proof. . . . .

To illustrate the applicability of the above notions, consider an $f \in \mathcal{A}$ that is a unit, so that $f(1) \neq 0$. We may then write

(4.4.8)
$$f = f(1)e - g$$

where $g \in \mathcal{A}$, and $g(1) = 0$. Then

$$f = f(1)\left[e - \frac{g}{f(1)} \right].$$

and for the inverse of f under convolution we have

$$f^{-1} = \frac{1}{f(1)} \cdot \left[e - \frac{g}{f(1)} \right]^{\langle -1 \rangle}.$$

Proceeding purely formally, one might conjecture that

(4.4.9)
$$\left[e - \frac{g}{f(1)} \right]^{\langle -1 \rangle} = \sum_{\nu=0}^{\infty} \left(\frac{1}{f(1)} \right)^{\nu} g^{\langle \nu \rangle}$$

where $g^{\langle 0 \rangle} = e$, and for $\nu \geq 1$,

$$g^{\langle \nu \rangle} = \underbrace{g * g * \cdots * g}_{\nu},$$

the convolution of g with itself ν times. Since $g(1) = 0$, it is easily seen that

$$\left\| \lim_{\nu \to \infty} \right\| g^{\langle \nu \rangle} = 0$$

so that the series on the right of (4.4.9) converges in the valuation. Finally, since

$$\left(e - \frac{g}{f(1)} \right) * \sum_{\nu=0}^{\infty} \left(\frac{1}{f(1)} \right)^{\nu} g^{\langle \nu \rangle} = \sum_{\nu=0}^{\infty} \left(\frac{1}{f(1)} \right)^{\nu} g^{\langle \nu \rangle} - \sum_{\nu=0}^{\infty} \left(\frac{1}{f(1)} \right)^{\nu+1} g^{\langle \nu+1 \rangle} = e,$$

the truth of (4.4.9) is established.

In general, all such formal manipulations with series that are convergent in the valuation may be validated. The above calculation is really an analysis of $(e - X)^{\langle -1 \rangle}$ for $X \in \mathcal{A}$ that shows that it has the series expansion

$$(e - X)^{\langle -1 \rangle} = \sum_{\nu=0}^{\infty} X^{\langle \nu \rangle}$$

valid in the domain $\mathcal{O}(X) > 1$, or $\|X\| < 1$.

Several other examples of a similar nature are included in the following exercises.

EXERCISES

1. Show that the definition

$$e^X = \sum_{\nu=0}^{\infty} \frac{1}{\nu!} X^{\langle \nu \rangle}$$

is valid for those $X \in \mathcal{A}$ such that $\|X\| < 1$. For such functions verify that

(*) $e^X * e^Y = e^{X+Y}.$

Can the definition of e^X be extended to *all* functions of $\mathcal{A}$ (with value in $\mathcal{A}$) so as to maintain the validity of (*)?

2. Determine the range of validity of the definition

$$\log(e + X) = \sum_{\nu=1}^{\infty} \frac{(-1)^{\nu-1}}{\nu} X^{\langle \nu \rangle}$$

Define $\log(\alpha e + X)$ for $\alpha \neq 0$ a complex constant. Show that if X and Y are units in $\mathcal{A}$,

$$\log(X * Y) = \log X + \log Y.$$

Is it true that $\text{Log } e^X = X$, or that $e^{\log X} = X$?

3. For k an integer ≥ 1, and $X \in \mathcal{A}$, define $X^{\langle 1/k \rangle}$ to be a function of $\mathcal{A}$ such that

$$(X^{\langle 1/k \rangle})^{\langle k \rangle} = X.$$

   Show that for X a unit, $X^{\langle 1/k \rangle}$ exists for all $k \geq 1$. Conversely, show that if $X^{\langle 1/k \rangle}$ exists for all $k \geq 1$, then X is a unit.

4. An *infinite convolution* of functions $f_n \in \mathcal{A}$ is said to converge to $f \in \mathcal{A}$, written $\ast_{n=1}^{\infty} f_n = f$, if and only if

$$\left\| \lim_{N \to \infty} \right\| (f_1 \ast f_2 \ast \cdots \ast f_N) = f.$$

   Show that if $f(n)$ is a completely multiplicative function of $\mathcal{A}$,

$$\underset{p}{\ast} \left(e - e\left(\frac{x}{p}\right) f(p) \right)^{\langle -1 \rangle} = f(x),$$

   where the infinite convolution runs over primes p. Use this to show that

$$\log 1_0 = \begin{cases} \dfrac{1}{\alpha} & \text{for } n = p^\alpha \text{ (a prime power)}, \\ 0 & \text{otherwise.} \end{cases}$$

5. Let $f \in \mathcal{A}$ be such that the values of f are real and nonnegative. Show for such f, that

$$\sup_{n,k} f^{(k)}(n) < \infty$$

   if and only if

$$\sum_{n=1}^{\infty} f(n) \leq 1.$$

6. Let g be any fixed completely additive arithmetic function such that $g(n) \geq 0$ for all integers n. For any $f \in \mathcal{A}$, $f \neq 0$, define $E_g(f)$ to be the largest non-negative integer k such that

$$g(n) < k \qquad \text{implies} \qquad f(n) = 0.$$

   Then set $\|f\|_g = 2^{-E_g(f)}$ for $f \neq 0$, and $\|f\|_g = 0$ for $f = 0$. Show that $\|f\|_g$ defines a nonarchimedian valuation over $\mathcal{A}$. What valuation results from the choice of $g(n) \equiv 0$?

7. Let P be any completely multiplicative arithmetic function such that $P(n) > 0$. For any $f \in \mathcal{A}$, define $\mathbb{O}_P(f)$ to be the value of $P(k)$ for the smallest positive integer k such that $f(k) \neq 0$. Define

$$\|f\|_P = \frac{1}{\mathbb{O}_P(f)}$$

for $f \neq 0$, $\|f\|_P = 0$ for $f = 0$; show that this provides a nonarchimedian valuation over $\mathcal{A}$.

§4.5 DERIVATIONS

A *derivation* over the ring $\mathcal{A}$ is a mapping D of the ring into itself such that

(4.5.1) $$D(f * g) = Df * g + f * Dg$$

and

(4.5.2) $$D(\alpha f + \beta g) = \alpha Df + \beta Dg$$

for all $f, g \in \mathcal{A}$, and complex constants α, β.
   For any such derivation, since $f * e = f$, from (4.5.1)

$$Df = D(f * e) = Df * e + f * De = Df + f * De$$

so that $f * De = 0$ for all f, and hence $De = 0$.
   There appears to be a richness of derivations over $\mathcal{A}$. In fact, a particularly large collection of them is provided by the following theorem.

Theorem 4.5.1.   For a given function $h \in \mathcal{A}$, a necessary and sufficient condition that

(4.5.3) $$Df = hf$$

constitutes a derivation over $\mathcal{A}$, is that h be completely additive.

   Proof. . . . .

   From given derivations others may be built in various ways. For example, if D is a given derivation and g a fixed function of $\mathcal{A}$,

(4.5.4) $$\hat{D}f = Df * g$$

is also a derivation. Also, taking linear combinations with constant coefficients of a finite number of derivations produces a derivation. The following lemma provides a more "local" process for producing a derivation.

Lemma 4.5.1.   For p a fixed prime, $\nu_p(m) = $ the exponent of the highest power of p dividing m, then

(4.5.5) $$(D_p f)(n) = f(np)\nu_p(np)$$

is a derivation.

Proof. . . . .

Combining the above constructions, for any fixed function $H_p \in \mathcal{A}$

$$(4.5.6) \qquad\qquad D_p f * H_p$$

is a derivation. Summing over a finite collection of primes (let p_i denote the ith prime)

$$Df = \sum_{i=1}^{N} D_{p_i} f * H_{p_i}$$

is also a derivation. Finally, if we let $N \to \infty$ in the above, we obtain the following lemma.

Lemma 4.5.2. If, for all $f \in \mathcal{A}$, the series

$$(4.5.7) \qquad\qquad \sum_{p} D_p f * H_p$$

(the primes are taken in their natural order) converges in the valuation, then

$$(4.5.8) \qquad\qquad Df = \sum_{p} D_p f * H_p$$

defines a derivation over $\mathcal{A}$.

Proof. . . . .

In view of the above lemma, it is natural to investigate under what circumstances the series (4.5.7) converges in the valuation for all $f \in \mathcal{A}$.

Lemma 4.5.3. The series (4.5.7) converges in the valuation for all $f \in \mathcal{A}$ if and only if

$$(4.5.9) \qquad\qquad \|\lim_{p \to \infty}\| H_p = 0.$$

Proof. For the necessity of (4.5.9), take $f = \sum_p e(x/p)$ and apply Theorem 4.4.3. . . . .

If we set

$$\mathcal{U}_m(x) = e\left(\frac{x}{m}\right) \in \mathcal{A}$$

we have the following lemma.

Lemma 4.5.4. If the derivation D is of the form (4.5.8), where the series converges in the valuation for all f, then for all primes p,

(4.5.10) $$H_p = D\,\mathcal{U}_p.$$

Proof. . . . .

The above lemmas shows that if Df is to have a representation such as (4.5.8), then it is *necessary* that

(4.5.11) $$\left\|\lim_{p\to\infty}\right\| D\,\mathcal{U}_p = 0.$$

The next step will be to translate this condition into some intrinsic property of the derivation D.

Definition 4.5.1. A derivation D over $\mathcal{A}$ will be called *continuous* if for every sequence $f_n \in \mathcal{A}$ such that

(4.5.12) $$\left\|\lim_{n\to\infty}\right\| f_n = f,$$

It follows that

(4.5.13) $$\left\|\lim_{n\to\infty}\right\| Df_n = Df.$$

Lemma 4.5.5. If D is a derivation of $\mathcal{A}$ of the form (4.5.8), then D is continuous.

Proof. Consider a sequence $f_n \in \mathcal{A}$ such that (4.5.12) holds. Then

(4.5.14) $$Df_n = \sum_p D_p f_n * D\,\mathcal{U}_p.$$

For a fixed integer m, (4.5.11) holds,.. . . . .
so that

(4.5.15) $$(Df_n)(m) = \sum_{p\le m^*} (D_p f_n * D\,\mathcal{U}_p)(m)$$

where we note that m^* depends only on m, and is independent of n, as well as of the function f_n. Next, since

$$D_p f_n =$$

and. . . . .
it follows that for all integers l dividing m

(4.5.16) $$\left\|\lim_{n\to\infty}\right\|(D_p f_n)(l) = (D_p f)(l).$$

From (4.5.15) and (4.5.16), we obtain

$$\|\lim_{n\to\infty}\| (Df_n)(m) = \sum_{p\leq m^*} (D_p f * D\mathcal{U}_p)(m)$$

$$= Df(m)$$

where the last equality follows from the uniform nature of m^*. This completes the proof of the lemma.

Lemma 4.5.6. If D is any derivation over $\mathcal{A}$, then for any integer $m \geq 1$,

$$(4.5.17) \qquad (D\mathcal{U}_m)(n) = \sum_{p|m} \nu_p(m)(D\mathcal{U}_p)\left(\frac{np}{m}\right),$$

where the sum on the right is over the distinct primes dividing m.

Proof.   . . . .

Theorem 4.5.2. A derivation D over $\mathcal{A}$ is of the form (4.5.8) if and only if it is continuous.

Proof.   Lemma 4.5.5 establishes half of the theorem, namely that a derivation of the form (4.5.8) is continuous. On the other hand, if we start with the hypothesis that D is continuous, it follows that

$$(4.5.18) \qquad \|\lim_{p\to\infty}\|D\mathcal{U}_p = 0.$$

For any given $f \in \mathcal{A}$, define for every integer $k \geq 1$,

$$(4.5.19) \qquad f_k(n) = \begin{cases} f(n) & \text{if } n \leq k \\ 0 & \text{if } n > k, \end{cases}$$

so that

$$(4.5.20) \qquad \|\lim_{k\to\infty}\| f_k = f.$$

Letting $c_m = f_k(m)$, we have

$$(4.5.21) \qquad f_k(n) = \sum_{m=1}^{k} c_m \mathcal{U}_m(n).$$

Next, we obtain

$$(4.5.22) \qquad Df_k(n) = \sum_{m=1}^{k} c_m D\mathcal{U}_m(n).$$

Using (4.5.17) on the right side of (4.5.22) we have. . . . .
This yields

$$(4.5.23) \qquad Df_k(n) = \sum_{p \le k} \sum_{t/n} v_p(pt) f_k(pt) D\mathcal{U}_p\left(\frac{n}{t}\right).$$

For n fixed, it follows from (4.5.18), that there exists a $k^* = k^*(n)$ such that for all $k \ge k^*$ the right side of (4.5.23) is the same. That is, (4.5.23) becomes

$$(4.5.24) \qquad Df_k(n) = \sum_{p \le k^*} \sum_{t/n} v_p(pt) f_k(pt) D\mathcal{U}_p\left(\frac{n}{t}\right).$$

Letting $k \to \infty$, via (4.5.19) and the continuity of D, (4.5.24) becomes

$$(Df)(n) = \sum_p f(pn)v_p(pn) * D\mathcal{U}_p$$

$$= \left(\sum_p D_p f * D\mathcal{U}_p\right)(n),$$

which completes the proof of the theorem.

Corollary. If H_p is any sequence of functions of $\mathcal{A}$ such that

$$(4.5.25) \qquad \|\lim_{p \to \infty}\|H_p = 0,$$

then

$$(4.5.26) \qquad Df = \sum_p D_p f * H_p$$

defines a continuous derivation.

Proof. . . . .

By way of illustration, if we take $H_p = \mathcal{U}_p$ for all primes p, (4.5.25) is satisfied and (4.5.26) gives

$$Df(n) = \sum_p \sum_{d/n} f(dp)v_p(dp)\mathcal{U}_p\left(\frac{n}{d}\right)$$

$$= \sum_{p/n} f(n)v_p(n) = \Omega(n)f(n).$$

That is, in this case, Df is the derivation that corresponds to multiplication by the completely additive function $\Omega(n)$.

As a second example, take $H_p = \mathcal{U}_{p^2}$ for all primes p. Again (4.5.25) is satisfied, and (4.5.26) gives

$$Df(n) = \sum_p \sum_{d/n} f(dp)\nu_p(dp)\mathcal{U}_{p^2}\left(\frac{n}{d}\right)$$

$$= \sum_{p^2/n} f\left(\frac{n}{p}\right)\nu_p\left(\frac{n}{p}\right),$$

so that since $\nu_p(n/p) = 0$ unless $p^2|n$, in this case,

$$Df(n) = \sum_{p/n} f\left(\frac{n}{p}\right)\nu_p\left(\frac{n}{p}\right)$$

In conclusion, we note that the representation

(4.5.27)
$$Df = \sum_p D_p f * D\mathcal{U}_p,$$

which is characteristic of continuous derivations, is, in form, analogous to the chain rule. In fact, if we introduce the somewhat fanciful notation $(\partial f/\partial \mathcal{U}_p)$ for $D_p f$, (4.5.27) may be rewritten as

$$Df = \sum_p \frac{\partial f}{\partial \mathcal{U}_p} * D\mathcal{U}_p.$$

§4.6 FORMAL TRANSFORMS, DIRICHLET SERIES, AND GENERATING FUNCTIONS

Given an arithmetic function $f \in \mathcal{A}$, we can form the symbolism

(4.6.1)
$$\sum_{n=1}^{\infty} \frac{f(n)}{n^s}.$$

Such an object is called a *formal Dirichlet series*. Naturally, for the ordinary numerical meaning to be associated with such a series, questions of convergence must be considered. For those values of s for which the series converges, we could then define a function of the real variable s by

(4.6.2)
$$F(s) = \sum_{n=1}^{\infty} \frac{f(n)}{n^s}.$$

However, for the purpose of this section, questions of convergence will be relatively unimportant and will enter only for guidance in handling the formal symbols. In the case where the Dirichlet series converge absolutely for a given s, rearrangement of terms is permissible and multiplication yields

$$\sum_{n=1}^{\infty} \frac{f(n)}{n^s} \cdot \sum_{n=1}^{\infty} \frac{g(n)}{n^s} = \sum_{n,m} \frac{f(n)g(m)}{(nm)^s} = \sum_{c=1}^{\infty} \frac{1}{c^s} \sum_{mn=c} f(n)g(m) = \sum_{c=1}^{\infty} \frac{(f * g)(c)}{c^s}.$$

This then is taken as a definition for multiplication of formal Dirichlet series, namely,

(4.6.3) $$\sum_{n=1}^{\infty} \frac{f(n)}{n^s} \cdot \sum_{n=1}^{\infty} \frac{g(n)}{n^s} = \sum_{n=1}^{\infty} \frac{(f * g)(n)}{n^s}.$$

The set $\mathcal{D}$ of formal Dirichlet series forms a ring under addition and multiplication. Let T be the map of $\mathcal{A}$ into $\mathcal{D}$ defined by

(4.6.4) $$f \rightarrow T(f) = \sum_{n=1}^{\infty} \frac{f(n)}{n^s}.$$

For this map we have the following theorem.

Theorem 4.6.1. The map T of (4.6.4) describes an isomorphism between the rings $\mathcal{A}$ and $\mathcal{D}$. Namely, it is 1–1 and both of the following hold:

(4.6.5) $T(f * g) = T(f)T(g)$

(4.6.6) $T(f + g) = T(f) + T(g).$

Proof. . . . .

As a functional, $T(f)$ may be thought of as a formal transform of f. For whatever s domain, the series

$$F(s) = T(f) = \sum_{n=1}^{\infty} \frac{f(n)}{n^s}$$

converges, the function $F(s)$ defined thereby is called the *generating function* of f. Thus we see that the statement and proof of arithmetic identities may have an isomorphic image in terms of generating functions. It should be noted, however, that not all $f \in \mathcal{A}$ have generating functions. For example, if $f(n) = 2^n$, the corresponding formal Dirichlet series converges nowhere. The familiar arithmetic functions have generating functions as follows [$\zeta(s)$ is defined in Ex. 4 of §2.8]:

$$(4.6.7) \qquad \sum_{n=1}^{\infty} \frac{\mu(n)}{n^s} = \frac{1}{\zeta(s)}, \qquad s > 1$$

$$(4.6.8) \qquad \sum_{n=1}^{\infty} \frac{\phi(n)}{n^s} = \frac{\zeta(s-1)}{\zeta(s)}, \qquad s > 2$$

$$(4.6.9) \qquad \sum_{n=1}^{\infty} \frac{d(n)}{n^s} = \zeta^2(s), \qquad s > 1$$

$$(4.6.10) \qquad \sum_{n=1}^{\infty} \frac{\sigma(n)}{n^s} = \zeta(s)\zeta(s-1), \qquad s > 2.$$

Verification of the above is left as an exercise.

The use of generating functions, when available, is convenient for finding identities, but the convolution is more facile (yet somewhat less routine). For example, from (4.6.8) and (4.6.10) we see that

$$(4.6.11) \qquad T(\phi)T(\sigma) = \zeta^2(s-1) = T(nd(n))$$

where the equality in (4.6.11) represents function value equality, valid for $s > 2$. This *suggests* that as transforms

$$T(\phi * \sigma) = T(\phi)T(\sigma) = T(nd(n)),$$

so that by the Isomorphism Theorem 4.6.1, we would get

$$(4.6.12) \qquad \phi * \sigma = nd(n).$$

Once noted, this is easily proved by means of the convolution arithmetic. Namely,

$$\phi * \sigma = (n * \mu) * (1_0 * n)$$
$$= (1_0 * \mu) * (n * n) = e * n * n$$
$$= (n * n) = n(1_0 * 1_0) = nd(n).$$

The conclusion that (4.6.12) is valid as a direct consequence of the function equality (4.6.11) follows once one has the following lemma.

Lemma 4.6.1. If $\sum_{n=1}^{\infty} f(n)/n^s$ and $\sum_{n=1}^{\infty} g(n)/n^s$ converge for $s > s_0$, and for such s

$$\sum_{n=1}^{\infty} \frac{f(n)}{n^s} = \sum_{n=1}^{\infty} \frac{g(n)}{n^s}$$

then for all integers $n \geq 1$, $f(n) = g(n)$.

Proof. . . . .

The key to many of the calculations involving generating functions is that when f is multiplicative and the generating function exists, an Euler factorization (see Ex. 4 of §2.8) is possible.

Theorem 4.6.2. If $\sum_{n=1}^{\infty} f(n)/n^s$ converges absolutely for $s > s_0$, $f(1) = 1$, and $f(n)$ is multiplicative, then for these s,

(4.6.13)
$$\prod_{p} \left(\sum_{\nu=0}^{\infty} \frac{f(p^\nu)}{p^{\nu s}} \right) = \sum_{n=1}^{\infty} \frac{f(n)}{n^s}.$$

Proof. . . . .

Corollary. Under the same hypothesis as the theorem, if $f(n)$ is completely multiplicative, then for s sufficiently large

(4.6.14)
$$\sum_{n=1}^{\infty} \frac{f(n)}{n^s} = \prod_{p} \frac{1}{1 - f(p)/p^s}.$$

Proof. . . . .

Consider next a generating function

$$F(s) = \sum_{n=1}^{\infty} \frac{f(n)}{n^s}$$

that converges for $s > s_0$. It is easily seen (prove this as an exercise) that in this interval the derivative $F'(s)$ exists and

(4.6.15)
$$F'(s) = -\sum_{n=1}^{\infty} \frac{f(n) \log n}{n^s}$$

Thus, if we let $L(n) = \log n \in \mathscr{A}$, (4.6.15) implies that

(4.6.16)
$$T((-L)f) = F'(s).$$

That is, differentiation in $\mathcal{D}$ corresponds under the isomorphism of Theorem 4.5.1 to multiplication by $-L$. But since $-L$ is completely additive, by Theorem 4.5.1, multiplication by $-L$ is a derivation. Thus, under the isomorphism T, differentiation in $\mathcal{D}$ corresponds to the derivation $D_0 f = -Lf$ in $\mathcal{A}$. This, in a sense, distinguishes the class of Dirichlet series $\mathcal{D}$ as being associated with this derivation D_0. For emphasis we write $\mathcal{D} = \mathcal{D}[D_0]$; (and correspondingly $T = T_{D_0}$).

The above discussion suggests the question of whether a generalization of the class of Dirichlet series may be formulated in correspondence to any given derivation D. If we had such a ring, denoted by $\mathcal{D}[D]$, we would require the existence of an isomorphism T_D between $\mathcal{A}$ and $\mathcal{D}[D]$ which is such that (whenever it makes sense) differentiation in $\mathcal{D}[D]$ corresponds to the application of the derivation D in $\mathcal{A}$. For the special derivations provided by Theorem 4.5.1, we have that

$$(4.6.17) \qquad\qquad Df = hf,$$

where $h(n)$ is a completely additive function. Then let $H(n) = \exp[h(n)]$, so that $H(n)$ is completely multiplicative, and form $\mathcal{D}[D]$ the class of formal sums of the form

$$(4.6.18) \qquad\qquad \sum_{n=1}^{\infty} \frac{a_n}{(H(n))^s}.$$

Then define

$$(4.6.19) \qquad\qquad T_D(f) = \sum_{n=1}^{\infty} \frac{f(n)}{(H(n))^s}$$

When the series on the right converges, we consider the generating function of f relative to D as

$$(4.6.20) \qquad\qquad F_D(s) = \sum_{n=1}^{\infty} \frac{f(n)}{(H(n))^s}$$

Such series are usually called *generalized Dirichlet series*.

Theorem 4.6.3. The map T_D of $\mathcal{A}$ into $\mathcal{D}[D]$ is an isomorphism (with an appropriate definition of multiplication in $\mathcal{D}[D]$) between $\mathcal{A}$ and $\mathcal{D}[D]$ such that differentiation in $\mathcal{D}[D]$ corresponds to the derivation $-D$ in $\mathcal{A}$.

Proof. . . . .

§4.7   UNITS, PRIMES, AND UNIQUE FACTORIZATION

As was seen in §4.2, those $f \in \mathcal{A}$ that are units are precisely those for which $f(1) \neq 0$. The nonunits are then those $f \in \mathcal{A}$ such that $f(1) = 0$.

Lemma 4.7.1.   For $f, g \in \mathcal{A}$

   (i)   If f and g are units, then $f * g$ is a unit;
   (ii)   If f is a nonunit, then $f * g$ is a nonunit.

Proof. . . . .

Letting $N(f)$ be the smallest positive integer m such that $f(m) \neq 0$ and $N(f) = 0$ if $f = 0$, we have the following lemma.

Lemma 4.7.2   For $f, g \in \mathcal{A}$

(4.7.1) $N(f * g) = N(f)N(g)$.

Proof. . . . .

Definition 4.7.1.   A function $f \in \mathcal{A}$ is called an $\mathcal{A}$-*prime* if and only if it is a nonunit, such that any equation of the form

$$f = h * g$$

implies that h or g is a unit.

Theorem 4.7.1.   Every $f \in \mathcal{A}$ ($f \neq 0$) has at least one representation of the form

(4.7.2) $f = \eta * g_1 * \cdots * g_r$

where η is a unit and $g_1, \ldots, g_r$ are $\mathcal{A}$-primes.

Proof. . . . .

In the particular case of the functions $\mathcal{U}_m$, m a positive integer, the $\mathcal{A}$-primes are $\mathcal{U}_p$, where p is a prime; and for $m = \prod_{i=1}^{s} p_i^{\alpha_i}$

$$\mathcal{U}_m = \mathcal{U}_{p_1}^{\langle \alpha_1 \rangle} * \mathcal{U}_{p_2}^{\langle \alpha_2 \rangle} * \cdots * \mathcal{U}_{p_s}^{\langle \alpha_s \rangle}.$$

Thus one sees that the unique factorization of m in the ring of integers corresponds to a factorization of $\mathcal{U}_m$ into a convolution of $\mathcal{A}$-primes. The natural question that follows is whether there is uniqueness of factorization in $\mathcal{A}$. The result is known, but the proof lies beyond the scope of this chapter. We content ourselves here with stating the theorem.

Definition 4.7.2. Let f, $g \in \mathcal{A}$. Then f, g are called *associates* if and only if $f = \epsilon * g$ where ϵ is a unit.

Theorem 4.7.2. If $f \in \mathcal{A}$ has two representations of the form (4.7.2)

$$f = \eta * g_1 * \cdots * g_r = \zeta * h_1 * \cdots * h_s,$$

where η, ζ are units, the g_i and h_j $\mathcal{A}$-primes, then $r = s$, and there is a 1–1 correspondence between the g_i and h_j, $g_i \leftrightarrow h_{j_i}$, such that g_i and h_{j_i} are associates.

§4.8 REMOVING THE ASYMMETRY

In §4.1 it was noted that the associative law fails to hold in general for convolutions of functions of $\mathcal{S}$. This stems essentially from the fact that in the definition

(4.8.1) $$(f * g)(x) = \sum_{n \leq x} f\left(\frac{x}{n}\right) g(n),$$

the role of f and g is highly asymetric. In particular, only the values of g at integer values of the argument are relevant, that is,

$$f * g = f * g_0$$

Of course, it is this same asymmetry that prevents the commutative law from holding in $\mathcal{S}$, but this is not as unnatural.

In order to reveal the source of this asymmetry and remove it, we deal first with a somewhat different structure. Let Ω denote the collection of functions $F(x, y)$ of two variables $x \geq 1$, $y \geq 1$, where x is restricted to rational numbers and y is restricted to integers. Addition of functions of Ω is defined as usual and a new multiplication (denoted by $\boxtimes$) is defined by

(4.8.2) $$(F \boxtimes G)(x, y) = \sum_{n \leq x} F(x, n) G(n, y).$$

This operation is noncommutative, but the distributive law holds between addition and $\boxtimes$. The associative law still does not hold in Ω. However, we now focus our attention on the subset $\Omega^* \subset \Omega$ consisting of those $F(x, y)$ that satisfy

(4.8.3) $$F(x, y) = 0 \quad \text{for} \quad y > x \geq 1.$$

Theorem 4.8.1. The set Ω^* is closed under the operation $\boxtimes$ (i.e., F, $G \in \Omega^*$ implies $F \boxtimes G \in \Omega^*$), and for functions of Ω^* the associative law holds for the operation $\boxtimes$.

Proof. . . . .

Within Ω^* we single out a subset $\mathcal{G}^*$ consisting of those functions $F(x, y)$ of the form

$$(4.8.4) \qquad\qquad F(x, y) = f\left(\frac{x}{y}\right), \qquad f \in \mathcal{G}.$$

In fact, we may view (4.8.4) as providing a 1—1 map between $\mathcal{G}$ and $\mathcal{G}^*$. Since $\mathcal{A} \subset \mathcal{G}$, $\mathcal{A}$ has an image in $\mathcal{G}^*$ under this map, that we denote by $\mathcal{A}^*$. Note that in interpreting (4.8.4) as a 1—1 map, we are tacitly assuming that the functions $f(x)$ of $\mathcal{G}$ are defined only for rational x. For two functions of $\mathcal{G}^*$,

$$F(x, y) = f\left(\frac{x}{y}\right), \qquad G(x, y) = g\left(\frac{x}{y}\right)$$

we have

$$(4.8.5) \qquad\qquad (F \boxtimes G)(x, y) = \sum_{n \le x} f\left(\frac{x}{n}\right) g\left(\frac{n}{y}\right).$$

Note also in this case that

$$(4.8.6) \qquad\qquad (F \boxtimes G)(x, 1) = (f * g)(x).$$

Next we ask whether $\mathcal{G}^*$ is closed under the $\boxtimes$ operation. Namely, in (4.8.5), is

$$(4.8.7) \qquad\qquad (F \boxtimes G)(x, y) = h\left(\frac{x}{y}\right)$$

for some $h \in \mathcal{G}$. This is equivalent to requiring that for every integer $t \ge 1$,

$$\sum_{n \le x} f\left(\frac{x}{n}\right) g\left(\frac{n}{y}\right) = \sum_{n \le tx} f\left(\frac{tx}{n}\right) g\left(\frac{n}{ty}\right)$$

or

$$(4.8.8) \qquad\qquad \sum_{y \le n \le x} f\left(\frac{x}{n}\right) g\left(\frac{n}{y}\right) = \sum_{ty \le n \le tx} f\left(\frac{tx}{n}\right) g\left(\frac{n}{ty}\right).$$

It is conceivable that a particular pair of $f, g \in \mathcal{G}$ would satisfy (4.8.8) without anything of interest emerging, for example, $f = 0$. Of greater pertinence is to consider those g such that (4.8.8) holds for all $f \in \mathcal{S}$. Indeed, the correct condition emerges by analyzing the case $f = e$. In this case, (4.8.8) reduces to

$$(4.8.9) \qquad e\left(\frac{x}{[x]}\right) g\left(\frac{[x]}{y}\right) = e\left(\frac{tx}{[tx]}\right) g\left(\frac{[tx]}{ty}\right).$$

Since. . . . .

This simplies that for x not an integer, we have

$$g(x) = 0.$$

In other words, $g \in \mathcal{A}$, which is essentially the necessity part of the following theorem.

Theorem 4.8.2. Let $G \in \mathcal{S}^*$, then a necessary and sufficient condition that $F \boxtimes G \in \mathcal{S}^*$ for all $F \in \mathcal{S}^*$ is that $G \in \mathcal{A}^*$.

Proof. . . . .

Thus we may view the source of the asymmetry which prevented associativity of the convolution operation in $\mathcal{S}$ from the point of view of Ω^*. Ω^* is associative under $\boxtimes$, but the image $\mathcal{S}^*$ of $\mathcal{S}$ in Ω^* is not closed under $\boxtimes$, and Theorem 4.8.2 reveals that the definition of * convolution corresponds to what must be done in order to remain in $\mathcal{S}^*$. More precisely, the set of elements G under which $\mathcal{S}^*$ is right invariant is precisely $\mathcal{A}^*$. This also accounts somewhat for the special role of the arithmetic functions.

NOTES

§4.1. The idea of a convolution arithmetic of arithmetic functions appears to have originated with M. Cipolla in 1915 [4.1, 4.2, 4.3].

This work is described by F. Pellegrino [4.4]. Cipolla defined the operation given in (4.1.4), and according to [4.5, 4.6], he referred to it as "multiplicazione (numerica) integrale." It has also been called "Cauchy product" [4.4] and "Dirichlet multiplication" [4.7]. However, it is now customary to refer to this operation as "convolution." Over the years notation has also varied widely, including fg [4.8], $f \circ g$ [4.9], and $f \hat{\ } g$ [4.10].

Cipolla gave a number of theorems about these convolutions and their relation with Dirichlet series. These results were quoted in [4.3], and the proofs reproduced in [4.4]. They deal with the algebraic structure of the ring of arithmetic functions.

The further devlopment of Cipolla's work was taken up by F. Succi in [4.11]. He considers the problem of finding f, where g and h are given, in the equation $f * g = h$. Other developments of these ideas, in Italy, are included in [4.12, 4.13, 4.14, 4.15].

There are many parallel presentations of the basic concepts of convolution airthmetic. A partial list of these includes [4.16, 4.17, 4.18, 4.19, 4.9].

The idea of this kind of convolution arithmetic has appeared in a variety of applications and generalizations. Among these are:

(i) S. Delsarte [4.20], where instead of Z the domain is the set of finite abelian groups.

(ii) L. Carlitz [4.21], where the algebraic independence of various arithmetic functions is treated.

(iii) S. Amitsur [4.22], where the convolution arithmetic is applied to the study of primes in arithmetic progression.

(iv) J. Popken [4.23], where the results of Carlitz noted above are generalized.

(v) W. Narkiewicz [4.24], where the convolution summation is restricted to a subset of the divisors of n, and studies the properties of the set of arithmetic functions with respect to addition and this convolution.

(vi) A. A. Gioia [4.25], where a generalized convolution is considered of the form

$$(f * g)(n) = \sum_{d/n} K(d, \, n/d) f(d) g(n/d),$$

where $K(u, v)$ is a given. The set of functions $K(u, v)$, that are functions of the greatest common divisor of u and v and for which convolution is associative, are characterized. T. M. K. Davison characterized those K of the form $K(uv, u)$ such that under convolution the subset of multiplicative arithmetic functions form a group [4.18]. In the case where K is symmetric [i.e., $K(u, v) = K(v, u)$], those for which convolution is associative are determined in [4.26] by S. Eilenberg and S. MacLane, and in [4.27] by B. Jessen, J. Karpf, and A. Thorup. I. Popa Fotino [4.28], obtains a complete characterization without assuming that K is symmetric.

(vii) G. C. Rota [4.29], where a general notion of convolution is introduced in the context of the lattice of a partial ordering.

§4.2. Many of the formalisms and properties of the ring of arithmetic functions are clarified if one views the ring $\mathcal{A}$ in a particular representation. Namely, $\mathcal{A}$ is isomorphic to the ring of formal power series $\mathcal{P}$ in an infinite sequence of variables. Letting $x_1, x_2, \ldots, x_n, \ldots$ denote these variables, the elements of $\mathcal{P}$ are the infinite sums

$$\Phi(x) = \Sigma \, a_{i_1, \ldots, i_r, \ldots} \, x_1^{i_1} x_2^{i_2} \ldots x_r^{i_r} \ldots$$

where the summation runs over all sequences of nonnegative integers that are zero from some point on, and the coefficients $a_{i_1, i_2, \ldots}$ are complex numbers. Addition and multiplication have natural definitions in $\mathcal{P}$. Then the isomorphism of the convolution ring $\mathcal{A}$ and $\mathcal{P}$ is established by mapping $f \in \mathcal{A}$ into

$$\Phi(X) = \sum_n f(n) x_{i_1}^{\alpha_1} \ldots x_{i_r}^{\alpha_r}$$

where $n = p_{i_1}^{\alpha_1} \ldots p_{i_r}^{\alpha_r}$ is the unique factorization of n, and p_i denotes the ith prime.

Thus, in particular, units of $\mathcal{A}$ must correspond to units of $\mathcal{P}$, which by Theorem 4.2.2 is equivalent to $\Phi(0, 0, \ldots) \neq 0$. In the context of the formal power series, this is quite well known.

§4.5. If the derivation D_{p_i}, given by (4.5.5), corresponding to the ith prime, is interpreted in the isomorphic formal power series ring, it is easily seen to represent formal partial differentiation with respect to the variable x_i. It is this that justifies the alternate notation $(\partial f / \partial \mathcal{U}_p)$ for $D_p f$.

§4.7. The Unique Factorization Theorem for the convolution ring of complex valued arithmetic functions was first proved by Cashwell and Everett [4.30]. The proof utilizes results concerning formal power series given by Krull [4.31], and Ruckert [4.32].

The question was posed by Ulam [4.8], of whether unique factorization remains true in the convolution ring when the arithmetic functions are restricted to rational integer values. Utilizing a result of P. Samuel [4.33] (obtained independently by D. Buchsbaum [4.34]), Cashwell and Everett [4.35] showed that the unique factorization theorem does remain true in the convolution ring of integer valued arithmetic functions. This was also obtained independently by Chin-Pi Lu [4.36].

In the case where a generalized convolution is defined with a kernel $K(u, v)$ via

$$(f * g)(n) = \sum_{d/n} K(d, \, n/d) f(d) g(n/d),$$

(and associativity is required), I. Popa Fotino has shown [4.28] that unique factorization in the resulting ring is connected with the symmetry of $K(u, v)$ [i.e., $K(u, v) = K(v, u)$].

REFERENCES

4.1. M. Cipolla, Specimen de Calculo Arithmetico Integrale, *Revista de Matematica*, **9**, 1908.

4.2. M. Cipolla, Sui Principi del Calculo Arithmetico Integrale, *Atti Accademia Gioenia Catonia*, **5** (8), 1915, 11.

4.3. S. Amante, Sulle funzioni analitiche numerico-integrale di una o piu funzioni numeriche, *Bolletino Unione Matematica Italiana*, **3** (2), 1947, 109–117.

4.4. F. Pellegrino, Sviluppi moderni del calculo numerico integrale di Michele Cipolla, *Atti del Quarto Congresso dell'Unione Matematica Italiana*, **2**, 1953, 161–168.

4.5. F. Pellegrino, Lineamenti di una teoria delle funzione arithmetiche I, *Rendiconti di Matematica e delle sue Applicazione*, **5** (15), 1956, 469–504.

4.6. F. Pellegrino, Sulle funzioni analitiche numerico-integrali, *Attidella Accademia Nazionale dei Lincei*, **8** (28), 1960, 32–36.

4.7. L. Carlitz, Rings of Arithmetic Functions, *Pacific Journal of Mathematics*, **14**, 1964, 1165–1171.

4.8. S. Ulam, *A Collection of Mathematical Problems*, New York: Interscience, 1960.

4.9. J. Lambek, Arithmetical Functions and Distributivity, *American Mathematical Monthly*, **73**, 1966, 969–973.

4.10. D. Yarden and T. Motzkin, The Dirichlet Convolution Theory of Numbers, *Riveon Lematematika*, **1**, 1946,1–7.

4.11. F. Succi, Sulla espressione del quoziente integrale di due funzione arithmetiche, *Rendiconti di Matematica e Applicazione*, **5**, (15), 1956, 80–92.

4.12. F. Succi, Una generalizzazione delle funzioni aritmetiche completamente moltiplicative, *Rendiconti di Matematica e delle sue Applicazione*, **5** (16), 1957, 255–280.

4.13. F. Succi, Divisibilita integrale locale delle funzioni aritmetiche, *Rendiconti di Matematica e delle sue Applicazione*, **5** (19), 1960, 174–192.

4.14. F. Succi, Sul gruppo moltiplicativo delle funzioni aritmetiche regolari, *Rendiconti di Matematica e delle sue Applicazione*, **5** (19),1960, 458–472.

4.15. F. Pellegrino, La divisione integrale, *Rendiconti di Matematica e delle sue Applicazione*, **5** (22), 1963, 489–497.

4.16. N. Kabaker, On the distributivity of Dirichlet Convolution, *Riveon Lematematika*, **1**, 1946, 29.

4.17. J. Popken, On convolutions in number theory, *Indigationes Mathematicae*, **17**, 1955, 10–15.

4.18. T. M.K. Davison, On Arithmetic Convolutions, *Canadian Math. Bulletin*, **9**, 1966, 287–296.

4.19. H. N. Shapiro, On the Convolution Ring of Arithmetic Functions, *Communications on Pure and Applied Mathematics*, **25**, 1972, 287–336.

4.20. S. Delsarte, Fonctions de Möbius sur les groupes abeliens finis, *Annals of Mathematics*, **2** (49), 1948, 600–609.

4.21. L. Carlitz, Independence of Arithmetic Functions, *Duke Mathematics Journal*, **19**, 1952, 65–70.

4.22. S. Amitsur, On Arithmetic Functions, Journal d'Analyse Mathematique, **5**, 1956–57, 273–314.

4.23. J. Popken, Algebraic dependence of arithmetic functions, Indagationes Mathematicae, **24**, 1962, 155–168.

4.24. W. Narkiewicz, On a Class of Arithmetical Convolutions, *Colloquium Mathematicum*, **10**, 1963, 81–94.

4.25. A. A. Gioia, The *K*-product of Arithmetic Functions, *Canadian Journal of Mathematics*, **17**, 1965, 970–976.

4.26. S. Eilenberg and S. MacLane, Group Extensions and Homology, *Annals of Mathematics*, **43**, 1942.

4.27. B. Jessen, J. Karpf, and A. Thorup, Some functional equations in groups and rings, *Math. Scand.*, **22**, 1968, 257–265.

4.28.   I. P. Fotino, Generalized Convolution Ring of Arithmetic Functions, *Pacific Journal of Mathematics*, **61**, 1975, 103–116.

4.29.   G. C. Rota, On The Foundations of Combinatorial Theory, I. Theory of Mobius Functions, *Zeitschrift für Wahrscheinlichkeitstheorie*, **2**, 1964, 340–368.

4.30.   E. Cashwell and C. Everett, The Ring of Number-theoretic Functions, *Pacific Journal of Mathematics*, **9**, 1959, 975–985.

4.31.   W. Krull, Beitrage zur Arithmetik kommutativer Integritätsbereiche, III. Zum dimensionsbegriff der Idealtheorie, *Mathematische Zeitschrift*, **42**, 1937, 745–766.

4.32.   W. Ruckert, Zum Eliminationsproblem der Potenzreihenideale, *Mathematische Annalen*, **107**, 1933, 259–281.

4.33.   P. Samuel, On Unique Factorization Domains, *Illinois Journal of Mathematics*, **5**, 1961, 1–17.

4.34.   D. A. Buchsbaum, Some remarks on factorization in power series rings, *Journal of Mathematics and Mechanics*, **10**, 1961, 749–753

4.35.   E. Cashwell and C. Everett, Formal Power Series, *Pacific Journal of Mathematics*, **13**, 1963, 45–64.

4.36.   C. P. Lu, On the Unique Factorization Theorem in the Ring of Arithmetic Functions, *Illinois Journal of Mathematics*, **9**, 1965, 40–46.

5

CONGRUENCES

§5.1 BASIC DEFINITIONS AND PROPERTIES

The processing of relations involving the divisibility of integers by a specified fixed integer is greatly facilitated by application of the idea of "congruence." The essence of this notion is a notation.

Definition 5.1.1. For a, b, m, integers, we say that *a is congruent to b modulo m*, written

$$(5.1.1) \qquad\qquad a \equiv b \pmod{m},$$

if and only if $a - b$ *is a multiple of m*. [The negation of (5.1.1) will be written as $a \not\equiv b \pmod{m}$.]

In the above definition, if $m = 1$, (5.1.1) reads

$$a \equiv b \pmod{1},$$

which is clearly true for any integers a, b. Another special case of note corresponds to $m = 0$,

$$a \equiv b \pmod{0},$$

which is clearly equivalent to the statement $a = b$. Thus, in a trivial sense, we see that congruence is a generalization of the notion of equality. However, for the remainder of this chapter it will be convenient to assume that, in all congruences, the modulus m is not zero.

The notion of congruence modulo a fixed integer m provides a natural equivalence relation in the following sense:

(i) *Reflexive Property:* $a \equiv a \pmod{m}$, since $a - a = 0$ is a multiple of m.
(ii) *Symmetric Property:* $a \equiv b \pmod{m}$ implies $b \equiv a \pmod{m}$, since if $a - b$ is a multiple of m, then $b - a$ is also a multiple of m.

(iii) *Transitive Property:* $a \equiv b$ (mod m) and $b \equiv c$ (mod m) implies $a \equiv c$ (mod m). This follows since

$$a - b = mq_1 \quad \text{and} \quad b - c = mq_2$$

implies

$$a - c = (a - b) + (b - c) = mq_1 + mq_2 = m(q_1 + q_2).$$

As an equivalence relation, congruence modulo m separates the integers into equivalence classes, where the integers of a given class are those that are congruent modulo m.

Given any integer a, applying the division algorithm yields

$$a = qm + r, \quad 0 \le r < m;$$

or

$$a - r = qm.$$

Thus we have $a \equiv r$ (mod m), where r is one of the integers $0, 1, 2, \ldots, m - 1$; that is, every integer is congruent modulo m to exactly one of these remainders. Denoting the congruence class modulo m that contains r by $\mathscr{S}_r$, we see that the m mutually disjoint sets

$$\mathscr{S}_0, \mathscr{S}_1, \ldots, \mathscr{S}_{m-1}$$

include all integers.

Definition 5.1.2. The classes $\mathscr{S}_0, \ldots, \mathscr{S}_{m-1}$ are called the *residue classes modulo m*. Any set of representatives, one from each of these classes, is called a *complete residue system modulo m*, (i.e., a set of integers r_i, $i = 0, \ldots, m - 1$, such that $r_i \in \mathscr{S}_i$).

Thus, for example, the integers $0, 1, \ldots, m - 1$ constitute a complete residue system modulo m. Clearly, a set such as

$$m, 2m + 1, 3m + 2, \ldots, (m - 2)m + (m - 1)$$

also constitutes a complete residue system modulo m.

Pursuing the analogy between congruences and equations we have the following theorem.

Theorem 5.1.1. If $a \equiv b$ (mod m) and $c \equiv d$ (mod m), then

(5.1.2) $a + c \equiv b + d$ (mod m)

(5.1.3) $ac \equiv bd$ (mod m).

Proof. The hypothesis provides that there are integers q_1, q_2 such that

$$a - b = q_1 m \quad \text{and} \quad c - d = q_2 m.$$

Then

$$(a + c) - (b + d) = (a - b) + (c - d) = q_1 m + q_2 m$$
$$= (q_1 + q_2)m,$$

which establishes (5.1.2).

Further, we have

$$ac - bd = (a - b)c + (c - d)b = q_1 mc + q_2 mb = (q_1 c + q_2 b)m$$

and this yields (5.1.3).

Corollary. If $a \equiv b$ (mod m) and $n \geq 0$, then

$$a^n \equiv b^n \pmod{m}.$$

Using the notion of congruence, it is quite easy to establish such assertions of divisibility as appear in Ex. 9 of §1.4. Since $13 \equiv -1$ (mod 7), it follows from the above corollary that for any integer $n \geq 0$ that

$$13^n \equiv (-1)^n \pmod{7},$$

and if n is even, this yields $13^n \equiv 1$ (mod 7) or 7 divides $13^n - 1$.

It is of interest to interpret Theorem 5.1.1 in terms of residue classes. If R_1 and R_2 are two residue classes modulo m, choose representatives $r_1 \in R_1$, $r_2 \in R_2$ and define the *sum* $R_1 + R_2$ to be the residue class that contains $r_1 + r_2$; and define the *product* $R_1 R_2$ to be the residue class that contains $r_1 r_2$. The essence of Theorem 5.1.1 is that these definitions of sum and product of residue classes are unambiguous (or well defined). That is, they are independent of the particular choice of the representatives r_1, r_2. It is easily verified that, with respect to these operations of "addition" and "multiplication," the residue classes modulo m satisfy the conditions (i)–(vi) of §1.1, and therefore constitute a commutative ring. The "0" of this ring is the residue class of 0 modulo m and the "1" of this ring is the residue class of 1 modulo m. In general, this ring need not be an integral domain. For if m is composite, say $m = ab$, $a \not\equiv 0$ (mod m), $b \not\equiv 0$ (mod m), the residue class of a times the residue class of b (the product of two nonzero residue classes) will be the zero class. If m is a prime, this cannot occur and it is easily seen that, in this case, the ring of residue classes is an integral domain.

Having investigated the behavior of congruences under addition and multiplication, it is natural to consider what can be done with respect to "division." A simple example shows that in general $ac \equiv bc$ (mod m) does not imply $a \equiv b$ (mod m), for example,

$$6 \equiv 3 \pmod{3} \quad \text{but} \quad 2 \not\equiv 1 \pmod{3}.$$

However, we do have the following theorem.

Theorem 5.1.2.   If $ac \equiv bc \pmod{m}$ and $d = (c, m)$, then

(5.1.4) $$a \equiv b \left(\operatorname{mod} \frac{m}{d} \right).$$

Proof.   The hypothesis provides that

$$(a - b)c = ac - bc = qm.$$

Dividing by $d = (c, m)$, this becomes

(5.1.5) $$(a - b)\frac{c}{d} = q \cdot \frac{m}{d},$$

where $(c/d, m/d) = 1$. Then since m/d divides the right side of (5.1.5), it follows from Theorem 1.4.3 that m/d divides $a - b$, and we have (5.1.4).

Corollary.   If $ac \equiv bc \pmod{m}$ and $(c, m) = 1$, then $a \equiv b \pmod{m}$.

Thus we see that the cancellation property, as with division, is available as long as the integer c (to be canceled) is relatively prime to the modulus m. In particular then, for $m > 0$, the positive integers less than m that are relatively prime to m are of special interest. We denote these by

(5.1.6) $$r_1, r_2, \ldots, r_{\phi(m)}.$$

Corresponding to these are the residue classes $\mathcal{S}_{r_i}$, $i = 1, \ldots, \phi(m)$.

Lemma 5.1.1.   The residue classes $\mathcal{S}_{r_i}$, $i = 1, \ldots, \phi(m)$, include all integers that are relatively prime to m.

Proof.   For any given integer n such that $(n, m) = 1$, we have $n \equiv r \pmod{m}$, $1 \leq r \leq m$. Then $n = r + qm$, and by Ex. 7 of §1.4,

$$1 = (n, m) = (r + qm, m) = (r, m).$$

Thus $n \in \mathcal{S}_r$, $(r, m) = 1$, and the lemma is proved.

A set of integers such as (5.1.6) is an example of a *reduced residue system modulo m*.

Definition 5.1.3.   A *reduced residue system* modulo m is a set of integers $a_1, \ldots, a_{\phi(m)}$ that constitutes a system of representatives for the residue classes $\mathcal{S}_{r_i}$, $i = 1, \ldots, \phi(m)$.

In more explicit terms, it follows that $a_1, \ldots, a_t$ is a reduced residue system modulo m if and only if

(i) $(a_i, m) = 1, i = 1, \ldots, t$

(ii) $a_i \not\equiv a_j \pmod{m}$ for $i \neq j$

(iii) For any integer n, $(n, m) = 1$, there is an a_j such that $n \equiv a_j \pmod{m}$.

If, in the above, we know a priori that $t = \phi(m)$, then $a_1, \ldots, a_{\phi(m)}$ constitutes a reduced residue system if and only if (i) and (ii) are satisfied. Indeed, under these circumstances, (iii) is a consequence of (i) and (ii).

The special role of "reduced residues" is further highlighted by the following theorem of Euler.

Theorem 5.1.3. If $(a, m) = 1$, $m > 0$, then

(5.1.7)
$$a^{\phi(m)} \equiv 1 \pmod{m}.$$

Proof. Let $a_1, a_2, \ldots, a_{\phi(m)}$ be a reduced residue system. Then

$$aa_1, aa_2, \ldots, aa_{\phi(m)}$$

is also a reduced residue system. From the discussion above, this follows if we verify conditions (i) and (ii). Condition (i) is clear since $(aa_i, m) = 1$. Also, if $aa_i \equiv aa_j$ $\pmod{m}$ the corollary to Theorem 5.1.2 would produce $a_i \equiv a_j \pmod{m}$, so that (ii) holds. Since the aa_i, $i = 1, \ldots, \phi(m)$ form a reduced residue system, they are congruent modulo m (in some order) to the a_j, $j = 1, \ldots, \phi(m)$. Thus

$$\prod_{i=1}^{\phi(m)} aa_i \equiv \prod_{i=1}^{\phi(m)} a_i \pmod{m}$$

or

$$\left(\prod_{i=1}^{\phi(m)} a_i \right) a^{\phi(m)} \equiv \prod_{i=1}^{\phi(m)} a_i \pmod{m}.$$

Then since $(m, \prod_{i=1}^{\phi(m)} a_i) = 1$, the corollary to Theorem 5.1.2 yields

$$a^{\phi(m)} \equiv 1 \pmod{m}.$$

The special case of the theorem, where m is a prime p, was given by Fermat and asserts the following.

Corollary. For p a prime, if $(a, p) = 1$, then

$$a^{p-1} \equiv 1 \pmod{p}.$$

At this point we give a simple example that illustrates one of the ways in which the above results may be applied. The problem in question is a natural generalization of the one posed in Ex. 7 of §2.1. It is asserted there that a nonconstant polynomial $f(x)$ with integer coefficients is composite infinitely often. This can now be extended to exponential polynomials as follows.

Theorem 5.1.4. Let $f(x)$ be an expression of the form

$$f(x) = \sum_{i=1}^{r} g_i(x)\, c_i^x,$$

where the c_i are nonzero integers, the $g_i(x)$ are polynomials with integer coefficients, and $|f(x)| \to \infty$ as $x \to \infty$ through the integers. Then, $f(x)$ is composite infinitely often.

Proof. Suppose the theorem false, so that $f(x)$ is prime for all large integral x. Choosing x_0 sufficiently large, $f(x_0) = p$ a prime, where

$$p > \max_{i=1,\ldots,r} |c_i|$$

and hence $(c_i, p) = 1$ for all i. Then, taking $x = x_0 + t \cdot p(p - 1)$, and applying Fermat's theorem, we have

$$f(x_0 + tp(p - 1)) = \sum_{i=1}^{r} g_i(x_0 + tp(p - 1))\, c_i^{x_0 + tp(p-1)}$$

$$\equiv \sum_{i=1}^{r} g_i(x_0) c_i^{x_0} = p \ (\mathrm{mod}\ p)$$

$$\equiv 0 \ (\mathrm{mod}\ p).$$

Thus, for t sufficiently large,

$$|f(x_0 + tp(p - 1))| > p,$$

and since p divides $f(x_0 + tp(p - 1))$, it cannot be a prime, which provides a contradiction.

If one attempts to generalize this one step further and allow another level of exponentiation, trouble is easily anticipated. For at this level would be included such simple functions as

$$2^{2^x} + 1,$$

the so-called Fermat numbers, and it is a difficult and as yet unsolved problem as to whether (for positive integers x) infinitely many of these are composite. However, apart from such specifiable exceptions, results can be obtained. An amusing illustration of this is provided by the following theorem.

Theorem 5.1.5. Let a, b, m, n be integers such that $a > 0$, $m > 1$, $n > 1$; and assume that

$$F(x) = am^{n^x} + b$$

fails to be composite infinitely often. Then, either

$$b = 1 \quad \text{or} \quad a = |1 - b|.$$

Proof. Assuming that $am^{n^x} + b$ is prime for all sufficiently large x, say $x \geq x^*$, it follows that $b \neq 0$. Further, we may choose $x_0 \geq x^*$ and large enough so that the prime $F(x_0)$ does not divide a, m, or b, and also that the following conditions are satisfied

(5.1.8) $$2^{x_o} > |1 - b|$$

(5.1.9) $$x_0 > \max_{q/n} \{\log |a^{q-1} - (1 - b)^{q-1}|/\log q\},$$

where the max on the right of (5.1.9) is taken over those primes q that divide n. We then have

(5.1.10) $$am^{n^{x_0}} + b = p, \text{ a prime.}$$

Let $q_1, \ldots, q_r$ be the distinct primes (if any) that divide $(n, p - 1)$, and write n as

$$n = \left(\prod_{i=1}^{r} q_i^{\mu_i}\right)\bar{n},$$

where $\mu_i > 0$ and $(\bar{n}, p - 1) = 1$. Also, write

$$p - 1 = \left(\prod_{i=1}^{r} q_i^{\lambda_i}\right)s$$

where $\lambda_i > 0$ and $(s, n) = 1$. Then by Euler's theorem, (Theorem 5.1.3), for $x = x_0 + t\phi(s)$, t an integer > 0,

$$\bar{n}^x = \bar{n}^{x_0}\bar{n}^{t\phi(s)} \equiv \bar{n}^{x_0} \pmod{s}.$$

Setting $n_0 = \prod_{i=1}^{r} q_i^{\mu_i}$,

$$n_0^{t\phi(s)} \equiv 1 \pmod{s}$$

so that

(5.1.11) $$n_0^{t\phi(s)} \cdot \bar{n}^x \equiv \bar{n}^{x_0} \pmod{s}.$$

Multiplying (5.1.11) by $n_0^{x_0}$ we obtain

(5.1.12) $n^x \equiv n^{x_0} \pmod{n_0^{x_0} s}$.

Since

$$n_0^{x_0} = \prod_{i=1}^{r} q_i^{\mu_i \cdot x_0},$$

if $\mu_i x_0 \geq \lambda_i$, $i = 1, \ldots, r$, it follows that

$$n_0^{x_0} s \equiv 0 \pmod{p - 1}$$

and hence from (5.1.12)

$$n^x \equiv n^{x_0} \pmod{p - 1}.$$

Then, in this case,

$$F(x) = am^{n^x} + b \equiv am^{n^{x_0}} + b \equiv 0 \pmod{p},$$

which would contradict the assumption that $F(x)$ is a prime.

Thus we are left with the possibility that for some i^*, $1 \leq i^* \leq r$

$$\mu_{i^*} x_0 < \lambda_{i^*}.$$

Then (5.1.10) implies

$$am^{n^{x_0}} \equiv 1 - b \pmod{q_{i^*}^{x_0 + 1}},$$

and since q_{i^*} divides n, $n^{x_0} = q_{i^*}^{x_0} v$, so that

(5.1.13) $am^{q_{i^*}^{x_0} v} \equiv (1 - b) \pmod{q_{i^*}^{x_0 + 1}}$.

We consider first the possibility that q_{i^*} divides m. This would imply that

$$q_{i^*}^{q_{i^*}^{x_0}} > q_{i^*}^{x_0 + 1}$$

divides $m^{q_{i^*}^{x_0} v}$ and therefore by (5.1.13),

$$1 - b \equiv 0 \pmod{q_{i^*}^{x_0 + 1}}.$$

Then [using (5.1.8)] since

$$q_{i^*}^{x_0 + 1} \geq 2^{x_0 + 1} > |1 - b|,$$

it would follow that $|1 - b| = 0$ or $b = 1$.

There remains the case where q_{i*} does not divide m. Raising both sides of (5.1.13) to the $q_{i*} - 1$ power, yields

$$a^{q_{i*}-1} m^{q_{i*}^{x_0}(q_{i*}-1)v} \equiv (1 - b)^{q_{i*}-1} \pmod{q_{i*}^{x_0+1}}.$$

Since Euler's theorem provides

$$m^{q_{i*}^{x_0}(q_{i*}-1)} \equiv 1 \pmod{q_{i*}^{x_0+1}},$$

the above yields

(5.1.14) $$a^{q_{i*}-1} - (1 - b)^{q_{i*}-1} \equiv 0 \pmod{q_{i*}^{x_0+1}}.$$

From (5.1.9) we have that

$$q_{i*}^{x_0+1} > |a^{q_{i*}-1} - (1 - b)^{q_{i*}-1}|,$$

so that (5.1.14) implies that

$$a^{q_{i*}-1} = (1 - b)^{q_{i*}-1}$$

or

$$a = |1 - b|,$$

which is the other possibility provided by the theorem.

EXERCISES

1. Show that for any integers m, n, and any prime p

$$(m + n)^p \equiv m^p + n^p \pmod{p}.$$

   Generalize this to $(m_1 + \cdots + m_r)^p \equiv m_1^p + \cdots + m_r^p \pmod{p}$. Taking all $m_i = 1$, give another proof of Fermat's theorem.

2. Prove that an integer is a multiple of nine if and only if the sum of its digits is a multiple of nine.

3. If a, b, m, n are positive integers such that $mn > 1$, show that

$$am^x + bn^x$$

   is composite for infinitely many integers $x > 0$.

4. Let $a > 0$, $m > 1$, $n > 1$, be integers such that $(m, n) = 1$. Show that $am^{n^x} + 1$ is composite for infinitely many positive integers x.

5. Let $m > 1$, $n > 1$, $b \neq 1$ be integers such that n divides m. Show that

$$|b - 1|m^{n^x} + b$$

is composite for infinitely many positive integers.

6. For $n = \prod p_i^{\alpha_i}$, let

$$\lambda(n) = 1.\text{c.m.} \{\phi(p_i^{\alpha_i})\}.$$

Show that for all a such that $(a, n) = 1$

$$a^{\lambda(n)} \equiv 1 \pmod{n}.$$

Note that $\lambda(n) \leq \phi(n)$ with equality holding only for n a power of a prime, or twice a power of an odd prime.

7. Show that if p is a prime, then $a \equiv b \pmod{p^\alpha}$ implies

$$a^{p^x} \equiv b^{p^x} \pmod{p^{x+\alpha}}.$$

8. Let p be a prime, a, b any integers. If $a^p - b^p$ is divisible by p, then it is also divisible by p^2.

9. Let a, $m > 0$, be integers such that $(a, m) = 1$, and denote by $\mathcal{S}$ the set of nonnegative intergers x such that

$$a^x \equiv 1 \pmod{m}.$$

Show that $\mathcal{S}$ satisfies the hypothesis of Theorem 1.3.2, so that if x^* is the smallest positive integer in $\mathcal{S}$, the integers of $\mathcal{S}$ are all multiples of x^*. This integer x^* is called *the order of a modulo m* and is denoted by $\mathbb{O}_m(a)$. Prove that $\mathbb{O}_m(a)$ divides $\phi(m)$.

10. Let $a > b > 0$ be integers. Show for every positive integer n that n divides $\phi(a^n - b^n)$.

11. Show that if there exists an integer a such that $\mathbb{O}_m(a) = \phi(m)$, then m must either be a prime power, or twice a power of an odd prime.

12. Show that for a odd, $a^2 \equiv 1 \pmod{8}$, and hence

$$a^{2^{\alpha-2}} \equiv 1 \pmod{2^\alpha}.$$

Combine this with Ex. 11 to conclude that if there exists an a such that $(a, m) = 1$ and $\mathbb{O}_m(a) = \phi(m)$, then m is either 2, 4, a power of an odd prime or twice such a power.

13. If there exists an integer a such that $(a, x) = 1$ and $\mathbb{O}_x(a) = x - 1$, show that x must be a prime.

§5.2   INTRODUCTION TO POLYNOMIAL CONGRUENCES

Euler's theorem was in reality a first encounter with a "polynomial congruence." Namely, if we consider the congruence

(5.2.1) $$x^{\phi(m)} \equiv 1 \pmod{m}$$

and ask for the number of incongruent solutions for x modulo m, Euler's theorem provides the answer $\phi(m)$. To see this, note first that an integer a such that $(a, m) > 1$ cannot be a solution of (5.2.1), so that the solutions of (5.2.1) correspond to the reduced residues modulo m. Thus the number of incongruent solutions equals the number of reduced residue classes modulo m, namely $\phi(m)$. We describe this situation by saying that (5.2.1) "has $\phi(m)$ solutions modulo m."

In general, the formulation of study of polynomial congruences is completely analogous to that of polynomial equations. (The results, however, are considerably different.) Given a polynomial

(5.2.2) $$f(x) = a_n x^n + \cdots + a_0$$

with integer coefficients, we have the following definition.

Definition 5.2.1. $f(x)$ is of degree n modulo m if and only if $a_n \not\equiv 0 \pmod{m}$. Implicit in the existence of the degree modulo m is that $f(x)$ does not have all of its coefficients congruent to zero modulo m.

Definition 5.2.2.   An integer a is a root of $f(x)$ modulo m if and only if

(5.2.3) $$f(a) \equiv 0 \pmod{m}.$$

In the study of polynomial congruences, there is another, but related, notion of congruence that plays a useful role.

Definition 5.2.3.   Two polynomials $f(x)$ and $g(x)$ are said to be *identically congruent modulo m* or *identical modulo m* if the coefficients of corresponding powers of x are congruent modulo m. Thus if $f(x)$ is given by (5.2.2) and

$$g(x) = b_n x^n + \cdots + b_0,$$

$f(x)$ is identical modulo m to $g(x)$ if and only if $a_i \equiv b_i \pmod{m}$, $i = 0, \ldots, n$. The notation utilized to describe the relationship is

(5.2.4) $$f(x) \cong g(x) \pmod{m}.$$

It is easily verified that the symbol $\cong$ possesses many of the same formal properties as the ordinary congruences.

In particular, a polynomial $f(x) \cong 0 \pmod{m}$ if and only if all of its coefficients are congruent to zero modulo m. It is important to note the distinction between the notions corresponding to $\equiv$ and $\cong$. If (5.2.4) holds, it follows that for all integers x,

(5.2.5)
$$f(x) \equiv g(x) \ (\text{mod } m).$$

However, it is *not* true that if (5.2.5) holds for all integers x, then (5.2.4) must hold. For example, for all integers x,

$$x^2 - x = x(x - 1) \equiv 0 \ (\text{mod } 2),$$

yet it is clearly not true that

$$x^2 - x \cong 0 \ (\text{mod } 2).$$

The investigation of polynomial congruences such as (5.2.3) focuses on two main problems. First is the problem of determining the solutions, if any, of a given polynomial congruence modulo a fixed integer m. The second problem is that of determining, for a given polynomial, the moduli m for which the corresponding polynomial congruence has solutions. For first degree polynomials, both problems are easily solved. For second degree polynomials, a solution may be formulated in terms of the "quadratic reciprocity law." For the general case of higher degree equations, the investigation belongs to the subject matter of algebraic number theory.

Since it is an implied objective to count the "number of roots" of the congruence

(5.2.6)
$$f(x) \equiv 0 \ (\text{mod } m),$$

we must reach some meaningful convention for making this count. First we note that if $x = k$ satisfies (5.2.6), then any integer $s \equiv k \ (\text{mod } m)$ is likewise a root, since

$$f(s) = \sum_{i=0}^{n} a_i s^i \equiv \sum_{i=0}^{n} a_i k^i = f(k) \equiv 0 \ (\text{mod } m).$$

Thus the property of an integer being a root of (5.2.6) is a function only of the residue class modulo m to which the integer belongs. Hence, two solutions are counted as distinct if and only if they are incongruent modulo m. In order to find all solutions of (5.2.6), it suffices to determine which of the integers

$$f(0), f(\pm 1), \ldots, f\left(\pm \left[\frac{m}{2}\right]\right)$$

is divisible by m.

As with the case of polynomial equations, care must be taken to provide for the possibility that a polynomial congruence has a given residue (or residue class) as a "multiple root." In order to elucidate this we begin with the following lemma.

Lemma 5.2.1.   If k is a root of the polynomial congruence (5.2.6), where $f(x)$ is of degree n, then there exists a polynomial $g_1(x)$ with integer coefficients, of degree $n - 1$, such that

$$(5.2.7) \qquad f(x) \cong (x - k)g_1(x) \ (\mathrm{mod}\ m).$$

   Proof.   We have

$$f(x) - f(k) = \sum_{i=1}^{n} a_i(x^i - k^i)$$

$$= (x - k) \sum_{i=1}^{n} a_i \sum_{\nu=0}^{i-1} x^{i-1-\nu}k^{\nu}.$$

Setting

$$g_1(x) = \sum_{i=1}^{n} a_i \sum_{\nu=0}^{i-1} x^{i-1-\nu}k^{\nu},$$

since $f(k) \equiv 0 \ (\mathrm{mod}\ m)$, (5.2.7) follows.

   If, in (5.2.7), the congruence $g_1(x) \equiv 0 \ (\mathrm{mod}\ m)$ has $x = k$ as a solution, we can write

$$(5.2.8) \qquad g_1(x) \cong (x - k)g_2(x) \ (\mathrm{mod}\ m),$$

where $g_2(x)$ is of degree $n - 2$. Equation (5.2.7) and (5.2.8) would then imply that

$$f(x) \cong (x - k)^2 g_2(x) \ (\mathrm{mod}\ m).$$

In general, we are thus led to the following definition.

Definition 5.2.4.   For a given integer k, let $\nu \geq 0$ be the largest integer such that there exists a polynomial $g(x)$ with integer coefficients, of degree $n - \nu$, such that

$$(5.2.9) \qquad f(x) \cong (x - k)^{\nu} g(x) \ (\mathrm{mod}\ m).$$

This integer ν is the *order* of k as a zero of $f(x)$ modulo m, or, in other words, its *multiplicity* as a root of (5.2.6). We shall use the notation

$$(5.2.10) \qquad \nu = \mathrm{mult}(f, k, m).$$

   It follows that if $f(x) \not\equiv 0 \ (\mathrm{mod}\ m)$, then mult (f, k, m) is well defined, the $g(x)$ in (5.2.9) is $\not\equiv 0 \ (\mathrm{mod}\ m)$, and

$$(5.2.11) \qquad \mathrm{mult}(f, k, m) \leq \deg(f, m)$$

where $\deg(f, m)$ denotes the degree of f modulo m. When $\text{mult}(f, k, m) > 1$, k is said to be a multiple root modulo m of $f(x)$.

The inequality (5.2.11) is very much analogous to what is true for polynomial equations. However, it should be noted that *unlike* the case of equations, the congruence (5.2.6) can have more solutions modulo m than $\deg(f, m)$. For example, we've already noted (Ex. 12 of §5.1) that

$$x^2 \equiv 1 \ (\text{mod } 8)$$

has the four solutions 1, 3, 5, 7 modulo 8.

Since, as noted above, a polynomial congruence can have more roots than the degree, it is of interest to see where the standard argument (for the case of equations) fails. If k is a root of order ν, we have the representation (5.2.9) where $g(k) \not\equiv 0$ (mod m). Now if $l \not\equiv k$ (mod m) and $f(l) \equiv 0$ (mod m), (5.2.9) yields

(5.2.12) $(l - k)^\nu g(l) \equiv 0 \ (\text{mod } m).$

Here we have $l - k \not\equiv 0$ (mod m), and the requisite next step (in order to proceed as with equations) would be to conclude that $g(l) \equiv 0$ (mod m). *But this cannot be done here!* For it is quite possible to have $a \not\equiv 0$ (mod m), $b \not\equiv 0$ (mod m), and yet have $ab \equiv 0$ (mod m) [e.g., $4 \cdot 2 \equiv 0$ (mod 8)]. However, as was noted previously, if m is a prime, the ring of residue classes modulo m is an integral domain; that is, $ab \equiv 0$ (mod p), p a prime, implies that p divides a or b. This then suggests that the usual statements concerning the number of zeros of a polynomial might carry over to congruences in the case of prime modulus. This is indeed the case.

Theorem 5.2.1.   If $f(x)$ is a polynomial of degree n modulo p, where p is a prime, then the number of roots of $f(x)$ modulo p, counted with their multiplicities, is at most n.

Proof.   Let $k_1, \ldots, k_r$ be the distinct (mod p) roots of $f(x) \equiv 0$ (mod p). If k_1 is of multiplicity ν_1, from (5.2.9)

$$f(x) \cong (x - k_1)^{\nu_1} g^{(1)}(x) \ (\text{mod } p),$$

where the degree of $g^{(1)}(x) = n - \nu_1$ and $g^{(1)}(k_1) \not\equiv 0$ (mod p). Since $f(k_2) \equiv 0$ (mod p), this yields

$$0 \equiv (k_2 - k_1)^{\nu_1} g^{(1)}(k_2) \ (\text{mod } p).$$

Since $(k_2 - k_1)^{\nu_1} \not\equiv 0$ (mod p), this, in turn, implies that k_2 is a zero of $g^{(1)}(x)$ mod p. Let ν_2 be the order of k_2 as a zero of $g^{(1)}(x)$, so that

$$g^{(1)}(x) \cong (x - k_2)^{\nu_2} g^{(2)}(x) (\text{mod } p),$$

where degree of $g^{(2)}(x) = n - \nu_1 - \nu_2$ and $g^{(2)}(k_2) \not\equiv 0$ (mod p). Also, since

$g^{(1)}(k_1) \not\equiv 0 \pmod{p}$, it follows that $g^{(2)}(k_1) \not\equiv 0 \pmod{p}$. Combining the above, we have

$$f(x) \cong (x - k_1)^{\nu_1}(x - k_2)^{\nu_2} g^{(2)}(x) \pmod{p},$$

and continuing this process produces

$$(5.2.13) \qquad f(x) \cong (x - k_1)^{\nu_1}(x - k_2)^{\nu_2} \cdots (x - k_r)^{\nu_r} g^{(r)}(x) \pmod{p},$$

where the degree of $g^{(r)}(x) = n - \nu_1 - \nu_2 \cdots - \nu_r$ and

$$g^{(r)}(k_i) \not\equiv 0 \pmod{p}, \qquad i = 1, \ldots, r.$$

Thus from (5.2.13) we see that $\nu_i = \text{mult}(f, k_i, p)$ for $i = 1, \ldots, r$. Further, since

$$0 \le \text{degree } g^{(r)}(x) = n - \nu_1 \cdots - \nu_r,$$

we obtain

$$\nu_1 + \cdots + \nu_r \le n.$$

Corollary.   Under the hypothesis of the above theorem, (5.2.13) follows, where the degree of $g^{(r)}(x)$ equals $n - \nu_1 \cdots - \nu_r$.

From Fermat's theorem we have that for p a prime, the congruence

$$x^{p-1} - 1 \equiv 0 \pmod{p}$$

has the $p - 1$ distinct solutions $1, 2, \ldots, p - 1$. Then from (5.2.13)

$$x^{p-1} - 1 \cong \prod_{i=1}^{p-1} (x - i)^{\nu_1} g^{(p-1)}(x) \pmod{p}$$

where degree $g^{(p-1)}(x) = p - 1 - \sum_{i=1}^{p-1} \nu_i \ge 0$. Since $\nu_i \ge 1$, this implies that all $\nu_1 = 1$, so that degree $g^{(r)}(x) = 0$, and

$$x^{p-1} - 1 \cong c \prod_{i=1}^{p-1} (x - i) \pmod{p},$$

where c is a fixed integer. Comparing coefficients of x^{p-1} yields that $c \equiv 1 \pmod{p}$, and hence

$$(5.2.14) \qquad\qquad x^{p-1} - 1 \cong \prod_{i=1}^{p-1} (x - i) \pmod{p}.$$

Comparing the constant terms on both sides of (5.2.14) gives

$$-1 \equiv \prod_{i=1}^{p-1} (-i) \equiv (p-1)! \pmod{p}$$

or

(5.2.15) $\qquad\qquad (p-1)! + 1 \equiv 0 \pmod{p}$

The assertion (5.2.15) for prime p is known as *Wilson's Theorem*.

EXERCISES

1. Prove the converse of Wilson's Theorem: If

$$(n-1)! + 1 \equiv 0 \pmod{n},$$

then n is a prime.

2. Prove that for p a prime, and $0 < i < p$

$$(i-1)! \, (p-i)! + (-1)^{i-1} \equiv 0 \pmod{p}.$$

3. Let $m_1, \ldots, m_p$ and $n_1, \ldots, n_p$ be two complete residue systems modulo the odd prime p. Show that $m_1 n_1, m_2 n_2, \ldots, m_p n_p$, *cannot* be a complete residue system modulo p.

4. Find all primes p such that $(p-1)! + 1$ is a power of p.

5. Let p be an odd prime, and $0 < i < p - 1$. Prove that

$$\binom{p-1}{i} \equiv (-1)^i \pmod{p}.$$

6. Prove that for p a prime, $p \equiv 1 \pmod{4}$,

$$\left(\left(\frac{p-1}{2} \right)! \right)^2 + 1 \equiv 0 \pmod{p}.$$

7. Find all primes $p \equiv 1 \pmod{4}$ such that

$$\left(\left(\frac{p-1}{2} \right)! \right)^2 + 1$$

is a power of p.

8. **For a polynomial** $f(x) = a_n x^n + \cdots + a_0$ recall that the definition of the derivative is $f'(x) = n a_n x^{n-1} + \cdots + a_1$. Show that k is a root of $f(x) \equiv 0$

(mod m) of order > 1 (i.e., a multiple root) if and only if k is a root of both $f(x) \equiv 0$ (mod m) and $f'(x) \equiv 0$ (mod m).

9. Show that the congruence

$$x^n - x + 1 \equiv 0 \ (\text{mod } m)$$

has a multiple root modulo m if and only if

a. $(n(1 - n), m) = 1$,   and

b. $n^n \equiv (n - 1)^{n-1}$ (mod m).

Further, prove in this case that there is exactly one multiple root modulo m, and this one is of order 2.

10. Show that for every fixed positive integer n, the congruence

$$\sum_{d|n} \mu\left(\frac{n}{d}\right) x^d \equiv 0 \ (\text{mod } n)$$

has exactly n distinct solutions modulo n.

§5.3 LINEAR CONGRUENCES

The simplest case of a polynomial congruence is a linear one, that is, the degree $n = 1$. This then has the form

(5.3.1) $f(x) = ax - b \equiv 0$ (mod m).

For this case, both of the basic problems delineated in §5.2 may be solved completely.

Lemma 5.3.1. If $d = (a, m)$ divides b, then every solution of (5.3.1) is a solution of

(5.3.2) $$\frac{a}{d} \cdot x \equiv \frac{b}{d} \left(\text{mod } \frac{m}{d}\right),$$

and conversely, every solution of (5.3.2) satisfies (5.3.1).

Proof. Equation 5.3.1 is equivalent to finding integers x, y such that $ax - b = my$. Dividing by d, this is equivalent to $(a/d)x - (b/d) = (m/d)y$, which is (5.3.2).

Since $(a/d, m/d) = 1$ in the above, Lemma 5.3.1 reduces the problem of finding solutions to (5.3.1) to the case where $(a, m) = 1$.

Lemma 5.3.2. If $(a, m) = 1$, the congruence $ax \equiv b \pmod{m}$ has a unique solution modulo m.

Proof. Since $(a, m) = 1$, there exist integers u and v such that

$$1 = (a, m) = ua + vm,$$

and multiplying both sides by b yields

$$b = (bu)a + (bv)m.$$

This, in turn, gives

$$a(bu) \equiv b \pmod{m},$$

so that the congruence $ax \equiv b \pmod{m}$ has $x = bu$ as a solution.

Next suppose we have two solutions

$$ax \equiv b \pmod{m}$$
$$ax' \equiv b \pmod{m}.$$

Subtracting gives $a(x - x') \equiv 0 \pmod{m}$, and since $(a, m) = 1$ this implies $x \equiv x' \pmod{m}$. Thus the uniqueness of the solution is established.

Putting these lemmas together will yield the following theorem.

Theorem 5.3.1. A necessary and sufficient condition for $ax \equiv b \pmod{m}$ to have a solution is that $d = (a, m)$ divides b. If this is satisfied, the congruence has exactly d solutions modulo m.

Proof. If $ax \equiv b \pmod{m}$ has a solution, then we have integers x, y such that $ax - b = my$. Since $d = (a, m)$ divides a and m, it follows that it divides b (i.e., the condition is necessary).

Assuming that $d = (a, m)$ divides b, Lemma 5.3.1 asserts that the solutions of $ax \equiv b \pmod{m}$ are the same as those of $(a/d)x \equiv (b/d) \pmod{(m/d)}$; and Lemma 5.3.2 asserts that the latter has a unique solution modulo m/d. Letting x_0 be the unique solution modulo m/d of

$$(5.3.3) \qquad \frac{a}{d}x_0 \equiv \frac{b}{d} \left(\bmod \frac{m}{d} \right),$$

then the integers $x = x_0 + t \cdot (m/d)$, t any integer, are also solutions (though not distinct modulo m/d) of

$$\frac{a}{d}x \equiv \frac{b}{d} \left(\bmod \frac{m}{d} \right).$$

Since they are also solutions of $ax \equiv b \pmod{m}$, and include all such solutions, the number of incongruent solutions modulo m of $ax \equiv b \pmod{m}$ equals the number of distinct residues modulo m represented by the integers

$$(5.3.4) \qquad\qquad x_0 + t\frac{m}{d}, \qquad t = 0, \pm 1, \pm 2, \ldots .$$

Consider

$$(5.3.5) \qquad\qquad x_0, \; x_0 + \frac{m}{d}, \; \ldots, \; x_0 + (d-1)\frac{m}{d}.$$

These are from the set (5.3.4) and are d incongruent integers modulo m. Also, for any integer t, the Division Algorithm gives

$$t = dq + r, \qquad 0 \le r < d,$$

so that

$$x_0 + t\frac{m}{d} = x_0 + r\frac{m}{d} + qm \equiv x_0 + r\frac{m}{d} \pmod{m}.$$

Hence (5.3.5) is a complete set of distinct residues mod m for the integers of (5.3.4).

Note that this theorem does indeed resolve both of the fundamental congruence problems in the linear case. On the one hand, the integers m for which $ax \equiv b \pmod{m}$ has solutions are those such that (a, m) divides b. On the other hand, if m has this property, the theorem describes the solutions.

As important special cases we have the following corollaries.

Corollary 1. If $(a, m) = 1$, then there exists an integer $\bar{a}$, unique modulo m, such that $a\bar{a} \equiv 1 \pmod{m}$. ($\bar{a}$ is called the inverse of a modulo m.).

Corollary 2. For m a prime, the ring of residue classes modulo m is a field.

EXERCISES

1. Consider the integers $1 \le a \le p - 1$, for p a prime. By "matching up" each such integer a with its inverse modulo p, give another proof of Wilson's Theorem.

2. Let $(a, m) = 1$ and use Euler's Theorem to produce the inverse of a modulo m.

3. Let m be any positive integer; show that

$$\left(\prod_{\substack{i=1 \\ (i,m)=1}}^{m} i \right)^2 \equiv 1 \pmod{m}.$$

4. Let m be an integer > 6, $(b, m) = 1$, and c any integer. Consider a sequence a_n which is defined recursively modulo m by

$$(*) \qquad\qquad a_{n-1}a_n \equiv ba_{n-1} + c \ (\mathrm{mod}\ m),$$

where the sequence is initiated with a_1, $(a_1, m) \neq 1$. This construction either stops when an a_{n-1} is reached such that the congruence $(*)$ has no solution for a_n, or continues indefinitely. In any event, show that the sequence a_n cannot contain $\phi(m)$ integers representing a reduced residue system modulo m.

5. Prove that the linear congruence

$$a_1x_1 + \cdots + a_nx_n + a_{n+1} \equiv 0 \ (\mathrm{mod}\ m)$$

in the variables $x_1, \ldots, x_n$ has dm^{n-1} sets of incongruent solutions modulo m, provided $d = (a_1, \ldots, a_n, m)$ divides a_{n+1}. If it does not divide a_{n+1}, there are no solutions.

§5.3A   AVERAGE OF THE DIVISOR FUNCTION OVER ARITHMETIC PROGRESSIONS

The result on linear congruences expressed by Theorem 5.3.1 is of fundamental importance in the execution of many number-theoretic calculations. In this section one illustrative example of this will be considered. The setting for this illustration is motivated by the Dirichlet divisor problem discussed earlier in §3.8. There the estimation of $\Sigma_{n \leq x}\, d(n)$ was considered, and in Ex. 5 of §3.8 this was extended to sums of the form $\Sigma_{n \leq x}\, d(An)$. This suggests the current application, namely, consideration of

$$(5.3A.1) \qquad\qquad D_{a,b}(x) = \sum_{n \leq x} d(an + b),$$

where $a > 0$.

We shall require the following lemma.

Lemma 5.3A.1. For A a given positive integer,

$$(5.3A.2) \qquad\qquad \sum_{\substack{n \leq x \\ (n,A)=1}} \frac{1}{n} = \alpha_1 \log x + \alpha_2 + 0\!\left(\frac{1}{x}\right),$$

where $\alpha_1 = \phi(A)/A$ and both α_2 and the $0(1/x)$ depend on A.

Proof. Using (3.7.8) we have

$$\sum_{\substack{n \leq x \\ (n,A)=1}} \frac{1}{n} = \sum_{n \leq x} \frac{1}{n} \sum_{d/(n,A)} \mu(d)$$

(5.3A.3)
$$= \sum_{d/A} \mu(d) \sum_{\substack{n \leq x \\ n \equiv 0 (\mathrm{mod}\, d)}} \frac{1}{n}.$$

Letting $n = d\tau$ in this last inner summation, we have, using Lemma 3.8.1,

$$\sum_{\substack{n \leq x \\ n \equiv 0 (d)}} \frac{1}{n} = \frac{1}{d} \sum_{\tau \leq x/d} \frac{1}{\tau}$$

$$= \frac{1}{d}\left(\log \frac{x}{d} + c_2 + 0\left(\frac{d}{x}\right)\right)$$

$$= \frac{1}{d} \log \frac{x}{d} + \frac{c_2}{d} + 0\left(\frac{1}{x}\right).$$

Inserting this in (5.3A.3) yields

$$\sum_{\substack{n \leq x \\ (n,A)=1}} \frac{1}{n} = \sum_{d/A} \frac{\mu(d)}{d} \log\frac{x}{d} + c_2 \sum_{d/A} \frac{\mu(d)}{d} + 0\left(\frac{1}{x}\right)$$

$$= \left(\sum_{d/A} \frac{\mu(d)}{d}\right) \log x + \left(c_2 \sum_{d/A} \frac{\mu(d)}{d} - \sum_{d/A} \frac{\mu(d)\log d}{d}\right) + 0\left(\frac{1}{x}\right)$$

and this establishes (5.3A.2).

Theorem 5.3A.1. For every pair of integers $a, b, a > 0$, we have

(5.3A.4)
$$D_{a,b}(x) = \beta_1 x \log x + 0(x).$$

where

(5.3A.5)
$$\beta_1 = \sum_{\tau/(a,b)} \frac{\phi(a/\tau)}{a/\tau},$$

and the $0(x)$ depends on a and b.

Proof.

$$D_{a,b}(x) = \sum_{n \leq x} d(an + b) = \sum_{n \leq x} \sum_{d/(an+b)} 1$$

(5.3A.6)
$$= \sum_{d \leq ax+b} \sum_{\substack{n \leq x \\ an+b \equiv 0 \,(\mathrm{mod}\, d)}} 1.$$

In the inner summation of (5.3A.6), we are summing over those integers $n \le x$ that satisfy the linear congruence $an + b \equiv 0 \pmod{d}$. From Theorem 5.3.1, we know that this has solutions only if (a, d) divides b. Thus we may restrict the outer summation to such values of d, so that

$$(5.3A.7) \qquad D_{a,b}(x) = \sum_{\substack{d \le ax+b \\ (d,a)/b}} \sum_{\substack{n \le x \\ an+b \equiv 0 \,(\mathrm{mod}\, d)}} 1.$$

Letting $\tau = (d, a)$, Theorem 5.3.1 provides that the congruence $an + b \equiv 0 \pmod{d}$ has exactly τ solutions for n modulo d. Let

$$r_1(d), r_2(d), \ldots, r_\tau(d)$$

be the smallest positive residue in the residue class modulo d of each of these solutions. Then

$$\sum_{\substack{n \le x \\ an+b \equiv 0 \,(\mathrm{mod}\, d)}} 1 = \sum_{i=1}^{\tau} \sum_{\substack{n \le x \\ n \equiv r_i(\mathrm{mod}\, d)}} 1.$$

But for each i,

$$\sum_{\substack{n \le x \\ n \equiv r_i(\mathrm{mod}\, d)}} 1 = \sum_{\substack{td+r_i \le x \\ t \ge 0}} 1 = \sum_{0 \le t \le (x-r_i)/d} 1$$

$$= \frac{x - r_i}{d} + 0(1)$$

$$= \frac{x}{d} + 0(1),$$

the last since $0 < r_i \le d$. Then we have

$$\sum_{\substack{n \le x \\ an+b \equiv 0(\mathrm{mod}\, d)}} 1 = \sum_{i=1}^{\tau} \left(\frac{x}{d} + 0(1) \right)$$

$$= \tau \frac{x}{d} + 0(\tau);$$

and inserting this in (5.3A.7) gives [since $\tau = (a, d)$ divides b, clearly τ divides (a, b)]

$$D_{a,b}(x) = \sum_{\substack{\tau/(a,b)}} \sum_{\substack{d \le ax+b \\ (a,d)=\tau}} \left(\tau \frac{x}{d} + 0(\tau) \right)$$

$$= x \sum_{\tau/(a,b)} \tau \sum_{\substack{d \le ax+b \\ (a,d)=\tau}} \frac{1}{d} + 0\left(\sum_{\tau/(a,b)} \tau \sum_{\substack{d \le ax+b \\ (a,d)=\tau}} 1 \right)$$

(5.3A.8)
$$= x \sum_{\tau/(a,b)} \tau \sum_{\substack{d \le ax+b \\ (a,d)=\tau}} \frac{1}{d} + 0(x).$$

Then from Lemma 5.3A.1, writing $a = \tau a'$, $d = \tau d'$, $(a', d') = 1$,

$$\sum_{\substack{d \le ax+b \\ (a,d)=\tau}} \frac{1}{d} = \frac{1}{\tau} \sum_{\substack{d' \le (ax+b)/\tau \\ (a',d')=1}} \frac{1}{d'}$$

$$= \frac{1}{\tau}\left(\frac{\phi(a')}{a'} \log \frac{ax+b}{\tau} + 0(1) \right),$$

and combining this with (5.3A.8) yields

$$D_{a,b}(x) = x\left(\sum_{\tau/(a,b)} \frac{\phi(a/\tau)}{a/\tau} \right) \log x + 0(x).$$

This is precisely (5.3A.4) with

$$\beta_1 = \sum_{\tau/(a,b)} \frac{\phi(a/\tau)}{a/\tau}.$$

EXERCISES

1. Combining the above method with the ideas of §3.8 prove that

$$D_{a,b}(x) = \beta_1 x \log x + \beta_2 x + 0(\sqrt{x})$$

where β_2 and the $0(\sqrt{x})$ depend on a, b.

2. Estimate

$$\sum_{n \le x} \frac{d(an+b)}{n}.$$

3. Using the result of Ex. 1 estimate

$$\sum_{n \le x} 2^{\nu(an+b)}.$$

4. Estimate

$$\sum_{n \leq x} \frac{\phi(an + b)}{an + b}.$$

5. Estimate

$$\sum_{n \leq x} \frac{\sigma(an + b)}{an + b}.$$

§5.4   THE CHINESE REMAINDER THEOREM AND SIMULTANEOUS CONGRUENCES

We next consider the question of solving systems of simultaneous congruences. In this connection, we begin with the following basic result which is known as the Chinese Remainder Theorem.

Theorem 5.4.1.   Let $m_1, \ldots, m_k$ be positive integers that are relatively prime in pairs (i.e., $(m_i, m_j) = 1$ for $i \neq j$). Then for any given integers $b_1, \ldots, b_k$, the system of congruences

$$(5.4.1) \qquad\qquad x \equiv b_i \ (\text{mod } m_i), \qquad i = 1, \ldots, k$$

has a unique solution for x modulo $m_1 m_2 \cdots m_k$.

   Proof.

   EXISTENCE OF A SOLUTION.   For $k = 1$, $x \equiv b \ (\text{mod } m)$ is a solution and the existence of a solution is established for this case. Proceeding by induction on the number of congruences, assume the existence of a solution to be true for the case of $k - 1$ simultaneous congruences of the form (5.4.1). Then the first $k - 1$ of the congruences (5.4.1) have a solution x_0 modulo $m_1 \cdots m_{k-1}$. It follows that for any integer t, $x_0 + tm_1 \cdots m_{k-1}$ is also a solution of these first $k - 1$ congruences. We therefore wish to choose t so that the kth congruence is satisfied. That is, we require

$$x_0 + tm_1 \cdots m_{k-1} \equiv b_k \ (\text{mod } m_k)$$

or

$$(5.4.2) \qquad\qquad tm_1 \cdots m_{k-1} \equiv b_k - x_0 \ (\text{mod } m_k).$$

Since the hypothesis of the theorem insures that in (5.4.2) $(m_1 \cdots m_{k-1}, m_k) = 1$, it follows from Theorem 5.3.1 that (5.4.2) has a solution $t = t_0$. Then $x_0 + t_0 m_1 \cdots m_{k-1}$ satisfies all of the congruences (5.4.1), and the induction is complete.

UNIQUENESS OF THE SOLUTION. Let $x = x_1$ and $x = x_2$ be two solutions of the system of congruences (5.4.1). Then for $i = 1, \ldots, k,$

$$x_1 \equiv b_i \ (\text{mod } m_i) \qquad \text{and} \qquad x_2 \equiv b_i \ (\text{mod } m_i).$$

Subtracting these equations we have

$$x_1 - x_2 \equiv 0 \ (\text{mod } m_i), \qquad i = 1, \ldots, k,$$

and since the m_i are relatively prime in pairs, it follows (Ex. 2 of §1.4) that $x_1 - x_2 \equiv 0 \ (\text{mod } m_1 \cdots m_k)$.

Next we consider the generalization of Theorem 5.4.1 to the case where the moduli m_i are not relatively prime by pairs.

Theorem 5.4.2. The simultaneous congruences (5.4.1) have a solution if and only if for all $i, j,$

$$(5.4.3) \qquad b_i - b_j \equiv 0 \ (\text{mod } (m_i, m_j)),$$

in which case the solution is unique modulo $\{m_1, \ldots, m_k\}$.

Proof.

NECESSITY. If there is a solution to (5.4.1), then for any $i, j,$

$$x \equiv b_i (\text{mod } (m_i, m_j))$$
$$x \equiv b_j (\text{mod } (m_i, m_j))$$

so that $b_i - b_j \equiv 0 \ (\text{mod } (m_i, m_j))$.

SUFFICIENCY. Again we proceed by induction on the number of congruences in (5.4.1). The result holds for $k = 1$ and we assume it for $k - 1$. The first $k - 1$ of the congruences (5.4.1) have a solution x_0 modulo $\{m_1, \ldots, m_{k-1}\}$. Then for any integer t, $x_0 + t\{m_1, \ldots, m_{k-1}\}$ is also a solution and we seek an integer t for which the kth congruence is also satisfied. That is,

$$x_0 + t\{m_1, \ldots, m_{k-1}\} \equiv b_k \ (\text{mod } m_k),$$

or

$$(5.4.4) \qquad t\{m_1, \ldots, m_{k-1}\} \equiv b_k - x_0 \ (\text{mod } m_k).$$

Theorem 5.3.1 can be applied to insure the existence of a solution for t, provided we verify

$$(m_k, \{m_1, \ldots, m_{k-1}\}) \text{ divides } b_k - x_0$$

But this, in turn, follows from the hypothesis since (Ex. 17 of §1.4)

$$(m_k, \{m_1, \ldots, m_{k-1}\}) = \{(m_1, m_k), \ldots, (m_{k-1}, m_k)\}$$

and for $i < k$, $x_0 \equiv b_i \pmod{(m_i, m_k)}$, so that

$$b_k - x_0 \equiv b_k - b_i \equiv 0 \pmod{(m_i, m_k)}.$$

Thus (5.4.4) has a solution $t = t_0$ and $t_0\{m_1, \ldots, m_{k-1}\} + x_0$ satisfies all k equations, which completes the induction.

UNIQUENESS.   Consider two solutions x_1, x_2 to (5.4.1), so that for $i = 1, \ldots, k$

$$x_1 \equiv b_i \pmod{m_i} \qquad \text{and} \qquad x_2 \equiv b_i \pmod{m_i}.$$

Subtracting, we obtain $x_1 - x_2 \equiv 0 \pmod{m_i}$, $i = 1, \ldots, k$, and hence

$$x_1 - x_2 \equiv 0 \pmod{\{m_1, \ldots, m_k\}},$$

which shows that the solution is unique modulo $\{m_1, \ldots, m_k\}$.

The results can be generalized one step further.

Theorem 5.4.3.   The system of congruences

$$(5.4.5) \qquad\qquad a_i x \equiv b_i \pmod{m_i}, \qquad i = 1, \ldots, k$$

has a solution if and only if for all i, j,

$$(5.4.6) \qquad\qquad d_i = (a_i, m_i) \text{ divides } b_i$$

and

$$(5.4.7) \qquad\qquad \Delta_{ij} = \frac{b_j a_i - b_i a_j}{d_i d_j} \equiv 0 \bmod \left(\left(\frac{m_i}{d_i}, \frac{m_j}{d_j}\right)\right),$$

in which case the solution is unique modulo

$$\left\{\frac{m_1}{d_1}, \ldots, \frac{m_k}{d_k}\right\}.$$

Proof.

SUFFICIENCY.   Dividing each of the congruences (5.4.5) by the corresponding $d_i = (a_i, m_i)$, and setting $a_i = d_i a_i'$, $b_i = d_i b_i'$, $m_i = d_i m_i'$, (5.4.5) is equivalent to the system

$$a_i' x \equiv b_i' \pmod{m_i'}, \qquad i = 1, \ldots, k,$$

where $(a_i', m_i') = 1$. Thus a_i' has an inverse $\bar{a}_i'$ modulo m_i', and these congruences are in turn equivalent to

$$(5.4.8) \qquad x \equiv \bar{a}_i' b_i' \pmod{m_i'}, \qquad i = 1, \ldots, k.$$

However, Theorem 5.4.2 gives the existence of a solution to the system (5.4.8) provided that for all i, j

$$(5.4.9) \qquad \bar{a}_i' b_i' - \bar{a}_j' b_j' \equiv 0 \pmod{(m_i', m_j')}.$$

Since $(a_i' a_j', (m_i', m_j')) = 1$, the conditions (5.4.9) are equivalent to those obtained by multiplying (5.4.9) by $a_i' a_j'$. These are

$$(5.4.10) \qquad a_j' b_i' - a_i' b_j' \equiv 0 \left(\mathrm{mod}\ \left(\frac{m_i}{d_i}, \frac{m_j}{d_j} \right) \right),$$

which is precisely (5.4.7).

NECESSITY. If (5.4.5) has a solution for x, then $a_i x \equiv b_i \pmod{m_i}$ implies that $d_i = (a_i, m_i)$ divides b_i. It follows by the same steps as in the sufficiency proof that x is a solution of (5.4.8). Since the conditions (5.4.9) are necessary for this, the necessity of (5.4.7) follows.

UNIQUENESS. Since x a solution of (5.4.5) is equivalent to its being a solution of (5.4.8), Theorem 5.4.2 yields that x is unique modulo $\{m_1/d_1, \ldots, m_k/d_k\}$.

Corollary. In the case where the m_i, $i = 1, \ldots, k$ are pairwise relatively prime, and $d_i = (a_i, m_i)$ divides b_i, the number of solutions to (5.4.5) modulo $m_1 \cdots m_k$ is $d_1 \cdots d_k$.

Though the existence of solutions in the above proofs was based on inductive arguments, it may also be executed on the basis of arguments involving basic properties of the greatest common divisor. For example, in Theorem 5.4.1, setting $M_i = m_1 \cdots m_k/m_i$, the hypothesis implies that $(M_1, \ldots, M_k) = 1$. Then by Ex. 2 of §1.5, there exist integers $u_1, \ldots, u_k$ such that

$$u_1 M_1 + \cdots + u_k M_k = 1,$$

so that

$$u_i M_i \equiv 1 \pmod{m_i},$$
$$M_j \equiv 0 \pmod{m_i}, \qquad \text{for } i \neq j.$$

Setting

$$x = \sum_{i=1}^{k} u_i M_i b_i,$$

we see that x is a solution of the congruences (5.4.1).

The results of the above theorems on systems of congruence have a highly "multiplicative appearance." The Chinese Remainder Theroem itself can be applied to give this some precision and, at the same time, yield another useful method for treating simultaneous congruences.

Definition 5.4.1. An integer valued function $G(x_1, \ldots, x_n)$ of n integer variables $x_1, \ldots, x_n$ will be called *mod-$\mathcal{S}$-periodic,* for $\mathcal{S}$ a subset of the integers, if and only if for each $m \in \mathcal{S}$, $x_i \equiv x_i' \pmod{m}$, $i = 1, \ldots, n$ implies

$$G(x_1, \ldots, x_n) \equiv G(x_1', \ldots, x_n') \pmod{m}.$$

If $\mathcal{S} = Z$, the set of all integers, $G(x_1, \ldots, x_n)$ will be called mod-Z-periodic. Note that if m_1 and m_2 are integers such that $(m_1, m_2) = 1$, and G is mod-m_i-periodic, $i = 1, 2$, it follows that it is mod-$m_1 m_2$-periodic.

If in the following, two vectors $(x_1, \ldots, x_n)$ and $(x_1', \ldots, x_n')$ are considered the same modulo m if corresponding components are congruent modulo m.

Theorem 5.4.4. Let $\mathcal{S} = [m_1, \ldots, m_k]$ be a set of positive integers which are relatively prime in pairs; and for $i = 1, \ldots, k$, let $G_i(x_1, \ldots, x_n)$ be mod-m_i-periodic. Assume that for each $i = 1, \ldots, k$, the congruence

$$(5.4.11) \qquad\qquad G_i(x_1, \ldots, x_n) \equiv 0 \pmod{m_i}$$

has N_i distinct solutions modulo m_i for the vector $(x_1, \ldots, x_n)$. Then, the *simultaneous* system of congruences

$$(5.4.12) \qquad G_i(x_1, \ldots, x_n) \equiv 0 \pmod{m_i}, \qquad i = 1, \ldots, k$$

has exactly $N_1 N_2 \cdots N_k$ distinct solutions modulo $m_1 \cdots m_k$.

Proof. Let $X^{(i)} = (x_1^{(i)}, \ldots, x_n^{(i)})$ be a solution of (5.4.11) for each $i = 1, \ldots, k$. We wish to map the k vectors $X^{(i)}$ onto a single vector $(x_1^*, \ldots, x_n^*)$ which satisfies (5.4.12). That is,

$$(5.4.13) \qquad\qquad (X^{(1)}, \ldots, X^{(k)}) \rightarrow (x_1^*, \ldots, x_n^*).$$

To do this, for each fixed j, $1 \leq j \leq n$, consider the system of congruences

$$(5.4.14) \qquad\qquad x \equiv x_j^{(i)} \pmod{m_i}, \qquad i = 1, \ldots, k.$$

The Chinese Remainder Theorem provides a solution $x = x_j^*$ of the system (5.4.14) that is unique modulo $m_1 \cdots m_k$. Further, since G_i is mod-m_i-periodic, for each $i = 1, \ldots, k$

$$G_i(x_1^*, \ldots, x_n^*) \equiv G_i(x_1^{(i)}, \ldots, x_n^{(i)}) \equiv 0 \pmod{m_i}$$

so that $(x_1^*, \ldots, x_n^*)$ is an admissible image in the map (5.4.13) and is uniquely determined modulo $m_1 \cdots m_k$.

Next, to complete the proof of the theorem, it suffices to show that the map (5.4.13) is 1–1 onto, that is, every solution of the system (5.4.12) is covered exactly once. Starting with $(x_1^*, \ldots, x_n^*)$ a solution of (5.4.12), define

$$X^{(i)} = (x_1^*, \ldots, x_n^*),$$

for all $i = 1, \ldots, k$. This provides a set of $X^{(i)}$ that is mapped onto $(x_1^*, \ldots, x_n^*)$ modulo $m_1 \cdots m_k$ by (5.4.13).

That the map is 1–1 is seen by considering

$$(X^{(1)}, \ldots, X^{(k)}) \to (0, 0, \ldots, 0)$$

so that for $i = 1, \ldots, k; j = 1, \ldots, n$;

$$x_j^{(i)} \equiv 0 \pmod{m_i},$$

which implies that $X^{(i)} \equiv (0, \ldots, 0)$ modulo m_i.

The above theorem may be used to extend itself.

Theorem 5.4.5. Let $m_{ij}, i = 1, \ldots, k, j = 1, \ldots, l$, be positive integers such that for $j \neq j'$, $(m_{ij}, m_{i'j'}) = 1$ for all i, i'. For each $j = 1, \ldots, l$ consider the system of congruences

$$(5.4.15) \qquad G_{ij}(x_1, \ldots, x_n) \equiv 0 \pmod{m_{ij}},$$

$i = 1, \ldots, k$, where G_{ij} is mod-m_{ij}-periodic. If the system corresponding to a fixed j has D_j distinct solutions modulo $M_j = \prod_{i=1}^{k} m_{ij}$, then the full system (5.4.15), $j = 1, \ldots, l$, $i = 1, \ldots, k$, has exactly $D_1 \ldots D_l$ distinct solutions modulo $M_1 \ldots M_l$.

Proof. For each fixed j define

$$g_j(x_1, \ldots, x_n) = \begin{cases} 0 & \text{if } (x_1, \ldots, x_n) \text{ is a solution of the} \\ & \text{system (5.4.15), } i = 1, \ldots, k \\ 1 & \text{otherwise.} \end{cases}$$

Then the function $g_j(x_1, \ldots, x_n)$ is mod-M_j-periodic, the integers $M_1, \ldots, M_l$ are relatively prime in pairs, and the full system (5.4.15) is equivalent to

$$(5.4.16) \qquad g_j(x_1, \ldots, x_n) \equiv 0 \pmod{M_j}$$

for $j = 1, \ldots, l$. Thus Theorem 5.4.4 may be applied to the system (5.4.16) and this yields Theorem 5.4.5.

The above theorem provides a general type of "multiplicative reduction" for certain systems of congruences. For suppose that $\mathcal{S}_i$ = the set of divisors of a positive integer T_i, and G_i is mod-$\mathcal{S}_i$-periodic. We consider the system of congruences

(5.4.17) $$G_i(x_1, \ldots, x_n) \equiv 0 \ (\text{mod } T_i).$$

(Note that we are *not* assuming that the T_i are relatively prime in pairs.) Writing canonical factorizations

$$T_i = \prod_{j=1}^{l} p_j^{\alpha_{ji}},$$

we may replace (5.4.17) by the equivalent set of congruences

(5.4.18) $$G_i(x_1, \ldots, x_n) \equiv 0 \ (\text{mod } p_j^{\alpha_{ji}})$$

for $i = 1, \ldots, k$ and $j = 1, \ldots, l$. Setting $G_{ij} = G_i$, $m_{ij} = p_j^{\alpha_{ji}}$, we have an instance of Theorem 5.4.5. Thus the determination of the number of solutions to the system (5.4.17) reduces to carrying this out for the special case where the T_i are all powers of the same prime, that is,

$$G_i(x_1, \ldots, x_n) \equiv 0 \ (\text{mod } p^{\nu_i}), \qquad i = 1, \ldots, k.$$

As an illustration of how this is applied, let us consider Ex. 5 of §5.3, where we desire to count the number of solutions of

(5.4.19) $$a_1 x_1 + \cdots + a_n x_n + a_{n+1} \equiv 0 \ (\text{mod } m),$$

provided $d = (a_1, \ldots, a_n, m)$ divides a_{n+1} (an obviously necessary condition for solutions to exist). Writing $a_i = d a_i'$, $i = 1, \ldots, m$, $m = dm'$, (5.4.19) is equivalent to

(5.4.20) $$a_1' x_1 + \cdots + a_n' x_n + a_{n+1}' \equiv 0 \ (\text{mod } m')$$

where

(5.4.21) $$(a_1', \ldots, a_n', m') = 1.$$

Since $G_i(x_1, \ldots, x_n) = a_1' x_1 + \cdots + a_n' x_n + a_{n+1}'$ is mod-Z-periodic, the technique discussed above applies. If the unique factorization of m' is

$$m' = \prod_{i=1}^{l} p_i^{\alpha_i}, \qquad \alpha_i > 0,$$

we consider for each i, the congruence

$$(5.4.22) \qquad a_1' x_1 + \cdots + a_n' x_n + a_{n+1}' \equiv 0 \pmod{p_i^{\alpha_i}}.$$

From (5.4.21) it follows that $(a_1', \ldots, a_n', p_i^{\alpha_i}) = 1$, and hence for some j^*, $(a_{j^*}, p_i) = 1$. Thus in (5.4.22) if the $x_j, j \neq j^*$, are fixed arbitrarily modulo $p_i^{\alpha_i}$, then x_{j^*} is uniquely determined modulo $p_i^{\alpha_i}$ by (5.4.22). Hence, (5.4.22) has exactly $(p_i^{\alpha_i})^{n-1}$ solutions modulo $p_i^{\alpha_i}$. Then from Theorem 5.4.5, since $(p_i^{\alpha_i}, p_j^{\alpha_j}) = 1$, it follows that the number of solutions modulo m' of (5.4.20) is

$$\prod_{i=1}^{l} (p_i^{\alpha_i})^{n-1} = (m')^{n-1} = \frac{m^{n-1}}{d^{n-1}}.$$

In each of these solutions, x_1 is determined modulo (m/d), and hence has d incongruent "extensions" modulo m (which are congruent modulo (m/d)). Thus each solution of (5.4.20) extends to d^n solutions of (5.4.19) modulo m. Finally then, the number of solutions, modulo m, of (5.4.19) equals

$$d^n \cdot \frac{m^{n-1}}{d^{n-1}} = d \cdot m^{n-1}.$$

A special case of Theorem 5.4.5 which is of particular interest arises from the fact that polynomials $f(x)$ with integer coefficients are mod-Z-periodic. We thus have the following immediately.

Corollary 1. Let $f_i(x)$, $i = 1, \ldots, l$ be polynomials in x with integer coefficients. Then if $m_1, \ldots, m_l$ are pairwise relatively prime integers such that each congruence

$$f_i(x) \equiv 0 \pmod{m_i}$$

has d_i distinct solutions modulo m_i, the system of simultaneous congruences

$$f_i(x) \equiv 0 \pmod{m_i}, \qquad i = 1, \ldots, l$$

has exactly $d_1, d_2 \ldots d_l$ distinct solutions modulo $m_1 \ldots m_l$.

Corollary 2. Let $f(x)$ be a polynomial with integer coefficients, and $m = \prod p_i^{\alpha_i}$ the unique factorization of m. Then if N_i is the number of distinct solutions modulo $p_i^{\alpha_i}$ of

$$f(x) \equiv 0 \pmod{p_i^{\alpha_i}},$$

the number of distinct solutions modulo m of

$$f(x) \equiv 0 \pmod{m}$$

is $N_1 N_2 \ldots N_l$, ($l =$ the number of distinct prime factors of m).

EXERCISES

1. Let $f(x)$ be a polynomial with integer coefficients such that $f(x) \to \infty$ as $x \to \infty$. Show that given any integers $k > 1, N \geq 1$, there exists an integer x_0 such that each of

$$f(x_0), f(x_0 + 1), \ldots, f(x_0 + k - 1)$$

has at least N different prime factors.

2. Given any positive integer n, prove that there exist n consecutive positive integers each divisible by a square > 1.

3. Given any integer $K \geq 1$, and real number $\epsilon > 0$, show there exists an integer N such that

$$\frac{\phi(N + i)}{N + i} < \epsilon$$

for $i = 1, \ldots, K$.

4. Given any real number M, however large, and any integer $K \geq 1$, show there exists an integer N such that

$$\frac{\sigma(N + i)}{N + i} > M$$

for $i = 1, \ldots, K$.

5. What is the number of solutions modulo m of the congruence $x^k \equiv 0 \pmod{m}$, for k a fixed integer ≥ 1?

6. Let $f(x)$ be a polynomial with integer coefficients and let $\psi(n)$ denote the number of integers among

$$f(0), f(1), \ldots, f(n - 1)$$

that are relatively prime to n. Show that if ν_p denotes the number of solutions modulo p (a prime) to the congruence $f(x) \equiv 0 \pmod p$, then

$$\psi(n) = n \prod_{p/n} \left(1 - \frac{\nu_p}{p} \right).$$

§5.4A SIMULTANEOUS CONGRUENCES FOR POLYNOMIALS IN SEVERAL VARIABLES

In general, very little can be said about the existence of solutions to a system of simultaneous congruences involving polynomials in several variables. More precisely, let $F_i(x_1, \ldots, x_n)$, $i = 1, \ldots, k$, be polynomials with integer coefficients in the n variables $x_1, \ldots, x_n$; and consider the system of congruences

$$(5.4A.1) \qquad F_i(x_1, \ldots, x_n) \equiv 0 \pmod{m_i},$$

where the m_i are given positive integers. The Chinese Remainder Theorem, discussed in §5.4, focused on the special case where the F_i were linear and $n = 1$, whereas the discussion of (5.4.19) covered the case of a single linear polynomial F in n variables.

For the case where all the m_i are equal to a prime p, since the ring of residue classes modulo a prime form a field, the usual theory of homogeneous linear equations over a field may be carried over to a system of linear congruences modulo p. Thus if we have for $i = 1, \ldots, k$,

$$(5.4A.3) \qquad F_i = a_{i1}x_1 + \cdots + a_{in}x_n \equiv 0 \pmod{p},$$

this system will have a "nontrivial" solution [i.e., not all of the $x_i \equiv 0 \pmod{p}$], if $n > k$. This result has a generalization to the case of general polynomials $F_i(x_1, \ldots, x_n)$ that was conjectured by Artin and proved by Chevalley.

In considering Chevalley's theorem, various notions will be required, and they are set forth as follows.

Definition 5.4A.1. A set of integers $x_1\ x_2, \ldots, x_n$ is said to be *nontrivial modulo* m, if at least one of the x_i is $\not\equiv 0$ modulo m.

Thus if a set of integers is nontrivial modulo m, it follows that it is nontrivial modulo every multiple of m. It is *not* true that a set of integers that is nontrivial modulo m is, as a consequence, nontrivial modulo every divisor of m.

Definition 5.4A.2. If, in the polynomial

$$F(x_1, \ldots, x_n) = \sum_{i_1, \ldots, i_n} a_{i_1, \ldots, i_n}\, x_1^{i_1} x_2^{i_2} \cdots x_n^{i_n},$$

d = the largest sum $i_1 + \cdots + i_n$ that occurs for a term with a nonzero coefficient (i.e., $a_{i_1, \ldots, i_n} \neq 0$), then d is called the *degree* of F.

The above notation should not be confused with the degree of F when F is viewed as a polynomial in one of its variables x_i (with coefficients that are polynomials in the other variables). This degree is referred to as *the degree of F in the variable x_i* and is simply the exponent of the highest power of x_i appearing in any term of F.

Definition 5.4A.3. If two polynomials $F(x_1, \ldots, x_n)$, $G(x_1, \ldots, x_n)$ with integer coefficients are such that corresponding coefficients are congruent modulo m, we write

$$(5.4A.3) \qquad F(x_1, \ldots, x_n) \cong G(x_1, \ldots, x_n) \pmod{m}.$$

The above is a generalization of Definition 5.2.3 in which exactly this notation was presented for polynomials in a single variable. As was noted, when (5.4A.3) holds, it follows that for any integers $a_1, \ldots, a_n$

(5.4A.4) $F(a_1, \ldots, a_n) \equiv G(a_1, \ldots, a_n) \pmod{m}.$

However, the truth of (5.4A.4), even for all $a_1, \ldots, a_n$, need not imply (5.4A.3). Nonetheless, when m is a prime p, there is a case where this implication is valid as shown in the following lemma.

Lemma 5.4A.1. Let $F(x_1, \ldots, x_n)$ and $G(x_1, \ldots, x_n)$ be polynomials with integer coefficients, such that for a given prime p, and all integers $a_1, \ldots, a_n$

(5.4A.5) $F(a_1, \ldots, a_n) \equiv G(a_1, \ldots, a_n) \pmod{p}.$

Then, if each of F and G has the additional property that its degree in each of the variables is less than p, it follows that

(5.4A.6) $F(x_1, \ldots, x_n) \cong G(x_1, \ldots, x_n) \pmod{p}.$

Proof. In the case $n = 1$, the hypothesis provides that

$$F(x_1) - G(x_1) \equiv 0 \pmod{p},$$

has p solutions modulo p. Since the degree is less than p Theorem 5.2.1 yields that

$$F(x) - G(x) \cong 0 \pmod{p},$$

which is the required conclusion of this lemma. Proceeding by induction on the number of variables, assume the truth of the lemma for the case of $n - 1$ variables. Then, writing $F - G$ as a polynomial in x_n, we have

$$F(x_1, \ldots, x_n) - G(x_1, \ldots, x_n) = \sum_{j=0}^{p-1} P_j(x_1, \ldots, x_{n-1}) x_n^j.$$

Fixing $x_1 = a_1, \ldots, x_{n-1} = a_{n-1}$ arbitrarily, the hypothesis of the lemma asserts that the congruence

$$\sum_{j=0}^{p-1} P_j(a_1, \ldots, a_{n-1}) x_n^j \equiv 0 \pmod{p}$$

has p solutions for x_n, modulo p. By Theorem 5.2.1, this implies that

(5.4A.7) $P_j(a_1, \ldots, a_{n-1}) \equiv 0 \pmod{p}, \qquad j = 0, \ldots, p - 1.$

Since (5.4A.7) has been established for all $a_1, \ldots, a_{n-1}$, the inductive hypothesis yields that

$$P_j(x_1, \ldots, x_{n-1}) \cong 0 \ (\mathrm{mod}\ p), \qquad j = 0, \ldots, p - 1.$$

This, in turn, gives that $F(x_1, \ldots, x_n) - G(x_1, \ldots, x_n) \cong 0 \ (\mathrm{mod}\ p)$, which completes the induction and the proof of the lemma.

The above lemma is the principal tool in proving the aforementioned theorem of Chevalley, which is as follows.

Theorem 5.4A.1. Let $F_i(x_1, \ldots, x_n)$, $i = 1, \ldots, k$, be polynomials in $x_1, \ldots, x_n$ with integer coefficients, such that F_i is of degree $d_i \geq 1$, and $F_i(0, 0, \ldots, 0) = 0$. Then if

$$(5.4A.8) \qquad\qquad n > \sum_{k=1}^{k} d_i,$$

for any given prime p the system of congruences

$$(5.4A.9) \qquad F_i(x_1, \ldots, x_n) \equiv 0 \ (\mathrm{mod}\ p), \qquad i = 1, \ldots, k$$

has a nontrivial solution for $x_1, \ldots, x_n$.

Proof. Assume that the system (5.4A.9) has no nontrivial solution. Then for any set $a_1, \ldots, a_n$ not all $\equiv 0$ modulo p, for at least one i, say $i = i^*$,

$$F_{i^*}(a_1, \ldots, a_n) \not\equiv 0 \ (\mathrm{mod}\ p).$$

Then, by Fermat's theorem,

$$1 - F_{i^*}^{p-1}(a_1, \ldots, a_n) \equiv 0 \ (\mathrm{mod}\ p)$$

and consequently

$$\prod_{i=1}^{k} [1 - F_i^{p-1}(a_1, \ldots, a_n)] \equiv 0 \ (\mathrm{mod}\ p).$$

Thus if we introduce the polynomial

$$(5.4A.10) \qquad G(x_1, \ldots, x_n) = \prod_{i=1}^{k} [1 - F_i^{p-1}(x_1, \ldots, x_n)],$$

we have that

(5.4A.11) $G(a_1, \ldots, a_n) \equiv \begin{cases} 1 & \text{if all } a_i \equiv 0 \ (\text{mod } p) \\ 0 & \text{otherwise} \end{cases}\Bigg\} \ (\text{mod } p).$

We next describe a transformation of the polynomial G into another one $G^*(x_1, \ldots, x_n)$, generally of smaller degree, such that for all $a_1, \ldots, a_n$,

(5.4A.12) $G(a_1, \ldots, a_n) \equiv G^*(a_1, \ldots, a_n) \ (\text{mod } p).$

This is done as follows: for any appearance (in any term) of a symbol such as

$$x_i^t$$

write $t = (p - 1)q + r$, where $1 \le r \le p - 1$, and replace x_i^t by

$$x_i^r.$$

Carrying this out changes $G(x_1, \ldots, x_n)$ into a polynomial $G^*(x_1, \ldots, x_n)$ that is not of larger degree; that is,

(5.4A.13)
            degree of $G(x_1, \ldots, x_n) \ge$ degree of $G^*(x_1, \ldots, x_n).$

Also, we have

(5.4A.14)
            [degree of $G^*(x_1, \ldots, x_n)$ in x_i] $\le p - 1,$ $i = 1, \ldots, n$

That (5.4A.12) holds follows since for all integers a,

$$a^t = a^{(p-1)q+r} \equiv a^r \ (\text{mod } p)$$

is a consequence of Fermat's theorem and the condition $r \ge 1$.
    From (5.4A.11) and (5.4A.12), we obtain

$$G^*(a_1, \ldots, a_n) \equiv \begin{cases} 1 & \text{if all } a_i \equiv 0 \ (\text{mod } p) \\ 0 & \text{otherwise} \end{cases}\Bigg\} \ (\text{mod } p),$$

so that for all $a_1, \ldots, a_n$,

(5.4A.15) $G^*(a_1, \ldots, a_n) \equiv \prod_{i=1}^{n} (1 - a_i^{p-1}) \ (\text{mod } p).$

Since both G^* and $\Pi(1 - x_i^{p-1})$ have the property that the degree in each x_i is less than p, Lemma 5.4A.1 yields that

$$(5.4A.16) \qquad G^*(x_1, \ldots, x_n) \cong \prod_{i=1}^{n}(1 - x_i^{p-1}) \ (\text{mod } p).$$

The degree modulo p of the right side of (5.4A.16) is $n(p - 1)$, and also, by (5.4A.16), this is the degree of G^* modulo p as well. Hence (5.4A.13) gives that

$$(5.4A.17) \qquad \text{degree of } G(x_1, \ldots, x_n) \geq n(p - 1).$$

From (5.4A.10) we note that

$$\text{degree of } G(x_1, \ldots, x_n) \leq (p - 1)\sum_{i=1}^{k} d_i,$$

which together with (5.4A.17) implies

$$n \leq \sum_{i=1}^{k} d_i.$$

Since this contradicts the hypothesis (5.4A.8), the theorem is proved.

By utilizing the ideas and results of §5.4, Theorem 5.4A.1 can be extended to provide an analogous result for the general system (5.4A.1).

Theorem 5.4A.2.   Under the same hypothesis as Theorem 5.4A.1, the system of congruences (5.4A.1), where the m_i, $i = 1, \ldots, k$ are any positive integers, has a nontrivial solution modulo $\{m_1, \ldots, m_k\}$.

Proof.   Consider first the case where the m_i are all powers of the same prime p. Since we are concerned only with proving the existence of a nontrivial solution modulo the least common multiple, it suffices to consider all $m_i = p^\alpha$, the same power of p. Theorem 5.4A.1 provides a nontrivial solution $(a_1, \ldots, a_n)$ for the case $\alpha = 1$. Next we consider any $\alpha > 1$, and look for a solution of the form

$$(5.4A.18) \qquad t_1 p^{\alpha-1}, t_2 p^{\alpha-1}, \ldots, t_n p^{\alpha-1}.$$

That is, we consider the system of congruences

$$(5.4A.19) \qquad F_i(t_1 p^{\alpha-1}, \ldots, t_n p^{\alpha-1}) \equiv 0 \ (\text{mod } p^\alpha),$$

$i = 1, \ldots, k$. Applying Taylor's theorem to expand the left side of (5.4A.19) and using the fact that $F_i(0, \ldots, 0) = 0$ we obtain for $i = 1, \ldots, k$

(5.4A.20)

$$p^{\alpha-1}t_1\frac{\partial F_i}{\partial x_1}(0, \ldots, 0) + \cdots + p^{\alpha-1}t_n\frac{\partial F_i}{\partial x_n}(0, \ldots, 0) \equiv 0 \ (\text{mod } p^\alpha),$$

or dividing through by $p^{\alpha-1}$,

(5.4A.21)

$$t_1\frac{\partial F_i}{\partial x_1}(0, \ldots, 0) + \cdots + t_n\frac{\partial F_i}{\partial x_n}(0, \ldots, 0) \equiv 0 \ (\text{mod } p).$$

This last system of simultaneous congruences in the variables $t_1, \ldots, t_n$, viewed in the field of residue classes modulo p, is a set of k linear homogeneous equations in n unknowns. Since by hypothesis

$$n > \sum_{i=1}^{k} d_i \geq k,$$

there are more unknowns than equations, and a nontrivial solution exists for $t_1, \ldots, t_n$. Then the $t_i p^{\alpha-1}$ given in (5.4A.18) yield a nontrivial solution of

$$F_i(x_1, \ldots, x_n) \equiv 0 \ (\text{mod } p^\alpha), \qquad i = 1, \ldots, k.$$

Consider next the general case of (5.4A.1), with given moduli m_j, $j = 1, \ldots, k$. To prove the existence of a nontrivial solution in this case, it suffices to consider the system in which the individual m_i are all replaced by the least common multiple $\{m_1, \ldots, m_k\}$. Thus we see we may assume that all the $m_i = m > 1$. Then, by the above, if $m = \prod_{\nu=1}^{l} p_\nu^{\alpha_\nu}$, each of the systems

$$F_i(x_1, \ldots, x_n) \equiv 0 \ (\text{mod } p_\nu^{\alpha_\nu}), \ i = 1, \ldots, k$$

for $\nu = 1, \ldots, l$, has a nontrivial solution $(a_1^{(\nu)}, \ldots, a_n^{(\nu)})$. Thus, if D_ν is the number of distinct solutions modulo $p_\nu^{\alpha_\nu}$ of the νth system, we have $D_\nu \geq 2$ (there is always the trivial solution). Applying Theorem 5.4.4, the combined system of all congruences corresponding to $i = 1, \ldots, k$ and $\nu = 1, \ldots, l$ has

$$D_1 D_2 \cdots D_l \geq 2^l \geq 2$$

distinct solutions modulo m. It follows that the system

$$F_i(x_1, \ldots, x_n) \equiv 0 \ (\text{mod } m), \qquad i = 1, \ldots, k$$

has a nontrivial solution.

EXERCISES

1. Under the hypothesis of the theorem, if l = the number of distinct prime factors of $\{m_1, \ldots, m_k\}$, the system (5.4A.1) has at least 2^l solutions modulo $\{m_1, \ldots, m_k\}$.

2. Under the hypothesis of the theorem, letting $m = \{m_1, \ldots, m_k\}$, assume that for each m_i, $i = 1, \ldots, k$, $\nu(m/m_i) < \nu(m)$. Then the system (5.4A.1) has a solution that is nontrivial modulo each m_i, $i = 1, \ldots, k$. In particular, this holds when the m_i are relatively prime by pairs.

3. If $m = \{m_1, \ldots, m_k\}$ is squareful and $n > k$, then the system (5.4A.1) has a nontrivial solution modulo m. [It is assumed that all $F_i(0, \ldots, 0) = 0$.]

4. Give a justification of the assertion that (5.4A.19) implies (5.4A.20).

§5.5   THE GENERAL POLYNOMIAL CONGRUENCE

For a large number of problems of number theory it is necessary to acquire various kinds of information about the solutions of the general polynomial congruence

$$(5.5.1) \qquad\qquad f(x) \equiv 0 \;(\text{mod } m),$$

where $f(x)$ is a polynomial with integer coefficients and $m > 1$. Corollary 2 of Theorem 5.4.5 provides a reduction of this problem (as relates to the number of solutions) to the special case $m = p^\alpha$, a power of a prime.

To begin, we will assume that $f(x)$ has no multiple roots (in the field of complex numbers). Recalling Ex. 5 of §1.5, this assumption is equivalent to

$$(5.5.2) \qquad\qquad (f(x), f'(x)) = 1.$$

Then by Ex. 4 of §1.5, there exist polynomials $A(x)$, $B(x)$, with integer coefficients, and an integer $c > 0$ such that

$$(5.5.3) \qquad\qquad A(x)f(x) + B(x)f'(x) = c.$$

For definiteness, of all such equations we define $c = c(f)$ as the smallest positive integer that is representable as in (5.5.3). This function of f, $c(f)$, provides a simple means for limiting the possibility of the occurrence of multiple roots in the polynomial congruence.

Lemma 5.5.1. Let $f(x)$ be a polynomial with integer coefficients that has no multiple roots. If m and k are integers such that m divides $f(k)$ and $f'(k)$, then m divides $c(f)$.

Proof.   Setting $x = k$ in (5.5.3) yields

$$A(k)f(k) + B(k)f'(k) = c(f)$$

and from the hypothesis that m divides both $f(k)$ and $f'(k)$, it follows that m divides $c(f)$.

Corollary.   If $f(x)$ has no multiple roots and the congruence (5.5.1) has a multiple root modulo m, then m divides $c(f)$.

Letting $N(f, m)$ denote the number of solutions to the congruence (5.5.1), counted modulo m, and including multiplicities. As was seen in (5.2.11), the multiplicity of a given root cannot exceed the degree of $f(x)$. However, as has also been noted previously, the quantity $N(f, m)$ may exceed the degree of $f(x)$.

Considering the case $m = p^\alpha$, a power of a prime, let x_1 be a solution of $f(x_1) \equiv 0$ (mod p^α). Then, if $1 \leq \beta < \alpha$, it follows that $f(x_1) \equiv 0$ (mod p^β). In this second congruence, x_1 represents a residue class of solutions modulo p^β. This suggests the following useful terminology: if x_0 is a solution of $f(x_0) \equiv 0$ (mod p^β); and x_1 satisfies $f(x_1) \equiv 0$ (mod p^α), where $\alpha > \beta$, and

$$x_1 \equiv x_0 \text{ (mod } p^\beta\text{)},$$

then x_1 *is said to extend the solution* x_0 *modulo* p^β *to a solution modulo* p^α, or that x_1 is an *extension* of x_0. Thus our earlier remarks assert that a root modulo p^α is the extension of a unique root mod p^β. In particular then, every root modulo p^α is the extension of a unique root modulo p. The following lemma notes that under such extension, multiplicities cannot be increased.

Lemma 5.5.2.   Let the root x_1 of $f(x) \equiv 0$ (mod p^α) extend the root x_0 of $f(x) \equiv 0$ (mod p). Then

$$(5.5.4) \qquad\qquad\qquad \text{mult } (x_1, p^\alpha) \leq \text{mult } (x_0, p).$$

Proof.   If $\nu = \text{mult } (x_1, p^\alpha)$, then

$$f(x) \cong (x - x_1)^\nu g(x) \text{ (mod } p^\alpha\text{)}.$$

Since $\alpha \geq 1$, this implies

$$f(x) \cong (x - x_0)^\nu g(x) \text{ (mod } p\text{)}$$

so that mult $(x_0, p) \geq \nu$.

With respect to extensions we have the following theorem.

Theorem 5.5.1.   If $f(x)$ is a polynomial with integer coefficients of degree $n > 0$ modulo p, and x_0 is a solution of $f(x_0) \equiv 0$ (mod p) such that $f'(x_0) \not\equiv 0$ (mod p), then for every integer $\alpha > 0$, x_0 extends uniquely to a solution of $f(x) \equiv 0$ (mod p^α).

Proof. The assertion of the theorem is trivially true for $\alpha = 1$. Assume then for $\alpha > 1$ that x_0 has been uniquely extended to a root x_1 of $f(x) \equiv 0$ (mod $p^{\alpha-1}$). To complete the induction we must show that x_1 has a unique extension to a root of $f(x) \equiv 0$ (mod p^α). Consider then $x_1 + tp^{\alpha-1}$ and look for an integer t such that

$$(5.5.5) \qquad f(x_1 + tp^{\alpha-1}) \equiv 0 \text{ (mod } p^\alpha).$$

Using the Taylor expansion of the left side of (5.5.5), we have

$$f(x_1) + tp^{\alpha-1}f'(x_1) + \tfrac{1}{2}t^2p^{2\alpha-2}f''(x_1) + \cdots \equiv 0 \text{ (mod } p^\alpha).$$

Since $2\alpha - 2 \geq \alpha$ for $\alpha > 1$, this becomes (note that $f''(x_1)$ is even)

$$f(x_1) + tp^{\alpha-1}f'(x_1) \equiv 0 \text{ (mod } p^\alpha).$$

But $f(x_1) \equiv 0$ (mod $p^{\alpha-1}$), and this is equivalent to

$$(5.5.6) \qquad tf'(x_1) \equiv -\frac{f(x_1)}{p^{\alpha-1}} \text{ (mod } p).$$

From $x_1 \equiv x_0$ (mod p) we have

$$f'(x_1) \equiv f'(x_0) \not\equiv 0 \text{ (mod } p),$$

and hence (5.5.6) has a unique solution for t modulo p. This then provides a solution $x_1 + tp^{\alpha-1}$ of (5.5.5) that is unique modulo p^α, and completes the proof of the theorem.

Corollary 1. Let $f(x)$ be a polynomial with integer coefficients that has no multiple roots. If p is any prime that does not divide $c(f)$, then all the roots of $f(x) \equiv 0$ (mod p) extend uniquely to roots of $f(x) \equiv 0$ (mod p^α), for every integer $\alpha \geq 1$.

Proof. Since p does not divide $c(f)$, if $f(x_0) \equiv 0$ (mod p), Lemma 5.1.1 provides that $f'(x_0) \not\equiv 0$ (mod p) and the theorem applies.

The construction used in the proof of the theorem yields the following corollary.

Corollary 2. Under the same hypothesis as Corollary 1, if $x_1, \ldots, x_k$ are the roots of $f(x)$ modulo p^α, we have

$$x_i \equiv u_{i0} + u_{i1}p + \cdots + u_{i,\alpha-1}p^{\alpha-1} \text{ (mod } p^\alpha)$$

where $0 \leq u_{ij} < p$ and $u_{i0} \neq u_{j0}$ for $i \neq j$.

Theorem 5.5.2. Let $f(x)$ be a polynomial with integer coefficients that is of degree n and that has no multiple roots. If p is a prime that does not divide $c(f)$, then for every positive integer α,

(5.5.7) $$N(f, p^{\alpha}) \le n.$$

Proof. The case $\alpha = 1$ of the theorem is a direct consequence of Theorem 5.2.1. Further, since p does not divide $c(f)$, the congruence $f(x) \equiv 0 \pmod{p}$ has no multiple roots modulo p. Then, by Corollary 1 of Theorem 5.5.1, each of these roots extends uniquely to a root of $f(x) \equiv 0 \pmod{p^{\alpha}}$ and hence $N(f, p^{\alpha}) = N(f, p)$ which yields the theorem.

There remains the problem of estimating the number of solutions of $f(x) \equiv 0 \pmod{p^{\alpha}}$ in cases where multiple roots modulo p do occur. For this we have the following theorem.

Theorem 5.5.3. Let $f(x)$ have no multiple roots and consider any root x_0 of $f(x) \equiv 0 \pmod{p}$. Then, if p^{σ} is the highest power of p that divides $c(f)$, the number of roots of $f(x) \equiv 0 \pmod{p^{\alpha}}$, $\alpha \ge 1$ (counting multiplicities), that extend x_0 is less than or equal to

(5.5.8) $$p^{2\sigma} \text{ mult } (x_0, p).$$

Proof. Assume $\alpha > 2\sigma$. Then, if x_0 has an extension to a solution modulo p^{α}, it must have an extension to a solution modulo $p^{\alpha-\sigma}$. Let x_1 be an extension of x_0 such that

(5.5.9) $$f(x_1) \equiv 0 \pmod{p^{\alpha-\sigma}},$$

where

(5.5.10) $$x_1 = x_0 + hp.$$

Since x_1 is determined modulo $p^{\alpha-\sigma}$, h is determined modulo $p^{\alpha-\sigma-1}$. We consider the extensions of x_1 to solutions of $f(x) \equiv 0 \pmod{p^{\alpha}}$, and to solutions of $f(x) \equiv 0 \pmod{p^{\alpha+1}}$. In fact, it will be shown that for the moduli p^{α}, $p^{\alpha+1}$ the number of extensions of x_1, is the same.

Let $\epsilon = 0$ or 1, and consider the extensions of x_1 to solutions of

(5.5.11) $$f(x) \equiv 0 \pmod{p^{\alpha+\epsilon}}.$$

Setting

$$x = x_1 + tp^{\alpha-\sigma},$$

(5.5.11) becomes

(5.5.12) $$f(x_1 + tp^{\alpha-\sigma}) \equiv 0 \pmod{p^{\alpha+\epsilon}}.$$

Since $2(\alpha - \sigma) = \alpha + (\alpha - 2\sigma) \geq \alpha + \epsilon$, application of the Taylor expansion to the left side of (5.5.12) shows it is equivalent to

$$f(x_1) + tp^{\alpha-\sigma}f'(x_1) \equiv 0 \;(\text{mod } p^{\alpha+\epsilon}).$$

Since $f(x_1) \equiv 0 \;(\text{mod } p^{\alpha-\sigma})$, this may be rewritten as

(5.5.13) $$tf'(x_1) \equiv -\frac{f(x_1)}{p^{\alpha-\sigma}} \;(\text{mod } p^{\sigma+\epsilon}).$$

Let $\delta \geq 0$ be the exponent of the highest power of p that divides $f'(x_1)$. Then (5.5.13) has a solution for t if and only if

(5.5.14) $$(f'(x_1), p^{\sigma+\epsilon}) \text{ divides } \frac{f(x_1)}{p^{\alpha-\sigma}}.$$

If $\delta > \sigma$, (5.5.14) would imply that p^σ divides $f(x_1)/p^{\alpha-\sigma}$, which, in turn, yields that p^α divides $f(x_1)$. Since $\alpha > \sigma$, this would show that $p^{\sigma+1}$ divides $(f'(x_1), f(x_1))$. But this would require that $p^{\sigma+1}$ divide $c(f)$, contradicting the definition of σ. Hence $\delta \leq \sigma$, so that

$$(f'(x_1), p^{\sigma+\epsilon}) = p^\delta$$

and from (5.5.14) we conclude that (5.5.13) has solutions for t if and only if

(5.5.15) $$p^\delta \text{ divides } \frac{f(x_1)}{p^{\alpha-\sigma}},$$

in which case the number of solutions for t modulo $p^{\sigma+\epsilon}$ is p^δ. Since this conclusion is independent of the value of ϵ (i.e., 0 or 1), it follows that the number of extensions of x_1 to solutions modulo p^α equals the number of extensions to solutions modulo $p^{\alpha+1}$.

From the above, we conclude that the number of extensions of x_0 to moduli p^α is the same for all $\alpha > 2\sigma$. Hence the number of such extensions is less than or equal the number of extensions of x_0 to the modulus $p^{2\sigma+1}$. Since such extensions are of the form $x_0 + hp$, h is determined modulo $p^{2\sigma}$, so that the number of these is $\leq p^{2\sigma}$. From Lemma 5.5.2, every extension of x_0 to the modulus p^α has multiplicity $\leq$ mult (x_0, p). Finally, then, the number of roots of $f(x) \equiv 0 \;(\text{mod } p^\alpha)$ (counting multiplicities) that extend x_0 is $\leq$ mult $(x_0, p)p^{2\sigma}$, which is precisely (5.5.8).

The above proof shows that for $\alpha > 2\sigma$, the number of extensions of x_0 to solutions of $f(x) \equiv 0 \;(\text{mod } p^\alpha)$ is the same for all α. For the case where $\sigma = 0$, Corollary 1 of Theorem 5.5.1 had already shown this to be the case, in that x_0 extends uniquely to each modulus p^α. Clearly, this is also the case if p does not divide $f'(x_0)$. However, in general, it is *not* true that (for $\alpha > 2\sigma$) each solution mod p^α extends uniquely to a solution modulo $p^{\alpha+1}$. Ordinarily, there are sets of p

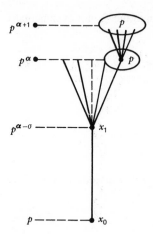

FIGURE 5.5.1

solutions modulo p^α such that only one of these extends to modulo $p^{\alpha+1}$, but this one has p such extensions. To see this, consider a case where $f'(x_0) \equiv 0 \pmod{p}$. Then, if x_0 extends to x_1 modulo $p^{\alpha-\sigma}$, we have $f'(x_1) \equiv 0 \pmod{p}$ so that if $p^\delta =$ the highest power of p dividing $f'(x_1)$, $\delta \geq 1$. The proof of the theorem provides that the extensions of x_1 to modulo p^α are given by

$$(5.5.16) \qquad x_1 + (t^* + \lambda p^{\sigma-\delta})p^{\alpha-\sigma}, \qquad \lambda = 0, 1, \ldots, p^\delta - 1$$

and the extensions of x_1 to modulo $p^{\alpha+1}$ are given by

$$(5.5.17) \qquad x_1 + (t^{**} + \gamma p^{\sigma-\delta+1})p^{\alpha-\sigma}, \qquad \gamma = 0, 1, \ldots, p^\delta - 1,$$

where $t^* \equiv t^{**} \pmod{p^{\sigma-\delta}}$. In asking for extensions from modulo p^α to modulo $p^{\alpha+1}$ of one of (5.5.16), we consider a fixed λ, and ask for γ such that

$$x_1 + (t^* + \lambda p^{\sigma-\delta})p^{\alpha-\sigma} \equiv x_1 + (t^{**} + \gamma p^{\sigma-\delta+1})p^{\alpha-\sigma} \pmod{p^\alpha}$$

or

$$(5.5.18) \qquad \frac{(t^* - t^{**})}{p^{\sigma-\delta}} + \lambda \equiv \gamma p \pmod{p^\delta}.$$

If the left side of (5.5.18) is not divisible by p, there are no solutions. For each of the $p^{\delta-1}$ values of λ such that the left side of (5.5.18) is divisible by p, (5.5.18) has exactly p solutions for γ modulo p^δ. Thus we see that one out of every p of the set (5.5.16) extends, and these have exactly p extensions among the set (5.5.17).

From Theorem 5.5.3 we obtain the following theorem.

Theorem 5.5.4. If $f(x)$ is a polynomial of degree n with no multiple roots, then

$$(5.5.19) \qquad N(f, m) \leq n^{\nu(m)}(c(f))^2.$$

Proof.   For any prime p, if σ_p is the power to which p divides $c(f)$, Theorem 5.5.3 gives that

$$N(f, p^\alpha) \leq \sum_{\substack{x_0 \\ f(x_0) \equiv 0 \,(\mathrm{mod}\, p)}} \mathrm{mult}(x_0, p) p^{2\sigma_p} \leq n p^{2\sigma_p}$$

(by Theorem 5.2.1). Then, from Corollary 2 of Theorem 5.4.5, for $m = \prod p_i^{\alpha_i}$,

$$N(f, m) \leq \prod N(f, p_i^{\alpha_i}) \leq \prod (n p^{2\sigma_p}) = n^{\nu(m)} (c(f))^2.$$

EXERCISES

1.   Let $f_i(x)$, $i = 1, \ldots, s$ be polynomials with integer coefficients of degree n_i, $i = 1, \ldots, s$, respectively, such that each $f_i(x)$ has no multiple roots. Prove that

$$N(f_1(x) \cdots f_s(x), m) \leq \left(\prod_{i=1}^s c^2(f_i) \right) \sum_{d_1 \cdots d_s = m} n_1^{\nu(d_1)} \cdots n_s^{\nu(d_s)}.$$

§5.5A   AVERAGE OF THE EULER FUNCTION OVER POLYNOMIAL SEQUENCES

Theorem 5.5.4 is extremely useful in many contexts. As a small illustration that can be carried out within the framework of the material developed thus far, we consider the problem of generalizing Theorem 3.8.1 to the case where the average is taken over a polynomial sequence. More precisely, we shall focus on the estimation of

(5.5A.1)
$$\sum_{n \leq x} \frac{\phi(f(n))}{f(n)}.$$

and

(5.5A.2)
$$\sum_{n \leq x} \phi(f(n)),$$

where $f(x)$ is a polynomial with integer coefficients [assume $f(n) > 0$] that has no multiple roots. Fixing f, we shall use the notation $B(m)$ to denote the number of solutions mod m, *not counting* multiplicities, of $f(x) \equiv 0$ (mod m). Thus

(5.5A.3)
$$B(m) \leq N(f, m),$$

and Theorem 5.4.5, Corollary 2, implies that $B(m)$ is multiplicative.

We shall require the following lemmas.

Lemma 5.5A.1. For $k \geq 1$, $L \geq 0$, any integers

$$(5.5A.4) \qquad \sum_{n \leq x} \frac{d(n)^L k^{\nu(n)}}{n} = 0((\log x)^{k \cdot 2^L}).$$

Proof. We observe first that

$$(5.5A.5) \qquad \sum_{n \leq x} \frac{d(n)^L k^{\nu(n)}}{n} \leq \left(\sum_{n \leq x} \frac{d(n)^L}{n} \right)^k$$

To see this, note that if the right side of (5.5A.5) were multiplied out, for each $n \leq x$, the term $1/n$ will be produced with a coefficient equal to

$$(5.5A.6) \qquad \sum_{u_1 \cdots u_k = n} d(u_1)^L d(u_2)^L \cdots d(u_k)^L \geq d(n)^L \sum_{u_1 \cdots u_k = n} 1.$$

Since n has $\nu(n)$ prime factors, the last sum on the right of (5.5A.6) is at least as big as the number of ways to partition $\nu(n)$ objects into $\leq k$ disjoint sets. This last number is $k^{\nu(n)}$ and (5.5A.5) follows. Next by a similar argument we see that

$$\left(\sum_{n \leq x} \frac{d(n)^L}{n} \right) \leq \left(\sum_{n \leq x} \frac{d(n)^{L-1}}{n} \right)^2$$

and when iterated, this yields

$$\left(\sum_{n \leq x} \frac{d(n)^L}{n} \right) \leq \left(\sum_{n \leq x} \frac{1}{n} \right)^{2^L} = 0((\log x)^{2^L}),$$

by Lemma 3.8.1. Inserting this last in (5.5A.5), we obtain (5.5A.4).

Lemma 5.5A.2. The series

$$(5.5A.7) \qquad \alpha = \sum_{d=1}^{\infty} \frac{\mu(d) B(d)}{d^2}$$

converges. Further,

$$(5.5A.8) \qquad \left| \sum_{d > x} \frac{\mu(d) B(d)}{d^2} \right| \leq \sum_{d > x} \frac{B(d)}{d^2} = 0\left(\frac{(\log x)^n}{x} \right),$$

where n is the degree of $f(x)$.

Proof. From Theorem 5.5.4 we have

(5.5A.9)
$$B(d) \leq (c(f))^2 n^{\nu(d)},$$

so that

(5.5A.10)
$$\sum_{d>z} \frac{B(d)}{d^2} = 0\left(\sum_{d>z} \frac{n^{\nu(d)}}{d^2}\right).$$

Setting

$$T(x) = \sum_{d \leq x} \frac{n^{\nu(d)}}{d},$$

we obtain after summing by parts:

$$\sum_{d>z} \frac{n^{\nu(d)}}{d^2} = \sum_{d>z} \frac{1}{d}(T(d) - T(d-1))$$

(5.5A.11)
$$= -\frac{T([z])}{[z]+1} + \sum_{d>z} \frac{T(d)}{d(d+1)}.$$

From (5.5A.4)

(5.5A.12)
$$T(x) = 0(\log^n x),$$

so that (5.5A.11) yields

$$\sum_{d>z} \frac{n^{\nu(d)}}{d^2} = 0\left(\sum_{d>z} \frac{\log^n d}{d^2}\right) + 0\left(\frac{\log^n z}{z}\right).$$

But

$$\sum_{d>z} \frac{\log^n d}{d^2} = 0\left(\frac{\log^n z}{z^{1/2}} \sum_{d>z} \frac{1}{d^{3/2}}\right) = 0\left(\frac{\log^n z}{z}\right),$$

where the last estimate uses (3.8.31). Finally, (5.5A.10) together with the above yields (5.5A.8).

Theorem 5.5A.1. Let $f(x)$ be a polynomial of degree n, with integer coefficients, that has no multiple roots $[f(m) > 0$ for $m \geq 1]$. Then

(5.5A.13) $$\sum_{m \leq x} \frac{\phi(f(m))}{f(m)} = \alpha x + 0(\log^n x),$$

where α is given by (5.5A.7).

Proof. Using (3.7.11), we have

$$\sum_{m \leq x} \frac{\phi(f(m))}{f(m)} = \sum_{m \leq x} \sum_{d/f(m)} \frac{\mu(d)}{d}$$

(5.5A.14)

$$= \sum_{d \leq ax^n} \frac{\mu(d)}{d} \sum_{\substack{m \leq x \\ f(m) \equiv 0 \,(\mathrm{mod}\, d)}} 1,$$

where the constant a is chosen such that $f(y) \leq ay^n$ for all $y \geq 1$. Since

$$\sum_{\substack{m \leq x \\ f(m) \equiv 0 \,(\mathrm{mod}\, d)}} 1 = B(d)\left(\frac{x}{d} + 0(1)\right),$$

(5.5A.14) becomes

$$\sum_{m \leq x} \frac{\phi(f(m))}{f(m)} = \sum_{d \leq ax^n} \frac{\mu(d)B(d)}{d}\left(\frac{x}{d} + 0(1)\right)$$

$$= x \sum_{d \leq ax^n} \frac{\mu(d)B(d)}{d^2} + 0\left(\sum_{d \leq ax^n} \frac{B(d)}{d}\right)$$

$$= \alpha x + 0\left(x \sum_{d > ax^n} \frac{B(d)}{d^2}\right) + 0\left(\sum_{d \leq ax^n} \frac{B(d)}{d}\right).$$

Using (5.5A.8) and (5.5A.12) to estimate the 0 terms, we obtain

$$\sum_{m \leq x} \frac{\phi(f(m))}{f(m)} = \alpha x + 0\left(\frac{\log^n x}{x^{n-1}}\right) + 0(\log^n x) = \alpha x + 0(\log^n x).$$

Using (5.5A.13) and a summation by parts argument, this transforms into the following theorem.

Theorem 5.5A.2. Under the same hypothesis as Theorem 5.5A.1,

(5.5A.15) $$\sum_{m \leq x} \phi(f(m)) = \frac{\alpha a_n}{n+1} x^{n+1} + 0(x^n \log^n x),$$

where a_n is the coefficient of x^n in $f(x)$.

Proof.   Setting

$$S(x) = \sum_{m \le x} \frac{\phi(f(m))}{f(m)},$$

we can write

$$\sum_{m \le x} \phi(f(m)) = \sum_{m \le x} f(m)(S(m) - S(m-1)).$$

Summing by parts, this gives [note that $S(0) = 0$]

(5.5A. 16) $$\sum_{m \le x} \phi(f(m)) = S([x])f([x]) - \sum_{m \le x-1} S(m)(f(m+1) - f(m)).$$

From (5.5A.13)

$$S(z) = \alpha z + 0(\log^n z)$$

so that

(5.5A.17)

$$S([x])f([x]) = (\alpha[x] + 0(\log^n x))f([x]) = \alpha a_n x^{n+1} + 0(x^n \log^n x).$$

Also,

(5.5A.18) $$\sum_{m \le x-1} S(m)(f(m+1) - f(m))$$

$$= \sum_{m \le x-1} (\alpha m + 0(\log^n m))(f(m+1) - f(m))$$

$$= \sum_{m \le x-1} \alpha m \left(f(m+1) - f(m) \right) + 0\left(\sum_{m \le x-1} (\log^n m)|f(m+1) - f(m)| \right)$$

Since

$$\sum_{m \le x-1} (\log^n m)|f(m+1) - f(m)| = 0(\log^n x) \sum_{m \le x-1} \left| f(m+1) - f(m) \right|$$

$$= 0(x^n \log^n x),$$

and

$$\sum_{m \le x-1} m(f(m+1) - f(m)) = \sum_{m \le x-1} m(na_n m^{n-1} + 0(m^{n-2}))$$

$$= na_n \sum_{m \le x-1} m^n + 0\left(\sum_{m \le x} m^{n-1}\right)$$

$$= \frac{n}{n+1} a_n x^{n+1} + 0(x^n),$$

(5.5A.18) becomes

(5.5A.19) $$\sum_{m \le x-1} S(m)(f(m+1) - f(m)) = \frac{an}{n+1} a_n x^{n+1} + 0(x^n \log^n x).$$

Inserting (5.5A.17) and (5.5A.19) into (5.5A.16) yields

$$\sum_{n \le x} \phi(f(m)) = \alpha a_n x^{n+1} - \frac{an}{n+1} a_n x^{n+1} + 0(x^n \log^n x)$$

$$= \frac{\alpha a_n}{n+1} x^{n+1} + 0(x^n \log^n x),$$

the desired result.

EXERCISES

1. a. Let $n_1, \ldots, n_s$ be positive integers. Prove that

$$\sum_{m \le x} \sum_{d_1 \cdots d_s = m} \frac{n_1^{\nu(d_1)} n_2^{\nu(d_2)}}{d_1 \; d_2} \cdots \frac{n_s^{\nu(d_s)}}{d_s} = 0 \; [(\log x)^{n_1 + \cdots + n_s}].$$

b. Using (a) and Ex. 1. of §5.5, show that if $f_i \, i = 1, \ldots, s$ are polynomials with integer coefficients, each with no multiple roots, degree $f_i = n_i$, then for $f = \prod_{i=1}^{s} f_i$ of degree $n = n_1 + \cdots + n_s$, prove that

$$\sum_{d > x} \frac{B(d)}{d^2} = 0\left(\frac{\log^n x}{x}\right),$$

where $B(d) = B(f, d)$.

c. Using the result of b, show that for the polynomial f of the type defined in b, (5.5A.15) holds; that is,

$$\sum_{n \le x} \phi(f(n)) = \frac{\alpha a_n}{n+1} x^{n+1} + O(x^n \log^n x).$$

d.   Using c and Ex. 2 of §2.5, and Ex. 6 of §1.5, show that (5.5A.15) holds for an arbitrary polynomial $f(x)$ with integer coefficients, provided only that $f(m) > 0$ for positive integers m.

2. Let $f(x)$ be any polynomial with integer coefficients, of degree n, and such that $f(m) > 0$ for all positive integers m. Prove that

$$\sum_{m \le x} \sigma(f(m)) = \frac{\beta a_n}{n+1} x^{n+1} + O(x^n \log^n x),$$

where

$$\beta = \sum_{d=1}^{\infty} \frac{B(d)}{d^2}$$

and a_n is the coefficient of x^n in $f(x)$.

§5.5B   AVERAGE OF THE DIVISOR FUNCTION OVER POLYNOMIAL SEQUENCES

The methods of the previous section do not apply as neatly to the study of the sum

$$\sum_{m \le x} d(f(m)),$$

where $f(x)$ is a polynomial of degree n with integer coefficients. The reason for this is easily seen. We have

$$\sum_{m \le x} d(f(m)) = \sum_{m \le x} \sum_{d / f(m)} 1$$

$$= \sum_{d \le ax^n} \sum_{\substack{m \le x \\ f(m) \equiv 0 \,(\mathrm{mod}\, d)}} 1$$

$$= \sum_{d \le ax^n} B(d)\left(\frac{x}{d} + O(1)\right)$$

$$= x \sum_{d \le ax^n} \frac{B(d)}{d} + O\left(\sum_{d \le ax^n} B(d)\right).$$

The sum in the 0-term is taken up to ax^n, and is in fact very large; indeed, large enough to dominate the first term completely. Thus we see that the difficulty stems from the large divisors of $f(m)$, for $m \leq x$. This difficulty as yet has not been completely overcome. What is known at present, for an irreducible $f(x)$, is that

$$(5.5B.1) \qquad \sum_{m \leq x} d(f(m)) = 0(x \log x).$$

The proof of (5.5B.1) is due to Erdos (Ref. 5.27) and requires methods and results that as yet have not been presented here. We will content ourselves with the following consideration of a weaker estimate attributed to Van der Corput.

Theorem 5.5B.1. For any polynomial $f(x)$ with integer coefficients and any given integer $r > 0$, there exists a constant $e = e(f, r)$ such that

$$(5.5B.2) \qquad \sum_{m \leq x} d^r(f(m)) = 0(x \log^e x).$$

Proof. We consider first the case where $f(x)$ has no multiple roots. Though there is no simple way to dispose of the big divisors of the $f(m)$, we can easily dispose of the large *prime* divisors of $f(m)$. For suppose that $f(m) = \prod p_i^{\alpha_i}$. Then, for $m \leq x$,

$$ax^n \geq f(m) \geq \prod_{p_i > x} p_i^{\alpha_i} > x^{\sum_{p_i > x} \alpha_i}$$

so that

$$(5.5B.3) \qquad \sum_{p_i > x} \alpha_i < K,$$

where K is a constant. Writing

$$f^*(m) = \prod_{p_i \leq x} p_i^{\alpha_i} \qquad \text{and} \qquad g(m) = \prod_{p_i > x} p_i^{\alpha_i},$$

we have

$$
\begin{aligned}
d(f(m)) &= d(f^*(m)g(m)) \\
&\leq d(f^*(m))d(g(m)) \\
&\leq 2^{\sum_{p_i > x} \alpha_i} d(f^*(m)) \\
&\leq 2^K d(f^*(m)).
\end{aligned}
$$

Thus we obtain

(5.5B.4)
$$\sum_{m \leq x} d^r(f(m)) \leq 2^{rK} \sum_{m \leq x} d^r(f^*(m)),$$

where $f^*(m)$ is that part of $f(m)$ that is composed of primes $\leq x$.

Since all the prime divisors of $f^*(m)$ are $\leq x$, $f^*(m)$ has at least one representation of the form

(5.5B.5)
$$f^*(m) = u_1 \cdots u_t,$$

where

(5.5B.6)
$$1 < u_i \leq x, \qquad i = 1, \ldots, t.$$

For given integer $m \leq x$, let t_m denote the smallest t such that $f^*(m)$ has a representation (5.5B.5) that satisfies (5.5B.6). For one of these representations of "minimal length," we write

(5.5B.7)
$$f^*(m) = u_1^{(m)} \cdots u_{t_m}^{(m)}.$$

If $t_m > 1$, we must have $u_i^{(m)} u_j^{(m)} > x$, for $i \neq j$, since otherwise the product could be "shortened" and still satisfy (5.5B.6). Thus

$$ax^n > f^*(m) = (u_1^{(m)} u_2^{(m)})(u_3^{(m)} u_4^{(m)}) \cdots > x^{(t_m - 1)/2},$$

and it follows for x sufficiently large that

$$t_m \leq 2n + 1 = L.$$

Augmenting the notation of (5.5B.7), we write

$$f^*(m) = u_1^{(m)} u_2^{(m)} \cdots u_L^{(m)},$$

where if $L > t_m$, we set $u_i^{(m)} = 1$ for $i > t_m$. In any event, in all cases $u_i^{(m)}$, which is $\leq x$, is a well defined divisor of $f(m)$.

Next we observe that

(5.5B.8)
$$d^r(f^*(m)) = d^r(u_1^{(m)} \cdots u_L^{(m)})$$
$$\leq d^r(u_1^{(m)}) \cdots d^r(u_L^{(m)})$$
$$\leq \sum_{i=1}^{L} d(u_i^{(m)})^{rL}.$$

(This last inequality, $x_1 \cdots x_L \leq \sum_{i=1}^{L} x_i^L$, $x_i \geq 1$, is left as an exercise.)

Combining (5.5B.8) and (5.5B.4), we obtain

$$(5.5B.9) \qquad \sum_{m \leq x} d^r(f(m)) \leq 2^{Kr} \sum_{i=1}^{L} \sum_{m \leq x} d(u_i^{(m)})^{rL}.$$

For each fixed i, since the $u_i^{(m)}$ are $\leq x$, we have

$$\sum_{m \leq x} d(u_i^{(m)})^{rL} = \sum_{u \leq x} d(u)^{rL} \sum_{\substack{u_i^{(m)} = u \\ m \leq x}} 1$$

$$\leq \sum_{u \leq x} d(u)^{rL} \sum_{\substack{m \leq x \\ f(m) \equiv 0 \,(\mathrm{mod}\, u)}} 1$$

$$\leq 2x \sum_{u \leq x} \frac{d(u)^{rL} B(u)}{u}$$

$$\leq 2(c(f))^2 x \sum_{u \leq x} \frac{d(u)^{rL} n^{\nu(u)}}{u}.$$

From Lemma 5.5A.1, this last is

$$0(x (\log x)^{2^{Lr} \cdot n}),$$

which completes the proof of (5.5B.2) in the case where $f(x)$ has no multiple roots.
For arbitrary $f(x)$, we can write

$$f(x) = f_1(x) f_2(x) \cdots f_s(x),$$

where each f_i is irreducible and hence has no multiple roots. Then

$$\sum_{m \leq x} d^r(f(m)) = \sum_{m \leq x} d^r(f_1(m) \cdots f_s(m))$$

$$\leq \sum_{m \leq x} d^r(f_1(m) \cdots d^r(f_s(m))$$

$$\leq \sum_{i=1}^{s} \sum_{m \leq x} d^{sr}(f_i(m)).$$

Using the fact that the result has been proved for each f_i, it follows that for some constants e_i,

$$\sum_{i=1}^{s} \sum_{m \leq x} d^{sr}(f_i(m)) = \sum_{i=1}^{s} 0(x (\log x)^{e_i}) = 0(x (\log x)^e).$$

EXERCISES

1. For some applications, it is required to have a generalization of Theorem 5.5B.1 to an estimation of

$$\sum_{\substack{m \le x \\ f(m) \equiv 0 \,(\mathrm{mod}\,A)}} d^r(f(m)),$$

that is uniform in A.

In this connection, show that if $f(x)$ has no multiple roots, there exists a constant $e = e(f, r)$, such that

$$\sum_{\substack{m \le x \\ f(m) \equiv 0 \,(\mathrm{mod}\,A)}} d^r(f(m)) = 0\left(\frac{B(A)d(A^e)}{A} x \,(\log x)^e\right)$$

uniformly in A, for $A \le x^{1-\epsilon}$ (ϵ fixed, $0 < \epsilon < 1$). In other words, for $A \le x^{1-\epsilon}$, the "0" in the above estimate depends only on ϵ, and in no other way on A than is expressed by the presence of the factor $[B(A)d(A^e)]/A$.

2. Show that for p a prime, we have an integer e such that

$$\sum_{\substack{m \le x \\ f(m) \equiv 0 \,(\mathrm{mod}\,p)}} d^r(f(m) = 0\left(\frac{x}{p^\alpha} \,(\log x)^e\right)$$

uniformly for $p^\alpha \le x^{1-\epsilon}$.

3. Prove that given any polynomial $f(x)$ with integer coefficients and an integer $D > 0$, there exists a number $C = C(D) > 0$ such that the number of integers $m \le x$ for which $f(m)$ has more than $C \log \log x$ distinct prime factors, is less than $x/(\log x)^D$.

NOTES

§5.1. The congruence notation was introduced by Gauss in his *Disquisitiones Arithmeticae* of 1801. However, the notion of congruence had already been employed by Euler, Lagrange, and Legendre (among others). In fact, the use of division remainders without retention of the quotient has recorded occurrences as early as 500 B.C. [5.1, 5.2].

Gauss developed the theory of congruences more extensively than had anyone before him, incorporating the work of Euler and others (which he acknowledged).

§5.1. The corollary to Theorem 5.1.4 is known as Fermat's (Little) Theorem. According to Dickson [5.1], it was known in 500 B.C. in China that if p is a prime, then p divides $2^p - 2$. However, the full statement of the theorem first appeared in a letter written by Fermat to Frenicle de Bessy on October 18, 1640 [5.3]. He did not include a proof in the letter since he felt it was too long.

The earliest recorded proof of Fermat's Theorem appears in a manuscript of Leibnitz (which must predate 1716, the year of Leibnitz' death). His proof utilized multinomial coefficients [5.1]. Exercise 1 suggests the method of Leibnitz' proof. Leibnitz' manuscript was the subject of a "famous" controversy in the eighteenth century, according to Gauss. A footnote to Article 50 in the *Disquisitiones Arithmeticae* indicates that König and Maupertius argued about the priority of Euler's proof of Fermat's Theorem. König claimed to have Leibnitz' manuscript with a proof that was very similar to Euler's first proof. Gauss himself stated, "We do not wish to deny this testimony, but certainly Leibnitz never published this discovery."

Euler gave three proofs of Fermat's Theorem. In the first one, given in 1736, he used the binomial theorem to prove that p divides $a^p - a$ for $a = 2$ and $a = 3$; then the general result by induction [5.4]. The second proof again utilized the binomial theorem. Euler showed that p divides $(a + b)^p - a^p - b^p$, so that for $b = 1$, this implies that

$$0 \equiv (a + 1)^p - a^p - 1 = (a + 1)^p - (a + 1) + a - a^p \pmod{p}.$$

Using this $a^p - a \equiv 0 \pmod{p}$ implies $(a + 1)^p - (a + 1) \equiv 0 \pmod{p}$, which enables an inductive proof [5.5].

Euler's third proof, published in 1758 (but mentioned in correspondence in 1742), did not use the binomial theorem. He showed that what we call $\mathbb{O}_p(a)$, the order of a modulo p (see Exercise 9), divides $p - 1$ for $a \not\equiv 0 \pmod{p}$. The desired result is then deduced from the fact that $a^{\mathbb{O}_p(a)} - 1$ divides $a^{p-1} - 1$. [5.6].

All of Euler's proofs were included by Gauss in Articles 49–50 of his *Disquisitiones Arithmeticae*. Gauss believed that Euler was motivated to produce his third demonstration because "the expansion of a binomial power seemed quite alien to the theory of numbers" [5.7]. It seems more likely that Euler was seeking a generalization of the theorem for composite moduli. In fact, shortly thereafter, in 1760, Euler obtained this generalization, which we now call Euler's Theorem.

§5.1. *Theorem 5.1.4, Euler's Theorem.* The statement and proof of Euler's Theorem first appeared in [3.5]. This is the same paper in which Euler developed the properties of the φ function (the generalization of Fermat's Theorem was the motivating factor). The method he used in his third proof of Fermat's Theorem [5.8] was also appropriate for the composite case [5.9].

Theorem 5.1.4 is proved in this text by a method first given by James Ivory in 1806 to establish Fermat's Theorem.

This method was rediscovered by Dirichlet. It is surprising that Gauss did not originate this proof, since it is very similar to the method he used in the part of his demonstration of the Quadratic Reciprocity Law that we now call Gauss' Lemma.

§5.1. *Theorem 5.1.5.* Functions that represent primes alone have long been sought. Euler, for example, mentions as "remarkable" the function $x^2 - x + 41$ because it is prime for $1 \le x \le 40$ (letter to Jean Bernouilli III in 1772 [5.10].) In 1743, C. Goldbach had observed in a letter to Euler [5.11] that a polynomial $f(x)$ cannot represent primes alone because the constant term would be 1 whereas it is $f(p)$ in $f(x + p)$. Euler seems to have forgotten this letter, because he later proposed the same theorem in a letter to Goldbach in 1752. Euler published his proof of this in 1762, noting that if $f(a) = A$, then $f(nA + a)$ is always divisible by A [5.12].

In the same paper, Euler indicated that the Fermat number $2^{2^5} + 1$ is divisible by 641. Fermat had believed that these numbers (of the form $2^{2^n} + 1$) always gave prime values.

Theorem 5.1.5 was once thought to represent a deep and difficult problem. It was first proved by M. Ward in 1930 [5.13].

§5.1. *Exercise 13.* This exercise gives a converse of Fermat's Theorem. In his *Théorie des Nombres*, Edouard Lucas proved that: "If $a^x - 1$ is divisible by n for $x = n - 1$, and it is not divisible by n for x equal to any other divisor of $n - 1$, the number n is prime." Lucas writes, "We announced this theorem for the first time in 1876 at the Congrès de l'Association Francaise pour l'Avancement des Sciences in a note entitled Sur la recherche des grands nombres premiers" [5.14].

A related problem that is still unsolved is D. H. Lehmer's 1932 conjecture that if $\varphi(n)$ divides $n - 1$ then n must be a prime.

§5.2. *Theorem 5.2.1.* This theorem is often called Lagrange's Theorem. In 1770, Lagrange proved by induction that a polynomial of degree n can have at most n roots between $-\frac{1}{2}p$ and $\frac{1}{2}p$ modulo p, p a prime [5.15]. Gauss also included an inductive proof of this theorem in his *Disquisitiones Arithmeticae* (Article 43).

§5.2. *Wilson's Theorem (5.2.15).* It is not clear how much Sir John Wilson had to do with the theorem named after him. The statement of the theorem was ascribed to him in 1770 by Edward Waring [5.16], but neither of them gave a proof. A 1682 manuscript of Leibnitz contains the equivalent statement: $(p - 2)! \equiv 1 \pmod{p}$ for p a prime [5.17].

The first published proof of Wilson's Theorem was by J. Lagrange in 1771, and was based on the binomial theorem [5.18]. Euler gave a proof in 1773 relying on the existence of primitive roots but the proof contained serious omissions. Dirichlet, in 1828, used Euler's associated numbers [a_i and a_j are associated if $a_i a_j \equiv 1 \pmod{n}$] to derive proofs of both Wilson's and Fermat's theorems [5.9].

§5.2. *Exercise 10.* In the case where $x = p^m$ is a power of a prime p, the given assertion is a consequence of a well known counting problem in algebra. Namely, in this case

$$\left(\frac{1}{n}\right) \sum_{d/n} \mu\left(\frac{n}{d}\right) p^{md}$$

equals the number of polynomials of degree n that are irreducible over the finite field that has p^m elements.

§5.4. The history of the Chinese Remainder Theorem is somewhat complicated by the inconsistency of the transliterations of Chinese names. We will follow the most commonly used forms.

The earliest appearance of a rule in the spirit of the Chinese Remainder Theorem is the "Ta yen," or "great expansion" in the "Sun-Tsze Swan-king" (*The Arithmetical Classic of Sun-Tsze*) [5.19, 5.20]. This contains a problem requiring a number whose remainder is 2, 3, 2, when divided by 3, 5, 7, respectively. Sun-Tsze describes his solution as follows: "The remainder divided by 3 is 2, so take 140. The remainder divided by 5 is 3, so take 63. The remainder divided by 7 is 2, so take 30. Adding these numbers gives 233 from which we subtract 210, giving the remainder 23" [5.21]. There are varying opinions about the age of Sun-Tsze's work. The best guess seems to be that he lived in the third century A.D.

The next recorded appearance of the "Ta yen" is in 717. A Buddhist priest, Yih-hing, developed the "Ta yen lei schu," or "great extension method of finding unity." He also considered the case where the moduli are not relatively prime. Yah-hing was motivated by applications of the method to astronomy [5.19, 5.20].

A more complete and detailed explanation of the "Ta yen" rule was given in 1247 by Tsin Kiu Tschaou in his "Su schu kiu tschang" (*Nine Sections of Mathematics*). It is from this that we learn of the work of Yih-hing [5.20, 5.21].

§5.4A. The assertion of Theorem 5.4A.1 was originally a conjecture of E. Artin, and was resolved by C. Chevalley in 1936 [5.22]. E. Warning made a more detailed study of the *number* of solutions, and in the context of Theorem 5.4A.1, showed that the number of solutions is divisible by p [5.23]. James Ax further refined Warning's result and proved: "Let $F = F(x_1, \ldots, x_n)$ be a polynomial of degree d over a finite field of characteristic p with q elements, and let $N(F)$ denote the number of solutions of $F(x_1, \ldots, x_n) = 0$ in this field; and let b be the largest integer less than n/d. Then $N(F)$ is divisible by q^b" [5.24].

§5.5. Theorem 5.5.4 was obtained independently in 1921 by O. Ore and T. Nagell [5.25].

§5.5A. Theorems 5.5A.1 and 5.5A.2 are very similar to those given by W. Schwarz in 1962 [5.26].

§5.5B. The assertion (5.5B.1) is credited to P. Erdos (1952) [5.27]. Erdos' proof used Theorem 5.5B.1 which was originally proved in 1939 by Van der Corput [5.28].

Erdos' result has been extended by F. Delmar, who proved the existence of positive constants c_1, c_2, such that

$$c_1 x (\log x)^L < \sum_{k \le x} \{d[f(k)]\}^s < c_2 x (\log x)^L$$

where $L = 2^s - 1$ [5.29].

REFERENCES

5.1. L. E. Dickson, Ref. 3.10, p. 59.

5.2. G. Peano, *Formulae Math.*, 3, Turin, 1901, p. 96.

5.3.   P. de Fermat, *Oeuvres de Fermat*, II, Paris: Gauthier-Villars et Fils, 1894.

5.4.   L. Euler, Theorematum quorandum ad numeros primos spectantium demonstratio, *Opera Omnia*, Leipzig: B. G. Teubner, 1915 (original: 1736).

5.5.   L. Euler, Theoremata circa divisores numerorum, *Opera Omnia*, I, 2, pp. 62–85 (original: 1747–1748).

5.6.   L. Euler, Demonstrationes circa residua ex divisione potestatum relicta, *Opera Omni*, I, 2, pp. 493–518 (original: 1758).

5.7.   C. F. Gauss, *Werke*, Göttingen: Königlichen Gesellschaft der Wissenschaften, I, p. 42.

5.8.   J. Ivory, *New Series of the Mathematical Repository*, **1**(2), 1806 6–8.

5.9.   G. L. Dirichlet, Démonstrations nouvelles de quelques théorèmes relatifs aux nombres, *Journal fur reine und angewandte Mathematik*, **3**, 1828, 390–393.

5.10. L. Euler, Extrait d'une Lettre de M. Euler le Père à M. Bernouilli concernant le mémoire imprimé parmi ceux de 1771, *Opera Omnia*, 1917 p. 318.

5.11. P. Fuss (Ed.), *Correspondance Mathématiques et Physique de quelques célèbres géomètres du XVIII-ème Siècle*, I. Johnson Reprint Company, 1968 (original: 1843).

5.12. L. Euler, De numeris primis valde magnis, *Opera Omnia*, I, 3, pp. 4–5 (original: 1762).

5.13. M. Ward, *Journal of the London Mathematical Society*, **5**, 1930, 106–107.

5.14. E. Lucas, *Theorie des Nombres*, Paris: Gauthier-Villars et Fils, 1891.

5.15. J. L. Lagrange, Nouvelle Méthode pour résoudre les Problèmes Indétermines en Nombres Entiers, *Oeuvres de Lagrange*, II, Paris: Gauthier-Villars, 1848 (original: 1770).

5.16. E. Waring, *Meditationes Algebraicae*, Cambridge, 1770, p. 218.

5.17. L. E. Dickson, Ref. 3.10, p. 60.

5.18. J. L. Lagrange, Demonstration d'un Théorème Nouveau concernant les Nombres Premiers, *Oeuvres de Lagrange*, Paris: Gauthier-Villars, 1869 (original: 1771).

5.19. L. Mattiessen, Ueber die sogenannte Restproblem in den chinesischen Werken *Swan-king* von *Sun-tsze* und *Tayen lei schu* von *Yih-hing*, *Crelle's Journal*, **91**, 1881, 254–261.

5.20. Y. Mikami, *The Development of Mathematics in China and Japan*, Leipzig: B. G. Teubner, 1913.

5.21. K. L. Biernatzki, Die Arithmetik der Chinesen, *Crelle's Journal*, **52**, 1856, 59–94.

5.22. C. Chevalley, Démonstration d'une hypothèse de M. Artin, Hamburg: Universität, Mathematisches Seminar, Abhandlungen **11**, 1936, 73–75.

5.23. E. Warning, Bemerkung zur vorstehenden Arbeit von Herrn Chevalley, Hamburg: Universität, Mathematisches Seminar, Abhandlungen **11**, 1936, 76–83.

5.24. J. Ax, Zeros of polynomials over finite fields, *American Journal of Mathematics*, **86**, 1964, 255–261.

5.25. T. Nagell, *Introduction to Number Theory*, New York: Wiley, 1951.

5.26. W. Schwarz, Uber die Summe $\sum_{n\leq x} \varphi(f(n))$ und verwandte Probleme, *Monatshefte für Mathematik*, **66**, 1962, 43–54.

5.27. P. Erdos, On the sum $\sum_{k=1}^{x} d(f(k))$, *Journal of the London Mathematical Society*, **27**, 1952, 7–15.

5.28. J. G. Van der Corput, Une inégalité relative au nombre des diviseures, *Indagationes Mathematicae*, **1**, 1939, 177–183.

5.29. F. Delmar, Sur la somme de diviseurs $\sum_{k\leq x} \{d(f(k))\}^s$, *Comptes Rendus de l'Académie des Sciences, Paris, Series A–B*, **272**, 1971, A849–A852.

6

STRUCTURE OF
THE REDUCED
RESIDUE CLASSES

§6.1 REDUCED RESIDUE CLASSES AS AN ABELIAN GROUP

Given a positive integer m, an integer r that is relatively prime to m determines a reduced residue class S_r modulo m. In §5.1 it was shown that the number of reduced residue classes modulo m is $\phi(m)$, in correspondence to the $\phi(m)$ integers of a reduced residue system. If S_{r_1} and S_{r_2} are two such reduced residue classes, the "product" of these classes is given by

$$(6.1.1) \qquad S_{r_1}S_{r_2} = S_{r_1r_2}.$$

Since $S_{r_1r_2}$ consists of all integers that are congruent to r_1r_2 modulo m, and $(r_1r_2, m) = 1$, it follows that $S_{r_1r_2}$ is a reduced residue class. In other words, the product of two reduced residue classes is a reduced residue class. It is also clear that such multiplication is commutative and associative, that is,

$$(6.1.2) \qquad S_{r_1}S_{r_2} = S_{r_2}S_{r_1},$$

$$(6.1.3) \qquad (S_{r_1}S_{r_2})S_{r_3} = S_{r_1}(S_{r_2}S_{r_3}).$$

For the residue class of 1 modulo m, it follows that

$$(6.1.4) \qquad S_r \cdot S_1 = S_1 \cdot S_r = S_r$$

for all reduced residues r.

Also, for a given integer r such that $(r, m) = 1$, Corollary 1 of Theorem 5.3.1 provides that there exists an $\bar{r}$, $(\bar{r}, m) = 1$ such that $r\bar{r} \equiv 1 \pmod{m}$. It is then easily seen that

$$(6.1.5) \qquad S_r \cdot S_{\bar{r}} = S_{\bar{r}} \cdot S_r = S_1.$$

These properties are typical of the structural notion of *group*, which may be defined as follows.

Definition 6.1.1. A collection of elements $G = \{a, b, \ldots\}$ together with an operation of composition $a \circ b$ (written like multiplication) is called a group if and only if:

(i) $a, b \in G$ implies $a \circ b \in G$;

(ii)   the associative law holds:

$$a \circ (b \circ c) = (a \circ b) \circ c;$$

(iii)   there exists an "identity" $e \in G$ such that for all $a \in G$,

$$a \circ e = e \circ a = a;$$

(iv)   for each $a \in G$ there exists an $\bar{a} \in G$ such that

$$a \circ \bar{a} = \bar{a} \circ a = e.$$

Thus letting $R(m)$ denote the collection of reduced residue classes modulo m, (6.1.1), (6.1.3), (6.1.4), and (6.1.5) are precisely the requisite conditions for $R(m)$ to constitute a group under the operation of multiplication of residue classes. If a group G satisfies the additional condition

$$a \circ b = b \circ a$$

for all $a, b \in G$, then it is called *abelian*. Thus (6.1.2) asserts that $R(m)$ is an abelian group. A group consisting of a finite number of elements is called a *finite* group. The number of elements in a finite group is called its *order*, so that $R(m)$ is of order $\phi(m)$.

The structure of $R(m)$ is to a large extent typical of finite abelian groups in general, with the additional feature that somewhat more explicit statements can be made. In any event, certain results of a group theoretic nature have already emerged in previous discussions. For example, Euler's theorem, Theorem 5.1.3, asserts for $R(m)$ that raising any element to a power equal to the order of $R(m)$ yields the identity. Indeed, this is a special case of the following theorem.

Theorem 6.1.1. If g is any element of a finite group of order N, then g^N equals the identity element of G.

For the case of G an abelian group, the proof of Theorem 6.1.1 is identical to the proof of Theorem 5.1.3. In the general case, where G is not necessarily abelian, the proof rests on other notions and is deferred to §6.1B.

If G is a finite group, the powers of a given element g cannot all be distinct, so that $g^{x_1} = g^{x_2}$ for integers $x_1 > x_2 \geq 0$. It follows that $g^{x_1 - x_2}$ is the identity. Since there is a positive power of g that equals the identity, there is a *smallest* such positive power that is called the *order of g*. We denote this by ord g. For the groups $R(m)$, this notion was presented in Ex. 9 of §5.1. In fact, that exercise is a special case of the following theorem.

Theorem 6.1.2. In a finite group, the order of each element divides the order of the group.

Proof. This result follows immediately from Theorem 1.3.1 and Theorem 6.1.1, for the integers x (positive or negative) such that g^x is the identity (g a fixed element of the group) form an ideal. This ideal is then the principal ideal generated by ord g and Theorem 6.1.1 asserts that the order of the group is in the ideal. From this the theorem follows.

The proof of the above theorem may be based directly on the Division Algorithm. Letting N denote the order of G, we have

$$N = (\text{ord } g)q + r, \qquad 0 \le r < \text{ord } g,$$

so that

$$g^r = g^N \cdot (g^{\text{ord } g})^{-q}.$$

Since g^N and $g^{\text{ord } g}$ are the identity, it follows that g^r is the identity; and from the definition of ord g we obtain $r = 0$. Hence, ord g divides N as claimed.

§6.1A   PRIMES OF THE FORM *km* + 1

As a consequence of Theorem 6.1.2, in the case of the residue groups, one can give a simple proof of the following theorem.

Theorem 6.1A.1. For any fixed positive integer m, the arithmetic progression $km + 1$, $k = 1, 2, \ldots$, contains infinitely many prime numbers.

Proof. Consider the polynomial $f(x) = x^m - 1$ and recall that by Ex. 6 of §2.1, the sequence of integers $f(1), f(2), \ldots$ contains, as divisors, an infinite number of distinct primes. Suppose that q is one of these primes and q divides $f(l)$, so that

(6.1A.1) $$l^m \equiv 1 \pmod{q}.$$

If we knew that m was the order of l in $R(q)$, Theorem 6.1.2 would provide that m divides $\phi(q) = q - 1$. Thus q would be of the form $km + 1$ and the theorem would be proved. In any event, we see that the theorem follows if we can provide infinitely many primes q such that (6.1A.1) holds and

(6.1A.2) $$m = \mathbb{O}_q(l)$$

[recall that $\mathbb{O}_q(l)$ is a notation for the order of l modulo q].

Achieving (6.1A.2) is equivalent to avoiding all the congruences $l^d \equiv 1 \pmod{q}$, for d a proper divisor of m (i.e., $d < m$). Thus we must seek primes q that divide a value of $x^m - 1$ but not $x^d - 1$ for $d < m$ that divide m. Since for d a divisor of m, $x^d - 1$ divides $x^m - 1$ algebraically, we are led to seek a *polynomial divisor* of $x^m - 1$ that has no common polynomial divisors with the $x^d - 1$, $d < m$, d a divisor of m. That there is such a nonconstant polynomial with integer coefficients is easily seen. For by Ex. 2 of §2.5, we have a factorization

$$(6.1A.3) \qquad\qquad x^m - 1 = \alpha g_1(x) g_2(x) \cdots g_k(x)$$

where the $g_i(x)$ are irreducible polynomials with integer coefficients of greatest common divisor 1, and α is a rational number. (Actually $\alpha = 1$, but this is not needed here.) Since $x = e^{2\pi i/m}$ is a complex root of the left side of (6.1A.3), it must be a root of some $g_i(x)$ on the right. Assume that $g_{i*}(x)$ has $x = e^{2\pi i/m}$ as a root. It follows that $g_{i*}(x)$ cannot divide $x^d - 1$ for any $d < m$, and since it is irreducible, $g_{i*}(x)$ has no common polynomial divisors with such an $x^d - 1$.

Furthermore, for every proper divisor d of m, $(x^m - 1)/(x^d - 1)$ is a polynomial that divides $x^m - 1$. Hence, from (6.1A.3), we have

$$(6.1A.4) \qquad\qquad \frac{x^m - 1}{x^d - 1} = \alpha_d g_{i*}(x) g_{i_1}(x) \cdots g_{i_\lambda}(x)$$

where α_d is some rational number and the $g_{i_\nu}(x)$, $\nu = 1, \ldots, \lambda$ are some subset of $g_1, \ldots, g_k$.

By Ex. 6 of §2.1, the values of $g_{i*}(x)$ for positive integral x contain infinitely many prime divisors. Let q be any one of these that does not divide m or the denominators of any of the α_d, d/m, $d < m$, or the denominator of α. If q divides $g_{i*}(l)$, (6.1A.1) then follows from (6.1A.3). It remains then to show that such a prime q also satisfies (6.1A.2).

Suppose, to the contrary, that there exists a divisor d of m, $d < m$, such that q divides $l^d - 1$; that is,

$$(6.1A.5) \qquad\qquad l^d \equiv 1 \pmod{q}.$$

Since $g_{i*}(l)$ is divisible by q, (6.1A.4) yields that

$$(6.1A.6) \qquad\qquad \frac{l^m - 1}{l^d - 1} \equiv 0 \pmod{q}.$$

But (6.1A.5) and (6.1A.6) together provide

$$0 \equiv \frac{l^m - 1}{l^d - 1} = \sum_{\nu=0}^{m/d-1} (l^d)^\nu \equiv \frac{m}{d} \pmod{q}.$$

Thus we get that q divides m, which is a contradiction, and the theorem is proved.

Note that the polynomial $g_{i*}(x)$, which occurs in the above argument only in the vague way needed for the proof, is in reality the *cyclotomic polynomial* for the *primitive mth roots of unity*. That is, its roots are the complex numbers

$$e^{2\pi i a/m}, \qquad (a, m) = 1, \qquad 1 \le a < m,$$

and it is given by the formula

(6.1A.7) $$F_m(x) = \prod_{d/m} (x^d - 1)^{\mu(m/d)}.$$

The proof of the fact that the cyclotomic polynomials are irreducible (over the rationals) would require a small digression and this is why explicit mention of them is avoided in the above proof of Theorem 6.1A.1. However, there is an alternative proof that uses the expression (6.1A.7) and avoids the issue of irreducibility. The outline of such a proof is provided in Exercises 1–3.

EXERCISES

1. Show that $F_m(x)$ is a polynomial with integer coefficients and that $F_m(x)$ divides the polynomial $x^m - 1$. Note that the degree of $F_m(x)$ is $\phi(m)$.

2. Show that for all d/m, $d < m$, the polynomial $x^d - 1$ has no common non-constant polynomial divisors with $F_m(x)$.

3. Let q be any prime that divides $F_m(x)$ for some positive integral value of x. Show that if q does not divide m, then q is of the form $km + 1$. From this fact deduce a proof of Theorem 6.1A.1.

4. Show that

$$x^m - 1 = \prod_{d/m} F_d(x).$$

5. Show that

$$F_m(x) = \prod_{\substack{1 \le a \le m \\ (a,m)=1}} (x - e^{2\pi i a/m}).$$

§6.1B   BASIC NOTIONS CONCERNING FINITE GROUPS

In this section we propose to review various basic notions that are relevant for the study of finite groups.

Definition 6.1B.1.   A subset H of a group G that also forms a group (with respect to the operation of composition of G) is called a *subgroup of G*. If H is not all of G, we will call it a *proper subgroup*.

Within a given group G, one can attempt to form a subgroup from any given element $g \in G$. One considers the set H consisting of all the powers of g (positive, negative, and zero) and easily verifies that H is a subgroup of G. In general, a group such as H that consists of the powers of a single element g is called *cyclic* and is said to be generated by g. A cyclic group is clearly abelian. If an element $g \in G$ is of finite order, the order of the subgroup that it generates equals the order of g.

Definition 6.1B.2.   Given any subgroup H of a group G, and any element $g \in G$, the collection of elements of the form hg, $h \in H$, is called a *right coset* of H, and is denoted by Hg. Similarly, the set of elements of the form gh, $h \in H$, is denoted by gH and is called a *left coset* of H.

The basic property of right (or left) cosets, which is the source of their usefulness, is that *if two of them intersect, then they must be identical*. For suppose $g_1, g_2 \in G$, H a subgroup of G, and

$$s \in Hg_1 \cap Hg_2.$$

We then have $s = h_1 g_1 = h_2 g_2$, where $h_1, h_2 \in H$, so that

$$g_1 = (h_1^{-1} h_2) g_2.$$

Thus for any element hg_1 of Hg_1 we have

$$hg_1 = h(h_1^{-1} h_2) g_2 \in Hg_2,$$

so that $Hg_1 \subset Hg_2$. Similarly $Hg_2 \subset Hg_1$ and hence $Hg_1 = Hg_2$, as claimed.

If we consider the collection of *distinct* right cosets of a subgroup H of G, denoting these by $\{Hg_\alpha\}$, these cosets are mutually disjoint. Further, their union includes all elements of G. (For if any element $g \in G$ were not contained in any of them, Hg would be a new right coset.) Thus

$$(6.1\text{B}.1) \qquad\qquad\qquad G = \cup Hg_\alpha$$

provides a decomposition of G into mutually disjoint sets, which is called a right coset decomposition of G.

In the case where H is a finite group, each coset Hg_α is a finite set, the number of whose elements equals ord H. If G is also of finite order, the number of cosets appearing in the decomposition (6.1B.1) is finite. Denoting this number by k, (6.1B.1) yields

$$(6.1\text{B}.2) \qquad\qquad\qquad \text{ord } G = k \text{ ord } H.$$

Theorem 6.1B.1.   The order of a subgroup of a finite group divides the order of the group.

From this theorem we see that the order of the cyclic subgroup generated by an element $g \in G$ must divide the order of G. But the order of this subgroup equals ord g. Since ord g divides ord G, it follows that for all $g \in G$, $g^{\text{ord}\,G}$ is the identity, which proves Theorem 6.1.1.

Clearly, in the case of abelian groups, a left coset is always a right coset, and conversely. More precisely, the sets Hg and gH are the same. In the general case, this property is the basis of an important notion given by the following definition.

Definition 6.1B.3. A subgroup H of a group G is called a *normal* or *invariant* subgroup if and only if the sets Hg and gH are the same for all $g \in H$. This may be written as

$$gH = Hg \qquad (\text{i.e., } gh = h'g, \; h, h' \in H)$$

or equivalently,

$$g^{-1}Hg = H.$$

We note then that every subgroup of an abelian group is an invariant subgroup.

An invariant subgroup H of a group G produces a kind of group theoretic analogue of division. We note first that in this case

$$(h_1 g_1)(h_2 g_2) = (h_1 h_2')g_1 g_2,$$

so that the product of an element of the coset Hg_1 with an element of the coset Hg_2 lies in the coset $H_{g_1 g_2}$, that is,

$$(Hg_1)(Hg_2) \subset Hg_1 g_2.$$

Since also (henceforth I will denote the identity element of the group)

$$hg_1 g_2 = (hg_1)(Ig_2) \subset (Hg_1)(Hg_2),$$

we obtain

(6.1B.3) $$(Hg_1)(Hg_2) = Hg_1 g_2.$$

Using (6.1B.3) to define multiplication of cosets, it is immediate that this definition is independent of the particular g_i used to represent the cosets. Also, with this multiplication, it is easily verified that the cosets form a group in which the subgroup H (it is one of the cosets) plays the role of identity, and the inverse of the coset Hg is the coset Hg^{-1}.

Definition 6.1B.4. Given an invariant subgroup H of a group G, the group consisting of the cosets of H under the operation of coset multiplication is called the *quotient* or *factor* group of G by H and is denoted by G/H.

In the case where G is a finite group, and H an invariant subgroup of G, (6.1B.2) may be rewritten as

(6.1B.4)   ord $G = (\text{ord } H)(\text{ord } G/H)$.

Just as factor or quotient groups introduce into group theory a spirit of "division," there is also an analogous concept in the spirit of "multiplication." This is the notion of *direct product* which arises as follows. Consider two subgroups H_1, H_2, of G such that

   (i)   for every pair $h_1 \in H_1$, $h_2 \in H_2$, $h_1 h_2 = h_2 h_1$ (i.e., the elements of H_1 and H_2 permute);
   (ii) the products $h_1 h_2$ are distinct for distinct such pairs (i.e., if $h_1 h_2 = h_1' h_2'$, $h_1, h_1' \in H_1$, $h_2, h_2' \in H_2$, then $h_1 = h_1'$ and $h_2 = h_2'$).

Then the set of elements $h_1 h_2$, $h_1 \in H_1$, $h_2 \in H_2$ is easily seen to form a subgroup of G.

Definition 6.1B.5.   The subgroup of G defined above is called the *direct product* of H_1 and H_2, and is denoted by $H_1 \times H_2$.
   Similarly, one can define the direct product of any finite number of subgroups of G, which would be notated as

$$H_1 \times H_2 \times \ \cdots \ \times H_t.$$

Note that the subgroup represented by this direct product has been provided with a sort of "unique factorization." In essence, it asserts that every element is *uniquely* expressible in the form

$$h_1 h_2 \cdots h_k$$

where $h_i \in H_i$, and order is unimportant since the h_i permute with each other.

EXERCISES

   1.   Prove that if a finite subset H of a group G is closed under multiplication, then it is a subgroup of G.

   2.   Prove that every subgroup of a cyclic group is cyclic.

   3.   As a consequence of Theorem 6.1B.1, show that a finite group whose order is a prime number must be cyclic.

   4.   Consider a cyclic group of order N generated by an element g. Prove that

$$\text{ord } g^i = \frac{N}{(i, N)}.$$

Deduce that the number of generators of such a group equals $\phi(N)$.

5. Show that in a cyclic group of order N, for each divisor d of N there are exactly $\phi(d)$ elements of order d. Use this to derive the identity

$$\sum_{d/N} \phi(d) = N.$$

6. Let G be a cyclic group of order N. For each divisor d of N, show that there exist exactly d solutions in G of the equation $x^d = I$ ($I =$ the identity of G), and these form a cyclic subgroup of order d. Deduce that it is the only subgroup of order d.

7. As a converse to Ex. 6, show that if a group G of order N has the property that for each d/N there are at most d solutions in G to $x^d = I$, then G is cyclic. Hence if for each d/N, G has at most one subgroup of order d, it must also be cyclic.

8. Given a subgroup H of a finite group G, for any element $g \in G$ define *the order of g relative to H*, which is denoted by $\operatorname{ord}_H g$, as the smallest positive power of g that lies in H. Note the case where H is the identity is the usual notion of order. Prove that

$$\operatorname{ord} g = (\operatorname{ord}_H g)(\operatorname{ord} g^{\operatorname{ord}_H g})$$

9. Verify in detail that for H an invariant subgroup of G, coset multiplication determines a group G/H as asserted in Definition 6.1B.4.

10. Let H be an invariant subgroup of G, and Hg an element of finite order in G/H. Show that the order of Hg as an element of G/H equals $\operatorname{ord}_H g$.

11. If G is abelian, H a subgroup of G, then G/H is abelian.

12. Verify in detail that the direct product $H_1 \times H_2$ is indeed a group, as claimed in Definition 6.1B.5. Further, if H_1 and H_2 are finite groups, show that

$$\operatorname{ord} (H_1 \times H_2) = (\operatorname{ord} H_1)(\operatorname{ord} H_2).$$

13. Show that the direct product of two subgroups H_1, H_2 of a group G is defined if the elements of H_1 permute with those of H_2, and $H_1 \cap H_2 = I$.

14. Prove that the direct product operation is associative, that is,

$$(H_1 \times H_2) \times H_3 = H_1 \times (H_2 \times H_3).$$

15. Prove that the direct product of abelian subgroups of a group G is necessarily abelian.

16. Two groups G and G' are said to be *isomorphic* if there exists a 1–1 correspon-

dence between their elements that preserves the group operation. In other words, if the correspondence is $g \leftrightarrow g'$, $g \in G$, $g' \in G'$ we have $(g_1 g_2)' = g_1' g_2'$. We use the notation $G \sim G'$ to denote that G is isomorphic to G'. The analogy between multiplication (division) with direct products (factor groups) extends to:

H_1 is an invariant subgroup of $H_1 \times H_2$ and

$$(H_1 \times H_2)/H_1 \sim H_2.$$

Prove this.

17. Show that if g and h are elements of a finite abelian group such that (ord g, ord h) = 1, then ord (gh) = (ord g)(ord h). By a simple example, show that the conclusion is not valid without the hypothesis (ord g, ord h) = 1.

§6.1C   DIRECT PRODUCTS IN ABELIAN GROUPS

Pursuing the analogy between direct products in groups and arithmetic factorization still further, we consider the question of expressibility as a direct product of subgroups. Certainly, some groups would have no such "factorizations," and would constitute a kind of generalization of the arithmetic notion of "prime." In this section, the case of finite abelian groups is considered and the analogue of the unique factorization theorem is proved.

The first theorem we will establish focuses on a simple case in which one easily ascertains the existence of a "factorization." As a preliminary, we require a lemma that provides the existence of certain subgroups.

Lemma 6.1C.1.   If a prime p divides the order of a finite abelian group G, then G contains at least one element of order p.

Proof.   We prove this by induction on the order of G. To begin, if ord $G = p$, the result follows from Ex. 3 of §6.1B. Assume it then for all abelian groups of order less than ord G, whose orders are divisble by p. Consider any fixed $g \in G$, $g \neq I$. We have two cases:

CASE (I). p divides ord g

By Ex. 4 of §6.1B we have that the ((ord g)/p) power of g has order p, and the lemma would be proved.

CASE (II). p does not divide ord g

Let H denote the cyclic subgroup generated by g, and form the factor group G/H. Since

$$\text{ord } (G/H) = \frac{\text{ord } G}{\text{ord } H} = \frac{\text{ord } G}{\text{ord } g}$$

it follows that the prime p divides ord G/H. Further, since ord $g > 1$, and ord $G/H < $ ord G, G/H is abelian and the inductive hypothesis applies. This yields that G/H contains an element of order p. This would be a coset Hs such that $\mathrm{ord}_H\, s = p$ (by Ex. 10 of §6.1B). But then from Ex. 8 of §6.1B, p divides ord s which returns us to Case (i) and again completes the proof of the lemma.

A case where a nontrivial "factorization" must exist is given as follows.

Lemma 6.1C.2. Let G be a finite abelian group of order mn where $(m, n) = 1$. Then G is expressible as a direct product of subgroups

$$G = H_1 \times H_2$$

where ord $H_1 = m$ and ord $H_2 = n$.

Proof. Let H_1 be the set of elements x of G such that $x^m = I$, and H_2 those such that $x^n = I$. Clearly H_1 and H_2 are subgroups of G. Further, if $g \in H_1 \cap H_2$ then ord g must divide both m and n and since $(m, n) = 1$, this yields ord $g = 1$ or $g = I$. Since $H_1 \cap H_2 = I$ and the group G is abelian, it follows that $G \supset H_1 \times H_2$. Also, since $(m, n) = 1$, there exist integers a, b such that $am + bn = 1$, so that for any $g \in G$

$$g = g^{am+bn} = (g^{bn})(g^{am}) \subset H_1 \times H_2.$$

Thus $G \subset H_1 \times H_2$ and consequently $G = H_1 \times H_2$.

Since the elements of H_1 satisfy $x^m = I$, it follows that $(\mathrm{ord}\ H_1, n) = 1$. For if a prime p divides n and ord H_1, Lemma 6.1C.1 provides that H_1 has an element of order p. Then p would also have to divide m, thus contradicting $(m, n) = 1$. Similarly $(\mathrm{ord}\ H_2, m) = 1$. From $G = H_1 \times H_2$, we obtain

$$(6.1C.1) \qquad\qquad mn = \mathrm{ord}\ G = (\mathrm{ord}\ H_1)(\mathrm{ord}\ H_2).$$

Since $(\mathrm{ord}\ H_1, n) = 1$, ord H_1 divides m and, similarly, ord H_2 divides n so that (6.1C.1) yields that ord $H_1 = m$ and ord $H_2 = n$ which completes the proof of the lemma.

From the above lemma we obtain immediately the following theorem.

Theorem 6.1C.1. Let G be a finite abelian group of order n, where the unique factorization of n as a product of powers of distinct primes is given by

$$(6.1C.2) \qquad\qquad n = p_1^{\alpha_1} p_2^{\alpha_2} \cdots p_r^{\alpha_r}.$$

Then G is expressible as a direct product of subgroups

$$(6.1C.3) \qquad\qquad G = H_1 \times \cdots \times H_r$$

where for $i = 1, \ldots, r$

$$(6.1C.4) \qquad\qquad \mathrm{ord}\ H_i = p_i^{\alpha_i}.$$

The above theorem reduces the problem of "factorization" in finite abelian groups to the case of those whose order is a power of a prime.

Theorem 6.1C.2. Let G be a finite abelian group of order p^α, $\alpha \geq 1$, where p is a prime. Then G is expressible as a direct product of subgroups

$$(6.1C.5) \qquad\qquad G = C_1 \times \cdots \times C_r$$

where each C_i is cyclic. Further, the order of C_i equals p^{ν_i}, and we can arrange the notation so that

$$(6.1C.6) \qquad\qquad \nu_1 \geq \nu_2 \geq \cdots \geq \nu_r > 0.$$

The proof of this theorem will be made to depend on the following simple lemma.

Lemma 6.1C.3. Let H denote the cyclic subgroup of an abelian group generated by an element g, where ord $g = p^\mu$, p a prime. Let s be another element of the group such that $\text{ord}_H s = p^\nu$ and ord $s \leq p^\mu$. Then the coset Hs contains an element $\hat{s}$ such that

$$(6.1C.7) \qquad\qquad \text{ord}_H \hat{s} = \text{ord } \hat{s} = p^\nu.$$

Proof. Since $\text{ord}_H s = p^\nu$, it follows that for some integer l, $0 \leq l < p^\mu$,

$$(6.1C.8) \qquad\qquad s^{p^\nu} = g^l.$$

We then equate the orders of the elements on each side of this equation, obtaining

$$\frac{\text{ord } s}{p^\nu} = \frac{p^\mu}{(l, p^\mu)}$$

Since ord $s \leq p^\mu$, this in turn yields

$$(l, p^\mu) \geq p^\nu,$$

which implies that p^ν divides l. Setting $l = p^\nu m$, (6.1C.8) may be rewritten as

$$(sg^{-m})^{p^\nu} = I.$$

Finally, letting $\hat{s} = sg^{-m}$, we clearly have (6.1C.7) and the lemma is proved.

Proof of Theorem 6.1C.2. The proof of the theorem will be carried out by induction on the order of the abelian group G. If ord G is a prime, G is cyclic and the theorem is trivially true, which initiates the induction. We proceed then by assuming the theorem for all abelian groups of prime power order less than the order of G.

If G is cyclic, the theorem is trivially true so that we may assume G not cyclic. Letting g be an element of G of maximum order, it follows that the cyclic subgroup H generated by g is not all of G. That is, $1 < \text{ord } H < \text{ord } G$. Further, since ord G is a power of the prime p,

$$(6.1C.9) \qquad \text{ord } H = \text{ord } g = p^\mu.$$

Forming the factor group G/H, we obtain an abelian group such that

$$1 < \text{ord } G/H < \text{ord } G$$

and ord G/H is a power of the prime p. Applying the inductive hypothesis to G/H yields that it is a direct product of cyclic subgroups. In other words, there exist cosets Hs_i, $i = 1, \ldots, r$ such that any coset Ht is uniquely expressible in the form

$$(6.1C.10) \qquad Ht = (Hs_1)^{u_1}(Hs_2)^{u_2} \cdots (Hs_r)^{u_r}$$

where for $i = 1, \ldots, r$, $0 \le u_i < p^{v_i}$, and

$$(6.1C.11) \qquad \text{ord}_H \, s_i = p^{v_i}.$$

The assertion (6.1C.10) is equivalent to

$$Ht = Hs_1^{u_1} s_2^{u_2} \cdots s_r^{u_r}.$$

This, in turn, gives that for any $t \in G$, there exist unique u_i, $0 \le u_i < p^{v_i}$ such that

$$(6.1C.12) \qquad t(s_1^{u_1} s_2^{u_2} \cdots s_r^{u_r})^{-1} \in H.$$

Since H is generated by g, it follows from (6.1C.12) that any $t \in G$ is uniquely representable in the form

$$(6.1C.13) \qquad t = g^{u_0} s_1^{u_1} \cdots s_r^{u_r}$$

where $0 \le u_i < p^{v_i}$, $i = 1, \ldots, r$, and $0 \le u_0 < p^\mu$.

Since $\text{ord}_H \, s_i \le \text{ord } g$, the hypothesis of Lemma 6.1C.3 is satisfied, so that each coset Hs_i contains an $\hat{s}_i$ such that

$$\text{ord}_H \, \hat{s}_i = \text{ord } \hat{s}_i = p^{v_i}.$$

Since $Hs_i = H\hat{s}_i$, the s_i may be replaced by the $\hat{s}_i$ in the preceding argument. Having done this, (6.1C.13) provides that G is the direct product of H and the cyclic groups generated by the $\hat{s}_i$, $i = 1, \ldots, r$, which completes the induction and the proof of the theorem.

The cyclic subgroups that appear in the representation (6.1C.5) of a given abelian group of order p^α, are not uniquely determined. In general, there may be many such representations. However, as stated in the following theorem, the orders of these cyclic subgroups are uniquely determined.

Theorem 6.1C.3.   In the representation (6.1C.5) of a finite abelian group of order p^α as a direct product of cyclic subgroups, the sequence of ν_i given in (6.1C.6) is uniquely determined.

Proof.   Let $r(u)$ denote the number of solutions in G of the equation

$$(6.1C.14) \qquad\qquad x^{p^u} = I.$$

If G has a representation such as (6.1C.5), the elements of G are uniquely representable in the form

$$x = s_1 \cdots s_r, \qquad s_i \in C_i.$$

Then (6.1C.14) is equivalent to

$$s_1^{p^u} \cdots s_r^{p^u} = I,$$

which, in turn, is equivalent to (because the product was direct)

$$(6.1C.15) \qquad\qquad s_i^{p^u} = I, \qquad i = 1, \ldots, r.$$

For a fixed i, the number of solutions of (6.1C.15) for $s_i \in C_i$ equals

$$p^{\min(u,\,\nu_i)}.$$

Thus the number of solutions of (6.1C.15) is

$$(6.1C.16) \qquad\qquad N(u) = p^{\sum\limits_{i=1}^{r} \min(u,\,\nu_i)}.$$

Since $N(u)$ is independent of the representation (6.1C.5), it follows that for each $u \geq 0$,

$$\sum_{i=1}^{r} \min(u, \nu_i)$$

is independent of the representation. Hence for each integer $u \geq 1$,

$$\sum_{i=1}^{r} (\min(u, \nu_i) - \min(u - 1, \nu_i))$$

is independent of the representation. Since

$$\min(u, v_i) - \min(u - 1, v_i) = \begin{cases} 1 & \text{if } v_i \geq u \\ 0 & \text{if } v_i < u \end{cases},$$

this says that for each $u \geq 1$, the number of $v_i \geq u$ is independent of the representation. Then finally, for each $u \geq 1$, the number of $v_i = u$ is also independent of the representation. But this is precisely the statement of the theorem.

Having seen that finite abelian groups are structurally describable as a direct product of cyclic subgroups of prime power order, it is natural to seek the explicit nature of this representation in the special case of the groups $R(m)$. This is the substance of the next three sections.

EXERCISES

1. Show that the subgroups $H_1, \ldots, H_r$ of (6.1C.3), provided by Theorem 6.1C.1, are unique.

2. Show that a direct product

$$G = C_1 \times \cdots \times C_r$$

of cyclic subgroups is cyclic if and only if the ord C_i, $i = 1, \ldots, r$ are relatively prime by pairs.

3. Let G be an abelian group of order n. For d any divisors of n, show that G has at least one subgroup of order d.

4. Let G be an abelian group that is a direct product of cyclic subgroups

$$G = C_1 \times \cdots \times C_r.$$

Show that a subgroup H of G need not be of the form

$$H = K_1 \times \cdots \times K_r$$

where K_i is a subgroup of C_i, $i = 1, \ldots, r$.

5. Let G be a finite abelian group, and H any subgroup of G. Show that G has a subgroup that is isomorphic to G/H (see §6.1B Exercise 16).

§6.2   **THE STRUCTURE OF** $R(2^\alpha)$

We consider the group $R(2^\alpha)$ of reduced residue classes modulo 2^α which is of order $\phi(2^\alpha) = 2^{\alpha-1}$. For $\alpha = 1$, $R(2)$ is the identity; for $\alpha = 2$, $R(4)$ is of order 2 and hence cyclic. For $\alpha > 2$, noting that a reduced residue r modulo 2^α is odd, we have

$$r^2 \equiv 1 \pmod{2^3}.$$

Then by Ex. 12 of §5.1 it follows that

(6.2.1) $r^{2^{\alpha-2}} \equiv 1 \pmod{2^\alpha}.$

Thus for $\alpha > 2$, the largest possible order of an element of $R(2^\alpha)$ is $2^{\alpha-2}$, and consequently $R(2^\alpha)$ is not cyclic in this case.

The analysis of $R(2^\alpha)$, in the case $\alpha > 2$, depends on a generalization of the result contained in Ex. 7 of §5.1.

Definition 6.2.1.   We say that "a is exactly congruent to b modulo p^α" and write

$$a \equiv b \pmod{p^\alpha}$$

if and only if $a \equiv b \pmod{p^\alpha}$, and it is *not* true that $a \equiv b \pmod{p^{\alpha+1}}$. The aforementioned generalization of Ex. 7 of §5.1 is

Lemma 6.2.1.   If $\beta \geq 2$, and

$$A \equiv 1 \pmod{2^\beta},$$

then for all integers $k \geq 0$,

$$A^{2^k} \equiv 1 \pmod{2^{\beta+k}}.$$

   Proof.   It clearly suffices to prove the lemma in the case $k = 1$. From the hypothesis we have that

$$A = 1 + 2^\beta s, \qquad (s, 2) = 1.$$

Then

$$A^2 = (1 + 2^\beta s)^2 = 1 + 2^{\beta+1} s (1 + 2^{\beta-1} s),$$

and since $\beta \geq 2$, $1 + 2^{\beta-1} s$ is odd, so that the above yields

$$A^2 \equiv 1 \pmod{2^{\beta+1}}.$$

Theorem 6.2.1.   The integers of the form $8t + 3$ and $8t + 5$ are of order $2^{\alpha-2}$ modulo 2^α, for $\alpha \geq 3$. For $\alpha = 3$ (i.e., modulo 2^3), the residue 7 is also of order $2^{\alpha-2}$. For $\alpha \geq 4$, the integers of the form $8t + 3$ and $8t + 5$ are the only ones of order $2^{\alpha-2}$ modulo 2^α.

Proof. Letting $c = 3$ or 5, we note that

$$(8t + c)^2 - 1 = 8(2t(c + 4t) + (c - 2)),$$

and hence

$$(8t + c)^2 \doteq 1 \pmod{2^3}.$$

Thus, from 6.2.1, for all $\gamma \geq 1$ we have

(6.2.2) $$(8t + c)^{2^\gamma} \doteq 1 \pmod{2^{\gamma+2}}$$

Let τ denote the order of $8t + c$ modulo 2^α, $\alpha \geq 3$. From (6.2.1) it follows that τ divides $2^{\alpha-2}$, and hence $\tau = 2^\lambda$, $1 \leq \lambda \leq \alpha - 2$. On the other hand, we must have

$$(8t + c)^\tau = (8t + c)^{2^\lambda} \equiv 1 \pmod{2^\alpha},$$

so that (6.2.2) with $\gamma = \lambda$ implies that $\lambda + 2 \geq \alpha$ or $\lambda \geq \alpha - 2$. Hence $\lambda = \alpha - 2$ and $\tau = 2^{\alpha-2}$, as claimed.

For $\alpha \geq 4$, the other integers (i.e., not of the form $8n \pm 3$) relatively prime to 2^α are of the form $8t \pm 1$. For these

$$(8t \pm 1)^2 = 1 + 16t(4t \pm 1)$$

so that

$$(8t \pm 1)^2 \equiv 1 \pmod{2^4}.$$

Applying Ex. 7 of §5.1, this yields that for $\alpha \geq 4$

$$(8t \pm 1)^{2^{\alpha-3}} \equiv 1 \pmod{2^\alpha},$$

and hence such integers are of order, at most, $2^{\alpha-3}$.

Theorem 6.2.2. For $\alpha \geq 3$, $R(2^\alpha)$ is the direct product of a cyclic subgroup of order 2 and a cyclic subgroup of order $2^{\alpha-2}$.

Proof. From Theorem 6.2.1, we note in particular that the residue class of 3 is of order $2^{\alpha-2}$ modulo 2^α, $\alpha \geq 3$. Thus 3 generates a cyclic subgroup H of order $2^{\alpha-2}$. Let K denote the cyclic subgroup of order 2, generated by the residue class of -1. The claim is that

(6.2.3) $$R(2^\alpha) = H \times K.$$

In order to verify this, we need only verify that the residues $\pm\, 3^i, 0 \leq i < 2^{\alpha-2}$, are all distinct modulo 2^α. Suppose that

(6.2.4) $\pm 3^i \equiv \pm 3^j \pmod{2^\alpha}$,

where $0 \leq j < i < 2^{\alpha-2}$. Then

$$3^{i-j} \equiv \pm 1 \pmod{2^\alpha}$$

and, by Ex. 7 of §5.1, this implies

(6.2.5) $3^{2(i-j)} \equiv 1 \pmod{2^{\alpha+1}}$.

Since 3 is of order $2^{\alpha-1}$ modulo $2^{\alpha+1}$, it follows from (6.2.5) that $2^{\alpha-1}$ divides $2(i - j)$. Hence $2^{\alpha-2}$ divides $i - j$ and since $|i - j| < 2^{\alpha-2}$, we must have $i = j$. Then in (6.2.4) the sign must be the same on both sides of the congruence.

The result of Theorem 6.2.1 enjoys a current mystical vogue in the world of computers. This relates to the problem of constructing "random number generators" which are used in computer simulations. Without involving ourselves in the details of this, the requirement is to have integers r such that the order of r modulo 2^α is large. The fact that the 2^α enters, stems from the binary nature of most computers. For similar reasons, it is convenient to specify r in terms of a small number of powers of 2. From Theorem 6.2.1, r may be taken as

$$r = 2^i + 2^\epsilon + 1, \qquad \epsilon = 1 \text{ or } 2$$

for any $i \geq 3$ and will be of order $2^{\alpha-2}$ modulo 2^α.

EXERCISES

1.   For each $\alpha \geq 3$, determine the integers that are of order $2^{\alpha-3}$ modulo 2^α.

2.   How many subgroups of order $2^{\alpha-2}$ are contained in $R(2^\alpha)$, $\alpha \geq 3$? Are they all isomorphic? Are there any isomorphic to $R(2^{\alpha-1})$?

§6.3   THE STRUCTURE OF $R(p^\alpha)$, p AN ODD PRIME

For p an odd prime, the group $R(p^\alpha)$ is an abelian group of order $\phi(p^\alpha) = p^{\alpha-1}(p - 1)$. Actually, for all odd primes p, and integers $\alpha \geq 1$, $R(p^\alpha)$ is a cyclic group. Thus from the point of view of its structure as an abelian group, if $\phi(p^\alpha)$, $= \Pi q_i^{\beta_i}$ is the unique factorization of $\phi(p^\alpha)$, $R(p^\alpha)$ is the direct product of cyclic groups of order $q_i^{\beta_i}$. This result may be derived in many ways. The first that we consider is based on the following extension of Lemma 6.2.1.

Lemma 6.3.1.   If p is an odd prime, and $\alpha \geq 1$ an integer,

$$A \doteq B \pmod{p^{\alpha}}$$

implies

$$A^{p^i} \doteq B^{p^i} \pmod{p^{\alpha+i}}$$

for every $i \geq 1$.

Proof. It clearly suffices to prove the lemma for $i = 1$. By the hypothesis

$$A = B + p^{\alpha}s, \qquad (s,p) = 1.$$

Then

$$A^p = (B + p^{\alpha}s)^p = B^p + \sum_{\nu=1}^{p} \binom{p}{\nu} p^{\alpha\nu} s^{\nu} B^{p-\nu}$$

so that

$$A^p = B^p + p^{\alpha+1}sB^{p-1} + \sum_{\nu=2}^{p} \binom{p}{\nu} p^{\alpha\nu} s^{\nu} B^{p-\nu}.$$

For $\alpha \geq 1$, $\alpha\nu > \alpha + 1$ for all $\nu > 2$ and for $\nu = 2$, since p is odd, $p^{\alpha+2}$ divides $\binom{p}{2}p^{2\alpha}$. Thus the above equation yields

$$A^p \doteq B^p \pmod{p^{\alpha+1}},$$

which proves the lemma.

The above lemma is in turn the source of the following lemma.

Lemma 6.3.2.   For any fixed integer B, $(B,p) = 1$, $\nu \geq 0$, $\alpha \geq 2$, the congruence

(6.3.1) $$x^{p^{\nu}} \equiv B \pmod{p^{\alpha}}$$

has at most p^{ν} distinct solutions modulo p^{α}.

Proof.   Consider the case $\nu = 1$ of (6.3.1), that is,

(6.3.2) $$x^p \equiv B \pmod{p^{\alpha}}.$$

If x_1, x_2 are two solutions, then

$$(x_1 x_2^{-1})^p \equiv 1 \pmod{p^{\alpha}},$$

so that the number of solutions of (6.3.2) is less than or equal to the number of solutions modulo p^α of

(6.3.3) $x^p \equiv 1 \pmod{p^\alpha}.$

Thus the proof of the case $\nu = 1$ of the lemma reduces to showing that (6.3.3) has, at most, p solutions.

Equation 6.3.3 implies $x^p \equiv 1 \pmod{p}$, which in turn implies $x \equiv 1 \pmod{p}$. Thus if $\alpha = 1$, (6.3.3) has exactly one solution modulo p, namely, $x \equiv 1 \pmod{p}$. If $\alpha = 2$, since the solutions are of the form $x = 1 + kp$, and, in fact, $x \equiv 1 \pmod{p}$ implies $x^p \equiv 1 \pmod{p^2}$, we see that there are exactly p solutions mod p^2.

For $\alpha \geq 3$, if $x \not\equiv 1 \pmod{p^2}$, then $x \overset{.}{\equiv} 1 \pmod{p}$, and Lemma 6.2.1 yields $x^p \overset{.}{\equiv} 1 \pmod{p^2}$ which contradicts (6.3.3). Hence we have $x \equiv 1 \pmod{p^2}$, so that

$$\frac{x^p - 1}{x - 1} = x^{p-1} + \cdots + 1 \equiv p \pmod{p^2}$$

and consequently

$$\frac{x^p - 1}{x - 1} \overset{.}{\equiv} 0 \pmod{p}.$$

Since

$$0 \equiv x^p - 1 = (x - 1)\left(\frac{x^p - 1}{x - 1}\right) \pmod{p^\alpha},$$

it follows that

$$x \equiv 1 \pmod{p^{\alpha-1}}$$

and there are exactly p such residues modulo p^α.

Having established the case $\nu = 1$, we proceed by induction on ν. The congruence (6.3.1) is equivalent to the system

$$\begin{cases} y^{p^{c-1}} \equiv B \pmod{p^\alpha} \\ x^p \equiv y \pmod{p^\alpha} \end{cases}.$$

The first of these has at most $p^{\nu-1}$ solutions for y modulo p^α, via the inductive hypothesis. For each fixed y, the second congruence has at most p solutions for x modulo p^α, by the case $\nu = 1$. Thus the system has at most $p \cdot p^{\nu-1} = p^\nu$ solutions for x.

Lemma 6.3.3.   For every d dividing $\phi(p^\alpha)$, the number of distinct solutions of

(6.3.4) $x^d - 1 \equiv 0 \pmod{p^\alpha}$

is less than or equal to d,

Proof. Since $\phi(p^\alpha) = p^{\alpha-1}(p - 1)$, $d = p^\nu d'$, $d'/(p - 1)$, and $0 \le \nu \le \alpha - 1$. First we note that, for $\alpha = 1$, the lemma follows immediately from Theorem 5.2.1. Thus, henceforth we may assume $\alpha \ge 2$.

We rewrite (6.3.4) as

$$(x^{p^\nu})^{d'} - 1 \equiv 0 \pmod{p^\alpha}$$

and replace this by the equivalent system

(6.3.5) $\begin{cases} z^{d'} - 1 \equiv 0 \pmod{p^\alpha} \\ x^{p^\nu} \equiv z \pmod{p^\alpha} \end{cases}$.

Since $(d', p) = 1$, the congruence $z^{d'} - 1 \equiv 0 \pmod{p^\alpha}$ has no multiple roots modulo p, so that from Theorem 5.5.1 it follows that the congruence

$$z^{d'} - 1 \equiv 0 \pmod{p^\alpha}$$

has at most d' solutions for z. For z any one of these, Lemma 6.3.2 provides that $x^{p^\nu} \equiv z \pmod{p^\alpha}$ has at most p^ν solutions for x. Thus the system (6.3.5) has, at most, $d'p^\nu = d$ solutions for x.

Theorem 6.3.1.   The group $R(p^\alpha)$ is cyclic for p an odd prime, $\alpha \ge 1$.

Proof.   Lemma 6.3.3 asserts that, for every d dividing the order of $R(p^\alpha)$, there are at most d solutions of $x^d = I$. The Theorem follows immediately from Ex. 7 of §6.1B. For completeness, a somewhat more primitive proof is provided below.

For d a divisor of $\phi(p^\alpha)$, let $\psi(d)$ be the number of elements of $R(p^\alpha)$ of order d. If $\psi(d) \ne 0$, there is an element r of order d that generates a cyclic subgroup of order d. Since these d elements satisfy $x^d = I$, they must be all of the solutions. In particular then, this subgroup contains all the elements of $R(p^\alpha)$ that are of order d. By Ex. 5. of §6.1B, the number of elements of order d in the subgroup equals $\phi(d)$. Thus in all cases

(6.3.6) $\psi(d) \le \phi(d).$

Then

$$\phi(p^\alpha) = \sum_{d/\phi(p^\alpha)} \psi(d) \le \sum_{d/\phi(p^\alpha)} \phi(d) = \phi(p^\alpha),$$

so that equality holds in (6.3.6) for all $d/\phi(p^\alpha)$. In particular

$$\psi(\phi(p^{\alpha})) = \phi(\phi(p^{\alpha})) > 1$$

so that $R(p^{\alpha})$ is cyclic and has $\phi(\phi(p^{\alpha}))$ generators.

The above proof of Theorem 6.3.1 has been based on the "extension" of congruences through powers of p developed in Chapter 5 and on the special result given in Lemma 6.3.1. The principal step that incorporated this material was Lemma 6.3.3. There is another route to the theorem in which use of the previous extension theory is minimized and a more critical role given to Lemma 6.3.1.

The first step in the type of proof proposed above is to provide a proof of the theorem in the case $\alpha = 1$. In this case, Lemma 6.3.3 is an immediate consequence of Theorem 5.2.1 and we may retain the elements of the previous argument. Alternatively, there is a simple direct proof that for $d/(p - 1)$, the congruence $x^d - 1 \equiv 0 \pmod{p}$ has exactly d solutions. For this, note the factorization

$$x^{p-1} - 1 = (x^d - 1)\left(\frac{x^{p-1} - 1}{x^d - 1}\right).$$

Since $d/(p - 1)$, the second factor on the right is a polynomial in x of degree $p - 1 - d$. From Theorem 5.2.1 it follows that $(x^{p-1} - 1)/(x^d - 1)$ has at most $p - 1 - d$ roots modulo p. However, since $x^{p-1} - 1$ has exactly $p - 1$ roots modulo p, it follows that $x^d - 1$ has at least d roots modulo p. Since Theorem 5.2.1 provides that it has at most d roots modulo p, it follows that $x^d - 1$ must have exactly d roots modulo p. With this information, the number of generators of $R(p)$ can be counted directly. Noting that

$$\sum_{d/(p-1)/\operatorname{ord} m} \mu(d) = \begin{cases} 1 & \text{if } \operatorname{ord} m = p - 1 \\ 0 & \text{otherwise} \end{cases}$$

we see that the number of generators of $R(p)$ equals

$$\sum_{m=1}^{p-1} \sum_{d/(p-1)/\operatorname{ord} m} \mu(d) = \sum_{d/(p-1)} \mu(d) \sum_{\operatorname{ord} m/(p-1/d)} 1.$$

Since $\sum_{\operatorname{ord} m/(p-1)/d} 1 = $ the number of solutions of $x^{(p-1)/d} \equiv 1 \pmod{p}$ which equals $(p - 1)/d$, the number of generators of $R(p)$ equals

$$\sum_{d/(p-1)} \mu(d)\frac{p - 1}{d} = \phi(p - 1).$$

The next step is to extend generators of $R(p)$ to generators of $R(p^{\alpha})$. Let g be a generator of $R(p)$. If $g^{p-1} \equiv 1 \pmod{p^2}$, as may happen, g could not be a generator modulo p^2, and in fact not for any p^{α}, $\alpha \geq 2$. Thus we replace g by $g + tp$

and attempt to choose t so that it is *not* a root of

(6.3.7) $$(g + tp)^{p-1} \equiv 1 \ (\text{mod } p^2).$$

Since this is equivalent to

$$(g + tp)^p \equiv (g + tp) \ (\text{mod } p^2)$$

and

$$(g + tp)^p \equiv g^p \ (\text{mod } p^2),$$

(6.3.7) is equivalent to

$$t \equiv \frac{g^p - g}{p} \ (\text{mod } p).$$

That is, there is exactly one residue class modulo p that t must avoid. Thus there are exactly $p - 1$ values of t mod p such that for $\hat{g} = g + tp$

$$\hat{g}^{p-1} \not\equiv 1 \ (\text{mod } p^2).$$

Of course, $\hat{g}$ still has order $p - 1$ mod p.

Such a $\hat{g}$ will be a generator modulo p^α for any fixed $\alpha \geq 2$. To see this, note first that the above preparations imply that

$$\hat{g}^{p-1} \overset{\cdot}{\equiv} 1 \ (\text{mod } p).$$

Then Lemma 6.3.1 implies that for any $\gamma \geq 1$,

(6.3.8) $$\hat{g}^{p^{\gamma-1}(p-1)} \overset{\cdot}{\equiv} 1 \ (\text{mod } p^\gamma).$$

Suppose then that $\hat{g}$ is of order τ dividing $\phi(p^\alpha)$ in $R(p^\alpha)$, for fixed $\alpha \geq 2$. Then

$$\hat{g}^\tau \equiv 1 \ (\text{mod } p)$$

so that $p - 1$ divides τ, and hence τ is of the form

$$\tau = p^{\gamma-1}(p - 1), \qquad 1 \leq \gamma \leq \alpha.$$

Further, since $\hat{g}^\tau \equiv 1 \ (\text{mod } p^\alpha)$, (6.3.8) implies that $\gamma \geq \alpha$ and hence $\gamma = \alpha$. Thus $\tau = \phi(p^\alpha)$ and $\hat{g}$ is a generator of $R(p^\alpha)$.

The group theoretic terminology of "generator" is not generally used in number theory. It is more customary to refer to integers whose residue classes modulo p^α

generate $R(p^\alpha)$ as *primitive roots modulo p^α*. Let g be a primitive root modulo p^α, so that for any integer n, $(n, p) = 1$, there is a unique integer i, $0 \le i < \phi(p^\alpha)$ such that

$$n \equiv g^i \pmod{p^\alpha}.$$

This integer i represents the analogue of a kind of "logarithm to the base g" and is called *the index of n to the base g* modulo p^α. When the modulus p^α is understood, we write for i,

$$\text{ind}_g n.$$

It is easily seen that if $(m, p) = (n, p) = 1$,

(6.3.9) $\text{ind}_g(mn) \equiv \text{ind}_g(m) + \text{ind}_g(n) \pmod{\phi(p^\alpha)}$.

From Ex. 4 of §6.1B, we see that the order of an integer modulo p^α (i.e., an integer relatively prime to p) is determined in terms of its index by

(6.3.10) $$\text{ord } n = \frac{\phi(p^\alpha)}{(\text{ind}_g n, \phi(p^\alpha))}$$

There is also a special number theoretic terminology and notation that is used instead of order. That "n has order d modulo m" is notated as

(6.3.11) $n \to d \pmod{m}$,

and is read "n belongs to d modulo m." [It is of course understood that $(n, m) = 1$.]

EXERCISES

1. If p is an odd prime, and $\alpha \ge 1$, show that

$$\prod_{\substack{1 \le a \le p^\alpha \\ (a, p) = 1}} a \equiv -1 \pmod{p^\alpha}.$$

(This is a generalization of Wilson's Theorem.)

2. For p a prime greater than 3, prove that the product of all primitive roots modulo p is congruent to 1 modulo p. Generalize this result to prime powers.

3. Show that if $a \to (p - 1)/2 \pmod{p}$ where p is a prime of the form $4n - 1$, then $-a$ is a primitive root modulo p.

4. Suppose that $a \to \tau \pmod{p}$ and that $a^\tau \stackrel{.}{\equiv} 1 \pmod{p^\lambda}$, $\lambda \ge 1$. Prove that for all $\alpha \ge \lambda$,

$$a \rightarrow \tau p^{\alpha - \lambda} \pmod{p^\alpha}.$$

5. Show that for $\alpha \geq 2$, the solutions of the congruence $x^p \equiv 1 \pmod{p^\alpha}$ are exactly the p residue classes of the form $x = 1 + kp^{\alpha - 1}$.

6. Verify (6.3.9) in detail.

7. Show that for each fixed positive integer u, the diophantine equation

$$u^r - u^s = n!$$

has, at most, a finite number of positive integer solutions for r, s, n. Determine all the solutions for $u = 2, 3, 5$. Some examples of such solutions are:

$2^2 - 2^1 = 2!$ $3^2 - 3 = 3!$ $5^3 - 5 = 5!$
$2^3 - 2^1 = 3!$ $3^3 - 3 = 4!$
$2^5 - 2^3 = 4!$
$2^7 - 2^3 = 5!$

8. Let R be a commutative ring with multiplicative identity 1, such that for a given prime p, $px = 0$ for all $x \in R$. Let $N(u) = $ the number of solutions $x \in R$ of $x^u = 1$. If for a given integer $M = p^\alpha s$, $(s,p) = 1$, we have
   a. $N(M) = M$,
   b. $N(u)N(s/u) \leq s$ for all u/s,
   c. $N(p) \leq p$,
   show that for all d that divide M, it follows that

$$N(d) = d.$$

9. (Unsolved) Prove that there exist infinitely many primes p such that 2 is a primitive root modulo p.

§6.4   THE GENERAL CASE OF $R(m)$

The group $R(m)$ is of order $\phi(m)$. If

$$\phi(m) = q_1^{\beta_1} \cdots q_s^{\beta_s}$$

is the unique factorization of $\phi(m)$ as a product of powers of distinct primes, Theorem 6.1C.1 provides that $R(m)$ is the direct product of subgroups of order $q_i^{\alpha_i}$. However, it is more revealing to examine how the direct product representation relates to the unique factorization of m as

$$m = p_1^{\alpha_1} \cdots p_r^{\alpha_r}.$$

Theorem 6.4.1. $R(m)$ is the direct product of subgroups H_i, $i = 1, \ldots, r$, where H_i is isomorphic to $R(p_i^{\alpha_i})$.

Proof. Let g be any integer such that $(g, p_i) = 1$; and consider the system of congruences

(6.4.1)
$$x \equiv g \pmod{p_i^{\alpha_i}}$$
$$x \equiv 1 \pmod{p_j^{\alpha_j}}, \qquad j = 1, \ldots, r, j \neq i.$$

The Chinese Remainder Theorem insures that this system has a unique solution modulo m. This solution will be an element of $R(m)$ and we denote it by $x^{(i)}(g)$. As g runs over the elements of $R(p_i^{\alpha_i})$, the $x^{(i)}(g)$ determine a set of distinct elements of $R(m)$ that we denote by H_i. Clearly, H_i is a subgroup of $R(m)$, and the correspondence

(6.4.2)
$$g \leftrightarrow x^{(i)}(g)$$

is an isomorphism between H_i and $R(p_i^{\alpha_i})$ (see §6.1B Exercise 16). Further, for any $g \in R(m)$

(6.4.3)
$$g \equiv x^{(1)}(g) \cdots x^{(r)}(g) \pmod{m}.$$

Since the number of $g \in R(m)$ is $\phi(m)$ and the number of products of the form

(6.4.4)
$$h_1 \cdots h_r, \qquad h_i \in H$$

equals $\phi(p_1^{\alpha_1}) \cdots \phi(p_r^{\alpha_r}) = \phi(m)$, we see that each g is congruent modulo m to a unique one of these products. This implies that $R(m)$ is the direct product of the H_i, as claimed.

Theorem 6.4.1 reduces the question of the structure of $R(m)$ to that of $R(p_i^{\alpha_i})$ and this, in turn, has been resolved in the previous sections.

It is natural to ask which of the $R(m)$ are cyclic. Thus far, of course, this has been obtained for $R(2)$, $R(4)$, and $R(p^\alpha)$, p an odd prime. Consider then $R(m)$ for $m = p_1^{\alpha_1} \cdots p_r^{\alpha_r}$. Since $R(p_i^{\alpha_i})$ is of order $\phi(p_i^{\alpha_i})$, it follows that for

$$\tau = \text{l.c.m.} \{\phi(p_1^{\alpha_1}), \ldots, \phi(p_r^{\alpha_r})\}$$

we have

$$x^\tau = I,$$

for all $x \in R(m)$. If $R(m)$ is cyclic, we must have

$$\tau \geq \phi(m) = \prod_{i=1}^{r} \phi(p_i^{\alpha_i}).$$

Since $\phi(2^\alpha) = 2^{\alpha-1}$ and for p_i odd, $\phi(p_i^{\alpha_i})$ is even, this implies that there can be, at most, one odd prime among the p_i. Consequently, the only possible cases are

$$m = 2^\alpha, p^\alpha, 2p^\alpha.$$

Among the $m = 2^\alpha$ we've seen that only $R(2)$ and $R(4)$ are cyclic. The $R(p^\alpha)$, p an odd prime, are all cyclic. The only new possibility is the $R(2p^\alpha)$, p an odd prime. These are, in fact, cyclic. For if g is a primitive root modulo p^α, so is $g + p^\alpha$, and at least one of these is odd. Then for g an odd primitive root modulo p^α, g is a reduced residue modulo $2p^\alpha$, and its order modulo $2p^\alpha$ equals $\phi(p^\alpha) = \phi(2p^\alpha)$. Thus g generates $R(2p^\alpha)$ which is, therefore, cyclic.

The residue class groups $R(m)$ are fairly varied in structure and, in fact, every finite abelian group appears as a subgroup of some of them.

Theorem 6.4.2. Given a finite abelian group H, there exists a residue class group $R(m)$ that contains a subgroup isomorphic to H.

Proof. Since H is abelian, it is a direct product of cyclic subgroups of prime power order (not necessarily the same prime)

$$H = C_1 \times C_2 \times \cdots \times C_r.$$

Choose p_i as an odd prime such that

$$p_i \equiv 1 \pmod{\text{ord } C_i}$$

$i = 1, \ldots, r$. Since, by Theorem 6.1A.1, there are infinitely many such primes (for each i), it can be arranged that no two of the p_i are the same. Set

$$m = p_1 p_2 \cdots p_r.$$

Then $R(m)$ is a direct product of subgroups

$$R(m) = H_1 \times \cdots \times H_r,$$

where H_i is isomorphic to $R(p_i)$, and hence of order $p_i - 1$. Since ord C_i divides $p_i - 1$, H_i contains a cyclic subgroup C_i' of order ord C_i. Thus $R(m)$ contains the direct product

$$C_1' \times C_2' \times \cdots \times C_r'$$

which is clearly isomorphic to H.

The fact that the group $R(m)$ has residue classes as elements is the source of additional special properties of these groups. In particular, as might be deduced from Theorem 6.4.1, if m divides N, then $R(N)$ contains a subgroup isomorphic to $R(m)$. However, there is a more natural description of the relationship between $R(m)$ and

$R(N)$ as an isomorphism between $R(m)$ and a factor group of $R(N)$ with respect to an appropriate subgroup.

Let $N = mt$ and consider the residue class modulo m, S_b, consisting of integers of the form $\lambda m + b$, where $(b, m) = 1$. There are certain reduced residues modulo N that are contained in S_b. These are the integers of the form $\lambda m + b$ that are relatively prime to $N = tm$. Since integers of the form $\lambda m + b$ are automatically relatively prime to m (for $(b, m) = 1$), the remaining condition of relative primeness is to the integer.

$$(6.4.5) \qquad\qquad t^* = \prod_{\substack{p/t \\ p \nmid m}} p.$$

Let c be relatively prime to t^*. Then since $(m, t^*) = 1$, the congruence

$$\lambda m + b \equiv c \ (\mathrm{mod}\ t^*)$$

always has a unique solution for λ modulo t^*, which we denote by λ_c. This implies that such integers form a subset of S_b consisting of those of the form

$$(6.4.6) \qquad\qquad \xi mt^* + (m\lambda_c + b).$$

The integers of (6.4.6) form a reduced residue class modulo mt^* (as ξ varies over all integers) which modulo mt breaks up into t/t^* reduced residue classes. Thus we see that the integers of S_b that are relatively prime to mt fill out precisely

$$\phi(t^*)\frac{t}{t^*}$$

reduced residue classes modulo mt. That is, each element of $R(m)$, viewed as a set, contains $\phi(t^*) \cdot t/t^*$ elements of $R(mt)$; and all elements of $R(mt)$ are so included.

This may be viewed somewhat differently. Consider the identity class S_1 of $R(m)$. The $\phi(t^*)\,t/t^*$ elements of $R(mt)$ (which it contains) form a subgroup $H_{mt,m}$ of $R(mt)$. The elements of $R(mt)$ contained in the element S_b of $R(m)$ are simply the elements of the coset

$$H_{mt,m}S_b.$$

From this it is easily seen that

$$(6.4.7) \qquad\qquad R(mt)/H_{mt,m} \sim R(m).$$

Hence, in particular, it follows that

$$(6.4.8) \qquad\qquad \mathrm{ord}\ H_{mt,m} = \frac{\phi(mt)}{\phi(m)}.$$

EXERCISES

1. Note that the proof of Theorem 6.4.2 utilizes Theorem 6.1A.1 only in the case
    of prime power modulus. That is, for p any prime, $\alpha > 0$, there are infinitely
    many primes of the form $kp^\alpha + 1$. For this case the proof may be simplified by
    considering any prime $q \neq p$ that divides

    $$\frac{a^{p^\alpha} - 1}{a^{p^{\alpha-1}} - 1},$$

    and showing that $q \equiv 1 \pmod{p^\alpha}$. Provide the details of this.

2. Show that if $(m_1, m_2) = 1$, then $R(m_1 m_2)$ is a direct product of two subgroups
    $H_1 \times H_2$ where H_1 is isomorphic to $R(m_1)$ and H_2 is isomorphic to $R(m_2)$.

3. Show that if N is a multiple of m, then $R(N)$ contains a subgroup isomorphic
    to $R(m)$. Show that there exist integers N relatively prime to m such that $R(N)$
    contains a subgroup isomorphic to $R(m)$.

4. Verify in detail that the sets H_i defined in the proof of Theorem 6.4.1 are indeed
    subgroups of $R(m)$. Also verify that (6.4.2) provides an isomorphism, as asser-
    ted.

5. Extend the assertion of Theorem 6.4.2 by showing that there exist infinitely
    many m such that $R(m)$ contains a subgroup isomorphic to a given abelian
    group.

6. Verify in detail that $H_{mt,m}$ is a subgroup of $R(mt)$, and that the isomorphism
    (6.4.7) is valid.

7. Show that for the integer t^* defined by (6.4.5),

    $$\phi(t^*)\frac{t}{t^*} = \frac{\phi(mt)}{\phi(m)}.$$

    Use this to give a direct proof of (6.4.8).

8. For a given odd integer $M > 1$, let $M^* = $ the product of the distinct primes
    dividing M. Prove that H_{M,M^*} is a cyclic group of order M/M^*. Further, show
    that there exists a fixed positive integer g^*, depending only on M^*, such that for
    all integers N such that $N^* = M^*$, the residue class of g^* mod N generates
    H_{N,M^*}.

§6.4A   THE VANDIVER–BIRKOFF THEOREM

Consider the relation (recall the notation (6.3.11))

(6.4A.1) $a \rightarrow n \pmod{p}$

where p is a prime number, and $(a, p) = 1$. There are three natural questions which may be posed:

(i)   Given a and p, $(a, p) = 1$, does there exist an integer n such that (6.4A.1) holds? On the basis of the previous discussion of $R(p)$, this is trivially true.

(ii) Given p, and the integer n that is a positive divisor of $p - 1$, does there exist an integer a such that (6.4A.1) holds? Since $R(p)$ is cyclic of order $p - 1$, there will exist $\phi(n)$ such residue classes modulo p.

(iii) Given a and n, does there exist a prime p such that (6.4A.1) holds?

The third question is somewhat more difficult than the first two, and its resolution is the substance of a theorem attributed to Vandiver and Birkhoff. Roughly speaking, the answer is yes, apart from certain exceptions that are completely determined.

An equivalent way of viewing question (iii) is to consider the sequence

$$(6.4A.2) \qquad a - 1, a^2 - 1, a^3 - 1, \ldots, a^n - 1, \cdots$$

and to ask whether $a^n - 1$ contains a prime divisor that has not appeared as a divisor of an earlier member of the sequence.

We will use the following notation:

(i) $E_p(m)$ = exponent of the highest power of p that divides m

(ii) $g_a(p)$ = order of a modulo p; when the a is understood we will omit it and write $g(p)$.

(iii) $F_m(x)$ = cyclotomic polynomial for the primitive mth roots of unity (see §6.1A).

The resolution of question (iii) is made to depend on the following lemmas which, in turn, lean strongly on Lemma 6.2.1 and Lemma 6.3.1.

Lemma 6.4A.1.   Let p be any prime, and $a \geq 2$ an integer relatively prime to p. Then if $a \equiv 1 \pmod{p}$ and $p \nmid n$, (i.e., p does not divide n)

$$(6.4A.3) \qquad\qquad E_p\left(\frac{a^n - 1}{a - 1}\right) = 0.$$

Also, for p odd, $a \equiv 1 \pmod{p}$ implies

$$(6.4A.4) \qquad\qquad E_p\left(\frac{a^p - 1}{a - 1}\right) = 1.$$

Proof.   If $a \equiv 1 \pmod{p}$ and $p \nmid n$, then

$$\frac{a^n - 1}{a - 1} = a^{n-1} + a^{n-2} + \cdots + 1 \equiv n \not\equiv 0 \pmod{p}$$

establishes (6.4A.3).

For p odd, if $\lambda = E_p(a - 1) > 0$, we have $a \stackrel{\cdot}{\equiv} 1 \pmod{p^\lambda}$. From Lemma 6.3.1, this yields $a^p \stackrel{\cdot}{\equiv} 1 \pmod{p^{\lambda+1}}$, which gives (6.4A.4).

The information of the above lemma is next translated into analogous information concerning cyclotomic polynomials.

Lemma 6.4A.2.   For p an odd prime, $a \geq 2$, $(a,p) = 1$, or if $p = 2$ and $E_2(a - 1) \geq 2$, we have

(6.4A.5) $E_p(F_{g(p)}(a)) = E_p(a^{g(p)} - 1)$

(6.4A.6) $E_p(F_{p^\alpha g(p)}(a)) = 1$ for $\alpha \geq 1$

(6.4A.7) $E_p(F_d(a)) = 0$ otherwise.

If $p = 2$ and $E_2(a - 1) = 1$, the only exception to the above occurs for the case $\alpha = 1$ of (6.4A.6), in which case

(6.4A.8) $E_2(F_2(a)) = E_2(a + 1)$.

Proof.   Since $a^n \equiv 1 \pmod{p}$ requires that $g(p)$ divide n, it follows that for $1 \leq m < g(p)$, $E_p(a^m - 1) = 0$. Thus since $F_m(a)$ divides $a^m - 1$, $E_p(F_m(a)) = 0$ for $1 \leq m < g(p)$. By Ex. 4 of §6.1A

$$a^{g(p)} - 1 = \left(\prod_{\substack{d/g(p) \\ d < g(p)}} F_d(a) \right) F_{g(p)}(a),$$

so that we obtain

$$E_p\left(a^{g(p)} - 1 \right) = E_p(F_{g(p)}(a)),$$

which is (6.4A.5).

Next, assuming that a prime p divides $F_n(a)$, it follows that p divides $a^n - 1$ and hence $g(p)$ divides n. We then write with $\alpha \geq 0$, $t \geq 1$, $(t,p) = 1$,

(6.4A.9) $n = p^\alpha g(p)t$.

Noting that $a^{p^\alpha g(p)} \equiv 1 \pmod{p}$ and $p \nmid t$, (6.4A.3) gives that

(6.4A.10) $\dfrac{a^n - 1}{a^{p^\alpha g(p)} - 1} \not\equiv 0 \pmod{p}$.

Then, if $t > 1$, $F_n(a)$ divides the left side of (6.4A.10), and since p divides $F_n(a)$ we have a contradiction. Thus we must have $t = 1$ in (6.4A.9) and this gives (6.4A.7).

If $\alpha \geq 1$, since

$$(6.4A.11) \qquad a^{p^{\alpha-1}g(p)} \equiv 1 \pmod{p},$$

for p odd, (6.4A.4) yields

$$(6.4A.12) \qquad E_p\left(\frac{a^{p^\alpha g(p)} - 1}{a^{p^{\alpha-1}g(p)} - 1}\right) = 1.$$

Since $F_{p^\alpha g(p)}(a)$ divides $a^{p^\alpha g(p)} - 1/a^{p^{\alpha-1}g(p)} - 1$, and is the only such $F_d(a)$ dividing this which is possibly divisible by p, this implies (6.4A.6).

Finally, for $p = 2$, we must have odd a, $g(2) = 1$, and from Lemma 6.2.1, (6.4A.12) holds for $\alpha \geq 2$. This leaves as the only possible exception the case where $\alpha = 1$, where (6.4A.8) holds since $F_2(a) = a + 1$.

In the next lemma we apply the above in order to examine the consequences of the assumption that we have an instance in which question (iii) has a negative answer.

Lemma 6.4A.3. Let $a \geq 2$, $n \geq 1$, and assume that there is no prime p such that

$$(6.4A.13) \qquad a \to n \pmod{p}.$$

Then either

$$(6.4A.14) \qquad a = 2, \qquad n = 1,$$

or there is an odd prime p such that

$$(6.4A.15) \qquad n = p^\alpha g(p), \qquad \alpha \geq 1, \quad \text{and} \quad |F_n(a)| = p$$

or

$$(6.4A.16) \qquad n = 2^\alpha, \qquad \alpha \geq 1, \quad \text{and} \quad |F_n(a)| = 2^\gamma$$

where $\gamma = 1$ if $\alpha \geq 2$; and for $\alpha = 1$,

$$\gamma = E_2(a + 1).$$

Proof. We consider first the possibility that $|F_n(a)| = 1$. For $n = 1$, $a \geq 2$, this is equivalent to $a - 1 = 1$ or $a = 2$, and this is the case listed under (6.4A.14). For $n = 2$, $|F_n(a)| = a + 1$ and the only solution would be $a = 0$ which cannot occur. For $n \geq 3$, the primitive nth roots of unity are all complex, and for ρ any one of these $|a - \rho| > 2 - 1 = 1$, so that

$$|F_n(a)| = \prod_\rho |a - \rho| > 1.$$

Thus, in all remaining cases there are primes that divide $F_n(a)$. Let p be any such prime. Since (6.4A.13) does not hold, $n \neq g(p)$, and by Lemma 6.4A.2 we see that

(6.4A.17) $n = p^\alpha g(p), \qquad \alpha \geq 1.$

Further, this same lemma yields that

(6.4A.18) $E_p(F_n(a)) = 1$

in all cases except $p = n = 2$, where

(6.4A.19) $E_2(F_2(a)) = E_2(a + 1).$

From the above we next conclude that there can be, at most, one prime dividing $F_n(a)$. For let p be the largest such and $p' < p$ any other. Then

$$n = (p')^{\alpha'} g(p') = p^\alpha g(p),$$

so that p must divide $g(p')$. But $g(p')$ divides $p' - 1$ which would imply $p < p'$, a contradiction.

If the unique prime divisor p of $F_n(a)$ is odd, (6.4A.18) yields (6.4A.15). If this $p = 2$, in all cases except $p = n = 2$, (6.4A.18) yields that $|F_n(a)| = 2$. If $p = n = 2$, it follows from (6.4A.19) that

$$|F_n(a)| = 2^{E_2(a+1)},$$

which establishes the last alternative, (6.4A.16).

We are now in a position to prove the following theorem.

Theorem 6.4A.1.   Let a be any given integer other than 0, 1, -1, and n any positive integer. Then there exists a prime p such that

(6.4A.20) $a \rightarrow n \pmod{p}$

in all except the following cases:

$$
\begin{array}{lll}
a = 2, & n = 1, & 6 \\
a = -2, & n = 2, & 3 \\
a = 2^\delta - 1, & n = 2, & \delta \geq 1 \\
a = -2^\delta - 1, & n = 2, & \delta \geq 1.
\end{array}
$$

Proof. To begin, we restrict ourselves to integers $a \geq 2$, and assume that we have an integer $n \geq 1$ such that (6.4A.20) does not hold for any prime p. Then Lemma 6.4A.3 produces the exception $a = 2$, $n = 1$ for which $a^n - 1 = 1$. Also, in (6.4A.16) it produces the possibility of $n = 2$, and

$$|F_2(a)| = 2^\gamma, \qquad \gamma \geq 1.$$

But this implies $a + 1 = 2^\gamma$ or $a = 2^\gamma - 1$. For such a, $a^2 - 1 = (a - 1)(a + 1) = 2^\gamma(a - 1)$, and since 2 divides $a - 1$, no prime divides $a^2 - 1$ which does not already divide $a - 1$. Thus another exception of our list is produced.

In all other cases, Lemma 6.4A.3 provides that $n = p^\alpha g(p)$, $\alpha \geq 1$, and $|F_n(a)| = p$. Thus $p \leq n$, and consequently

(6.4A.21) $$|F_n(a)| \leq n.$$

Thus we see that our assumptions produce restrictions on the order of magnitude of $F_n(a)$. It is the analysis of these restrictions on which the proof of the theorem rests.

At this point we distinguish two cases.

CASE I. $a \geq 3$, $n > 1$
Since

$$|F_n(a)| = \prod |a - \rho|,$$

where the product on the right is taken over the $\phi(n)$ primitive nth roots of unity, it follows immediately that

(6.4A.22) $$(a - 1)^{\phi(n)} < |F_n(a)| \leq (a + 1)^{\phi(n)}.$$

From $|F_n(a)| = p$ which divides $|F_{g(p)}(a)|$, we obtain

$$|F_n(a)| \leq |F_{g(p)}(a)| \leq (a + 1)^{\phi(g(p))},$$

so that from the lower inequality in (6.4A.22),

(6.4A.23) $$(a - 1)^{\phi(n)} < (a + 1)^{\phi(g(p))}.$$

However, since $n = p^\alpha g(p)$, $(p, g(p)) = 1$,

$$\phi(n) \geq (p - 1)\phi(g(p)),$$

so that (6.4A.23) yields

$$(a - 1)^{(p-1)\phi(g(p))} < (a + 1)^{\phi(g(p))}$$

or

$$(a - 1)^{p-1} < (a + 1).$$

For $p \geq 3$ this implies $(a - 1)^2 < a + 1$ or $a < 3$, a contradiction. For $p = 2$, $n = 2^\alpha$, $g(2) = 1$, so that (6.4A.23) becomes

$$(a - 1)^{2^{\alpha-1}} < (a + 1),$$

which again gives the contradiction $a < 3$ for $\alpha > 1$. Also for $\alpha = 1$, this yields $F_2(a) = a + 1 = 2$ or $a = 1$, which is precluded.

Note at this point that for $a \geq 3$, $n = 1$, the theorem is trivially true.

CASE II. $a = 2, n > 1$.

Since $2^n - 1$ is odd, p is odd. In this case, we propose to compare the upper bound of (6.4A.21) with a lower bound for $|F_n(2)|$. To obtain this lower bound we resort to the explicit formula

$$F_n(2) = \prod_{d/n} (2^d - 1)^{\mu(n/d)}.$$

From this

$$\log |F_n(2)| = \sum_{d/n} \mu\left(\frac{n}{d}\right) \log (2^d - 1)$$

$$= \sum_{d/n} \mu\left(\frac{n}{d}\right) \log 2^d + \sum_{d/n} \mu\left(\frac{n}{d}\right) \log \left(1 - \frac{1}{2^d}\right)$$

or

(6.4A.24) $$\log |F_n(2)| = \phi(n) \log 2 + \sum_{d/n} \mu\left(\frac{n}{d}\right) \log \left(1 - \frac{1}{2^d}\right).$$

For $0 < \theta \leq \frac{1}{2}$, it is easily seen that

$$\log (1 - \theta) \geq -\theta - \theta^2,$$

so that

$$\sum_{d/n} \mu\left(\frac{n}{d}\right) \log \left(1 - \frac{1}{2^d}\right) \geq -\sum_{d=1}^{\infty} \left(\frac{1}{2^d} + \frac{1}{4^d}\right) = -\frac{4}{3}.$$

Thus from (6.4A.24) we obtain

$$\log |F_n(2)| \geq \phi(n) \log 2 - \tfrac{4}{3}.$$

Combining this with $|F_n(2)| \leq n$ yields

$$\log n \geq \phi(n) \log 2 - \tfrac{4}{3}$$

or

$$e^{4/3} n \geq 2^{\phi(n)}.$$

Since $e^{4/3} < 4$, this, in turn, implies

$$(6.4A.25) \qquad\qquad\qquad 4n > 2^{\phi(n)}.$$

We propose now to determine, within the current context, all the solutions of (6.4A.25) (they are finite in number), and examine them for possible exceptions to the theorem. Note first that for $n = \prod p_i^{\alpha_i}$,

$$\frac{\phi(n)}{n^{1/2}} = \prod p_i^{\frac{1}{2}(\alpha_i - 1)}\left(\sqrt{p_i} - \frac{1}{\sqrt{p_i}}\right).$$

For $p_i \geq 3$, $\sqrt{p_i} - (1/\sqrt{p_i}) \geq \sqrt{3} - (1/\sqrt{3}) > 1$. If $p_j = 2$ occurs, since $n = p^{\alpha} g(p)$, p odd, another odd p_i must occur, and

$$\left(\sqrt{p_j} - \frac{1}{\sqrt{p_j}}\right)\left(\sqrt{p_i} - \frac{1}{\sqrt{p_i}}\right) \geq \left(\sqrt{2} - \frac{1}{\sqrt{2}}\right)\left(\sqrt{3} - \frac{1}{\sqrt{3}}\right) = \sqrt{\frac{2}{3}}.$$

Thus for these n we have in all cases

$$(6.4A.26) \qquad\qquad\qquad \phi(n) \geq \sqrt{2/3}\,\sqrt{n}.$$

Introducing (6.4A.26) into (6.4A.25) yields

$$4n > 2^{\sqrt{(2/3)n}},$$

which is clearly false for $n \geq 3/2 \cdot 9^2$ or $n > 121$.

For $n \leq 121$, (6.4A.25) implies

$$484 > 2^{\phi(n)}$$

or

$$\phi(n) \leq 8.$$

But among the integers $n \leq 121$, the only ones such that $\phi(n) \leq 8$ are

$$(6.4A.27) \quad n = 1, 2, 3, 4, 5, 6, 7, 8, 9, 10, 12, 14, 15, 16, 18, 24, 30.$$

For $n = 2, 3, 4, 5$ a solution for p in (6.4A.20) with $a = 2$, is seen to be, respectively, 3, 7, 5, 31. $n = 1$ is an exception already noted, and $n = 6$, (i.e., $2^6 - 1 = 3^2 \cdot 7$) also provides an exception. Since $n = p^\alpha g(p)$, $g(p) \neq 1$, and $g(p)$ a divisor of $p - 1$, n cannot be a power of a prime. Thus 7, 8, 9, 16 cannot be exceptions. This leaves for consideration

$$n = 10, 12, 14, 15, 18, 20, 24, 30.$$

If $n = 10$, then we would have $p = 5$, $g(5) = 2$, which is false. Similarly, $n = 12$ implies $p = 3$, $g(3) = 4$, which is false; $n = 14$ implies $p = 7$, $g(7) = 2$, which is false; and $n = 15$ implies that either $p = 3$, $g(3) = 5$ or $p = 5$, $g(5) = 3$, both of which are false. For $n = 24$, the $p = 3$, $g(p) = 8$, which is impossible. Also, for $n = 30$ we would have to have $p = 5$, $g(5) = 6$, which is false. This leaves only the cases

$$n = 18, 20$$

which can be disposed of as follows:

(i) $n = 18$ would imply $p = 3$, so that

$$|F_{18}(2)| = 3.$$

Since

$$F_{18}(2) = \frac{(2^{18} - 1)(2^3 - 1)}{(2^9 - 1)(2^6 - 1)} = \frac{2^9 + 1}{2^3 + 1},$$

this would imply

$$(64)(2^3) + 1 = 2^9 + 1 = 3(2^3 + 1)$$

or $(61) \cdot 2^3 = 2$, which is nonsense.

(ii) $n = 20$ would imply $p = 5$, so that

$$|F_{20}(2)| = 5.$$

Since

$$F_{20}(2) = \frac{(2^{20} - 1)(2^2 - 1)}{(2^{10} - 1)(2^4 - 1)} = \frac{2^{10} + 1}{2^2 + 1},$$

this would imply

$$2^8(2^2) + 1 = 2^{10} + 1 = 5(2^2 + 1)$$

or $(251)(2^2) = 4$, which is also nonsense. (This case can also be disposed of by noting that 6.4A.25 does not hold.)

The above cases complete the determination of all possible exceptions to the theorem, when $a \geq 2$. There remain the cases with $a \leq -2$. Writing $-a$, $a \geq 2$, we consider the sequence

$$(-a)^n - 1$$

which is equivalent to treating

$$a^n + 1 \text{ for } n \text{ odd,} \qquad a^n - 1 \text{ for } n \text{ even.}$$

For n odd, choose a prime p such tht $a \rightarrow 2n \pmod{p}$. Then since $a^{2n} - 1 = (a^n - 1)(a^n + 1)$ and p does not divide $a^n - 1$, it follows that p divides $a^n + 1$. For $t < n$, t even, it is clear that $a^t - 1$ is not divisible by p. For $t < n$, t odd, $a^t + 1 \equiv 0 \pmod{p}$ implies

$$a^{2t} - 1 \equiv 0 \pmod{p}$$

so that $2n$ would have to divide $2t < 2n$, a contradiction. Thus for n odd we've found a prime p that makes its first appearance as a divisor of an element of the sequence $(-a)^n - 1$. Note that in doing this we used the case $2n$ of the earlier discussion. Any exception there might produce an exception here. One possibility is $a = 2^\delta - 1$, $2n = 2$ which corresponds to $n = 1$, $a^n + 1 = 2^\delta$, which is not an exception since $\delta \geq 1$. The only other possibility is $2n = 6$, $a = 2$, or $n = 3$, $a = -2$ in the original notation. This is indeed an exception since

$$2 + 1 = 3, \qquad 2^2 - 1 = 3, \qquad 2^3 + 1 = 3^2.$$

For n even, $n = 2^\alpha s$, s odd, the corresponding term under consideration is $a^{2^\alpha s} - 1$. If $\alpha > 1$, by our previous discussion we can choose a prime p such that $a \rightarrow 2^\alpha s \pmod{p}$. Then if $t < n$, t even, p certainly cannot divide $a^t - 1$. If $t < n$, t odd, and $a^t + 1 \equiv 0 \pmod{p}$, then $a^{2t} - 1 \equiv 0 \pmod{p}$. This in turn implies that $2^\alpha s$ divides $2t$ or $2^{\alpha-1} s$ divides t. Since t is odd and we've assumed $\alpha > 1$, we have a contradiction.

If $\alpha = 1$, we proceed to make a different choice of p, choosing it so that $a \rightarrow s \pmod{p}$. Then if $t < n = 2s$, t even, and $a^t - 1 \equiv 0 \pmod{p}$, it follows that s divides t and since s is odd, $s \neq t$, so that $t \geq 2s$, which is a contradiction. Also, if $t < n = 2s$, t odd, and $a^t + 1 \equiv 0 \pmod{p}$, $a^{2t} - 1 \equiv 0 \pmod{p}$ so that s divides $2t$. Since s is odd, s divides t hence $s = t$. But then $a^s + 1 \equiv 0 \pmod{p}$ and $a^s - 1 \equiv 0 \pmod{p}$, so that there is a contradiction unless $p = 2$. This implies that a is odd and $s = 1$, so that we are through unless for $\delta \geq 1$,

$$a - 1 = 2^\delta \qquad \text{or} \qquad a = 2^\delta + 1.$$

This is an exception for $n = 2$ since for $\delta \geq 1$,

$$a^2 - 1 = (a + 1) \cdot 2^\delta$$

has no prime divisor not appearing in $(a + 1)$. For $\delta = 0$, $a = 2$, $a + 1 = 3$, $a^2 - 1 = 3$, and we get the exception corresponding to $a = 2$, $n = 1$. This is the only exception that can arise in this way since it is the only one with $a \geq 2$, and n odd.

This completes the proof of the theorem.

EXERCISES

1. Show that the integers given in (6.4A.27) constitute all the solutions of the inequality $\phi(n) \leq 8$.

2. Extend Theorem 6.4A.1 so as to cover the first appearance of primes in a sequence of the form $a^n - b^n$.

3. Let $F(x)$ be a given cyclotomic polynomial and consider the sequence $F(A^n)$, $n = 1, 2, \ldots$, where $A > 1$ is an integer. Show that apart from a finite number of "exceptions," there is always a prime dividing $F(A^n)$ that does not divide any earlier term of the sequence. Determine the exceptions.

§6.5 CHARACTERS OF $R(m)$

Of special importance for number theoretic applications are the complex valued completely multiplicative functions defined over $R(m)$.

Definition 6.5.1. A complex valued function $\chi(s)$ defined over the elements of $R(m)$, $\chi(s)$ not identically zero, is called a *character of $R(m)$*, if for all s_1, $s_2 \in R(m)$,

(6.5.1) $$\chi(s_1 s_2) = \chi(s_1)\chi(s_2).$$

Since the elements of $R(m)$ are residue classes modulo m, we can extend the definition of a character to integers n such that $(n, m) = 1$, by defining $\chi(n)$ to be the value of χ on the residue class of $R(m)$ to which n belongs. We may further extend it to all integers by adding $\chi(n) = 0$ if $(m, n) > 1$. This extended function retains the property (6.5.1) on the integers and will be called a *character modulo m*.

Definition 6.5.2. A complex valued function $\chi(n)$, not identically zero, defined over the integers, is called a *character modulo m* if it satisfies

(i) $\chi(n_1) = \chi(n_2)$ for $n_1 \equiv n_2 \pmod{m}$;

(ii) $\chi(n) = 0$ for $(n, m) > 1$;

(iii) $\chi(n_1 n_2) = \chi(n_1)\chi(n_2)$.

Thus we see that a character of $R(m)$ "extends" uniquely to a character modulo m. Conversely, by restricting a character modulo m to the residues relatively prime to m, condition (i) asserts that a common value is assigned to the integers of a given reduced residue class. Using this as the value of the character on the reduced residue class, condition (iii) insures that we have defined in this way a character $R(m)$. With this relationship understood, we shall use the notation χ both for a character of $R(m)$ and for its related character modulo m. In particular, we note that once all the characters of $R(m)$ are determined, so are all the characters modulo m.

$R(m)$ has one easily distinguished character defined by

$$(6.5.2) \qquad\qquad \chi_0(s) = 1 \qquad \text{for all} \qquad s \in R(m).$$

Corresponding to this is the character modulo m for which

$$(6.5.3) \qquad\qquad \chi_0(n) = \begin{cases} 1 & \text{if } (n, m) = 1 \\ 0 & \text{if } (n, m) > 1 \end{cases}.$$

In either case, the character χ_0 is called the *principal character*.

If I denotes the identity of $R(m)$ (i.e., the residue class of 1 modulo m), then

$$\chi(I) = \chi(I \cdot I) = \chi(I) \cdot \chi(I),$$

so that $\chi(I) = 0$ or 1. If $\chi(I) = 0$, then for any $s \in R(m)$

$$\chi(s) = \chi(s \cdot I) = \chi(s)\chi(I) = 0$$

which asserts that χ is identically zero. Thus we conclude that for all characters χ, $\chi(I) = 1$.

Given two characters of $R(m)$, χ and ψ, one can define a function over $R(m)$, which we denote by $\chi\psi$ as follows:

$$(6.5.4) \qquad\qquad (\chi\psi)(s) = \chi(s)\psi(s).$$

Since $(\chi\psi)(I) = \chi(I)\psi(I) = 1$, $\chi\psi$ is not identically zero. Also, for $s_1, s_2 \in R(m)$,

$$(\chi\psi)(s_1 s_2) = \chi(s_1 s_2)\psi(s_1 s_2) = \chi(s_1)\psi(s_1)\chi(s_2)\psi(s_2)$$

$$= (\chi\psi)(s_1)(\chi\psi)(s_2).$$

Thus we see that $\chi\psi$ is a character of $R(m)$, which we call "the product of the characters χ and ψ." Clearly the definition is such that the product of two characters of $R(m)$ is a commutative operation, that is, $\chi\psi = \psi\chi$.

With respect to the multiplication of characters of $R(m)$, since

$$(\chi_0\chi)(s) = \chi_0(s)\chi(s) = \chi(s),$$

it follows that $\chi_0\chi = \chi\chi_0 = \chi$ (i.e., χ_0 serves as an identity element).

Given a character χ of $R(m)$, for any element s of $R(m)$, $s^{\phi(m)} = I$ implies

$$1 = \chi(s^{\phi(m)}) = (\chi(s))^{\phi(m)}.$$

Hence the values of $\chi(s)$ must be $\phi(m)$th roots of unity, [i.e., complex numbers of the form $e^{2\pi i a/\phi(m)}$]. Consequently, of course, we have that $|\chi(s)| = 1$. Further, we can define the function $\overline{\chi}$ over $R(m)$ by

(6.5.5) $$\overline{\chi}(s) = \overline{\chi(s)}$$

where $\overline{\chi(s)}$ denotes the complex conjugate of $\chi(s)$. Since $\overline{\chi}(I) = \overline{\chi(I)} = 1$, and

$$\overline{\chi}(s_1 s_2) = \overline{\chi(s_1)\chi(s_2)} = \overline{\chi(s_1)}\,\overline{\chi(s_2)} = \overline{\chi}(s_1)\overline{\chi}(s_2),$$

it follows that $\overline{\chi}$ is also a character of $R(m)$.

With respect to multiplication of characters, $\overline{\chi}$ provides an inverse for χ (i.e., $\chi\overline{\chi} = \chi_0$). This is verified by

$$(\chi\overline{\chi})(s) = \chi(s)\overline{\chi}(s) = \chi(s)\overline{\chi(s)} = |\chi(s)|^2 = 1.$$

Since the values of characters are complex numbers wherein multiplication is associative, it is easily seen that the multiplication of characters is associative.

Gathering together the above observations, we see that they imply that the set of characters of $R(m)$, which we denote by $\Gamma(m)$, forms an abelian group with respect to the operation of multiplying characters, as given by (6.5.4). We give this formally as the following lemma.

Lemma 6.5.1.   The set of characters of $R(m)$, denoted by $\Gamma(m)$, is a finite abelian group with respect to multiplication of characters.

Proof.   The only detail that remains to be proved is the finiteness of $\Gamma(m)$. But as was seen above, the value of $\chi(s)$, for $\chi\epsilon\Gamma(m)$, must be among the $\phi(m)$, $\phi(m)$th roots of unity, and $R(m)$ has $\phi(m)$ elements. Consequently, the number of such functions over $R(m)$ is, at most,

$$\phi(m)^{\phi(m)}.$$

The proof of the finiteness of $\Gamma(m)$, given above, uses only that a character of $R(m)$ is a function over $R(m)$ whose values lie in a finite set. This is much too superficial to shed much light on how characters of $R(m)$ are manufactured. Rather, we must turn to the fact that $R(m)$ is abelian and may be represented as a direct product of cyclic subgroups. To see this, let

$$R(m) = H_1 \times \cdots \times H_r,$$

where H_i is a cyclic subgroup of $R(m)$ generated by an element h_i. Then each element

s of $R(m)$ has a unique representation of the form

(6.5.6) $$s = h_1^{u_1} h_2^{u_2} \cdots h_r^{u_r}, \qquad 0 \le u_i < \operatorname{ord} h_i.$$

Since a character χ of $R(m)$ satisfies

(6.5.7) $$\chi(s) = (\chi(h_1))^{u_1} \cdots (\chi(h_r))^{u_r},$$

it is clear that the character χ is determined by the values of $\chi(h_i)$, $i = 1, \ldots, r$. The value of $\chi(h_i)$ is a root of unity of an order restricted by the order of h_i, for

$$1 = \chi(I) = \chi(h_i^{\operatorname{ord} h_i}) = (\chi(h_i))^{\operatorname{ord} h_i}$$

[i.e., $\chi(h_i)$ must be an ord h_ith root of unity]. In fact, these are the only restrictions, and any choice of the values of $\chi(h_i)$ as ord h_ith roots of unity generates a character via (6.5.7). We shall not pursue this route here, but following more classical arguments, will use the above construction to prove the following lemma.

Lemma 6.5.2. If t is any element of $R(m)$ other than the identity there is at least one character χ of $R(m)$ such that $\chi(t) \ne 1$.

Proof. Choosing a fixed representation of $R(m)$ as a direct product of cyclic groups, every element of $R(m)$ has a unique representation of the form (6.5.6). In particular,

$$t = h_1^{v_1} \cdots h_r^{v_r}, \qquad 0 \le v_j < \operatorname{ord} h_j.$$

Since $t \ne I$, it follows that at least one of the v_j, say v_{j^*}, is not 0. Then set

(6.5.8) $$\chi(h_j) = \begin{cases} e^{2\pi i / \operatorname{ord} h_j}, & j = j^* \\ 1, & j \ne j^* \end{cases}$$

and using (6.5.8), define $\chi(s)$ in general (for $s = h_1^{\mu_1} \cdots h_r^{\mu_r}$) as

(6.5.9) $$\chi(s) = e^{2\pi i / \operatorname{ord} h_{j^*} \cdot u_{j^*}}.$$

Quite clearly, (6.5.9) defines a character of $R(m)$, and

$$\chi(t) = e^{2\pi i / \operatorname{ord} h_{j^*} \cdot v_{j^*}} \ne 1.$$

The properties of the characters of $R(m)$ that are most useful in number theory are given in the following theorem.

Theorem 6.5.1. The order of $\Gamma(m)$ is $\phi(m)$, and

(6.5.10) $$\frac{1}{\phi(m)} \sum_{s \in R(m)} \chi(s) = \begin{cases} 1 & \text{if } \chi = \chi_0 \\ 0 & \text{if } \chi \neq \chi_0 \end{cases};$$

(6.5.11) $$\frac{1}{\phi(m)} \sum_{\chi \in \Gamma(m)} \chi(s) = \begin{cases} 1 & \text{if } s = I \\ 0 & \text{if } s \neq I \end{cases}.$$

Proof. The assertion (6.5.10) is immediate for $\chi = \chi_0$, since

$$\sum_{s \in R(m)} \chi_0(s) = \sum_{s \in R(m)} 1 = \phi(m).$$

For $\chi \neq \chi_0$, there is an $\hat{s} \in R(m)$ such that $\chi(\hat{s}) \neq 1$. As s ranges over the elements of $R(m)$ so does $s\hat{s}$, so that

$$\sum_{s \in R(m)} \chi(s) = \sum_{s \in R(m)} \chi(s\hat{s}) = \chi(\hat{s}) \sum_{s \in R(m)} \chi(s).$$

Then

$$(1 - \chi(\hat{s})) \sum_{s \in R(m)} \chi(s) = 0$$

and hence

$$\sum_{s \in R(m)} \chi(s) = 0,$$

which completes the verification of (6.5.10).

For the case of (6.5.11) where $s = I$,

(6.5.12) $$\sum_{\chi \in \Gamma(m)} \chi(s) = \sum_{\chi \in \Gamma(m)} 1 = \text{ord } \Gamma(m).$$

If $s \neq I$, Lemma 6.5.2 provides the existence of a character $\hat{\chi} \in \Gamma(m)$ such that $\hat{\chi}(s) \neq 1$. Since $\Gamma(m)$ is a group, as χ ranges over the elements of $\Gamma(m)$, so does $\hat{\chi}\chi$. Thus

$$\sum_{\chi \in \Gamma(m)} \chi(s) = \sum_{\chi \in \Gamma(m)} (\hat{\chi}\chi)(s) = \hat{\chi}(s) \sum_{\chi \in \Gamma(m)} \chi(s)$$

or

$$(1 - \hat{\chi}(s)) \sum_{\chi \in \Gamma(m)} \chi(s) = 0,$$

and hence in this case

$$\sum_{\chi \in \Gamma(m)} \chi(s) = 0.$$

Summarizing then, instead of (6.5.11) we have

$$(6.5.13) \qquad \sum_{\chi \in \Gamma(m)} \chi(s) = \begin{cases} \text{ord } \Gamma(m) & \text{if } s = I \\ 0 & \text{if } s \neq I \end{cases},$$

so that (6.5.11) follows once we show that

$$(6.5.14) \qquad \text{ord } \Gamma(m) = \phi(m).$$

Using (6.5.10) and (6.5.13), we have

$$\sum_{s \in R(m)} \sum_{\chi \in \Gamma(m)} \chi(s) = \text{ord } \Gamma(m)$$

and

$$\sum_{\chi \in \Gamma(m)} \sum_{s \in R(m)} \chi(s) = \phi(m).$$

But the two double sums are clearly equal and (6.5.14) follows.

Translating Theorem 6.5.1 into a statement concerning characters modulo m, we have the following theorem.

Theorem 6.5.2.   In the following sums, let n run over a complete residue system modulo m, and the character χ over all characters modulo m. Then

$$(6.5.15) \qquad \frac{1}{\phi(m)} \sum_n \overline{\psi}(n)\chi(n) = \begin{cases} 1 & \text{if } \chi = \psi \\ 0 & \text{if } \chi \neq \psi \end{cases}$$

for each character ψ modulo m, and

$$(6.5.16) \qquad \frac{1}{\phi(m)} \sum_\chi \overline{\chi}(B)\chi(n) = \begin{cases} 1 & \text{if } n \equiv B \pmod{m} \\ 0 & \text{if } n \not\equiv B \pmod{m} \end{cases}$$

for each integer B that is relatively prime to m.

Proof.   We note first that in the sum on the left side of equation (6.5.15), integers n that are not relatively prime to m make no contribution. Thus

$$\frac{1}{\phi(m)} \sum_n \overline{\psi}(n)\chi(n) = \frac{1}{\phi(m)} \sum_{s \in R(m)} \overline{\psi}(s)\chi(s) = \frac{1}{\phi(m)} \sum_{s \in R(m)} (\overline{\psi}\chi)(s),$$

and, by (6.5.10), this last expression is 1 or 0 according as $\overline{\psi}\chi = \chi_0$ or $\overline{\psi}\chi \neq \chi_0$. Since $\overline{\psi}\chi = \chi_0$ is equivalent to $\psi = \chi$, this establishes (6.5.15).

Let $\overline{B}$ be such that $B\overline{B} \equiv 1 \pmod{m}$. Then $1 = \chi(1) = \chi(B)\chi(\overline{B})$ so that $\chi(\overline{B}) = \overline{\chi}(B)$. Thus (6.5.16) may be rewritten as

$$\frac{1}{\phi(m)} \sum_\chi \chi(\overline{B}n),$$

and (6.5.11) provides that this is 1 or 0 according as $\overline{B}n \equiv 1 \pmod{m}$ or not. Since $\overline{B}n \equiv 1 \pmod{m}$ is equivalent to $n \equiv B \pmod{m}$, this establishes (6.5.16).

The relation (6.5.11) may be viewed as a numerical test for whether an element of $R(m)$ is the identity element. There is a useful generalization of (6.5.11) that provides a test of whether an element of $R(m)$ is in a given subgroup of $R(m)$. This requires the following definition.

Definition 6.5.3. For a given subgroup H of $R(m)$, let $H^\#$ denote the subset of characters $\chi \in \Gamma(m)$ that are such that

$$\chi(h) = 1 \qquad \text{for all } h \in H.$$

Clearly $H^\#$ is a subgroup of $\Gamma(m)$.

We also require a generalization of Lemma 6.5.2 which is as follows.

Lemma 6.5.3. Let H be a given subgroup of $R(m)$, and t an element of $R(m)$ that is not in H. Then there exists a character χ in $H^\#$ such that $\chi(t) \neq 1$.

Proof. This lemma will be proved by direct construction as in the proof of Lemma 6.5.2. Note that

$$R(m)/H = C_1 \times \cdots \times C_r$$

is a direct product of cyclic subgroups $C_j = \{Hs_j\}$. For the given $t \in R(m)$, we have a representation of the form

$$(6.5.17) \qquad t = c_1 \cdots c_r h, \quad Hc_j \in C_j, \quad j = 1, \ldots, r, \quad h \in H.$$

Further, since $t \notin H$, it follows that for some j^*,

$$(6.5.18) \qquad c_{j^*} \notin H.$$

Since C_j is cyclic as noted above, we have

(6.5.19) $$c_j = s_j^{t_j}$$

for $j = 1, \ldots, r$, (Hs_j is a generator of C_j). Then every element g of $R(m)$ is expressible in the form

(6.5.20) $$g = s_1^{u_1} s_2^{u_2} \cdots s_r^{u_r} h'$$

where $h' \in H$, and the u_j are uniquely determined in the range $0 \leq u_j < \text{ord}_H s_j$. Thus the value of $\chi(s_j), j = 1, \ldots, r$, and $\chi(h'), h' \in H$, determine a character χ on $R(m)$. Taking $\chi(h') = 1$ for all $h' \in H$, $\chi(s_j) = 1, j \neq j^*, j = 1, \ldots, r$, and

$$\chi(s_{j^*}) = e^{(2\pi i/\sigma)}$$

where $\sigma = \text{ord}_H s_{j^*}$, it is easily verified that χ is a character on $R(m)$, $\chi \in H^{\#}$. We see from (6.5.17), (6.5.19), that

(6.5.21) $$\chi(t) = e^{(2\pi i/\sigma)t_{j^*}}.$$

Then (6.5.18) and (6.5.19) provide $s_{j^*}^{t_{j^*}} \notin H$, so that t_{j^*} is not divisible by $\sigma = \text{ord}_H s_{j^*}$. Thus from (6.5.21) we see that $\chi(t) \neq 1$, as desired.

The aforementioned generalization of (6.5.11) is given by the following theorem.

Theorem 6.5.3.   For H a subgroup of $R(m)$, $s \in R(m)$,

(6.5.22) $$\frac{1}{\text{ord } H^{\#}} \sum_{\chi \in H^{\#}} \chi(s) = \begin{cases} 1 & \text{if } s \in H \\ 0 & \text{if } s \notin H. \end{cases}$$

Proof.   If $s \in H$, $\chi(s) = 1$ for all $\chi \in H^{\#}$ and (6.5.22) is immediate in this case. On the other hand, if $s \notin H$, Lemma 6.5.3 provides the existence of a character $\psi \in H^{\#}$ such that $\psi(s) \neq 1$. Then as χ ranges over all the characters of $H^{\#}$, so does $\psi\chi$, so that

$$\sum_{\chi \in H^{\#}} \chi(s) = \sum_{\chi \in H^{\#}} (\psi\chi)(s) = \psi(s) \sum_{\chi \in H^{\#}} \chi(s)$$

or

$$(1 - \psi(s)) \sum_{\chi \in H^{\#}} \chi(s) = 0,$$

and hence

$$\sum_{\chi \in H^{\#}} \chi(s) = 0.$$

Corollary. For H any subgroup of $R(m)$, we have

$$(\text{ord } H)(\text{ord } H^\#) = \phi(m).$$

Proof. By (6.5.10),

$$\sum_{\chi \in H^\#} \sum_{s \in R(m)} \chi(s) = \phi(m),$$

and by Theorem 6.5.3,

$$\sum_{s \in R(m)} \sum_{\chi \in H^\#} \chi(s) = (\text{ord } H)(\text{ord } H^\#).$$

Since both of the above double sums are the same, the corollary follows.

Note that for $H = $ the identity, $H^\# = \Gamma(m)$ so that the assertion (6.5.22) becomes (6.5.11).

EXERCISES

1. For any finite abelian group G, define a character χ of G to be a complex valued function $\chi(g)$ defined over $g \in G$ that is not identically zero and such that

$$\chi(g_1 g_2) = \chi(g_1)\chi(g_2).$$

   a. Show that the function $\chi_0(g) = 1$ for all g is a character. This is called the principal character.

   b. Show that for all characters of G, $\chi(I) = 1$ (where I is the identity of G).

   c. For χ a character of G, define $\bar{\chi}(n) = \overline{\chi(n)}$. Show that $\bar{\chi}$ is also a character of G.

   d. For two characters χ_1, χ_2 of G, define the product $\chi_1\chi_2$ by

$$\chi_1\chi_2(g) = \chi_1(g)\chi_2(g).$$

   Show that $\chi_1\chi_2$ is a character, and that with respect to this notion of character multiplication, the characters of G, which we denote by $\Gamma(G)$, form a finite abelian group.

2. a. For G a finite abelian group, H a subgroup of G, let $H^\#$ denote the set of characters of G such that $\chi(g) = 1$ for all $g \in H$. Show that $H^\#$ is a subgroup of $\Gamma(G)$. Note, in particular, that for $H = $ the identity, $H^\# = \Gamma(G)$, and also $G^\# = \chi_0$.

   b. For Λ a subgroup of the character group $\Gamma(G)$, let $\Lambda^b = $ the set of elements of G that are such that $\chi(g) = 1$ for all $\chi \in \Lambda$. Show that Λ^b is a subgroup of G.

c. Show that for H any subset of G,

$$(H^\#)^b \supset H.$$

d. Show that for Λ any subset of Γ,

$$(\Lambda^b)^\# \supset \Lambda$$

e. For H any subset of G, Λ any subset of Γ, show that we have

$$((H^\#)^b)^\# = H^\#$$
$$((\Lambda^b)^\#)^b = \Lambda^b.$$

3. Given a finite abelian group G and subgroup H, show that for $s \in G$, $s \notin H$, there exists a character $\chi \in H^\#$ such that $\chi(s) \neq 1$. (Note that after establishing the case $H = I$, the general case follows by applying this to the factor group G/H.)

4. For H a subgroup of a finite abelian group G, show that

$$\frac{1}{\operatorname{ord} H^\#} \sum_{\chi \in H^\#} \chi(s) = \begin{cases} 1 & \text{if } s \in H \\ 0 & \text{if } s \notin H \end{cases}.$$

5. For Λ a subgroup of $\Gamma(G)$, G a finite abelian group, show that

$$\frac{1}{\operatorname{ord} \Lambda^b} \sum_{s \in \Lambda^b} \chi(s) = \begin{cases} 1 & \text{if } \chi \in \Lambda \\ 0 & \text{if } \chi \notin \Lambda \end{cases}.$$

6. Show that for H a subgroup of a finite abelian group G,

$$(H^\#)^b = H;$$

also, for Λ a subgroup of $\Gamma(G)$,

$$(\Lambda^b)^\# = \Lambda.$$

7. Show that for H a subgroup of a finite abelian group G, there exists a "natural identification" between $\Gamma(G/H)$ (i.e., the characters of G/H and the elements of $H^\#$). From this deduce that $\Gamma(G/H) \sim H^\#$.

8. For H a subgroup of a finite abelian group G, show that
   a. $G/H \sim H^\#$
   b. $\Gamma(G)/H^\# \sim H.$

Thus, in particular, $G \sim \Gamma(G)$.

9. For Λ a subgroup of $\Gamma(G)$, G a finite abelian group, show that $G/\Lambda^b \sim \Lambda$ and $\Gamma(G)/\Lambda \sim \Lambda^b$.

10. A character of an abelian group is called *real* if all of its values are real. Determine the number of real characters of $R(m)$.

§6.5A PRIMITIVE CHARACTERS

In §6.4, it was seen that there is a natural isomorphism of the group $R(m)$ with a factor group of the group $R(N)$ for N a multiple of m. this stemmed from the fact that the residue classes of $R(N)$ that are included as sets of integers in the identity class of $R(m)$ form a subgroup $H_{N,m}$ of $R(N)$ of order $\phi(N)/\phi(m)$. The cosets of $H_{N,m}$ in $R(N)$ are then precisely the collections of reduced residues modulo N that are contained in each residue class modulo m, and it is easily seen that

$$R(N)/H_{N,m} \sim R(m).$$

With this structural relationship in view, there is a simple construction for "extending" a character of $R(m)$ to a character of $R(N)$ (where N is a multiple of m). Starting with a character χ of $R(m)$, define the character $\hat\chi$ of $R(N)$ by defining for $\hat s \in R(N)$,

$$\hat\chi(\hat s) = \chi(s),$$

where s is the residue class of $R(m)$ that contains $\hat s$. We will then say that $\hat\chi$ *is an extension of* χ from $R(m)$ to $R(N)$. This then leads to the following definition.

Definition 6.5A.1. A character of $R(N)$ is called *primitive* modulo N if and only if it is not the extension of a character of $R(m)$ for m any proper divisor of N. A character modulo N is called primitive if the character of $R(N)$ to which it corresponds is primitive.

The principal character χ_0 is primitive modulo 1, since every principal character is an extension of the principal character of $R(1)$. For $n = p$ a prime, the only divisor $m < p$ of p is 1, so that modulo a prime every nonprincipal character is primitive. Thus for p an odd prime, $R(p)$ has exactly $p - 2$ primitive characters.

The first question that arises is the characterization of those characters of $R(N)$ that are extensions of a character of $R(m)$ for m a divisor of N.

Theorem 6.5A.1. For m a divisor of N, the characters of $R(N)$ that are extensions of characters of $R(m)$ are precisely those in the subgroup $H_{N,m}^{\#}$ of the character group of $R(N)$. Hence, their number equals $\phi(m)$.

Proof. The characters of $R(N)$ that extend those of $R(m)$ must be constant on each reduced residue class modulo m. Since $H_{N,m}$ contains the elements of $R(N)$ that are contained in the identity class of $R(m)$, and the inclusion in the other classes is by cosets of $H_{N,m}$, it follows that an extension of a character from $R(m)$ to $R(N)$ is recognized by its being constant on the elements of $H_{N,m}$; hence equal to 1 on these. Thus the extended characters are precisely $H_{N,m}^{\#}$.

Since

$$\operatorname{ord} H_{N,m}^{\#} = \frac{\operatorname{ord} R(N)}{\operatorname{ord} H_{N,m}} = \frac{\operatorname{ord} R(N)}{\operatorname{ord} R(N)/\operatorname{ord} R(m)} = \operatorname{ord} R(m) = \phi(m),$$

the theorem is proved.

Since $\phi(m)$ is the total number of characters of $R(m)$, as might have been easily anticipated, all the characters of $R(m)$ extend to characters of $R(N)$, for N a multiple of m.

Corollary 1. A character χ modulo N is primitive if and only if N is its smallest period, that is, the smallest positive integer such that for all integers n,

(6.5A.1) $$\chi(n + N) = \chi(n).$$

Proof. Let m be the smallest positive period of χ. Since the periods form an ideal, it follows that they are all multiples of m; since N is a period, m must divide N. Then, for $n \equiv 1 \pmod m$, $(n, N) = 1$, we have $\chi(n) = \chi(1) = 1$. Viewing χ as a character of $R(N)$, this asserts that $\chi \in H_{N,m}^{\#}$, so that χ extends a character of $R(m)$. Thus χ is primitive if and only if $m = N$, which is the statement of the corollary.

Corollary 2. If χ is a character of $R(N)$ (or a character modulo N), there is a unique divisor m of N and primitive character χ^* of $R(m)$ such that χ extends χ^*.

Proof. Viewing χ as a character modulo N, let m be its smallest period. The proof of Corollary 1 then provides that χ is the extension of a primitive character χ^* modulo m. If χ were the extension of a primitive character ψ modulo m_1, then m_1 is a period of χ so that m divides m_1, which in turn divides N. Also, for $n \equiv 1 \pmod m$, $(n, N) = 1$,

$$\psi(n) = \chi(n) = \chi^*(n) = 1.$$

Since every reduced residue class mod m_1 contains integers relatively prime to N, it follows that for $n \equiv 1 \pmod m$, $(n, m_1) = 1$, we must have $\psi(n) = 1$. Thus ψ is an extension of χ^*, and since they are both primitive, it follows that $m = m_1$, $\psi = \chi^*$.

Theorem 6.5A.2. The number of primitive characters modulo N equals

(6.5A.2) $$\sum_{d/N} \mu(d)\phi\left(\frac{N}{d}\right)$$

where ϕ is the Euler function.

Proof. Let $p(u)$ denote the number of primitive characters modulo u. Every character modulo N is the extension of a primitive character modulo some divisor d of N. Further, for each divisor d of N, all of the primitive characters modulo d extend to characters modulo N. Since there are $\phi(N)$ characters modulo N, we obtain for all N that

$$\phi(N) = \sum_{d/N} p(d).$$

By the Möbius inversion formula, (6.5A.2) follows immediately.

For $N = p$, a prime, the formula (6.5A.2) equals $p - 2$, as was seen previously. For $N = p^\alpha$, p a prime, $\alpha \geq 2$, this equals

$$p^\alpha - 2p^{\alpha-1} + p^{\alpha-2} = p^{\alpha-2}(p - 1)^2.$$

Since the expression (6.5A.2) is multiplicative, we see that for $N = \prod p_i^{\alpha_i}$, the number of primitive characters modulo N equals

(6.5A.3) $$\prod_{\alpha_i > 1} p_i^{\alpha_i-2}(p_i - 1)^2 \prod_{\alpha_i = 1} (p_i - 2).$$

EXERCISES

1. Show that there exist primitive characters modulo m if and only if m is either odd or divisible by 4.

2. Show that the only prime power moduli for which there exist real primitive characters are 4, 8, and odd primes p. For modulo 8 there are two such characters and in each of the other cases, exactly one.

3. For a fixed positive integer d, derive a formula for the number of primitive characters modulo N that are solutions of $\chi^d = \chi_0$.

§6.6 POWER RESIDUES

If the congruence

(6.6.1) $$x^d \equiv b \pmod{m}$$

has a solution for x, then b is called a *dth power residue* modulo m. For $d = 2, 3, 4$, the special terminology quadratic, cubic, and quartic residue, respectively, is used.

Special interest has focused on the case where m is a prime p, $(b, p) = 1$, and d divides $p - 1$. For this case, a simple criterion for *dth* power residues is given by the following theorem.

Theorem 6.6.1. An integer b, $(b,p) = 1$, p a prime, is a dth power residue modulo p, where $d/(p-1)$, if and only if

$$(6.6.2) \qquad\qquad b^{(p-1)/d} \equiv 1 \pmod{p}.$$

Proof. If (6.6.1) holds, then raising both sides to the $(p-1)/d$th power yields

$$b^{(p-1)/d} \equiv x^{p-1} \equiv 1 \pmod{p}$$

since $(x,p) = 1$ (by Fermat's theorem).

Conversely, if (6.6.2) holds, for g a primitive root modulo p, $b \equiv g^u \pmod{p}$, so that

$$g^{u[(p-1)/d]} \equiv 1 \pmod{p}.$$

Since the order of g modulo p is $p-1$, it follows that $p-1$ divides $u((p-1)/d)$. This in turn implies that u/d is an integer, so that

$$b \equiv g^u = (g^{u/d})^d \pmod{p}$$

and b is a *d*th power residue modulo p.

The dth power residues clearly form a subgroup H_d of $R(p)$. From Theorem 6.1.1, it follows that this subgroup is the cyclic subgroup of order $(p-1)/d$. Thus the character subgroup $H_d^\#$ is or order d. From this we may formulate a numerical test for dth power residues, as follows.

Theorem 6.6.2. Let p be a prime, d a divisor of $p-1$, then

$$(6.6.3) \qquad \frac{1}{d} \sum_{\chi^d = \chi_0} \chi(b) = \begin{cases} 1 & \text{if } b \text{ is a } d\text{th power residue mod } p \\ 0 & \text{otherwise,} \end{cases}$$

where the summation is over all characters modulo p such that $\chi^d = \chi_0$.

Proof. the assertion (6.6.3) is precisely (6.5.22) provided we can show that $\chi \in H_d^\#$ is equivalent to $\chi^d = \chi_0$. Since for g a primitive root modulo p, $g^d \in H_d$, $\chi \in H_d^\#$ implies that $\chi(g)^d = 1$ which, in turn, implies that $\chi^d = \chi_0$. Conversely, if $\chi^d = \chi_0$ and $b \equiv u^d \pmod{p}$, then

$$\chi(b) = \chi(u^d) = \chi^d(u) = 1$$

so that $\chi \in H_d^\#$.

EXERCISES

1. Give a direct proof of (6.6.3) without making use of (6.5.22).

2. Formulate and prove a generalization of Theorem 6.6.1 in which the hypothesis

$d/(p - 1)$ is removed. Generalize this, in turn, to the case of a prime power modulus.

NOTES

§6.1. The earliest proponents of a group theoretic approach to reduced residue classes were F. Frobenius and G. Stickelberger [6.1]. In 1878 they investigated the $\varphi(M)$ classes of rational integers modulo M and indicated that they formed a finite group under multiplication. In 1888, H. W. Lloyd Tanner [6.2] determined all the subgroups of this multiplicative group and also expressed it as a direct product of cyclic groups.

Tanner was motivated by a remark of J. J. Sylvester concerning primitive roots in an 1879 paper on "cyclotomic functions" [6.3]. Sylvester himself did not use group theory. In fact, he used the word "group" to mean a set. Sylvester gave several examples involving powers of reduced residues modulo n. To treat this type of system, Tanner adapted the methods of group theory as given by E. Netto [6.4]. Tanner cites Netto as the source of the tabular arrangement in which the cosets of a subgroup are used to obtain factor groups. Netto, in turn, attributes this to Cayley [6.5].

Other early explicit occurrences of the multiplicative group properties of reduced residues are found in:

(i) K. Zsigmondy, 1896 [6.6], who investigated the number, sum, and product of the roots of $x^m \equiv 1$ (mod n).

(ii) H. Weber, 1899 [6.7], who used it to deduce Euler's Theorem. Weber called reduced residue systems the most important example of finite abelian groups.

(iii) G. A. Miller, 1903 [6.8, 6.9], used these ideas to generalize Wilson's Theorem and to study the Euler Function.

(iv) M. Cipolla, 1907 [6.10], used group properties of reduced residues to deal with congruences of the form $x^n \equiv a$ (mod p^m), p a prime, where n divides $\varphi(p^m)$.

§6.1. Theorem 6.1.2 is generally attributed to A. Cauchy who proved a specialized version of it in 1844 [6.11].

§6.1A. Euler, in 1775, seems to have been the first to state that a progression of the form $km + 1$ contains infinitely many primes [6.12].

In 1808, Legendre gave an incorrect proof that $2am + b$ represents infinitely many primes if $(2a, b) = 1$ [6.13].

The general theorem on the infinitude of primes in arithmetic progressions $am + b$ with $(a, b) = 1$ was proved by Dirichlet in 1837 [6.14].

Following Dirichlet, more elementary proofs were sought for the cases $km + 1$. The earliest such proofs involved the special case where m is a prime. (W. Emerson has pointed out that this is, in fact, an easy consequence of the assertion of Exercise 10 of §5.1.) For this case, V. A. Lebsgue gave a proof in 1843 using the fact that all prime factors of $(x^m - y^m)/(x - y)$ are of the form $2km + 1$ [6.15]. The first simple complete proof, for arbitrary m, (that $km - 1$ contains infinitely many primes), was given by A. Genocchi in 1868 [6.16]. He utilized the functions $A_n(x)$ and $B_n(x)$ defined by: $(x + \sqrt{b})^n = A_n(x) + B_n(x) \sqrt{b}$. Numbers of this form had been used by Euler in 1773 [6.17]. The divisibility properties of such functions had been previously studied by Gauss [6.18] and Lagrange [6.19]. Also Dirichlet used them in 1827 to prove that all prime divisors of $(x^p - 1)/(x - 1)$ have the form $2 mp + 1$ [6.20]. He also proved that every divisor of $x^{2^L} + 1$ must have the form $2^{L+1}m + 1$.

Genocchi's paper seems to have been overlooked. As late as 1875, L. Kronecker again gives a simple proof for the case of m a prime [6.21]. E. Wendt gave still another proof in 1895 [6.22]. Other proofs of the general case $km + 1$ were given by E. Lucas, 1878 [6.23], A. S. Bang, 1886 [6.24], J. J. Sylvester, 1888 [6.25], and G. D. Birkhoff and H. S. Vandiver, 1903 [6.26].

Lucas used certain sequences of functions u_n and v_n in his proof. Genocchi noted in 1875 [6.27] that these were related to his A_n and B_n via

$$u_n = \frac{(a + \sqrt{b})^n - (a - \sqrt{b})^n}{2b} = B_n$$

$$v_n = (a + \sqrt{b})^n + (a - \sqrt{b})^n = 2A_n.$$

Genocchi's report at the scientific congress at Turin in 1875 preceded Lucas' own presentation of divisibility properties of u_n and v_n [6.28]. These properties were similar to those given for A_n and B_n by Genocchi in 1868 [6.29].

The most readily available of the early proofs of the $km + 1$ case is that given by Landau [6.30]. Landau must have been aware of Genocchi's work since his related papers are listed in the bibliography of the "Primzahlen."

Landau also gives an elementary proof of the case $km - 1$. The argument is very similar to that given by M. Bauer in 1905 [6.31]. Though the main argument and notation of Bauer's paper is identical to that given by Landau, Bauer's paper is not referenced by Landau.

§6.1B. Since we have restricted ourselves to the absolute basics of the subject of group theory, the interested reader is referred for more details to one or more of the following references: [6.3, 6.33, 6.34, 6.35, 6.36, 6.37].

As for the origins of the concept of group, certainly that of abelian group is implicit in the work of Euler on power residues and primitive roots [6.38]. In fact, in his generalization of Fermat's Theorem, Euler used properties of the reduced residue system modulo m that are essentially group theoretic properties. However, in spite of this, no abstraction of the notion of group occurred till more than a century later.

Early interest in groups was rooted in the study of substitution groups and the solvability of equations. This led Lagrange in 1760 to Theorem 6.1B.1, which is often called Lagrange's Theorem [6.39]. The other important figure in eighteenth century group theory was P. Ruffini [6.40, 6.37].

Gauss, the greatest nineteenth century mathematician, never explicitly employed the concept of group, but many group theoretic results are found in his work [6.41].

Two significant nineteenth century figures in the development of group theory were Niels Henrik Abel, 1802–1829, and Evariste Galois, 1811–1832, both of whom applied group theory to the solution of fundamental problems in the theory of equations [6.42, 6.43]. In particular, it was Galois who introduced the concepts of coset and invariant subgroup.

A. Cauchy, who was already an established mathematician during the lifetime of Abel and Galois, is often considered "the father of group theory" [6.33]. He was the first to attempt to organize the subject into a cohesive theory. In 1845, he proved Lemma 6.1C.1 [6.44], which had previously been stated by Galois [6.43].

According to Burnside [6.35], the first "connected exposition" of group theory was given by J. A. Serret in 1866 [6.45]. Another major contributor to the theory was L. Kronecker who, in 1870, initiated the formulation of sets of postulates for an abstract group [6.46]. In that same year, C. Jordan published the first separate text devoted to group theory [6.47]. Jordan developed Galois' ideas and introduced the notion of "factorgruppe" or quotient group, along with the G/H notation, in 1873 [6.36].

Equation (6.1B.2) was given by O. Holder in 1889 [6.48]. He, in turn, attributed it to W. Dyck, 1882 [6.49]. However, the notion of the coset decomposition of a group already appears (at least in spirit) in earlier work of C. Gauss and E. Schering [6.41, 6.50].

§6.1C. Lemma 6.1C.1 is given by Frobenius and Stickelberger [6.1] (for the case of substitution groups), and they credit it to Cauchy [6.44].

Theorem 6.1C.1 is contained implicitly in the works of Gauss and Shering, but the first explicit proof is attributed to Kronecker, 1870 [6.46]. Kronecker also gives what is essentially Theorem 6.1C.2.

§6.2. Theorem 6.2.1 was given by Gauss in Article 90 of his *Disquisitiones Arithmeticae*, 1801.

Theorem 6.2.2 is given in Weber [6.7].

§6.3. Lemma 6.3.3 may be found in Gauss' *Disquisitiones Arithmeticae*, Articles 85–88. Article 92 contains the essence of Theorem 6.3.1.

§6.3. Exercise 1, the generalized Wilson Theorem, was discussed by Gauss in Article 78 of the *Disquisitiones Arithmeticae*. Though he omitted the complete proof, he sketched the main idea. G. A. Miller later gave a proof using the elementary theorems of substitution groups [6.51].

The special case $u = 2$ of Exercise 7 first appeared in a seminar of H. Rademacher.

§6.4. That $R(2)$, $R(4)$, $R(p^m)$ and $R(2p^m)$ (p an odd prime) are the only cyclic $R(m)$, was given by

Gauss in Article 92 of the *Disquisitiones Arithmeticae*. Later G. A. Miller gave a group theoretic treatment of this [6.52].

Theorem 6.4.2 is closely related to results obtained by Miller concerning isomorphisms between a group and a subgroup of its group of isomorphisms [6.53].

§6.4A. Vandiver and Birkhoff gave Theorem 6.4A.1 in 1903 [6.26]. Unknown to them, a similar theorem had been proved previously by A. S. Bang [6.24, 6.54].

§6.5. One of the earliest appearances of characters modulo m is in connection with Dirichlet's work concerning the existence of infinitely many primes in arithmetic progressions $am + b$ with $(a, b) = 1$. Dirichlet himself did not explicitly use "characters" in his 1837 proof, but employed the relevant roots of unity in his series [6.14]. However, in Dedekind's edition of Dirichlet's *Vorlesungen über Zahlentheorie* in 1894, Dedekind included a footnote in which he singled out the notion of "character," defined it explicitly, and denoted it by $\chi(n)$ [6.55]. However, he did not give the function a name.

Weber's *Lehrbuch der Algebra*, II, 1899, defined the function $\chi(A)$ as a "Gruppencharakter," and developed some of its elementary properties. Also, a proof of Theorem 6.5.1 is given [6.7].

Kronecker's *Vorlesungen über Zahlentheorie*, I, 1901 (edited by Hensel), also contains a development of the elementary theory of characters. This material was appended by Hensel.

E. Landau's use of the symbol $\chi(n)$ in his texts, together with the terms "charakter" and "Hauptcharakter" most probably led to the subsequent widespread acceptance of the notation and terminology [6.30].

Landau credited G. Torelli, 1901, with playing a major role in applying the theory of functions to the study of prime numbers [6.56]. Landau's treatment of characters in 1905 [6.57] suggests that it was Torelli's use of notation that led to Landau's. This is further supported by a 1918 paper of Landau [6.58], where $\chi(n)$ is introduced in connection with a discussion of Torelli's results.

§6.6. Theorem 6.6.1 is often called Euler's criterion. In 1748, Euler proved that if a is a dth power residue modulo p, then $a^{(p-1)/d} - 1$ is divisible by the prime p. [6.59]. He proved the converse in 1758 [6.60].

REFERENCES

6.1. F. Frobenius and G. Stickelberger, Ueber Gruppen von vertauschbaren Elementen, *Crelle's Journal*, **86**, 1878, 217–262.

6.2. H. W. L. Tanner, On Cyclotomic Functions, *Proceedings of the London Mathematical Society*, **20**, 1888, 63–83.

6.3. J. J. Sylvester, On the Divisors of Cyclotomic Functions, *Americal Journal of Mathematics*, **2**, 1879, 357–393.

6.4. E. Netto, *Gruppen und Substitutionentheorie*, Leipzig: G. J. Goschen'sche Verlagshandlung, 1908.

6.5. E. Netto, *Substitutionentheorie und ihre Anwendung auf die Algebra*, Leipzig: G. J. Goschen'sche Verlagshandlung, 1908 (original: 1882).

6.6. K. Zsigmondy, Beitrage zur theorie Abel'scher Gruppen und ihre Anwendung auf die Zahlentheorie, *Monatshefte Math. Phys.*, **7**, 1896, 185–289.

6.7. H. Weber, *Lehrbuch der Algebra*, 2 vols., 2nd edition, Braunschweig: F. Vieweg, 1899.

6.8. G. A. Miller, A New Proof of the Generalized Wilson's Theorem, *Annals of Mathematics*, **4**, 1903, 188–190.

6.9. G. A. Miller, Note on the Totient of a Number, *American Mathematical Monthly*, **12**, 1905, 41–43.

6.10. M. Cipolla, *Atti della Reale Accademia di Lincei, Rendiconti*, **5**, (16), I, 1907, 732–741.

6.11. A. Cauchy, Mémoire sur les arrangements que l'on peut former avec des lettres données et sur les permutations ou substitutions a l'aide desquelles on passe d'un arrangement a un autre, *Oevres Completes*, Paris: Gauthier-Villars, II, 13, 1932, pp. 171–282 (original: 1844).

6.12.   L. Euler, De summa seriei ex numeris primis formitatae ubi numeri primi formae $4n - 1$ habent signum positivum formae autem, $4n + 1$ signum negativum, *Opera Omnia,* Leipzig: B. G. Teubner, I, 4, pp. 146–162 (original: 1775).

6.13.   A. M. Legendre, *Théorie des Nombres,* 2 vols., 4th edition, Paris: A. Blanchard, 1955 (2nd edition: 1830).

6.14.   G. L. Dirichlet, *Beweis der Satzes dass jede unbegrenzte arithmetische Progression deren erstes Glied und Differenz ganze Zahlen ohne gemeinschaftlichen Factor sind unendliche viele Primzahlen enthalt, Werke,* Leipzig: G. Reimer, 1889, I, pp. 313–342 (original: 1837).

6.15.   V. A. Lebsgue, *Journal de Mathematique,* **2** (7), 1862, 417.

6.16.   A. Genocchi, Intorno ad alcune forme di numeri primi, *Annali di Matematichi,* **2** (2), 1868–1869, 256–267.

6.17.   L. Euler, Resolutio Aequationis $Ax^2 + 2Bxy + Cy^2 + 2Dx + 2Ey + F = 0$ per numeros tam rationales quam integros, *Opera Omnia,* 1917, I, 3, pp. 297–309 (original: 1773).

6.18.   C. F. Gauss, *Disquisitiones Arithmeticae,* Göttingen: Königlichen Gesellschaft der Wissenschaften, 1863, I, pp. 1–466 (original: 1801).

6.19.   J. L. Lagrange, Recherches d'Arithmétique, *Oeuvres,* Paris: Gauthier-Villars, III, pp. 693–795 (original: 1775).

6.20.   G. L. Dirichlet, *De formis linearibus in quibus continentur divisores primi quarundam formularum graduum superiorum, Werke,* Leipzig: G. Riemer, 1889, I, pp. 47–62 (original: 1827).

6.21   L. Kronecker, *Vorlesungen uber Zahlentheorie,* (K. Hensel, Ed.) Leipzig: B. G. Teubner, 1901.

6.22   E. Wendt, Elementarer Beweis des Satzes dass in jeder unbegrenzen arithmetischen Progression $my + 1$ unendlich viele Primzahlen vorkommen, *Journal für die Reine und angewandte mathematik,* **115,** 1895, 85–88.

6.23   E. Lucas, Théorie des Fonctions Numériques Simplement Périodiques, *American Journal of Mathematics,* **1,** 1978, 289–321.

6.24.   A. S. Bang, Taltheoretiske Undersogelser, *Tidsskrift for Mathematik,* **5,** (4), 1886, 70–80, 130–137.

6.25.   J. J. Sylvester, Preuve Elémentaire du Théorème de Dirichlet sur les Progression arithmétiques dans les cas où la Raison est 8 ou 12, *Comptes Rendus,* **106,** 1888, 1278–1281, 1385–1386.

6.26.   G. D. Birkhoff and H. S. Vandiver, On The Integral Divisors of $a^n - b^n$, *Annals of Mathematics,* **5,** 1903–1904, 173–182.

6.27.   A. Genocchi, Cenni di Ricerche intorno di numeri primi, *Atti della Reale Accademia delle Scienze di Torino,* **11,** 1875–1876, 924–927.

6.28.   E. Lucas, Sur la théorie des nombres premiers, *Atti della Reale Accademia delle Scienze di Torino,* **11,** 1875–1896, 928–937.

6.29.   A. Genocchi, Sur les diviseurs de certains polynômes et l'existence de certains nombres premiers, *Comptes Rendus,* **98,** 1884, 411–413.

6.30.   E. Landau, *Handbuch der Lehre von der Verteilung der Primzahlen,* Leipzig: B. G. Teubner, 1909.

6.31.   M. Bauer, Uber die arithmetische Reihe, *Journal für die reine und angewandte Mathematik,* **131,** 1906, 265–267.

6.32.   A. G. Kurosh, *The Theory of Groups,* I, 2nd edition, New York: Chelsea Publishing Company, 1960.

6.33.   G. A. Miller, H. F. Blichfeld, and L. E. Dickson, *Theory and Application of Finite Groups,* 2nd edition, New York: Chelsea Publishing Company, 1958.

6.34.   B. L. van der Waerden, *Modern Algebra,* I, New York: Frederick Ungar Publishing Company, 1953.

6.35.   W. Burnside, *Theory of Groups of Finite Order,* 2nd edition, Cambridge University Press, 1911.

6.36. W. Magnus, Allgemeine Gruppentheorie, *Enzyklopadie der Mathematischer Wissenschaften*, **1**, (2), Heft 4, Teil I, No. 9, 1939, pp. 1–51.

6.37. G. A. Miller, History of the Theory of Groups to 1900, *Collected Works,* Urbana: University of Illinois Press, 1935, pp. 427–467.

6.38. L. Euler, Theoremata circa residua ex divisione potestatum relicta, *Opera Omnia*, 1915, I, 2, pp. 493–518 (original: 1758).

6.39. J. L. Lagrange, Réflexions sur la resolution algébrique des équations, *Oeuvres de Lagrange*, Paris: Gauthier-Villars, IV, 1896, pp. 203–421 (original: 1770).

6.40. P. Ruffini, Teoria generale delle equazioni in cui si dimonstra impossibile la soluzione algebrica delle equazioni generali di grado superiore al quarto, *Tipografia Matematica di Palermo,* 1915 (original: 1799).

6.41. C. F. Gauss, *Demonstration de quelques Théoremes concernant les periodes des classes de formes binaires du second degré, Werke*, Göttingen: Königlichen Gesellschaft der Wissenschaften, 1876, II, pp. 266–268 (original: 1801).

6.42. N. Abel, Beweis der Unmoglichkeit algebraische Gleichungen von hoheren Graden als dem vierten allgemeis ausszulosen, *Crelle's Journal,* **1,** 1826, 65–84.

6.43. E. Galois, *Oevres Mathematiques d'Evariste, Galois* (Emile Picard ed.) Paris: Gauthier-Villars, 1897.

6.44. A. Cauchy, Mémoire sur diverses propriétés remarquable des substitutions régulières ou irrégulières et des systèmes de substitutions conjugées, *Oeuvres,* Paris: Gauthier-Villars, 1896, I, 9, pp. 342–360 (original: 1845).

6.45. J. A. Serret, *Cours d'Algèbre,* 6th edition, Paris: Gauthier-Villars, 1910 (1st edition: 1866).

6.46. L. Kronecker, *Auseinandersetzung einiger eigenschaften der klassenanzahl idealer complexen Zahlen, Werke,* Leipzig: B. G. Teubner, 1895, I, pp. 273–282.

6.47. C. Jordan, *Traité des Substitutions et des Equations Algébriques,* Paris: Gauthier-Villars, 1957, (original: 1870).

6.48. O. Holder, Zuruckfuhrung einer belieben algebraischen Gleichung auf eine Kette von Gleichungen, *Mathematische Annalen,* **34,** 1889, 26–56.

6.49. W. Dyck, Gruppentheorie Studien I, *Mathematische Annalen,* **20,** 1882, 1–44.

6.50. E. Schering, *Die fundamental klassen der zusammensetzbaren arithmetischen Formen, Gesammelte Mathematische Werke,* Berlin: Mayer und Müller, 1902, I, pp. 135–148 (original: 1868).

6.51. G. A. Miller, A New Proof of the Generalized Wilson's Theorem, *Annals of Mathematics,* **4,** 1903, 188—190.

6.52. G. A. Miller, On Holomorphisms and Primitive Roots, *Bulletin of the American Mathematical Society,* **7,** 1901, 350–354.

6.53. G. A. Miller, Note on the Group of Isomorphisms, *Bulletin of the American Mathematical Society,* **6,** 1900, 337–339.

6.54. A. S. Bang, Om Primtal af bestemte Former, *Nyt Tidsskrift for Mathematik,* **2B,** 1891, 73–82.

6.55. G. L. Dirichlet, *Vorlesungen uber Zahlentheorie,* (R. Dedekind, ed.) Braunschweig: F. Vieweg (2nd edition: 1871; 4th edition: 1894).

6.56. G. Torelli, Sulla totalita dei numeri primi fino a un limite assegnato, *Atti della Reale Accademia delle Scienze Fisiche e Matematiche di Napoli, ser. 2,* **11** (1), 1901.

6.57. E. Landau, Uber ein Satz von Tchebyschev, *Mathematische Annalen,* **61,** 1905, 527–550.

6.58. E. Landau, Uber einige altere Vermatungen und Behauptungen in der Primzahltheorie, *Mathematische Zeitschrift,* **1,** 1918, 1–24.

6.59. L. Euler, Theoremata circa divisores numerorum, *Opera Omnia,* I, 2, pp. 62–85 (original: 1747).

6.60. L. Euler, Demonstrationes circa residua ex divisione potestatum per numeros primos resultantia, *Opera Omnia,* I, 3, pp. 240–281 (original: 1773).

7

QUADRATIC CONGRUENCES

§7.1 THE GENERAL QUADRATIC CONGRUENCE

In §5.2 we have already noted the general polynomial congruence as given in (5.2.6). In this chapter we wish to focus on the special case where $f(x)$ is quadratic; that is, a congruence of the form

$$(7.1.1) \qquad ax^2 + bx + c \equiv 0 \ (\text{mod } m),$$

where a, b, c are integers and $a \not\equiv 0 (\text{mod } m)$. We've already seen in §5.5 that the solution of (7.1.1) reduces to consideration of the case where m is a prime power. (This, of course, stems from the Chinese Remainder Theorem.) For the moment then, let us take $m = p^\alpha$, p a prime, so that (7.1.1) becomes

$$(7.1.2) \qquad ax^2 + bx + c \equiv 0 \ (\text{mod } p^\alpha),$$

where we assume that p is odd and does not divide a. (We postpone for later consideration the cases where either $p = 2$ or p divides a.)

We process (7.1.2) by completing the square. Multiplying by $4a$ yields the equivalent form

$$(2ax + b)^2 \equiv b^2 - 4ac \ (\text{mod } p^\alpha),$$

or setting $D = b^2 - 4ac$, $y = 2ax + b$, we have

$$(7.1.3) \qquad y^2 \equiv D \ (\text{mod } p^\alpha).$$

Clearly, the solutions of (7.1.3) yield the solutions of (7.1.2) via the relation

$$x \equiv (2a)^{-1}(y - b) \ (\text{mod } p^\alpha).$$

By the familiar process of §5.5, the solution of (7.1.3) may be reduced to the case of prime modulus. Namely, each solution of (7.1.3) is congruent mod p to a solution of

$$(7.1.4) \qquad\qquad y^2 \equiv D \pmod{p};$$

so that we may focus on extending solutions of (7.1.4) to solutions of (7.1.3). Assume that p does not divide D, and that we have already achieved such an extension to a solution y_0 of

$$y_0^2 \equiv D \pmod{p^{\alpha-1}},$$

where $\alpha > 1$ (and $p \nmid y_0$). An extension of y_0 to a solution mod p^α would be of the form $y_0 + tp^{\alpha-1}$, $0 \le t < p$, so that we require

$$(y_0 + tp^{\alpha-1})^2 \equiv D \pmod{p^\alpha}.$$

This, in turn, is equivalent to

$$y_0^2 + 2tp^{\alpha-1}y_0 \equiv D \pmod{p^\alpha}$$

or

$$2ty_0 \equiv \frac{D - y_0^2}{p^{\alpha-1}} \pmod{p},$$

which has a unique solution for t modulo p (since $(2y_0, p) = 1$). We conclude that in this case if we have a solution to (7.1.4) (recall our assumption that $p \ne 2$), then it extends uniquely to a solution of (7.1.3) for all p^α. Thus we see that, in this case, the number of solutions of (7.1.3) is the same as the number of solutions modulo p of (7.1.4). Since we've assumed that p does not divide D, the number of solutions of (7.1.4) is either zero or two. Clearly, the two solutions correspond to the fact that if y_0 is one solution mod p, then $-y_0$ is another.

Returning to the case where p divides D, we see that the number of solutions to (7.1.3) depends on the exact power of p that divides D. For example, if $D = 0$, (7.1.3) becomes $y^2 \equiv 0 \pmod{p^\alpha}$; the solutions of this are those integers of the form $tp^{\lfloor(\alpha+1)/2\rfloor}$ in the range 0 to $p^\alpha - 1$. That is, $0 \le t < p^{\alpha - \lfloor(\alpha+1)/2\rfloor}$, so that the number of solutions modulo p^α equals $p^{\alpha - \lfloor(\alpha+1)/2\rfloor} = p^{\lfloor\alpha/2\rfloor}$.

If $D \ne 0$, assume that $p^\nu \| D$, $D = p^\nu \overline{D}$ $(\overline{D}, p) = 1$, and $\nu \ge 1$. (Recall that $p^\nu \| D$ denotes the fact that p^ν is the highest power of p dividing D). Then for $\alpha \le \nu$, (7.1.3) reduces to $y^2 \equiv 0 \pmod{p^\alpha}$, which we've seen has $p^{\lfloor\alpha/2\rfloor}$ solutions modulo p^α. For $\alpha > \nu$, (7.1.3) implies that $y^2 \equiv 0 \pmod{p^\nu}$, so that any solution y must be of the form $y = tp^{\lfloor(\nu+1)/2\rfloor}$. Thus we require

$$t^2 p^{2\lfloor(\nu+1)/2\rfloor} \equiv p^\nu \overline{D} \pmod{p^\alpha}$$

or

(7.1.5) $$t^2 p^{2[(\nu+1)/2]-\nu} \equiv \overline{D} \pmod{p^{\alpha-\nu}}.$$

If ν is odd, $2[(\nu + 1)/2] - \nu > 0$, so that if (7.1.5) had a solution for t, it would follow that $\overline{D} \equiv 0 \pmod{p}$, a contradiction. Thus, for ν odd, $\alpha > \nu$, *there are no solutions* to (7.1.3). For ν even, $2[(\nu + 1)/2] - \nu = 0$, and (7.1.5) reduces to $t^2 \equiv \overline{D} \pmod{p^{\alpha-\nu}}$, which has either zero or two solutions modulo $p^{\alpha-\nu}$. In this case, it follows that the number of solutions of (7.1.3) modulo p^α is correspondingly either 0 or $2p^{[\nu/2]}$.

We turn next to the case where the modulus is a power of 2; that is,

(7.1.6) $$ax^2 + bx + c \equiv 0 \pmod{2^\alpha}.$$

To begin, as before, we assume that $(a, 2) = 1$ and, in addition, that $(b, 2) = 1$. If x_0 is a solution modulo $2^{\alpha-1}$, $\alpha > 1$ [i.e., $ax_0^2 + bx_0 + c \equiv 0 \pmod{2^{\alpha-1}}$], we set $x = x_0 + t \cdot 2^{\alpha-1}$, so that

$$a(x_0 + t \cdot 2^{\alpha-1})^2 + b(x_0 + t \cdot 2^{\alpha-1}) + c \equiv 0 \pmod{2^\alpha}$$

or

$$(ax_0^2 + bx_0 + c) + bt \cdot 2^{\alpha-1} \equiv 0 \pmod{2^\alpha}$$

$$bt = -\frac{(ax_0^2 + bx_0 + c)}{2^{\alpha-1}} \pmod{2}.$$

Since we've assumed that b is odd, this has a unique solution for t modulo 2. Thus here each solution of (7.1.6) is the unique extension of a solution for $\alpha = 1$. Since $(a, 2) = (b, 2) = 1$, this, in turn, is

$$x^2 + x + c \equiv c \equiv 0 \pmod{2}$$

which gives that there are either two or zero solutions to (7.1.6) according as c is even or odd.

Next we treat (7.1.6) under the assumption that $(a, 2) = 1$, but that b is even. In this case, we have equivalently

$$\left(ax + \frac{b}{2}\right)^2 \equiv \frac{b^2}{4} - ac \pmod{2^\alpha}$$

which is of the form (7.1.3) with $y = (ax + b/2)$ and $D = b^2/4 - ac$. Consider first the case where D is odd. Then for $\alpha = 1$, $y^2 \equiv D \pmod{2}$ always has one solution; for $\alpha = 2$, $y^2 \equiv D \pmod{4}$ can have solutions only if $D \equiv 1 \pmod{4}$, and then there are two solutions modulo 4. In the case $\alpha = 3$, $y^2 \equiv D$ (mod 8), since the odd squares are $\equiv 1$ (mod 8), there are solutions only if $D \equiv 1$ (mod 8), and, in this case, there are four solutions modulo 8. In the general case, D odd, with $\alpha \geq 3$, we consider the solutions of $y^2 \equiv D$ (mod 2^α) and investigate the possibility of extending these to solutions modulo $2^{\alpha+1}$. the solutions of $y^2 \equiv D$ (mod 2^α) occur in pairs, solutions y_0, $0 < y_0 < 2^{\alpha-1}$, and $y_1 = y_0 + 2^{\alpha-1}$. Moreover, we have

$$y_1^2 - D = y_0^2 - D + 2^\alpha(y_0 + 2^{\alpha-2}) = y_0^2 - D + 2^\alpha s_0,$$

where $s_0 = y_0 + 2^{\alpha-2}$ is odd, since y_0 is odd and $\alpha \geq 3$. Thus

$$\frac{y_1^2 - D}{2^\alpha} = \frac{y_0^2 - D}{2^\alpha} + s_0,$$

so that exactly one of the quantities $(y_0^2 - D)/2^\alpha$, $(y_1^2 - D)/2^\alpha$, is even [i.e., gives a solution of $y^2 \equiv D$ (mod $2^{\alpha+1}$)]. The extensions of a $\hat{y}$ such that $\hat{y}^2 \equiv D$ (mod 2^α) to a solution of $y^2 \equiv D$ (mod $2^{\alpha+1}$) are of the form $\hat{y} + 2^\alpha t$, where $t = 0, 1$. We require

$$(\hat{y} + 2^\alpha t)^2 \equiv D \ (\text{mod } 2^{\alpha+1}),$$

and since $2\alpha \geq \alpha + 1$, this is the same as

(7.1.7) $\qquad\qquad\qquad\qquad \hat{y}^2 \equiv D \ (\text{mod } 2^{\alpha+1}).$

But this is independent of t and we have seen that in the pairing of solutions, y_0, y_1 described above, exactly one of the pair satisfied (7.1.7). Hence exactly one of each such pair has two extensions.

Summarizing the above, for the case where $(a, 2) = 1$, $(b, 2) = 2$, $D = (b^2/4) - ac$ odd, for $\alpha \geq 3$, the number of solutions modulo 2^α of (7.1.6) is 4 if $D \equiv 1$ (mod 8), and 0 otherwise.

If D is even, the argument proceeds much as in the case of an odd prime p. Setting $D = 2^\nu \overline{D}$, $\overline{D}$ odd, $\nu \geq 1$, the congruence $y^2 \equiv D$ (mod 2^α) becomes $y^2 \equiv 2^\nu \overline{D}$ (mod 2^α). Then, for $\alpha \leq \nu$, this asserts that $y^2 \equiv 0$ (mod 2^α) which has $2^{\lfloor \alpha/2 \rfloor}$ solutions modulo 2^α. On the other hand, for $\alpha > \nu$, y must have the form $y = t2^{\lfloor(\nu+1)/2\rfloor}$, so that

(7.1.8) $\qquad\qquad\qquad t^2 2^{2\lfloor(\nu+1)/2\rfloor - \nu} \equiv \overline{D} \ (\text{mod } 2^{\alpha-\nu}).$

Then we see that for ν odd, since $2\lfloor(\nu + 1)/2\rfloor - \nu > 0$, (7.1.8) can have no solutions for t. For ν even, it becomes $t^2 \equiv \overline{D}$ (mod $2^{\alpha-\nu}$), $\alpha - \nu > 0$, $\overline{D}$ odd, a case that has been treated previously.

The remaining detail of treating all the cases of (7.1.1) in which the prime p divides a is given in the following exercises (as Exercise 1).

EXERCISES

1. For p a prime, consider the congruence

    $$(7.1.9) \qquad\qquad p^s \overline{a} x^2 + bx + c \equiv 0 \pmod{p^\alpha},$$

    where $s \geq 1$ and $(\overline{a}, p) = 1$. For $\alpha \leq s$, this reduces to the linear congruence $bx + c \equiv 0 \pmod{p^\alpha}$. For $\alpha > s$, relate the solutions of (7.1.9) to those of $bx + c \equiv 0 \pmod{p^s}$.

2. Analyze the existence and number of solutions modulo p^α, p a prime, for x and y in the congruence $ax^2 + bxy + y^2 \equiv 0 \pmod{p^\alpha}$.

3. Show that for $\alpha \geq 3$, $x^2 \equiv a \pmod{2^\alpha}$ has a solution if and only if $a \equiv 1 \pmod 8$.

§7.2   QUADRATIC RESIDUES

The above discussion justifies the narrowing of our focus to solutions of quadratic congruences of the form (7.1.4). For a given prime p, this is equivalent to considering those integers D that are "squares modulo p."

Definition 7.4.1. For p a prime, an integer b, not divisible by p, is called a *quadratic residue modulo p* if the congruence $x^2 \equiv b \pmod p$ has a solution. Otherwise it is called a *quadratic nonresidue modulo p*. The *Legendre symbol* $\left(\dfrac{b}{p}\right)$ is defined for all b prime to p by

$$(7.2.1) \qquad\qquad \left(\frac{b}{p}\right) = \begin{cases} 1 & \text{if } b \text{ is a quadratic residue mod } p, \\ -1 & \text{if } b \text{ is a quadratic nonresidue mod } p. \end{cases}$$

Given a prime p, it is easy to find the quadratic residues, for we need only consider the squares of a reduced residue system modulo p. For example, for $p = 11$, we have $(\pm 1)^2, (\pm 2)^2, (\pm 3)^2, (\pm 4)^2, (\pm 5)^2$ congruent mod 11 to $1, 4, 9, 5, 3$. Thus these are the quadratic residue classes modulo 11; $2, 6, 7, 8, 10$, represent the quadratic nonresidues modulo 11. Hence, for example, we have $(\frac{5}{11}) = 1$ and $(\frac{2}{11}) = -1$.

Applying Theorem 6.6.1 with $d = 2$ yields *Euler's criterion* to the effect that for p an odd prime, $(b, p) = 1$,

$$(7.2.2) \qquad\qquad b^{(p-1)/2} \equiv \left(\frac{b}{p}\right) \pmod p.$$

We give here a quick direct proof. If $\left(\dfrac{b}{p}\right) = 1$, $b \equiv a^2 \pmod p$, implying $b^{(p-1)/2} \equiv a^{p-1} \equiv 1 \pmod p$. Conversely, if $b^{(p-1)/2} \equiv 1 \pmod p$, for g a primitive root modulo p, $b \equiv g^l \pmod{p}$, hence $b^{(p-1)/2} \equiv g^{l \cdot (p-1)/2} \equiv 1 \pmod{p}$. This implies that 2 divides l, so that $b \equiv (g^{l/2})^2 \pmod{p}$; that is, $\left(\dfrac{b}{p}\right) = 1$.

From (7.2.2) it follows for $(b_1, p) = (b_2, p) = 1$ that

$$\left(\frac{b_1 b_2}{p}\right) \equiv (b_1 b_2)^{(p-1)/2} \equiv b_1^{(p-1)/2} \cdot b_2^{(p-1)/2} \equiv \left(\frac{b_1}{p}\right)\left(\frac{b_2}{p}\right) \pmod{p}$$

Since both sides of the above congruence equal ± 1 and $p > 2$, it follows that

(7.2.3)
$$\left(\frac{b_1 b_2}{p}\right) = \left(\frac{b_1}{p}\right)\left(\frac{b_2}{p}\right).$$

[It is trivial that (7.2.3) also holds for $p = 2$.]

Translating (7.2.3) into words yields:

   (i)   A quadratic residue times a quadratic residue is a quadratic residue.

  (ii)   A quadratic residue times a quadratic nonresidue is a quadratic nonresidue.

 (iii)   A quadratic nonresidue times a quadratic nonresidue equals a quadratic residue.

These can also be proved directly without the use of Euler's criterion.

Other properties of the Legendre symbol that are obvious include: for $(a, p) = 1$

(7.2.4)
$$\left(\frac{a^2}{p}\right) = 1,$$

and for $a \equiv a' \pmod{p}$

(7.2.5)
$$\left(\frac{a}{p}\right) = \left(\frac{a'}{p}\right).$$

Thus we see that the basic problem of determining for a given integer b, those primes p not dividing b for which

$$x^2 \equiv b \pmod{p}$$

has a solution, is that of determining p so that $\left(\dfrac{b}{p}\right) = 1$. This problem is solved by a theorem of Gauss which is as follows.

Theorem 7.2.1. (Quadratic Reciprocity Law)

For p an odd prime we have

(7.2.6) $$\left(\frac{-1}{p}\right) = (-1)^{(p-1)/2}$$

and

(7.2.7) $$\left(\frac{2}{p}\right) = (-1)^{(p^2-1)/8}.$$

Further, for any distinct odd primes p, q,

(7.2.8) $$\left(\frac{p}{q}\right)\left(\frac{q}{p}\right) = (-1)^{(p-1)(q-1)/4}.$$

We note that the assertion (7.2.6), included in the statement of the Quadratic Reciprocity law, is an immediate consequence of Euler's criterion (7.2.2). For applying it with $b = -1$ yields

$$\left(\frac{-1}{p}\right) \equiv (-1)^{(p-1)/2} \pmod{p},$$

and since the prime $p > 2$, (7.2.6) follows.

The Quadratic Reciprocity Law, together with the properties (7.2.3), (7.2.4), and (7.2.5), provides an algorithm for evaluating the Legendre symbol $\left(\frac{a}{p}\right)$. Before illustrating this by an example, note that (7.2.8) is equivalent to

(i) $\left(\frac{p}{q}\right) = \left(\frac{q}{p}\right)$ if either p or q is $\equiv 1 \pmod{4}$

(ii) $\left(\frac{p}{q}\right) = -\left(\frac{q}{p}\right)$ if both p and q are $\equiv -1 \pmod{4}$.

We then have, for example,

$$
\begin{aligned}
\left(\tfrac{30}{199}\right) &= \left(\tfrac{2}{199}\right)\left(\tfrac{3}{199}\right)\left(\tfrac{5}{199}\right) && \text{by (7.2.3)} \\
&= \left(\tfrac{3}{199}\right)\left(\tfrac{5}{199}\right) && \text{by (7.2.7)} \\
&= -\left(\tfrac{199}{3}\right)\left(\tfrac{199}{5}\right) && \text{by (7.2.8)} \\
&= -\left(\tfrac{1}{3}\right)\left(\tfrac{4}{5}\right) && \text{by (7.2.5)} \\
&= -1 && \text{by (7.2.4).}
\end{aligned}
$$

This process is by no means unique and for this example a quicker route would be

$$\left(\tfrac{30}{199}\right) = \left(\tfrac{-169}{199}\right) \qquad \text{by (7.2.5)}$$

$$= (\tfrac{-1}{199})(\tfrac{13^2}{199}) \qquad \text{by (7.2.3)}$$

$$= -1 \qquad\qquad \text{by (7.2.6) and (7.2.4).}$$

In any event, we conclude that 30 is a nonresidue of 199.

This algorithm can be applied in a more generic way to determine, for a given integer $a \neq 0$, all the odd primes p not dividing a such that $\left(\dfrac{a}{p}\right) = 1$. We illustrate this with $a = 15$. Since

$$\left(\frac{15}{p}\right) = \left(\frac{3}{p}\right)\left(\frac{5}{p}\right)$$

we must analyze $\left(\dfrac{3}{p}\right)$ and $\left(\dfrac{5}{p}\right)$ separately. From (7.2.8) we have

$$\left(\frac{5}{p}\right) = \left(\frac{p}{5}\right) = \begin{cases} 1 & \text{if } p \equiv 1,\ 4 \ (\mathrm{mod}\ 5) \\ -1 & \text{if } p \equiv 2,\ 3 \ (\mathrm{mod}\ 5) \end{cases}$$

On the other hand,

$$\left(\frac{3}{p}\right) = (-1)^{(p-1)/2}\left(\frac{p}{3}\right) = \begin{cases} (-1)^{(p-1)/2} & \text{if } p \equiv 1 \ (\mathrm{mod}\ 3) \\ -(-1)^{(p-1)/2} & \text{if } p \equiv -1 \ (\mathrm{mod}\ 3). \end{cases}$$

Putting these together, we find that $\left(\dfrac{15}{p}\right) = 1$ in the following cases:

(i) $p \equiv 1 \ (\mathrm{mod}\ 4)$, and $p \equiv 1 \ (\mathrm{mod}\ 3)$, and $p \equiv 1,\ 4 \ (\mathrm{mod}\ 5)$

(ii) $p \equiv 1 \ (\mathrm{mod}\ 4)$, and $p \equiv -1 \ (\mathrm{mod}\ 3)$, and $p \equiv 2,\ 3 \ (\mathrm{mod}\ 5)$

(iii) $p \equiv -1 \ (\mathrm{mod}\ 4)$, and $p \equiv -1 \ (\mathrm{mod}\ 3)$, and $p \equiv 1,\ 4 \ (\mathrm{mod}\ 5)$

(iv) $p \equiv -1 \ (\mathrm{mod}\ 4)$, and $p \equiv 1 \ (\mathrm{mod}\ 3)$, and $p \equiv 2,\ 3 \ (\mathrm{mod}\ 5)$.

These correspond to $p \equiv \pm 1, \pm 7, \pm 11, \pm 17 \ (\mathrm{mod}\ 60)$. Thus these are precisely the odd primes for which the congruence $x^2 \equiv 15 \ (\mathrm{mod}\ p)$ has a solution.

EXERCISES

1. Solve the following congruences:

    a. $x^2 \equiv 3 \ (\mathrm{mod}\ 23)$

    b. $x^2 \equiv 3 \ (\mathrm{mod}\ 11)$

    c. $x^2 \equiv 5 \ (\mathrm{mod}\ 89)$

    d. $x^2 \equiv 11 \ (\mathrm{mod}\ 157)$

    e. $x^2 \equiv 37 \ (\mathrm{mod}\ 137)$.

2. Prove (7.2.3) directly from the definitions (i.e., without using Euler's criterion).

3. Give a direct proof that for p an odd prime, modulo p, there are $(p - 1)/2$ quadratic residues and $(p - 1)/2$ nonresidues.

4. Prove that for any odd prime p,

$$\sum_{n=1}^{p-1} \left(\frac{n}{p}\right) = 0.$$

5. For what primes p is $\left(\dfrac{11}{p}\right) = 1$?

6. Show that the product of the positive quadratic residues mod p that are less than p is congruent to $(-1)^{(p+1)/2}$ modulo p.

7. Prove that any quadratic residue of an odd prime p is a quadratic residue modulo p^n for any positive integer n.

8. Prove that for $p > 3$, the sum of the positive quadratic residues mod p (that are less than p) is divisible by p.

9. Let the reduced residues modulo an odd prime p be divided into two nonempty sets S and T such that the product of two classes in S or two in T is again in S, whereas the product of one in S and one in T is in T. Prove that S consists of the quadratic residues mod p, and T the nonresidues.

10. Show that for any odd prime p, the congruence $x^2 + y^2 + 1 \equiv 0 \pmod{p}$ always has integer solutions x, y.

11. For any given integer $n > 2$, show that there exist infinitely many primes p such that the congruence $x^2 \equiv p \pmod{n}$ has no solutions.

    $\Bigg[$ *Hint.* dispose first of the case where n is a power of 2. For remaining cases, it suffices to deal with $n = q$ an odd prime. Use the fact that for any fixed integer b, $(b, q) = 1$, t can be chosen so that $tb + 1 = m$ satisfies $\left(\dfrac{m}{q}\right) = -1. \Bigg]$

§7.3 GAUSS' LEMMA

Perhaps the most popular of all proofs of the Quadratic Reciprocity Law is based on a result known as Gauss' Lemma. Motivated by (7.2.2), to the effect that for $(a, p) = 1$,

(7.3.1) $$a^{(p-1)/2} \equiv \left(\frac{a}{p}\right) \pmod{p},$$

and noting its superficial resemblance to Fermat's result (the corollary to Theorem 5.1.3), which states

$$(7.3.2) \qquad\qquad a^{p-1} \equiv 1 \ (\text{mod } p),$$

suggests that the proof of (7.3.2) might shed some light on $\left(\dfrac{a}{p}\right)$. More precisely, since the proof of (7.3.2) deals with a permutation of a complete reduced residue system modulo p consisting of $p - 1$ residues, one is led to consider a similar processing of "half" of a system of such representatives and this, in fact, leads to the desired result.

Theorem 7.3.1. (Gauss' Lemma)

For p an odd prime and q any prime distinct from p, consider the integers $q, 2q, \ldots, ((p - 1)/2)q$, and let $r = $ the number of these whose least positive residue modulo p is greater than $p/2$. Then

$$(7.3.3) \qquad\qquad \left(\frac{q}{p}\right) = (-1)^r.$$

Proof. For the least positive residues mod p of the integers $q, 2q, \ldots,$ $((p - 1)/2)q$, let $a_1, \ldots, a_l$ be those $< p/2$ and $b_1, \ldots, b_r$, those $> p/2$. Hence $l + r = (p - 1)/2$ and $a_1, \ldots, a_l, b_1, \ldots, b_r$ are a permutation mod p of $q, 2q, \ldots, ((p - 1)/2)q$. Multiplying the integers of each set, we obtain

$$q(2q) \cdots \left(\frac{p - 1}{2}\right)q \equiv a_1 \cdots a_l b_1 \cdots b_r \ (\text{mod } p)$$

or

$$(7.3.4) \qquad\qquad q^{(p-1)/2}\left(\frac{p - 1}{2}\right)! \equiv \prod_{i=1}^{l} a_i \prod_{j=1}^{r} b_j \ (\text{mod } p).$$

Since $p/2 < b_j < p$, we have $0 < p - b_j < p/2$. Further, we cannot have $p - b_j = a_i$ for any i. For then

$$0 \equiv a_i + b_j \equiv m_i \cdot q + m_j q = (m_i + m_j)q \ (\text{mod } p)$$

where $0 < m_i < p/2, 0 < m_j < p/2$; which implies $m_i + m_j \equiv 0 \ (\text{mod } p)$, impossible since $0 < m_i + m_j < p$. Thus the integers

$$a_1, \ldots, a_l, p - b_1, \ldots, p - b_r$$

are a permutation of $1, 2, \ldots, (p - 1)/2$, implying

(7.3.5) $$\left(\frac{p-1}{2}\right)! \equiv \prod_{i=1}^{l} a_i \prod_{j=1}^{r}(p - b_j) \pmod{p}.$$

Since $p - b_j \equiv -b_j \pmod{p}$, this in turn gives

(7.3.6) $$\left(\frac{p-1}{2}\right)! \equiv (-1)^r \prod_{i=1}^{l} a_i \prod_{j=1}^{r} b_j \pmod{p}.$$

Multiplying (7.3.4) by $(-1)^r$ and using (7.3.6) yields

$$q^{(p-1)/2}\left(\frac{p-1}{2}\right)! (-1)^r \equiv \left(\frac{p-1}{2}\right)! \pmod{p}.$$

Then since p does not divide $((p - 1)/2)!$, it follows that $q^{(p-1)/2} \equiv (-1)^r \pmod{p}$ and because $p > 2$, (7.3.3) follows from (7.3.1).

Gauss' Lemma is itself a kind of algorithm for evaluating the Legendre symbol. In particular, the case of $\left(\frac{2}{p}\right)$, p is an odd prime, is now quite simple. For here $q = 2$, and $r = $ the number of $2, 4, \ldots, p - 1$ that are $> p/2$. That is, $r = $ the number of integers t such that $p/2 < 2t \le p - 1$ or $p/4 < t \le (p - 1)/2$. Hence

(7.3.7) $$r = \frac{p-1}{2} - \left[\frac{p}{4}\right]$$

and from this it follows easily that

(7.3.8) $$r = \begin{cases} \dfrac{p-1}{4} & \text{for } p \equiv 1 \pmod 4 \\[2mm] \dfrac{p+1}{4} & \text{for } p \equiv -1 \pmod 4 \end{cases}$$

Since $(p^2 - 1)/8 = (p - 1)(p + 1)/(2)(4)$, (7.3.8) implies $(p^2 - 1)/8 \equiv r \pmod 2$, so that (7.3.3) gives (7.2.7). That is,

$$\left(\frac{2}{p}\right) = (-1)^{(p^2-1)/8}.$$

Broken down into cases, this last asserts that $\left(\frac{2}{p}\right) = 1$ for $p \equiv \pm 1 \pmod 8$, and $\left(\frac{2}{p}\right) = -1$ for $p \equiv \pm 3 \pmod 8$.

EXERCISES

1. For any prime $p > 3$, use Gauss' Lemma directly to show that

$$\left(\frac{3}{p}\right) = (-1)^{[p/3]-[p/6]}.$$

2. For any prime $p > 5$, use Gauss' Lemma directly to show that

$$\left(\frac{5}{p}\right) = (-1)^{\left[\frac{p}{5}\right]-\left[\frac{p}{10}\right]+\left[\frac{2p}{5}\right]-\left[\frac{3p}{10}\right]}.$$

§7.4 PROOF OF THE QUADRATIC RECIPROCITY LAW

The assertions (7.2.6) and (7.2.7) of the Quadratic Reciprocity Law have been established and there remains the main statement (7.2.8). A route to this via Gauss' Lemma is provided by extending the argument used above for $q = 2$ to the general case where both p and q are odd primes. In doing this, the idea is to apply the same procedure used to prove Gauss' Lemma itself, but this time focusing on addition instead of multiplication.

We recall that

(i) $r =$ the number of least positive residues modulo p, which are $> p/2$, of the integers $q, 2q, \ldots, ((p-1)/2)q$.

(ii) $a_1, \ldots, a_l$ are the least positive residues modulo p of q, $2q, \ldots, ((p-1)/2)q$, which are $< p/2$.

(iii) $b_1, \ldots, b_r$ are the least positive residues modulo p, of q, $2q, \ldots, ((p-1)/2)q$, which are $> p/2$. Writing

$$(7.4.1) \qquad kq = p\left[\frac{kq}{p}\right] + t_k,$$

we note that as k runs over to 1 to $(p-1)/2$, the t_k sweeps out the set of all a_i and b_j. Thus summing (7.4.1) over i yields

$$q \sum_{k=1}^{(p-1)/2} k = p \sum_{k=1}^{(p-1)/2} \left[\frac{kq}{p}\right] + \sum_{i=1}^{l} a_i + \sum_{j=1}^{r} b_j$$

so that

$$(7.4.2) \qquad \sum_{k=1}^{(p-1)/2} k \equiv \sum_{k=1}^{(p-1)/2} \left[\frac{kq}{p}\right] + \sum_{i=1}^{l} a_i + \sum_{j=1}^{r} b_j \pmod{2}.$$

In the proof of Gauss' Lemma we saw that the a_i together with the $p - b_j$ constitute a permutation of $1, 2, \ldots, (p - 1)/2$. Hence

$$\sum_{K=1}^{(p-1)/2} k = \sum_{i=1}^{l} a_i + \sum_{j=1}^{r} (p - b_j)$$

so that since $p \equiv 1 \pmod 2$,

(7.4.3) $$\sum_{k=1}^{(p-1)/2} k \equiv \sum_{i=1}^{l} a_i - \sum_{j=1}^{r} b_j + r \pmod 2.$$

Adding (7.4.2) and (7.4.3) yields

(7.4.4) $$r \equiv \sum_{k=1}^{(p-1)/2} \left[\frac{kq}{p} \right] \pmod 2.$$

As noted earlier, this may be viewed as an additive version of Gauss' Lemma. From (7.3.3) and (7.4.4), it follows that

$$\left(\frac{q}{p} \right)\left(\frac{p}{q} \right) = (-1)^s$$

where

(7.4.5) $$s = \sum_{i=1}^{(p-1)/2} \left[\frac{iq}{p} \right] + \sum_{j=1}^{(q-1)/2} \left[\frac{jp}{q} \right].$$

Thus our objective, (7.2.8), will be achieved if we can show that

(7.4.6) $$s \equiv \left(\frac{p - 1}{2} \right)\left(\frac{q - 1}{2} \right) \pmod 2.$$

We have

$$\sum_{1 \le i < p/2} \left[\frac{iq}{p} \right] = \sum_{1 \le i < p/2} \sum_{1 \le k \le iq/p} 1$$

and interchanging the order of the i and k summations this

$$= \sum_{1 \le k < q/2} \sum_{pk/q \le i < p/2} 1 = \sum_{1 \le k < q/2} \left(\frac{p-1}{2} - \sum_{i < pk/q} 1 \right).$$

Since for $1 \le k < q/2$, pk/q is never an integer, this last

$$= \frac{p-1}{2} \cdot \frac{q-1}{2} - \sum_{1 \le k < q/2} \sum_{i \le pk/q} 1$$

$$= \frac{p-1}{2} \cdot \frac{q-1}{2} - \sum_{k=1}^{(q-1)/2} \left[\frac{kp}{q} \right],$$

which yields that

(7.4.7) $$s = \sum_{i < p/2} \left[\frac{iq}{p} \right] + \sum_{k < q/2} \left[\frac{kp}{q} \right] = \frac{p-1}{2} \cdot \frac{q-1}{2},$$

so that certainly (7.4.6) holds. This completes the proof of (7.2.8) and of the Quadratic Reciprocity Law.

EXERCISES

1. Show that an equivalent formulation of the Quadratic Reciprocity Law is given by

$$\left(\frac{p}{q} \right) = \left(\frac{(-1)^{(q-1)/2} q}{p} \right)$$

    (original form of Gauss).

2. A popular proof of the identity (7.4.7), attributed to Eisenstein, views $\sum_{1 \le i < p/2} [iq/p]$ as the number of points (u, v) with integer coordinates such that $1 \le v \le uq/p$, $1 \le u < p/2$. Similarly, $\sum_{1 \le j < q/2} [jp/q]$ is the set of such (u, v) with $1 \le u \le vp/q$ and $1 \le v < q/2$. Deduce (7.4.7) by noting that these two sets of (u, v) are disjoint and together comprise exactly those (u, v) such that $1 \le u < p/2$, $1 \le v < q/2$.

3. Give a direct evaluation of $\left(\frac{2}{p} \right)$, p an odd prime, via the following steps:

    a. $\left(\frac{2}{p} \right) = -1$ if $p \equiv \pm 3 \pmod 8$. This is true for $p = 3$. Assume it is false and let $p > 3$ be smallest prime $p \equiv \pm 3 \pmod 8$ such that $\left(\frac{2}{p} \right) = 1$. Then there is a positive odd integer $r < p$ satisfying $r^2 - 2 = p \cdot s$. Show that this yields an odd prime $p' \equiv \pm 3 \pmod 8$, $p' < p$, such that $\left(\dfrac{2}{p}\right) = 1$.

b. $\left(\dfrac{2}{p}\right) = 1$ if $p \equiv -1 \pmod 8$. Since here $\left(\dfrac{-1}{p}\right) = -1$, it suffices to show that $\left(\dfrac{-2}{p}\right) = -1$. We claim that $\left(\dfrac{-2}{p}\right) = -1$ for $p \equiv -1$ or $-3 \pmod 8$. [The case of $p \equiv 3 \pmod 8$ already follows from (a).] This claim is true for $p = 5$. Assume it is false and let $p > 5$ be smallest prime $p \equiv -1$ or $-3 \pmod 8$ such that $\left(\dfrac{-2}{p}\right) = 1$. Then there is an odd positive integer $r < p$ such that $r^2 + 2 = ps$. Show that this yields an odd prime $p' < p$, $p' \equiv -1$ or $-3 \pmod 8$ such that $\left(\dfrac{-2}{p}\right) = 1$.

c. $\left(\dfrac{2}{p}\right) = 1$ for $p \equiv 1 \pmod 8$. Use $x^4 + 1 = (x^2 + 1)^2 - 2x^2$.

4. Determine those primes p of which
   a. -4 is a fourth power residue,
   b. -1 is a fourth power residue.

5. For m a nonzero integer, prove that
   a. $12m^2 - 1$ is divisible by at least one prime of the form $12n + 11$;
   b. if $(m, 3) = 1$, $4m^2 + 3$ is divisible by at least one prime of the form $12n + 7$;
   c. if $(m, 6) = 1$, $4m^2 + 1$ is divisible by at least one prime of the form $12n + 1$;
   d. all prime divisors of $m^4 - m^2 + 1$ are of the form $12n + 1$.

6. Show that each of the arithmetic progressions $12n + 1$, $12n + 5$, $12n + 7$, $12n + 11$, contains infinitely many primes.

§7.4A GAUSS SUMS AND THE QUADRATIC RECIPROCITY LAW

Though the proof of the Quadratic Reciprocity Law given above is the most primitive from a pedagogical point of view, it is perhaps the least penetrating. It is certainly the least suggestive of generalizations. Of the many other proofs that have been given, apart from variations on the theme of Gauss' Lemma, the next most basic proof is one utilizing exponential sums. For a given odd prime p, these so-called Gauss sums are given by

(7.4A. 1) $$\tau(a) = \sum_{h=1}^{p-1} \left(\frac{h}{p}\right) e^{2\pi i a h/p},$$

for any integer a.

If p divides a, then $\tau(a) = 0$. This is easily seen directly since the number of quadratic residues modulo p equals the number of nonresidues. It also follows since the Legendre symbol determines a nonprincipal character mod p (simply extend its definition so that $\left(\dfrac{a}{p}\right) = 0$ if p divides a).

For the case where $(a, p) = 1$, we have

$$\left(\frac{a}{p}\right)\tau(a) = \sum_{h=1}^{p-1} \left(\frac{ah}{p}\right) e^{2\pi i ah/p} = \tau(1)$$

(since ah runs through a reduced residue system) and hence

(7.4A.2) $$\tau(a) = \left(\frac{a}{p}\right)\tau(1).$$

Upon squaring, this yields for $(a, p) = 1$ that

(7.4A.3) $$\tau^2(a) = \tau^2(1).$$

We next focus on the evaluation of $\tau^2(1)$.

Lemma 7.4A.1.

(7.4A.4) $$\tau^2(1) = (-1)^{(p-1)/2} \cdot p$$

Proof.   Note first that the complex conjugate $\overline{\tau(a)}$ satisfies

$$\overline{\tau(a)} = \sum_{h=1}^{p-1} \left(\frac{h}{p}\right) e^{-2\pi i ah/p} = \sum_{h=1}^{p-1} \left(\frac{-h}{p}\right) \cdot e^{2\pi i ah/p} = \left(\frac{-1}{p}\right)\tau(a).$$

Then from (7.4A.3), for any fixed a, $(a, p) = 1$,

$$\tau^2(1) = \tau^2(a) = \tau(a)\overline{\tau(a)}\left(\frac{-1}{p}\right).$$

Thus

$$\tau^2(1) = \left(\frac{-1}{p}\right) \sum_{h=1}^{p-1} \left(\frac{h}{p}\right) e^{2\pi i ah/p} \sum_{h'=1}^{p-1} \left(\frac{h'}{p}\right) e^{-2\pi i ah'/p}$$

or

(7.4A.5) $$\tau^2(1) = \left(\frac{-1}{p}\right) \sum_{h,h'=1}^{p-1} \left(\frac{h}{p}\right)\left(\frac{h'}{p}\right) e^{2\pi i a(h-h')}$$

Noting that the sum on the right of (7.4A.5) vanishes for $a = 0$, and summing (7.4A.5) over $a = 1, \ldots, p - 1$ yields

$$(7.4A.6) \qquad (p - 1)\tau^2(1) = \left(\frac{-1}{p}\right) \sum_{h,h'=1}^{p-1} \left(\frac{h}{p}\right)\left(\frac{h'}{p}\right) \sum_{a=0}^{p-1} e^{2\pi i a(h-h')/p}.$$

Then since

$$\sum_{a=0}^{p-1} e^{2\pi i a t/p} = \begin{cases} p & \text{if } p \text{ divides } t \\ 0 & \text{otherwise}, \end{cases}$$

in (7.4A.6), only the sums where $h \equiv h' \pmod{p}$ contribute. But, in this range, this requires $h = h'$, so that (7.4A.6) becomes

$$(p - 1)\tau^2(1) = p\left(\frac{-1}{p}\right) \sum_{h=1}^{p-1} \left(\frac{h}{p}\right)^2 = p(p - 1)\left(\frac{-1}{p}\right);$$

hence

$$\tau^2(1) = p \cdot \left(\frac{-1}{p}\right) = (-1)^{(p-1)/2} \cdot p.$$

Writing $\tau = \tau(1)$, a proof of the Quadratic Reciprocity Law is achieved by evaluating τ^{q+1} modulo q in two different ways. First, we have from (7.4A.4) that for q an odd prime,

$$\tau^{q+1} = \tau^2 \cdot \tau^{q-1} = \tau^2 \cdot ((-1)^{(p-1)/2} \cdot p)^{(q-1)/2}$$

$$= \tau^2 \cdot (-1)^{(p-1)/2 \cdot (q-1)/2} \cdot p^{(q-1)/2}$$

so that, from (7.2.2), this yields

$$(7.4A.7) \qquad \tau^{q+1} \equiv \tau^2 \cdot (-1)^{(p-1)/2 \cdot (q-1)/2}\left(\frac{p}{q}\right) \pmod{q}.$$

On the other hand,

$$\tau^{q+1} = \tau\left(\sum_{h=1}^{p-1} \left(\frac{h}{p}\right) e^{2\pi i h/p}\right)^q$$

$$= \tau \sum_{h=1}^{p-1} \left(\frac{h}{p}\right)^q e^{2\pi i h q/p} + q\tau J$$

$$= \tau \sum_{h=1}^{p-1} \left(\frac{h}{p}\right) e^{2\pi i h q/p} + q\tau J$$

$$= \tau\tau(q) + \tau qJ,$$

so that

(7.4A.8) $$\tau^{q+1} = \tau^2 \cdot \left(\frac{q}{p}\right) + \tau qJ.$$

Since τ^2 is a rational integer, both $\tau^{q+1} = (\tau^2)^{(q+1)/2}$ and $\tau^2 \cdot \left(\frac{q}{p}\right)$ are rational integers. It then follows from (7.4A.8) that τJ is a rational number whose denominator divides q. Actually, τJ is a rational integer, a fact easily seen within the context of algebraic number theory. However, not having considered such material to this point, we will have to detour around it. For the moment, *assume* that τJ is a rational integer. Then (7.4A.8) yields

$$\tau^{q+1} \equiv \tau^2 \cdot \left(\frac{q}{p}\right) \pmod{q},$$

and inserting this in (7.4A.7) gives

(7.4A.9) $$\tau^2 \cdot \left(\frac{q}{p}\right) \equiv \tau^2 (-1)^{(p-1)/2 \cdot (q-1)/2} \left(\frac{p}{q}\right) \pmod{q}.$$

Since $\tau^2 = (-1)^{(p-1)/2} p$ is prime to q, (7.4A.9) implies

$$\left(\frac{p}{q}\right)\left(\frac{q}{p}\right) \equiv (-1)^{(p-1)/2 \cdot (q-1)/2} \pmod{q},$$

and since $q > 2$, this yields

$$\left(\frac{p}{q}\right)\left(\frac{q}{p}\right) = (-1)^{(p-1)/2 \cdot (q-1)/2}.$$

To complete the proof as given above requires the following lemma.

Lemma 7.4A.2.   The quantity J defined above is such that τJ is a rational integer.

Proof.   To show that τJ is an integer, we've seen that it suffices to prove that its denominator (it is rational) cannot be q. Suppose the contrary, that is, that q does appear in the denominator of τJ. Then q would also appear in the denominator of

$$(\tau J)^2 = \tau^2 J^2 = (-1)^{(p-1)/2} p J^2,$$

implying that it appeared in the denominator of the rational number J^2. We will obtain a contradiction by showing that, in fact, J^2 is a rational integer.

J is defined by the first equation prior to (7.4A.8) in which it appears. Thus we see that

$$qJ = \left(\sum_{h=1}^{p-1} \left(\frac{h}{p} \right) e^{2\pi i h/p} \right)^q - \sum_{h=1}^{p-1} \left(\frac{h}{p} \right)^q e^{2\pi i h q/p}$$

so that by the multinomial theorem,

$$(7.4A.10) \quad J = \sum_{\substack{j_i+\cdots+j_m=q \\ j_\nu > 0 \\ m > 1}} \frac{(q-1)!}{j_1! \cdots j_m!} \sum_{1 \le h_1 < \cdots < h_m \le p-1} \left(\frac{h_1}{p} \right)^{j_1} \cdots \left(\frac{h_m}{p} \right)^{j_m}$$

$$\times \, e^{(2\pi i/p)(j_1 h_1 + \cdots + j_m h_m)}.$$

Setting $\rho = e^{2\pi i/p}$ and using that $\rho^p = 1$, it follows from (7.4A.10) that J^2 has the form

$$(7.4A.11) \qquad\qquad J^2 = \sum_{k=0}^{p-1} A_k \rho^k$$

where the A_k are rational integers. We then see that since $\rho^p = 1$

$$\rho J^2 = \sum_{k=0}^{p-1} A_k \rho^{k+1} = A_{p-1} + \sum_{k=1}^{p-1} A_{k-1} \rho^k.$$

By repeating this process of multiplying by ρ, we obtain that for all j, $0 \le j \le p - 1$, there are integers a_{kj} such that

$$\rho^j J^2 = \sum_{k=0}^{p-1} a_{kj} \rho^k$$

or

$$(7.4A.12) \qquad\qquad -\sum_{\substack{k=0 \\ k \ne j}}^{p-1} a_{kj} \rho^k + (J^2 - a_{jj}) \rho^j = 0.$$

Viewing (7.4A.12) as a system of p linear equations, $j = 0, \ldots, p - 1$, with the nontrivial solution $1, \rho, \rho^2, \ldots, \rho^{p-1}$, it follows that the determinant of the coefficients must vanish. This yields a relation of the form

(7.4A.13) $(J^2)^p + B_{p-1}(J^2)^{p-1} + \cdots + B_1(J^2) + B_0 = 0$

with rational integral coefficients B_ν. Then if $J^2 = r/s$, r, s, positive integers such that $(r, s) = 1$, (7.4A.13) implies

$$r^p + B_{p-1}r^{p-1}s + \cdots + B_1rs^{p-1} + B_0s^p = 0.$$

This, in turn, gives that s divides r^p, and since $(r, s) = 1$ we must have $s = 1$ (i.e., J^2 is an integer as claimed).

§7.4B   THE LEAST POSITIVE NONRESIDUE

A special problem of interest is to obtain an upper bound for the smallest positive quadratic nonresidue of a prime p. If we assume that all the positive integers $\leq x$ are quadratic residues of an odd prime p, we have that

$$[x] = \sum_{n \leq x} \left(\frac{n}{p} \right).$$

Thus if we can derive an upper bound for the sum on the right that conflicts with this equation, we deduce the presence of a quadratic nonresidue $\leq x$. The derivation of such estimates depends on the consideration of the Gauss sums introduced in the previous section. Here we use them to obtain the following simple lemma.

Lemma 7.4B.1.   For a given odd prime p, let $\mathscr{S}$ and $\mathscr{T}$ be sets of integers such that within each set no two of the integers are congruent modulo p. Then

(7.4B.1) $$\left| \sum_{u \in \mathscr{S}, v \in \mathscr{T}} \left(\frac{u + v}{p} \right) \right| \leq \sqrt{pN(\mathscr{S})N(\mathscr{T})},$$

where $N(\mathscr{S})$ and $N(\mathscr{T})$ denote the number of integers in $\mathscr{S}$ and $\mathscr{T}$, respectively.

   Proof.   We recall (7.4A.2) which asserts that

(7.4B.2) $$\tau(1)\left(\frac{a}{p} \right) = \sum_{h=0}^{p-1} \left(\frac{h}{p} \right) e^{2\pi i ha/p}$$

[where we define $\left(\dfrac{h}{p} \right) = 0$ if p divides h]. Setting $a = u + v$, and summing over all $u \in \mathscr{S}$, $v \in \mathscr{T}$ yields

$$\tau(1) \sum_{u \in \mathscr{S}, v \in \mathscr{T}} \left(\frac{u+v}{p}\right) = \sum_{h=0}^{p-1} \left(\frac{h}{p}\right) \sum_{u \in \mathscr{S}} e^{2\pi i h u/p} \sum_{v \in \mathscr{T}} e^{2\pi i h v/p}.$$

Taking absolute values and using Schwarz's Inequality, we have

(7.4B.3) $\displaystyle |\tau(1)| \left| \sum_{u \in \mathscr{S}, v \in \mathscr{T}} \left(\frac{u+v}{p}\right) \right|$

$$\le \sum_{h=0}^{p-1} \left| \sum_{u \in \mathscr{S}} e^{2\pi i h u/p} \right| \left| \sum_{v \in \mathscr{T}} e^{2\pi i h v/p} \right|$$

$$\le \left\{ \sum_{h=0}^{p-1} \left| \sum_{u \in \mathscr{S}} e^{2\pi i h u/p} \right|^2 \cdot \sum_{h=0}^{p-1} \left| \sum_{v \in \mathscr{T}} e^{2\pi i h v/p} \right|^2 \right\}^{1/2}.$$

We note next that

$$\sum_{h=0}^{p-1} \left| \sum_{u \in \mathscr{S}} e^{2\pi i h u/p} \right|^2 = \sum_{h=0}^{p-1} \left(\sum_{u \in \mathscr{S}} e^{2\pi i h u/p} \right) \overline{\left(\sum_{u \in \mathscr{S}} e^{2\pi i h u/p} \right)}$$

$$= \sum_{h=0}^{p-1} \sum_{\substack{u \in \mathscr{S} \\ u' \in \mathscr{S}}} e^{2\pi i h(u-u')/p}$$

$$= \sum_{\substack{u \in \mathscr{S} \\ u' \in \mathscr{S}}} \sum_{h=0}^{p-1} e^{2\pi i h(u-u')/p}.$$

But the inner sum above equals p or 0 according as $u - u' \equiv 0 \pmod{p}$ or not. Since the integers of $\mathscr{S}$ are incongruent mod p, we conclude that

(7.4B.4) $\displaystyle \sum_{h=0}^{p-1} \left| \sum_{u \in \mathscr{S}} e^{2\pi i h u/p} \right|^2 = pN(\mathscr{S})$

and, similarly,

(7.4B.5) $\displaystyle \sum_{h=0}^{p-1} \left| \sum_{v \in \mathscr{T}} e^{2\pi i h v/p} \right|^2 = pN(\mathscr{T}).$

Inserting (7.4B.4) and (7.4B.5) in (7.4B.3) gives

(7.4B.6) $\displaystyle |\tau(1)| \left| \sum_{u \in \mathscr{S}, v \in \mathscr{T}} \left(\frac{u+v}{p}\right) \right| \le p\{N(\mathscr{S})N(\mathscr{T})\}^{1/2}.$

Since $\tau^2(1) = (-1)^{(p-1)/2}p$, it follows that $|\tau(1)| = p^{1/2}$, so that (7.4B.6) yields (7.4B.1).

Using the lemma we can easily prove the following theorem.

Theorem 7.4B.1.   The least positive quadratic nonresidue of an odd prime p is less than $2\sqrt{p} + 1$.

Proof.   Let x be the largest positive integer such that $x < p$ and all the positive integers $\leq x$ are quadratic residues modulo p. Thus if we let $\mathcal{S}$ denote the set of integers from 1 to $[x/2]$, we have

$$(7.4B.7) \qquad \left[\frac{x}{2}\right]^2 = \sum_{\substack{u \in \mathcal{S} \\ v \in \mathcal{S}}} \left(\frac{u + v}{p}\right).$$

On the other hand, applying Lemma 7.4B.1 with $\mathcal{T} = \mathcal{S}$, we obtain

$$(7.4B.8) \qquad \left| \sum_{\substack{u \in \mathcal{S} \\ v \in \mathcal{S}}} \left(\frac{u + v}{p}\right) \right| \leq p^{1/2}\left[\frac{x}{2}\right]$$

Combining (7.4B.7) and (7.4B.8) gives $[x/2]^2 \leq p^{1/2}[x/2]$ or $[x/2] \leq p^{1/2}$.

If x is even, this becomes $x \leq 2p^{1/2}$ and hence $x < 2p^{1/2}$. In fact, this must be the case; that is, except for one trivial possibility, x cannot be odd. For if x were odd, $x + 1$ would be even and the smallest positive nonresidue. If $x + 1 > 2$, then $\left(\frac{2}{p}\right) = 1$ and

$$-1 = \left(\frac{x + 1}{p}\right) = \left(\frac{2}{p}\right)\left(\frac{(x + 1)/2}{p}\right) = \left(\frac{(x + 1)/2}{p}\right)$$

would produce a smaller positive nonresidue than $x + 1$. This leaves only the possibility that $x + 1 = 2$, in which case, $x = 1 < 2\sqrt{p}$.

Note that the smallest nonresidue $n < 2\sqrt{p} + 1$, modulo p, provided by Theorem 7.4B.1 must itself be a prime. For if n were composite, $\left(\frac{n}{p}\right) = -1$ implies the existence of a prime q dividing n (hence $q < n$), such that $\left(\frac{q}{p}\right) = -1$.

There is a "dual problem" in which given a prime q, one asks for an upper estimate on the smallest prime p such that

$$(7.4B.9) \qquad \left(\frac{q}{p}\right) = -1.$$

If $q \equiv 1 \pmod 4$, by the Quadratic Reciprocity Law $\left(\frac{q}{p}\right) = \left(\frac{p}{q}\right)$, so that this new question would be answered by Theorem 7.4B.1. Namely, we would have a prime $p < 2\sqrt{q} + 1$ such that $\left(\dfrac{q}{p}\right) = -1$. In the case $q \equiv -1 \pmod 4$, if we knew that the prime p that is the smallest nonresidue of q were such that $p \equiv 1 \pmod 4$, we could again come through by the reciprocity law. However, this would still leave many cases to be handled and so we use this occasion to once again illustrate the use of Lemma 7.4B.1.

Theorem 7.4B.2. For any given odd prime $q > 3$, there exists an odd prime $p < 8\sqrt{q} + 8$ such that $\left(\dfrac{q}{p}\right) = -1$.

Proof. From the above remarks, we see that for $q \equiv 1 \pmod 4$ this is an immediate consequence of Theorem 7.4B.1 and the Quadratic Reciprocity Law. Assume now that $q \equiv -1 \pmod 4$.

Suppose that the positive integer $x < q$ is such that for all integers $1 \le n \le x$ such that $n \equiv 1 \pmod 4$ we have $\left(\dfrac{n}{q}\right) = 1$. Letting $\mathcal{S} = \{n, 1 \le n \le x/2, n \equiv 1 \pmod 4\}$, and $\mathcal{T} = \{n, 1 \le n \le x/2, n \equiv 0 \pmod 4\}$, it follows that

$$(7.4B.10) \qquad\qquad N(\mathcal{S})N(\mathcal{T}) = \sum_{\substack{u \in \mathcal{S} \\ v \in \mathcal{T}}} \left(\frac{u+v}{q}\right)$$

(since all these $u + v \equiv 1 \pmod 4$, positive, and $\le x$). Then using (7.4B.1) to estimate the right side of (7.4B.10), we obtain

$$N(\mathcal{S})N(\mathcal{T}) \le \sqrt{q N(\mathcal{S})N(\mathcal{T})}$$

or

$$(7.4B.11) \qquad\qquad N(\mathcal{S})N(\mathcal{T}) \le q.$$

But here $N(\mathcal{S}) = [(x+6)/8]$ and $N(\mathcal{T}) = [x/8]$, so that (7.4B.11) implies

$$\left[\frac{x+6}{8}\right]\left[\frac{x}{8}\right] \le q$$

or

$$\left(\frac{x}{8} - \frac{1}{4}\right)\left(\frac{x}{8} - 1\right) < q,$$

which easily gives that $x < 8\sqrt{q} + 7$.

Hence there must be a positive integer $n < 8\sqrt{q} + 8$, such that $n \equiv 1 \pmod 4$ and $\left(\dfrac{n}{q}\right) = -1$, provided of course $8\sqrt{q} + 8 < q$. This last requirement holds for all primes $q > 79$. For the finite number of primes where $8\sqrt{q} + 8 \geq q$, $q \equiv -1$ (mod 4), $q \leq 79$, and it is easily verified that except for $q = 3$; $(\frac{11}{3}) = -1$, and the rest have 5, 13, or 17 as a nonresidue modulo q. Proceeding with $q > 79$, we write $n = n'n''$, where n' contains the primes dividing n that are $\equiv 1$ (mod 4), and n'' those that are $\equiv -1$ (mod 4). We then have the following cases:

(i) for *some* p/n'', $\left(\frac{p}{q}\right) = 1$. Then $\left(\frac{q}{p}\right) = -1$ by the Quadratic Reciprocity Law;

(ii) for *all* p/n'', $\left(\frac{p}{q}\right) = -1$. Since $n \equiv n'' \equiv 1$ (mod 4), the number of primes dividing n'' must be even, implying that $\left(\frac{n''}{q}\right) = 1$. Hence, $\left(\frac{n}{q}\right) = -1$ yields $\left(\frac{n'}{q}\right) = -1$, so that for some prime p/n', $\left(\frac{p}{q}\right) = -1$. This p is $\equiv 1$ (mod 4) and the Quadratic Reciprocity Law gives $\left(\frac{q}{p}\right) = -1$. (Note that this includes the case where $n'' = 1$.)

This completes the proof of the theorem.

Gauss observed that in the case where $q \equiv 1$ (mod 8), one could prove the existence of a prime $< 2\sqrt{q} + 1$ such that $\left(\frac{q}{p}\right) = -1$ *without* using the Quadratic Reciprocity Law. (Note that $2\sqrt{q} + 1 < q$ for $q \geq 7$, and since $q \equiv 1$ (mod 8), $q \geq 17$.) His method proceeded from the assumption that this was false for all positive odd primes $p < 2\sqrt{q} + 1$. That is, for these primes $\left(\frac{q}{p}\right) = 1$. Then for any prime $p < 2\sqrt{q} + 1$, the congruence $x^2 \equiv q$ (mod p) has a solution for x, which in turn implies that $x^2 \equiv q$ (mod p^α) has a solution for any integer $\alpha \geq 1$. Further, since $q \equiv 1$ (mod 8), we know that $x^2 \equiv q$ (mod 2^α) also has a solution for any $\alpha \geq 1$ (Exercise 3 of §7.1). It then follows from the Chinese Remainder Theorem that for M any positive integer all of whose prime factors are $< 2\sqrt{q} + 1$, the congruence

$$x^2 \equiv q \ (\text{mod } M)$$

has a solution for x. In particular, such an M is

$$M = (2m + 1)!, \qquad m = [\sqrt{q}].$$

Then there exists a positive odd integer k such that

(7.4B.12) $q \equiv k^2 \ (\text{mod } M)$

(k is odd since q is odd and M is even). Then from (7.4B.12),

$$k(q - 1^2)(q - 2^2) \cdots (q - m^2) \equiv k(k^2 - 1^2) \cdots (k^2 - m^2) \ (\text{mod } M).$$

But the right side equals

$$(k + m)(k + m - 1) \cdots k \cdots (k - m) = (2m + 1)! \binom{k + m}{2m + 1}$$

and is therefore divisible by M. Since $q > 2m + 1 = 2[\sqrt{q}] + 1$ for all $q \geq 7$, $(q, M) = 1$, and hence $(k, M) = 1$. Thus we obtain that

$$(q - 1^2)(q - 2^2) \cdots (q - m^2) \equiv 0 \pmod{M}.$$

This asserts that

$$\frac{(q - 1^2) \cdots (q - m^2)}{M} = I$$

where I is a rational integer. But since

$$M = (2m + 1)! = (m + 1) \prod_{i=1}^{m} [(m + 1)^2 - i^2],$$

this may be rewritten as

$$(7.4B.13) \qquad \frac{1}{m + 1} \prod_{i=1}^{m} \frac{q - i^2}{(m + 1)^2 - i^2} = I.$$

Since $(m + 1)^2 = ([\sqrt{q}] + 1)^2 > q$, the m factors in the product are each less than 1, implying that $0 < I < 1/(m + 1) < 1$, a contradiction.

EXERCISES

1. As asserted in the proof of Theorem 7.4B.2, verify that for all primes q, $11 < q \leq 103$, $q \equiv -1 \pmod 4$, at least one of 5, 13, or 17 is a nonresidue modulo q.

2. Summing (7.4B.2) over $a = 1, 2, \ldots, x$ yields

$$p^{1/2} \left| \sum_{a \leq x} \left(\frac{a}{p} \right) \right| \leq \left| \sum_{h=0}^{p-1} e^{2\pi i h a / p} \right|.$$

Use this inequality to prove that for a prime p,

$$\left| \sum_{a \leq x} \left(\frac{a}{p} \right) \right| \leq \sqrt{p} \log p.$$

3. Prove that the sum of the quadratic residues modulo an odd prime p of the form $4n + 1$ is $p(p - 1)/4$.

4. For q an odd prime, and $A > B \geq 0$ given integers, obtain an upper estimate for the smallest positive integer n, $n \equiv B \pmod{A}$, such that $\left(\dfrac{n}{q}\right) = -1$.

§7.4C APPLICATION TO A DIOPHANTINE EQUATION

The Quadratic Reciprocity Law arises naturally in many arithmetic problems. Not at all surprising is the fact that this is usually connected to the presence of a "square" somewhere in the problem. An illustration of such an application is in the consideration of various special cases of the diophantine equation

$$(7.4C.1) \qquad\qquad bx^m + ay^2 = k.$$

That is, for given integers a, b, k, m ($m > 0$), we consider the solutions to (7.4C.1) in integers x, y. The method we will pursue is particularly effective for demonstrating that there are no such solutions. The simple idea is that of assuming that k has a representation of the form

$$(7.4C.2) \qquad\qquad bv^m + c = k.$$

Then, under suitable circumstances, the assumption that (7.4C.1) also has integer solutions x, y either leads to a contradiction or "reduces" to a similar diophantine equation with a smaller k. We content ourselves here with several illustrations of this method.

First we consider an example that depends only on the quadratic character of -1 and 2.

Theorem 7.4C.1. The diophantine equation

$$(7.4C.3) \qquad\qquad x^3 - y^2 = 24$$

has no integer solutions.

Proof. This is the special case $b = 1$, $a = -1$, $m = 3$, $k = 24$, of (7.4C.1). Assuming then that (7.4C.3) has integer solutions for x, y, we note that $24 = 2^3 + 4^2$ is of type (7.4C.2) with $b = 1$, $v = 2$, $c = 4^2$. We have several cases.

CASE I. x odd and hence y odd.
  We have $x^3 - y^2 = 2^3 + 4^2$, so that

$$(7.4C.4) \qquad\qquad x^3 - 2^3 = y^2 + 4^2.$$

Since y is odd, every prime dividing the right side of (7.4C.4) must be $\equiv 1 \pmod 4$.

On the other hand, $x^2 + 2x + 4$ divides the left side, and x odd implies (note that $x^2 + 2x + 4 > 1$)

$$x^2 + 2x + 4 \equiv -1 \pmod 4,$$

so that the left side of (7.4C.4) must be divisible by a prime $\equiv -1 \pmod 4$. This contradiction disposes of this case.

CASE II. x even and hence y even.
   Setting $x = 2\bar{x}$, $y = 2\bar{y}$, we obtain from (7.4C.3) that

$$2\bar{x}^3 - \bar{y}^2 = 6.$$

But then $\bar{y}$ is even, and setting $\bar{y} = 2\hat{y}$ gives

(7.4C.5)
$$\bar{x}^3 - 2\hat{y}^2 = 3,$$

where $\bar{x}$ must be odd. We now treat this equation in the same way by using the representation $3 = (-1)^3 + 2^2$. We then have $\bar{x}^3 - 2\hat{y}^2 = (-1)^3 + 2^2$, or

(7.4C.6)
$$\bar{x}^3 + 1 = 2\hat{y}^2 + 2^2.$$

Note first that if p is any odd prime dividing $\bar{x}^3 + 1$ we have $2\hat{y}^2 + 2^2 \equiv 0 \pmod p$, $p \nmid \hat{y}$, and $\left(\dfrac{-2}{p}\right) = 1$. From the reciprocity law, this implies that either $p \equiv 1 \pmod 8$ or $p \equiv 3 \pmod 8$. We will obtain a contradiction by showing that $\bar{x}^3 + 1$ must have an odd prime divisor that is *not* in either of these admissible progressions modulo 8.
   First note that since $\bar{x}$ is odd, (7.4C.5) gives

$$\bar{x} \equiv 2\hat{y}^2 + 3 \equiv \pm 3 \pmod 8.$$

Then $\bar{x}^2 - \bar{x} + 1$ divides $\bar{x}^3 + 1$, and

$$\bar{x}^2 - \bar{x} + 1 \equiv 5 \text{ or } 7 \pmod 8.$$

Since $\bar{x}^2 - \bar{x} + 1$ is odd, and the residue classes $\{1, 3\}$ form a subgroup modulo 8, it follows that $\bar{x}^2 - \bar{x} + 1 > 0$ has at least one odd prime factor p such that $p \equiv 5$ or $7 \pmod 8$. This is a contradiction and completes the proof of the theorem.
   A second example of a slightly different nature is given by the following theorem.

Theorem 7.4C.2. The diophantine equation

(7.4C.7)
$$x^5 - y^2 = 52$$

has no integer solutions.

Proof.   This is a case of (7.4C.1) with $b = 1$, $a = -1$, $k = 52$, $m = 5$. We note the representation $52 = 2^5 + 5 \cdot 2^2$, and assuming that (7.4C.7) has an integer solution x, y, we have

$$(7.4C.8) \qquad\qquad x^5 - 2^5 = y^2 + 5 \cdot 2^2.$$

CASE I. x odd and hence y odd.
    Any prime p dividing $y^2 + 5 \cdot 2^2$ is odd, and will have (for $p \neq 5$)

$$(7.4C.9) \qquad\qquad \left(\frac{-5}{p}\right) = 1.$$

On the other hand, from (7.4C.8), $x \equiv x^5 \equiv 1 + 5 \cdot 2^2 \equiv 5 \pmod 8$; then

$$\frac{x^5 - 2^5}{x - 2} = x^4 + 2x^3 + 4x^2 + 8x + 16 \equiv -1 \pmod 8.$$

Hence $[(x^5 - 2^5)/(x - 2)] > 1$ is divisible by at least one odd prime $p \equiv -1$ (mod 4), which also satisfies (7.4C.9). (Note that this $p \neq 5$.) But then by the reciprocity law,

$$1 = \left(\frac{-5}{p}\right) = -\left(\frac{5}{p}\right) = -\left(\frac{p}{5}\right) = -\left(\frac{1}{5}\right) = -1,$$

a contradiction {since for all $p \,|\, [(x^5 - 2^5)/(x - 2)]$, $p \equiv 1 \pmod 5$ if $p \neq 5$}. This contradiction disposes of this case.

CASE II. x even and hence y even.
    Setting $x = 2\bar{x}$, $y = 2\bar{y}$, yields that $8\bar{x}^5 - \bar{y}^2 = 13$. This, in turn, implies $\bar{y}^2 \equiv 3$ (mod 8), which is impossible.

EXERCISES

1. For any prime $p \equiv 3 \pmod 4$, show that $x^p - y^2 = 3$ has no integer solutions for x, y.

2. Show that $y^2 - x^3 = 5^n$ has an integer solution if and only if $n \not\equiv 5 \pmod 6$.

3. Show that all integer solutions x, y, of $x^3 - y^2 = 6860$ are of the form $x = 7\bar{x}$, $y = 7\bar{y}$, where $\bar{x}$, $\bar{y}$, satisfy $\bar{x}^3 - 7\bar{y}^2 = 20$. Use this to find a solution of the original diophantine equation.

4. Let p be any prime $p \equiv 3 \pmod 8$. Show that not both of the diophantine equations, $x^p - y^2 = p$, $u^p + v^2 = p$, can have a solution.

§7.5 THE JACOBI SYMBOL

For certain conveniences of calculation, the Legendre symbol is extended to a more general notation called the *Jacobi symbol*.

Definition 7.5.1. For $m = p_1, \ldots, p_r$, a product of positive odd primes (not necessarily distinct), and an integer a, $(a, m) = 1$, the Jacobi symbol $\left(\dfrac{a}{m}\right)$ is defined by

$$(7.5.1) \qquad\qquad \left(\frac{a}{m}\right) = \prod_{i=1}^{r} \left(\frac{a}{p_i}\right),$$

where the factors on the right of (7.5.1) are Legendre symbols.

Clearly, in the case where m is a prime, the Jacobi symbol $\left(\dfrac{a}{m}\right)$ is a Legendre symbol. In general, of coure, it is not. However, if $\left(\dfrac{a}{m}\right) = -1$, we see that, for at least one of the factors on the right of (7.5.1), we must have $\left(\dfrac{a}{p_i}\right) = -1$. Hence, in such a case the congruence $x^2 \equiv a$ (mod p_i) would have no solution for x, implying that $x^2 \equiv a$ (mod m) has no solutions. Thus we see that if the Jacobi symbol equals -1, the related quadratic congruence has no solutions (just as with the Legendre symbol). However, if $\left(\dfrac{a}{m}\right) = 1$, this *does not* imply that $x^2 \equiv a$ (mod m) need have a solution, as we see by choosing a prime p such that $\left(\dfrac{a}{p}\right) = -1$ and taking $m = p^2$.

The Jacobi symbol does inherit many of the properties of the Legendre symbol. These we set forth as the following theorem.

Theorem 7.5.1.

(i) If $a \equiv a'$ (mod m), $(a, m) = 1$, then

$$(7.5.2) \qquad\qquad \left(\frac{a}{m}\right) = \left(\frac{a'}{m}\right).$$

(ii) For $(a, m) = (b, m) = 1$, we have

$$(7.5.3) \qquad\qquad \left(\frac{a}{m}\right)\left(\frac{b}{m}\right) = \left(\frac{ab}{m}\right);$$

and, in particular, $\left(\dfrac{a^2}{m}\right) = 1$.

(iii) For $(a, m_1) = (a, m_2) = 1$, we have

$$(7.5.4) \qquad\qquad \left(\frac{a}{m_1 m_2}\right) = \left(\frac{a}{m_1}\right)\left(\frac{a}{m_2}\right).$$

(iv)   The reciprocity law for the Jacobi symbol, including

(7.5.5) $$\left(\frac{-1}{m}\right) = (-1)^{(m-1)/2},$$

(7.5.6) $$\left(\frac{2}{m}\right) = (-1)^{(m^2-1)/8},$$

and for m, n, positive odd integers such that $(m, n) = 1$,

(7.5.7) $$\left(\frac{m}{n}\right)\left(\frac{n}{m}\right) = (-1)^{(m-1)(n-1)/4}.$$

Proof:

(i)   Since $a \equiv a'$ (mod m), $a \equiv a'$ (mod p_i) for every prime p_i dividing m, and hence $(a/p_i) = (a'/p_i)$. Thus

$$\left(\frac{a}{m}\right) = \Pi\left(\frac{a}{p_i}\right) = \Pi\left(\frac{a'}{p_i}\right) = \left(\frac{a'}{m}\right).$$

(ii)   For $m = p_1 \cdots p_r$

$$\left(\frac{ab}{m}\right) = \prod_{i=1}^{r}\left(\frac{ab}{p_i}\right) = \prod_{i=1}^{r}\left(\frac{a}{p_i}\right)\left(\frac{b}{p_i}\right) = \left(\frac{a}{m}\right)\left(\frac{b}{m}\right).$$

(iii)   For $m_1 = p_1 \cdots p_r$, $m_2 = q_1 \cdots q_s$,

$$\left(\frac{a}{m_1 m_2}\right) = \prod_{i=1}^{r}\left(\frac{a}{p_i}\right)\prod_{j=1}^{s}\left(\frac{a}{q_j}\right) = \left(\frac{a}{m_1}\right)\left(\frac{a}{m_2}\right).$$

(iv)   For $m = p_1 \cdots p_r$ we have

$$\left(\frac{-1}{m}\right) = \prod_{i=1}^{r}\left(\frac{-1}{p_i}\right) = \prod_{i=1}^{r}(-1)^{(p_i-1)/2} = (-1)^{\sum_{i=1}^{r}(p_i-1)/2};$$

but

$$m = \prod_{i=1}^{r} p_i = \prod_{i=1}^{r} [(p_i - 1) + 1] \equiv 1 + \sum_{i=1}^{r}(p_i - 1) \ (\text{mod } 4),$$

since each factor $p_i - 1$ is even. Thus

$$\frac{m-1}{2} \equiv \sum_{i=1}^{r} \frac{p_i - 1}{2} \pmod{2}$$

so that (7.5.5) follows.
  Similarly we have

$$\left(\frac{2}{m}\right) = \prod_{i=1}^{r}\left(\frac{2}{p_i}\right) = \prod_{i=1}^{r} (-1)^{(p_i^2-1)/8} = (-1)^{\sum_{i=1}^{r}(p_i^2-1)/8}.$$

But

$$m^2 = \prod_{i=1}^{r} p_i^2 = \prod_{i=1}^{r} (1 + (p_i^2 - 1)) \equiv 1 + \sum_{i=1}^{r} (p_i^2 - 1) \pmod{64},$$

since each factor $p_i^2 - 1$ is divisible by 8. Thus

$$\frac{1}{8}(m^2 - 1) \equiv \frac{1}{8}\sum_{i=1}^{r} (p_i^2 - 1) \pmod{8}$$

and (7.5.6) results.

  As for (7.5.7), let $m = \prod_{i=1}^{r} p_i$, $n = \prod_{j=1}^{s} q_j$, $(m, n) = 1$, both m and n odd and positive. Then

$$\left(\frac{m}{n}\right)\left(\frac{n}{m}\right) = \prod_{i=1}^{r}\prod_{j=1}^{s} \left(\frac{p_i}{q_j}\right)\left(\frac{q_j}{p_i}\right) = \prod_{i=1}^{r}\prod_{j=1}^{s} (-1)^{(p_i-1)/2 \cdot (q_j-1)/2}$$

$$= (-1)^{\sum_{i=1}^{r}\sum_{j=1}^{s} [(p_i-1)/2][(q_j-1)/2]}.$$

But we've already seen that $(m - 1)/2 \equiv \sum_{i=1}^{r} [(p_i - 1)/2]$ (mod 2) and $(n - 1)/2 \equiv \sum_{j=1}^{s} [(q_j - 1)/2]$ (mod 2), which implies

$$\left(\frac{m-1}{2}\right)\left(\frac{n-1}{2}\right) \equiv \sum_{i=1}^{r}\sum_{j=1}^{s} \left(\frac{p_i-1}{2}\right)\left(\frac{q_j-1}{2}\right) \pmod{2}$$

and hence yields (7.5.7).
  The above rules allow for rapid algorithmic evaluation of the Jacobi symbol. In particular, we may start with a Legendre symbol and evaluate it by treating it as a Jacobi symbol. For example, the illustration in §7.2 may now be carried out as

$$\left(\tfrac{30}{199}\right) = \left(\tfrac{2}{199}\right)\left(\tfrac{15}{199}\right) = \left(\tfrac{15}{199}\right) = -\left(\tfrac{199}{15}\right) = -\left(\tfrac{4}{15}\right) = -1.$$

This process can be speeded up further if the property (7.5.7) is translated into a kind of "periodicity" statement for the "lower half" of the symbol. Namely, consider $(a, m) = 1$, $a > 0$, $m > 0$, $m' > 0$, $m \equiv m'$ (mod a), where a, m, and m' are odd. Then

$$\left(\frac{a}{m}\right) = (-1)^{(a-1)/2 \cdot (m-1)/2}\left(\frac{m}{a}\right) = (-1)^{(a-1)/2 \cdot (m-1)/2}\left(\frac{m'}{a}\right)$$

$$= (-1)^{((a-1)/2)((m-1)/2+(m'-1)/2)}\left(\frac{a}{m'}\right).$$

Since

$$\frac{m-1}{2} + \frac{m'-1}{2} = \frac{m-m'}{2} + m' - 1 \equiv \frac{m-m'}{2} \text{ (mod 2)},$$

this may be rewritten as

(7.5.8) $$\left(\frac{a}{m}\right) = (-1)^{(a-1)/2 \cdot (m-m')/2}\left(\frac{a}{m'}\right).$$

EXERCISES

1. Extend the definition of the Jacobi symbol $\left(\dfrac{a}{m}\right)$ to include the case $m < 0$, by defining $\left(\dfrac{a}{m}\right) = \left(\dfrac{a}{-m}\right)$ in this case. Also define $\left(\dfrac{a}{\pm 1}\right) = 1$. Show that (7.5.7) extends unless both m and n are negative.

2. Prove that for positive integers a, b, c, the expression $4abc - b - c$ cannot be a square.

3. For m and n odd relatively prime positive integers, show that

    $$\left(\frac{m}{n}\right) = (-1)^{(1/2)(m-1)}\left(\frac{m}{2m-n}\right).$$

4. Verify that if the Quadratic Reciprocity Law

    $$\left(\frac{p}{q}\right)\left(\frac{q}{p}\right) = (-1)^{(p-1)/2 \cdot (q-1)/2}$$

    for the Legendre symbol is known for all odd primes p and q less than x, then (7.5.7) holds for all positive odd integers m, n, $(m, n) = 1$, that are composed of primes less than x. (Also if one but not both of these m, n, is negative.)

5. If n is a nonzero integer such that $\left(\dfrac{n}{p}\right) = 1$ for all primes p not dividing n, show that n must be a square.

6. Show that for $\epsilon = \pm 1$ we have $\left(\dfrac{\epsilon}{f}\right) = (-1)^{(f-1)/2 \cdot (\epsilon-1)/2}$.

§7.5A INDUCTIVE PROOF OF THE RECIPROCITY LAW

Gauss was the first to prove the Quadratic Reciprocity Law. Though he ultimately provided many proofs, his very first proof was by means of a remarkable induction over the primes. Today, this proof may seem somewhat unnatural, but actually it was a very natural crystallization of the special cases that had been discovered by Euler and Legendre. Moreover, it reflected the spirit of the time in that in proving this theorem concerning integers, there is a direct and intimate handling of integers in relation to divisibility properties. We follow here Gauss' proof as modified by Dirichlet.

To begin, we assume that (7.2.6) and (7.2.7) have been established, providing the quadratic character of -1 and 2, respectively. Our focus will be on establishing (7.2.8) for any two distinct positive odd primes p, q. For the very "smallest" possible case, $p = 3$, $q = 5$, (7.2.8) is easily verified. Proceeding by induction, we assume that (7.2.8) holds for all pairs of odd primes p and p' less than the prime q (hence $q \geq 7$). It follows then (Exercise 4 of §7.5) that the reciprocity law for the Jacobi symbol holds for integers composed solely of primes less than q.

To complete the induction requires that we prove (7.2.8) for any odd prime $p < q$ and q. This proceeds via several cases, as follows.

CASE I. $q \equiv -1 \pmod 4$

In this case, by Exercise 1 of §7.4 our objective becomes that of verifying

(7.5A.1)
$$\left(\frac{p}{q}\right) = \left(\frac{-q}{p}\right).$$

Let $\epsilon = \left(\dfrac{p}{q}\right)$, so that since $q \equiv -1 \pmod 4$, we have $\left(\dfrac{\epsilon p}{q}\right) = 1$. Thus there is an even integer e, $0 < e < q$ such that

(7.5A.2)
$$e^2 - \epsilon p = f \cdot q.$$

Since $p < q$, $|e^2 - \epsilon p| \leq (q - 1)^2 + p < (q - 1)^2 + q$, and it follows from (7.5A.2) that $0 < f < q$, and f is odd. ($f > 0$ follows from $p < q$, the choice of e even from the fact that $q - e$ is also a solution.)

From (7.5A.2) we have $\left(\dfrac{\epsilon p}{f}\right) = 1$, implying

(7.5A.3)
$$\left(\frac{\epsilon}{f}\right) = \left(\frac{p}{f}\right).$$

Also from (7.5A.2), $1 = \left(\dfrac{fq}{p}\right) = \left(\dfrac{-f}{p}\right)\left(\dfrac{-q}{p}\right)$, implying that

(7.5A.4)
$$\left(\dfrac{-q}{p}\right) = \left(\dfrac{-f}{p}\right).$$

Then since all the prime factors of f are less than q,

$$\left(\dfrac{-q}{p}\right) = \left(\dfrac{-f}{p}\right) = (-1)^{(p-1)/2\cdot(f+1)/2}\left(\dfrac{p}{f}\right) = (-1)^{(p-1)/2\cdot(f+1)/2}\left(\dfrac{\epsilon}{f}\right) \qquad \text{by (7.5A.3)}$$

or

(7.5A.5) $\qquad \left(\dfrac{-q}{p}\right) = (-1)^{(p-1)/2\cdot(f+1)/2+(\epsilon-1)/2\cdot(f-1)/2} \qquad$ by Exercise 6 of §7.5.

But since e is even and $q \equiv -1 \pmod 4$, (7.5A.2) yields $\epsilon p \equiv f \pmod 4$; hence

$$\frac{p-1}{2}\cdot\frac{f+1}{2} + \frac{\epsilon-1}{2}\cdot\frac{f-1}{2} \equiv \frac{p-1}{2}\cdot\frac{\epsilon p+1}{2} + \frac{\epsilon-1}{2}\cdot\frac{\epsilon p-1}{2} \pmod 2$$

$$\equiv \left(\frac{\epsilon-1}{2}\right)^2 \pmod 2.$$

Thus from (7.5A.5),

$$\left(\frac{-q}{p}\right) = (-1)^{[(\epsilon-1)/2]^2} = \epsilon,$$

which is precisely (7.5A.1).

CASE II. $q \equiv 1 \pmod 4$ and $\left(\dfrac{p}{q}\right) = 1$

In this case, by Exercise 1 of §7.4 our objective is to verify that

(7.5A.6)
$$\left(\dfrac{q}{p}\right) = \left(\dfrac{p}{q}\right) = 1.$$

As in Case I, there is an even integer e such that $0 < e < q$, and

(7.5A.7)
$$e^2 - p = f\cdot q.$$

Again, since $p < q$, we can set $0 < f < q$, and f odd. From (7.5A.7), $1 = \left(\dfrac{fq}{p}\right) = \left(\dfrac{f}{p}\right)\left(\dfrac{q}{p}\right)$, so that

(7.5A.8)
$$\left(\frac{q}{p}\right) = \left(\frac{f}{p}\right).$$

Since f is composed solely of primes less than q, we get (if it occurs, $\left(\frac{p}{1}\right) = 1$)

$$\left(\frac{q}{p}\right) = \left(\frac{f}{p}\right) = (-1)^{(p-1)/2 \cdot (f-1)/2}\left(\frac{p}{f}\right);$$

from (7.5A.7), $\left(\frac{p}{f}\right) = 1$ (since p is a quadratic residue of every prime dividing f), so that

(7.5A.9)
$$\left(\frac{q}{p}\right) = (-1)^{(p-1)/2 \cdot (f-1)/2}.$$

Since e is even and $q \equiv 1 \pmod 4$, (7.5A.7) implies $f \equiv -p \pmod 4$, so that

$$\frac{p-1}{2} \cdot \frac{f-1}{2} \equiv \frac{p^2 - 1}{4} \equiv 0 \pmod 2.$$

Thus (7.5A.9) implies $\left(\frac{q}{p}\right) = 1$, as desired.

CASE III. $q \equiv 1 \pmod 4$, $\left(\frac{p}{q}\right) = -1$

Our objective here is to derive that $\left(\frac{q}{p}\right) = -1$.

If $q \equiv 1 \pmod 8$, we've seen in §7.4B how Gauss provided a proof (not using the Quadratic Reciprocity Law) of the existence of an odd prime $p' < 2\sqrt{q} + 1$ such that $\left(\frac{q}{p'}\right) = -1$. (Note $p' < q$, since $2\sqrt{q} + 1 < q$ for $q \geq 7$.) For $q \equiv 5 \pmod 8$, $\frac{1}{2}(q + 1) \equiv 3 \pmod 8$, so that there exists a $p' \equiv -1 \pmod 4$ dividing $\frac{1}{2}(q + 1)$, and clearly $p' < q$. But then $\left(\frac{q}{p'}\right) = \left(\frac{-1}{p'}\right) = -1$. Thus in all cases where $q \equiv 1 \pmod 4$, we have a positive odd prime $p' < q$ such that $\left(\frac{q}{p'}\right) = -1$. For this prime we must have $\left(\frac{p'}{q}\right) = -1$, since if $\left(\frac{p'}{q}\right) = +1$, the argument of Case II would imply that $\left(\frac{q}{p'}\right) = 1$.

Assuming that our objective is false $\left[\text{i.e., that } \left(\frac{q}{p}\right) = 1\right]$, it follows that $p \neq p'$ and since $\left(\frac{p'}{q}\right) = -1$, we have

$$\left(\frac{pp'}{q}\right) = \left(\frac{p}{q}\right)\left(\frac{p'}{q}\right) = 1.$$

Thus there exists a positive even integer e, $0 < e < q$ such that

(7.5A.10) $e^2 - pp' = fq,$

where f is odd. Since $0 < pp' < q^2$, $0 < e^2 < q^2$, it follows that $0 < |f| < q$. From (7.5A.10) and $\left(\frac{q}{p'}\right) = -1$,

$$1 = \left(\frac{fq}{pp'}\right) = \left(\frac{f}{pp'}\right)\left(\frac{q}{pp'}\right) = \left(\frac{f}{pp'}\right)\left(\frac{q}{p}\right)\left(\frac{q}{p'}\right) = -\left(\frac{f}{pp'}\right)\left(\frac{q}{p}\right),$$

so that

(7.5A.11) $\left(\frac{q}{p}\right) = -\left(\frac{f}{pp'}\right).$

From (7.5A.10) and the fact that p, p' and all the primes dividing f are less than q, we obtain (see Exercise 4, §7.5)

$$1 = \left(\frac{pp'}{f}\right) = (-1)^{(pp'-1)/2 \cdot (f-1)/2}\left(\frac{f}{pp'}\right);$$

combining this with (7.5A.11) gives

(7.5A.12) $\left(\frac{q}{p}\right) = -(-1)^{(pp'-1)/2 \cdot (f-1)/2}.$

But from (7.5A.10), since e is even and $q \equiv 1 \pmod 4$, it follows that $pp' \equiv -f \pmod 4$. Thus

$$\frac{pp'-1}{2} \cdot \frac{f-1}{2} \equiv \frac{-(pp')^2 + 1}{4} \equiv 0 \pmod 2,$$

so that (7.5A.12) asserts the desired result $\left(\frac{q}{p}\right) = -1$.

This completes the induction in all cases.

NOTES

§7.1. In article 152 of the *Disquisitiones Arithmeticae*, Gauss describes the reduction of the question of solutions to congruences of the form (7.1.1) to those of the form (7.1.4). He sketches the main ideas but omits many details. Others treated this subject before Gauss, most notably Euler, who also investigated quadratic residues and quadratic forms. However, Gauss' introduction of the congruence notation considerably clarified the presentation.

§7.2. Definition 7.2.1. Euler was the first to define residues and nonresidues and systematically investigate their properties [7.1, 7.2].

The "Legendre symbol" was first introduced by Legendre in 1798 [7.3].

Euler applied the more general condition for dth power residues (see §6.6) to the case of quadratic residues (i.e., $d = 2$). [7.4].

§7.2. Theorem 7.2.1. The Quadratic Reciprocity Law was given the name "Loi de réciprocité" by Legendre, who stated it in a form equivalent to (7.2.8) in 1785 [7.5]. The first complete proof was given by Gauss in 1801 in his *Disquisitiones Arithmeticae* (Articles 125–145).

There is some belief that Euler preceded Gauss in proving the Quadratic Reciprocity Law. However, though he established many special cases, Euler apparently never published a complete proof. According to Kummer [7.6], Euler did enunciate a proposition that is equivalent to the Quadratic Reciprocity Law (1783) [7.4]. Kronecker [7.7] pointed out that Euler's formulation asserted: ". . . a positive prime s is a quadratic residue or non-residue of a positive prime p according as $(-1)^{(1/2)(p-1)}p$ is a quadratic residue or non-residue of s." Kronecker expressed some surprise that Gauss was unaware of this formulation. Kronecker also indicated that the reciprocity law could be derived from material present in an earlier work of Euler [7.8] with the addition of a "simple observation."

Legendre attempted to give a proof of the Quadratic Reciprocity Law in 1785 [7.5] and in 1798 [7.3]. His proof required the "assumption" that given a prime p of the form $4n + 1$ there exists a prime q of the form $4n + 3$ such the $\left(\dfrac{p}{q}\right) = -1$. Since Legendre could not prove this, his proof was properly considered incomplete. In 1859, Kummer noted that Dirichlet's theorem on primes in arithmetic progressions easily fills the gap [7.6].

Notwithstanding all of the above, we must reiterate that the *first proof* of the Quadratic Reciprocity Law was given by Gauss in 1801. A second proof also appears in the *Disquisitiones Arithmeticae* in Article 262 and involves quadratic forms. In all, Gauss gave six different proofs of the Quadratic Reciprocity Law. The third and fourth proofs appeared in 1808 [7.9, 7.10] and the fifth and sixth proofs in 1817 [7.11]. Variations on Gauss' sixth proof were later given independently by Jacobi, Eisenstein, and Cauchy [7.12].

Gauss' motivation in seeking new proofs of the Quadratic Reciprocity Law was to develop methods for treating higher reciprocity laws [7.11].

§7.3. Theorem 7.3.1. Gauss' Lemma first appeared in connection with Gauss' third proof of the Quadratic Reciprocity Law in 1808 [7.9]. This proof was Gauss' favorite.

§7.4. This is the method employed by Gauss in his third proof [7.9].

§7.4A. Gauss' fourth and sixth proofs both depend on properties of finite sums of certain roots of unity. The fourth proof utilizes $w = 1 + r + r^4 + r^9 + \cdots + r^{(n-1)^2}$ where $r = e^{2\pi i/n}$, n odd; and $w' = 1 + r^h + r^{4h} + r^{9h} + \cdots + r^{h(n-1)^2}$. Gauss then observes that $w' = \left(\dfrac{h}{p}\right)w$ (where $n = p$ a prime not dividing h) and uses this as the basis of the proof.

Gauss' sixth proof utilizes

$$\xi = \sum_{s=0}^{p-2} (-1)^s x^{As}$$

where A is a primitive root modulo p.

This last most closely resembles the proof given in this section. Actually, the proof given here is an adaptation of Gauss' sixth proof as given by Eisenstein in 1844. [7.13].

§7.4B. In Article 129 of the *Disquisitiones Arithmeticae*, Gauss proves that if $q \equiv 1 \pmod 8$ is a given prime, there must exist a prime $p < 2\sqrt{q} + 1$ such that q is a nonresidue of p.

§7.4C. The method presented here is given in more general form in [7.17].

§7.5. Definition 7.5.1. The Jacobi symbol was introduced by K. Jacobi in 1846 [7.14]. In this paper, Jacobi gave and proved Theorem 7.5.1.

§7.5A. As noted above, the first proof of the Quadratic Reciprocity Law given by Gauss was by induction. H. J. Smith called it "repulsive." (He was wrong. It is really quite beautiful.) Dirichlet modified this proof in 1854 so as to reduce the number of separate cases [7.15], and in this form it was included in Dedekind's edition of Dirichlet's *Vorlesungen* [7.16].

REFERENCES

7.1. L. Euler, Demonstratio Theorematis Fermationi omnem numerum sine integrum sine fractum esse summam quatuor pauciorumue quadratorum, *Opera Omnia*, Leipzig: B. G. Teubner, I, 2, pp. 338–372 (original: 1754).

7.2. L. Euler, Disquisitio accuratior circa residua ex divisione quadratorum altiorumque potestatum per numeros primos relicta, *Opera Omnia*, I, 3, pp. 513–543 (original: 1783).

7.3. A. M. Legendre, *Essai sur la Théorie des Nombres*, 1798.

7.4. L. Euler, Observationes circa divisionem quadratorum per numeros primos, *Opera Omnia*, I, 3, pp. 497–512 (original: 1783).

7.5. A. M. Legendre, *Recherches d'analyse indeterminée*, Paris: Histoire de l'Académie des Sciences, 1785, p. 465.

7.6. E. Kummer, Uber die allgemeinen Reciprocitatsgesetz unter den Resten und Nichtresten der Potenzen deren Grad eine Primzahl ist, *Collected Papers*, New York: Springer Verlag, 1975, I, pp. 699–838 (original: 1859).

7.7. L. Kronecker, Ref. 6.21.

7.8. L. Kronecker, *Zur Geschichte des Reciprocitätsgesetzes, Werke*, Leipzig: B. G. Teubner, 1895, II, pp. 3–10 (original: 1875).

7.9. C. F. Gauss, *Theorematis arithmetici demonstratio nova, Werke*, 2nd edition, Göttingen: Königlichen Gesellschaft der Wissenschaften, 1876, II, pp. 1–8 (original: 1808).

7.10. C. F. Gauss, *Summatio quarumdam serierum singularium, Werke*, 2nd edition, 1876, II, pp. 9–45 (original: 1808).

7.11. C. F. Gauss, *Theorematis fundamentalis in doctrina de residuis quadraticis deminstrationes et ampliationes novae, Werke*, 2nd edition, 1876, II, pp. 47–64 (original: 1817).

7.12. H. J. Smith, *History of the Theory of Numbers*, p. 67.

7.13. F. Eisenstein, La loi de réciprocité tirée des formules de M. Gauss, sans avoir déterminée preablement le signe du radical, *Crelle's Journal*, **28**, 1844, 41–43.

7.14. K. Jacobi, Ueber die Kreistheilung und ihre Anwendung auf die Zahlentheorie, *Crelle's Journal*, **30**, 1846, 166–182.

7.15. G. L. Dirichlet, Uber den ersten der von Gauss gegebenen Beweise des Reciprocitatsgesetzes in der Theorie der quadratischen Reste, *Crelle's Journal*, **47**, 1854, 139–150.

7.16. G. L. Dirichlet, Ref. 6.55.

7.17. H. N. Shapiro and G. H. Sparer, Power Quadratic Diophantine Equations and Descent, *Communications on Pure and Applied Mathematics*, **31**, 1978, 185–203. (Also, **32**, 1979, 277–279.)

8

COUNTING PROBLEMS
A Do-It-Yourself Chapter

§8.1 FORMULATION OF THE PROBLEMS

A great deal of the subject matter of the theory of numbers is concerned with counting the number of elements (usually integers) in a given set $\mathscr{S}$. More generally, the formulation of a counting problem may be formally described as follows.

Definition 8.1.1. The *formulation of a counting problem* consists of a finite set $\mathscr{S}$ and a property P, with the objective of evaluating (or estimating) $N(\mathscr{S}, P)$ = the number of elements of $\mathscr{S}$ with the property P. The counting problem itself will be denoted by $(\mathscr{S}, P)$. Two such problems, $(\mathscr{S}, P)$ and $(\mathscr{S}', P')$, are called equivalent if $N(\mathscr{S}, P) = N(\mathscr{S}', P')$.

For example, if $\mathscr{S}$ = the positive integers $\leq n$, and P is the property of being "prime to n", the counting problem $(\mathscr{S}, P)$ is that of determining the number of positive integers less than or equal to n that are prime to n. Then $N(\mathscr{S}, P) = \phi(n)$, the Euler function, and the essence of the problem is that of finding a formula for $\phi(n)$ (a problem that was solved in §3.7).

In general, there are many equivalent formulations of a given counting problem. An easy illustration arises by letting $n = n_1 n_2$, $(n_1, n_2) = 1$, $\mathscr{S}_1$ = the positive integers $\leq n$ that are prime to n_1, and P_2 the property of being prime to n_2. Then $(\mathscr{S}_1, P_2)$ and the $(\mathscr{S}, P)$ given above are equivalent formulations in that $N(\mathscr{S}_1, P_2) = N(\mathscr{S}, P)$.

The basic procedure for setting up the machinery for approaching a counting problem $(\mathscr{S}, P)$ is to introduce the *characteristic function* $C_P(x)$ of the property P, defined for all $x \in \mathscr{S}$ and such that

$$(8.1.1) \qquad C_P(x) = \begin{cases} 1 & \text{if } x \text{ has the property } P \\ 0 & \text{otherwise.} \end{cases}$$

Then summing $C_P(x)$ over all $x \in \mathscr{S}$ yields $N(\mathscr{S}, P)$. That is

$$(8.1.2) \qquad N(\mathcal{S}, P) = \sum_{x \in \mathcal{S}} C_P(x).$$

This is quite tautological in a very trivial way and achieves little until one constructs and introduces some representation of the characteristic function $C_P(x)$. One starting point for the building of such representations is the Möbius function $\mu(n)$ and the property (3.5.3) that asserts

$$(8.1.3) \qquad \sum_{d/n} \mu(d) = \begin{cases} 1 & \text{if } n = 1 \\ 0 & \text{otherwise.} \end{cases}$$

This is, in fact, clearly a representation of the characteristic function of the property of the integer n: *that it equals one*. One can transform this into representations of many other characteristic functions. We recall the second proof of Theorem 3.7.1 in which exactly this route was followed. There the property $\hat{P}$ of the integer m is that of being prime to a fixed integer n. For this property $\hat{P}$,

$$(8.1.4) \qquad C_{\hat{P}}(m) = \sum_{d/(m,n)} \mu(d).$$

If the set $\hat{\mathcal{S}} = $ the positive integers $\leq n$, (8.1.2) then becomes

$$(8.1.5) \qquad \phi(n) = N(\hat{\mathcal{S}}, \hat{P}) = \sum_{m=1}^{n} \sum_{d/(m,n)} \mu(d).$$

More generally, if $\hat{\mathcal{S}} = $ the positive integers $\leq x$, a given integer,

$$(8.1.6) \qquad \phi(x, n) = N(\hat{\mathcal{S}}, \hat{P}) = \sum_{m=1}^{x} \sum_{d/(m,n)} \mu(d).$$

As was seen in the second proof of Theorem 3.7.1, the "processing" of (8.1.5) as a "productive piece of machinery" involves the interchange of the order of summation on the right. For the case of (8.1.6), this gives (recall that $[z] = $ the largest integer $\leq z$)

$$\phi(x, n) = \sum_{d/n} \mu(d) \sum_{\substack{m \leq x \\ m \equiv 0 \,(\mathrm{mod}\, d)}} 1$$

$$= \sum_{d/n} \mu(d) \left[\frac{x}{d} \right] = \sum_{d/n} \mu(d) \left(\frac{x}{d} + 0(1) \right),$$

so that

(8.1.7)
$$\phi(x, n) = \frac{\phi(n)}{n} \cdot x + O(2^{\nu(n)}),$$

where $\nu(n) = $ the number of distinct prime factors of n and, as given, the O is uniform in n and x [i.e., the term is $\leq c 2^{\nu(n)}$ in absolute value, where c is a constant independent of both n and x].

From the above illustration, we see that the actualization of "how you count" has two distinct but interactive components: first, of all the equivalent formulations of a counting problem, the choice of one $(\mathcal{S}, P)$; and second, for this choice, an appropriate representation of the characteristic function $C_P(x)$. For example, suppose that one were concerned with the counting of integers m, $1 \leq m \leq x$, that are prime to a fixed integer A, and in a given arithmetic progression $Bt + C$ ($t = 0, 1, 2, \ldots$). Following are the natural and equivalent formulations of this counting problem.

    (i) $\mathcal{S}_1 = \{m, 1 \leq m \leq x, (m, A) = 1\}$
           $P_1 = $ property of being in the progression $Bt + C$
   (ii) $\mathcal{S}_2 = \{m, 1 \leq m \leq x, m \equiv C \pmod{B}\}$
           $P_2 = $ property of being prime to A
  (iii) $\mathcal{S}_3 = \{m, 1 \leq m \leq x\}$
           $P_3 = $ property of being in progression $Bt + C$ and prime to A.

For this specific problem, the formulation (ii) is an a priori preferable choice simply because the characteristic function of P_2 has the natural representation provided by (8.1.4). Using (ii), (8.1.2) and (8.1.4) yield

(8.1.8)
$$N(\mathcal{S}_2, P_2) = \sum_{\substack{m \leq x \\ m \equiv C \,(\mathrm{mod}\, B)}} \sum_{d | (m, A)} \mu(d).$$

We will return to this problem in the next section. It is, in fact, atypical in that there is a clear a priori preference for the problem formulation. In general, though, there is usually a clearly defined *collection* of preferable formulations; the one that is *best* emerges only after trying them all.

The above illustration suggests the introduction of a simple "arithmetic of properties" (all assumed defined over a common domain). We let $P_1 \cdot P_2$ denote the property that requires that both the property P_1 and the property P_2 hold. (In the above example $P_3 = P_1 \cdot P_2$.) Similarly, $P_1 + P_2$ denotes the property that at least one of the properties P_1 or P_2 holds. Let $\overline{P}$ be the property that P does not hold. This translates into characteristic functions as shown in the following lemma.

Lemma 8.1.1. For properties P_1, P_2, P,

(8.1.9)
$$C_{P_1 P_2}(x) = C_{P_1}(x) C_{P_2}(x)$$

(8.1.10) $C_{P_1 + P_2}(x) = C_{P_1}(x) + C_{P_2}(x) - C_{P_1}(x)C_{P_2}(x)$

(8.1.11) $C_{\bar{P}}(x) = 1 - C_P(x).$

Proof.   . . . .

This enhances the building up of representations of more complex properties out of representations of the component properties. Of course, the question remains as to what are the available building blocks for representations of characteristic functions. Actually, there is no complete list and new ideas are always evolving. Among the principal tools, we have (i) the Möbius function, (ii) characters (and roots of unity in general), (iii) binomial coefficients, (iv) Fourier series and transforms, and (v) complex integrals. In what follows here, we will touch mainly on (i), (ii), and (iii). In any event, it is all meant to be purely illustrative and *not in any sense* a complete methodology for all counting problems.

§8.1A   PRIME TO A GIVEN INTEGER AND IN A GIVEN PROGRESSION

We continue here with a detailed analysis of the problem defined above where

$$\mathcal{S}_2 = \{m, \ 1 \le m \le x, \ m \equiv C \ (\mathrm{mod} \ B)\},$$

$$P_2 = \text{relatively prime to } A,$$

and A, B, C are integers, $A > 0$, $B > C \ge 0$. Our objective is to estimate $N(x) = N(\mathcal{S}_2, P_2) = $ the number of positive integers $1 \le m \le x$ such that $m \equiv C$ (mod B) and $(m, A) = 1$.

Note first the following lemma.

Lemma 8.1A.1.   If $(A, B, C) > 1$, then $N(x) = 0$ for all $x \ge 1$.

Proof.   . . . .

Theorem 8.1A.1.   For $N(x) = $ the number of positive integers $m \le x$ such that $m \equiv C$ (mod B) and $(m, A) = 1$, we have that $N(x) = 0$ if $(A, B, C) > 1$; if $(A, B, C) = 1$,

(8.1A.1) $$N(x) = \frac{x}{B} \prod_{\substack{p/A \\ p \nmid B}} \left(1 - \frac{1}{p}\right) + O(1),$$

where the $O(1)$ depends on A but is uniform in B and C.

Proof. By Lemma 8.1A.1, it suffices to consider $(A, B, C) = 1$. From (8.1.8),

(8.1A.2)
$$N(x) = \sum_{\substack{m \leq x \\ m \equiv C \,(\mathrm{mod}\, B)}} \sum_{d/(m,A)} \mu(d).$$

Interchanging the summations on the right of (8.1A.2) yields

(8.1A.3)
$$N(x) = \sum_{d/A} \mu(d) \sum_{\substack{m \leq x \\ m \equiv C(\mathrm{mod}\, B) \\ m \equiv 0(\mathrm{mod}\, d)}} 1.$$

Here the system of congruences $m \equiv C \,(\mathrm{mod}\, B)$, $m \equiv 0 \,(\mathrm{mod}\, d)$, has a solution if and only if $(d, B) = 1$ . . . .
Thus writing $m = td$, (8.1A.3) becomes

(8.1A.4)
$$N(x) = \sum_{\substack{d/A \\ (d,B)=1}} \mu(d) \sum_{\substack{t \leq (x/d) \\ td \equiv C(\mathrm{mod}\, B)}} 1.$$

From this we derive. . . . .

(8.1A.5)
$$N(x) = \sum_{\substack{d/A \\ (d,B)=1}} \mu(d)\left(\frac{x}{Bd} + O(1)\right),$$

which yields (8.1A.1), since . . . .
  If we wish a closed formula without error, as in the case of $\phi(n)$, we must work with an integer x that is "commensurable" with both the property of being prime to A and that of being $\equiv C(\mathrm{mod}\, B)$. Since the property of being prime to A has A as a period, and that of being $\equiv C(\mathrm{mod}\, B)$ has B as a period, we should consider x at some common period point. A natural choice is $x = \{A, B\} =$ the least common multiple of A and B.

Theorem 8.1A.2. If $(A, B, C) = 1$, we have

(8.1A.6)
$$N(\{A, B\}) = \frac{\phi(A)}{\phi((A, B))}.$$

Proof. For an arbitrary positive integer k, let $x = k\{A, B\}$. From the periodicity of the properties involved we have

(8.1A.7)
$$N(x) = kN(\{A, B\}).$$

since . . . .

Inserting this in (8.1A.1), dividing by k, and letting k tend to infinity yields (8.1A.6) because . . . .

§8.1B SUM OF INTEGERS EACH PRIME TO A GIVEN INTEGER

There is a tendency for the solutions of counting problems to propagate themselves as potential tools in the solution of other problems. In this section we illustrate such a situation with respect to the result of §8.1A. The problem we consider here is that of counting the number of representations of a positive integer N as the sum of two positive integers, each prime to a given positive integer A. That is,

$$(8.1B.1) \qquad N = a_1 + a_2, \qquad (a_1, A) = (a_2, A) = 1,$$

where $a_1 > 0$, $a_2 > 0$. We note that there is one trivial case, namely A even and N odd. For then the integers prime to A are all odd and (8.1B.1) would have no solutions. However, in general, we have the following theorem.

Theorem 8.1B.1. Given a fixed positive integer A, let $\gamma(N, A) = $ the number of representations of N in the form (8.1B.1). Then

$$(8.1B.2) \qquad \gamma(N, A) = \frac{\phi(A)}{A}\left(\prod_{\substack{p/A \\ p \nmid N}} \frac{p-2}{p-1}\right)N + O(1),$$

where the $O(1)$ depends on A but is uniform in N.

Proof. We have

$$\gamma(N, A) = \sum_{\substack{1 \le a < N \\ (a,A)=1}} \sum_{d/(N-a,A)} \mu(d)$$

so that

$$(8.1B.3) \qquad \gamma(N, A) = \sum_{\substack{d/A \\ (d,N)=1}} \mu(d) \sum_{\substack{1 \le a < N \\ (a,A)=1 \\ a \equiv N(\bmod d)}} 1,$$

where the condition $(d, N) = 1$ is acquired by virtue of the fact that otherwise the inner sum above is zero. We next evaluate the inner sum by applying (8.1A.1) with $B = d$, $C = N$, $x = N - 1$. Thus

$$(8.1B.4) \qquad \sum_{\substack{1 \le a < N \\ (a,A)=1 \\ a \equiv N(\bmod d)}} 1 = \frac{N}{d} \prod_{\substack{p/A \\ p \nmid d}} \left(1 - \frac{1}{p}\right) + O(1),$$

where the O depends on A but is uniform in d and N. Inserting (8.1B.4) in (8.1B.3) yields

$$(8.1\text{B}.5) \qquad \gamma(N, A) = N \sum_{\substack{d/A \\ (d,N)=1}} \frac{\mu(d)}{d} \prod_{\substack{p/A \\ p\nmid d}} \left(1 - \frac{1}{p}\right) + O(1),$$

where the O depends on A but is independent of N. Then, transform the sum in (8.1B.5), ... and (8.1B.2) follows.

Note that for the case where A is even and N odd, (8.1B.2) gives $\gamma(N, A) = O(1)$, in agreement with our observation that $\gamma(N, A) = 0$ in this case.

Corollary. Except for the case where A is even and N odd, every integer N that is sufficiently large with respect to A has a representation of the form (8.1B.1).

Proof. . . . .

§8.2 POWERFREE INTEGERS

In §2.7 we've already met the squarefree integers, which are those integers whose only square factor is one. All primes are squarefree and, in addition, all products of distinct primes such as 6, 10, 14, 30, etc. Since each positive integer m is uniquely expressible in the form $m = s^2 q$ where q is squarefree, we see that

$$\sum_{d^2/m} \mu(d) = \sum_{d/s} \mu(d) = \begin{cases} 1 & \text{if } s = 1 \\ 0 & \text{otherwise.} \end{cases}$$

(The first sum is over those d whose square divides m.) Hence

$$(8.2.1) \qquad \sum_{d^2/m} \mu(d) = \begin{cases} 1 & \text{if } m \text{ is squarefree} \\ 0 & \text{otherwise.} \end{cases}$$

That is, (8.2.1) gives a representation of the characteristic function of the property that an integer be squarefree.

Recall also from §2.7 that, for each integer $r \geq 2$, an "rth powerfree integer" is defined as one such that the only rth power dividing it is 1. (Thus the case $r = 2$ is that of the squarefree integers.) Let $C_r(m)$ denote the characteristic function of the property that an integer be rth powerfree.

Lemma 8.2.1. For $r \geq 2$ a fixed integer, we have

$$(8.2.2) \qquad C_r(m) = \sum_{d^r/m} \mu(d).$$

Proof. . . . .

Theorem 8.2.1. For $Q_r(x)$ = the number of positive rth powerfree integers $\leq x$, we have

(8.2.3)
$$Q_r(x) = \alpha_r x + O(x^{1/r}),$$

where α_r is a constant given by

(8.2.4)
$$\alpha_r = \sum_{d=1}^{\infty} \frac{\mu(d)}{d^r}.$$

Proof. Using (8.2.2) in (8.1.2), for this problem we have

(8.2.5)
$$Q_r(x) = \sum_{m \leq x} \sum_{d^r \mid m} \mu(d).$$

Interchanging the order of summation in (8.2.5) . . . . and we get

(8.2.6)
$$Q_r(x) = x \sum_{d \leq x^{1/r}} \frac{\mu(d)}{d} + O(x^{1/r}).$$

Extending the sum to infinity . . . .

Q.E.D.

For the case $r = 2$, it can be shown that $\alpha_2 = 6/\pi^2$. In general, for $r \geq 2$, note that (8.2.4) implies

(8.2.7)
$$\alpha_r = \prod_p \left(1 - \frac{1}{p^r}\right).$$

This agrees with a simple "probabilistic heuristic" argument. Namely, the "fraction" of the integers divisible by p^r is $(1/p^r)$; hence those not divisible by p^r are $1 - (1/p^r)$ of the whole set. Assuming a kind of statistical independence between the primes, the fraction not divisible by the rth power of any prime would be $\prod_p (1 - (1/p^r))$, which agrees with (8.2.7) and (8.2.3).

Next we consider the problem of counting the number of rth powerfree integers $\leq x$ that are in the arithmetic progression $kt + \ell$ ($t = 0, 1, 2, \ldots$), where $k > \ell \geq 0$. We denote this number by $Q_r(x, k, \ell)$. The same kind of heuristic argument as that given above may be used to guess at the answer. In the progression $kt + \ell$, for any prime p not dividing k or ℓ, there is no interaction between the progression and not being divisible by p^r. For such a prime a "factor" of $1 - (1/p^r)$ is anticipated. If $v_p(k) > v_p(\ell)$, we have

$$kt + \ell = p^{\nu_p(\ell)}\left(t\frac{k}{p^{\nu_p(\ell)}} + \frac{\ell}{p^{\nu_p(\ell)}}\right),$$

so that $\nu_p(kt + \ell) = \nu_p(\ell)$. For any such p, if $\nu_p(\ell) \geq r$, we have $Q_r(x, k, \ell) = 0$. On the other hand, if $\nu_p(\ell) < r$ for such a p, $kt + \ell$ cannot be divisible by p^r, and this prime does not condition the problem at all. If $\nu_p(k) \leq \nu_p(\ell)$

$$kt + \ell = p^{\nu_p(k)}\left(t\frac{k}{p^{\nu_p(k)}} + \frac{\ell}{p^{\nu_p(k)}}\right),$$

and again if $\nu_p(k) \geq r$, $Q_r(x, k, \ell) = 0$. If $\nu_p(k) < r$, we are only counting the fraction such that the second factor is $[r - \nu_p(k)]$th powerfree. This all suggests the overall fraction (the product is taken over all primes p which satisfy the indicated condition)

$$\prod_{\min(\nu_p(k),r)\leq\nu_p(\ell)}\left(1 - \frac{1}{p^{\max(r-\nu_p(k),0)}}\right).$$

Theorem 8.2.2.   For given integers $k > \ell \geq 0$, and $r \geq 2$,

$$(8.2.8) \qquad Q_r(x, k, \ell) = \frac{x}{k}\prod_{\min(\nu_p(k),r)\leq\nu_p(\ell)}\left(1 - \frac{1}{p^{\max(r-\nu_p(k),0)}}\right) + O(x^{1/r}).$$

Proof.   We choose the formulation of this counting problem corresponding to

$$\mathscr{S} = \{m, 1 \leq m \leq x, m \equiv \ell(\text{mod } k)\}$$

$$P = \text{property of being } r\text{th powerfree},$$

so that

$$(8.2.9) \qquad\qquad Q_r(x, k, \ell) = \sum_{\substack{m\leq x \\ m\equiv\ell(\text{mod } k)}} \sum_{d^r|m} \mu(d).$$

Interchanging the order of summation yields

$$(8.2.10) \qquad\qquad Q_r(x, k, \ell) = \sum_{d\leq x^{1/r}} \mu(d) \sum_{\substack{m\leq x \\ m\equiv\ell(\text{mod } k) \\ m\equiv 0(\text{mod } d^r)}} 1.$$

Since . . . .
The simultaneous congruences $m \equiv \ell(\text{mod } k)$, $m \equiv 0(\text{mod } d^r)$ have solutions if and only if (k, d^r) divides ℓ, in which case there is a unique solution modulo $\{d^r, k\}$. Then

from (8.2.10) . . . .
gives

(8.2.11) $$Q_r(x, k, \ell) = \frac{x}{k} \sum_{\substack{d \leq x^{1/r} \\ (d^r, k)/\ell}} \frac{(d^r, k)\mu(d)}{d^r} + O(x^{1/r}).$$

Extending the sum to infinity . . . .
and we get

(8.2.12) $$Q_r(x, k, \ell) = \frac{x}{k} \sum_{\substack{d \\ (d^r, k)/\ell}} \frac{(d^r, k)\mu(d)}{d^r} + O(x^{1/r}).$$

Since the summand of the infinite sum as well as (d^r, k) is multiplicative as a function of d, we need only evaluate it summing over $d = p^\alpha$, p a prime (and then multiply). In this case, the sum becomes

$$1 - \frac{(p^r, k)}{p^r}$$

provided $\min(\nu_p(k), r) \leq \nu_p(\ell)$, and 1 otherwise. Thus the series on the right of (8.2.12) equals

$$\prod_{\min(\nu_p(k), r) \leq \nu_p(\ell)} \left(1 - \frac{(p^r, k)}{p^r}\right),$$

and since . . . . ,
this gives (8.2.8), as desired.

Note that (8.2.8) asserts that [the notation $\sim$ is introduced in (3.8.28)]

$$Q_r(x, k, \ell) \sim C_{r,k,\ell} x,$$

For $C_{r,k,\ell}$ a constant depending on r, k, and ℓ, such that $C_{r,k,\ell} \geq 0$. It is of some interest to determine when $C_{r,k,\ell} > 0$.

Corollary. $C_{r,k,\ell} > 0$ if and only if (k, ℓ) is rth powerfree.

Proof.   . . . .

EXERCISES

1. Show that the error term $O(x^{1/r})$ in (8.2.8) is uniform in k.

2. Consider the problem of evaluating $Q_A^{(r)}(x) = $ the number of rth powerfree integers $\leq x$ that are prime to A. Show that

$$Q_A^{(r)}(x) = x \prod_{p/A} \left(1 - \frac{1}{p}\right) \prod_{p\nmid A} \left(1 - \frac{1}{p^r}\right) + O(x^{1/r}).$$

Give two solutions, using two different formulations of the problem.

3. Prove that every integer greater than one is expressible as the sum of two positive squarefree integers.

4. For each fixed $r \geq 3$, estimate the number of representations of the positive integer N as a sum of two positive rth powerfree integers.

5. As a generalization of Ex. 4, for given positive integers r, s, such that $(1/r) + (1/s) < 1$, estimate the number of representations of the positive integer N as a sum of an rth powerfree integer and an sth powerfree integer.

§8.2A   SQUAREFREE INTEGERS IN SMALL INTERVALS

There is a class of counting problems in which the set to be counted lies in an interval whose length is a function of the lower end. Such a problem arises naturally, for example, if we ask for any given $x > 0$, how far above x we must go before finding the next squarefree integer. In other words, we are looking for a positive h as a function of x such that we can assert that the interval $x < n \leq x + h$ contains a squarefree integer n. Symbolically, this is the requirement that

(8.2A.1) $Q_2(x + h) - Q_2(x) > 0.$

From (8.2.3) we have

$$Q_2(x + h) - Q_2(x) = \alpha_2 h + O(\sqrt{x})$$

so that (8.2A.1) follows for $h = c\sqrt{x}$ where c is an appropriate positive constant. It is somewhat surprising that this result can be improved easily to an h of smaller order. This is the following result of K. F. Roth.

Theorem 8.2A.1. There exists a positive constant c such that for all $x \geq 1$, the interval $x < n \leq x + cx^{1/3}$ always contains a squarefree integer n.

Proof.   We have for $h = cx^{1/3}$, ($c > 1$ to be fixed later),

$$Q_2(x + h) - Q_2(x) = \sum_{x < n \leq x + h} \sum_{d^2/n} \mu(d)$$

or

(8.2A.2) $Q_2(x + h) - Q_2(x) = \sum_{d \leq \sqrt{x+h}} \mu(d) \sum_{\substack{x < n \leq x+h \\ n \equiv 0 (\mathrm{mod}\, d^2)}} 1.$

The d-summation on the right is broken into two parts:

$$\Sigma_1 = \sum_{d \le \sqrt[3]{x}} \mu(d) \sum_{\substack{x < n \le x+h \\ n \equiv 0 (\mathrm{mod}\, d^2)}} 1$$

and

$$\Sigma_2 = \sum_{\sqrt[3]{x} < d \le \sqrt{x+h}} \mu(d) \sum_{\substack{x < n \le x+h \\ n \equiv 0 (\mathrm{mod}\, d^2)}} 1.$$

For Σ_1 we have

$$\Sigma_1 = \sum_{d \le \sqrt[3]{x}} \mu(d) \left(\frac{h}{d^2} + O(1) \right),$$

. . . .
so that (recall $\alpha_2 = (6/\pi^2)$) with a O uniform in c (for x large)

$$(8.2A.3) \qquad\qquad \Sigma_1 = \alpha_2 h + O(\sqrt[3]{x}).$$

As for Σ_2, we note that $|\Sigma_2| \le$ the number of positive integer pairs (d, e) such that

$$x < d^2 e \le x + h \qquad \text{and} \qquad \sqrt[3]{x} < d \le \sqrt{x + h}.$$

Since $d^2 > x^{2/3}$ it follows that $e \le [(x + h)/x^{2/3}] \le 2x^{1/3}$ for x large. Further for any two such $d, d', d \ne d'$, we have

$$|d^2 - d'^2| = |d - d'|(d + d') > \sqrt[3]{x}.$$

Thus for a fixed e, the number of d such that $d^2 e$ satisfies $x < d^2 e \le x + h$ is less than $1 + (h/e\sqrt[3]{x}) = 1 + (c/e)$. . . . .
It follows that $|\Sigma_2| \le 3x^{1/3}$ for x large, ...
$$\text{Q.E.D.}$$

EXERCISES

1. Prove that for all $x \ge 1$ there is always a squarefree integer between x and $x + (\pi^2/3)\sqrt[3]{x}$.

2. Generalize Theorem 8.2A.1 to the case of rth powerfree integers.

§8.3 POWERFUL INTEGERS

On occasion, the appropriate representation of the characteristic function of the property is nothing more than an equivalent arithmetic statement. This is illustrated by the counting problems involving a class of integers that are defined as follows.

Definition 8.3.1. An rth powerful integer m is one having the proprerty that for any prime p that divides m, p^r divides m. (For $r = 2$ these integers are called squareful.)

An arithmetic formulation of this notion is given by the following lemma.

Lemma 8.3.1. For r any integer ≥ 2, an integer m is rth powerful if and only if m has a representation in the form

$$(8.3.1) \qquad m = u^r q_1^{r+1} q_2^{r+2} \cdots q_{r-1}^{2r-1},$$

where the q_i, $i = 1, \ldots, r - 1$ are squarefree and relatively prime by pairs. Furthermore, this representation of m is unique.

Proof. Consider the unique factorization of m as

$$(8.3.2) \qquad m = p_1^{\alpha_1} p_2^{\alpha_2} \cdots p_\ell^{\alpha_\ell},$$

where the p_j are distinct primes . . . .

Note that for the case of squareful integers (i.e., $r = 2$), the above lemma states that m is squareful if and only if it has a representation in the form $m = u^2 q^3$, q squarefree. Also, such a representation is unique.

Let $\mathrm{pow}_r(x) =$ the number of positive rth powerful integers $\leq x$, we begin by considering the case $r = 2$. Essentially using the lemma to represent the characteristic function of the property of being squareful, we have

$$(8.3.3) \qquad \mathrm{pow}_2(x) = \sum_{m \leq x} \sum_{m = u^2 q^3} 1$$

(where q denotes a squarefree integer). Then

$$\mathrm{pow}_2(x) = \sum_{u^2 q^3 \leq x} 1 = \sum_{q \leq \sqrt[3]{x}} \sum_{u \leq \sqrt{x/q^3}} 1$$

or

$$\mathrm{pow}_2(x) = \sum_{q \leq \sqrt[3]{x}} \left(\frac{\sqrt{x}}{q^{3/2}} + O(1) \right)$$

. . . .

which yields

$$(8.3.4) \qquad \mathrm{pow}_2(x) = c_2 \sqrt{x} + O(x^{1/3})$$

where c_2 is a positive constant given by

$$(8.3.5) \qquad c_2 = \sum_q \frac{1}{q^{3/2}} = \prod_p \left(1 + \frac{1}{p^{3/2}} \right).$$

In deriving the corresponding result for $\mathrm{pow}_r(x)$ for general $r \geq 2$, we will require the following lemma.

Lemma 8.3.2. Let r and k be any fixed positive integers. Then

(8.3.6)
$$\sum_{v_1^{r+1} \cdots v_k^{r+k} \leq z} 1 = O(z^{1/(r+1)}).$$

(Note that the O depends on k and r.)

Proof. For $k = 1$ we have

$$\sum_{v_1^{r+1} \leq z} 1 = \sum_{v_1 \leq z^{1/(r+1)}} 1 = O(z^{1/(r+1)}),$$

which is (8.3.6). Proceeding by induction on k, assume (8.3.6) for $k - 1$, and all $z > 0$. Then for $k > 1$,

$$\sum_{v_1^{r+1} \cdots v_k^{r+k} \leq z} 1 = \sum_{v_k \leq z^{1/(r+k)}} \sum_{v_1^{r+1} \cdots v_{k-1}^{r+k-1} \leq z/v_k^{r+k}} 1$$

. . . .

We now have the following theorem.

Theorem 8.3.1. For any fixed integer $r \geq 1$,

(8.3.7)
$$\mathrm{pow}_r(x) = \beta_r x^{1/r} + O(x^{1/(r+1)})$$

where $\beta_r > 0$ is a constant.

Proof. From (8.3.1),

$$\mathrm{pow}_r(x) = \sum_{\substack{u^r q_1^{r+1} \cdots q_{r-1}^{2r-1} \leq x \\ (q_i, q_j) = 1}} 1$$

where the q_i are all squarefree. This may be rewritten as

$$\mathrm{pow}_r(x) = \sum_{\substack{q_1^{r+1} \cdots q_{r-1}^{2r-1} \leq x \\ (q_i, q_j) = 1}} \sum_{u \leq \left(\frac{x}{q_1^{r+1} \cdots q_{r-1}^{2r-1}}\right)^{1/r}} 1.$$

Then using summation by parts and Lemma 8.3.2 . . . .

EXERCISES

1. Estimate the number of positive integers $\leq x$ that are both rth powerful and prime to a given integer A.

2. Estimate the number of positive integers $\leq x$ that are both rth powerful and in a given arithmetic progression $kt + \ell$, $t = 0, 1, 2, \ldots$.

3. An integer m will be called (r, s)-powerful if it can be written in the form $m = m_1 m_2$ where m_1 is rth powerful and m_2 is sth powerful. Estimate the number of positive (r, s)-powerful integers $\leq x$.

4. Call an integer m, "rth powerful relative to A" if for any prime p dividing m that *does not* divide A, we have that p^r divides m. Count the number of positive integers $\leq x$ that are rth powerful relative to A.

§8.4 POWER RESIDUES MODULO A PRIME

For quadratic residues modulo an odd prime p, we see that

$$\frac{1}{2}\left(1 + \left(\frac{m}{p}\right)\right) = \begin{cases} 1 & \text{if } \left(\frac{m}{p}\right) = 1 \\ 0 & \text{if } \left(\frac{m}{p}\right) = -1 \end{cases}$$

gives a representation of the characteristic function of the property of being a quadratic residue modulo p. More generally, for dth power residues, $d/(p-1)$, we have from (6.6.3) that

$$(8.4.1) \qquad \frac{1}{d} \sum_{\chi^d = \chi_0} \chi(m) = \begin{cases} 1 & \text{if } m \text{ is a } d\text{th power residue mod } p \\ 0 & \text{otherwise} \end{cases}$$

Thus if $N_d(x) = $ the number of dth power residues $\leq x$ a positive integer $< p$, for $d/(p-1)$ we have

$$N_d(x) = \sum_{m \leq x} \frac{1}{d} \sum_{\chi^d = \chi_0} \chi(m)$$

$$= \frac{x}{d} + \sum_{m \leq x} \frac{1}{d} \sum_{\substack{\chi^d = \chi_0 \\ \chi \neq \chi_0}} \chi(m)$$

or

$$(8.4.2) \qquad N_d(x) = \frac{x}{d} + \frac{1}{d} \sum_{\substack{\chi^d = \chi_0 \\ \chi \neq \chi_0}} \sum_{m \leq x} \chi(m).$$

Hence we are led to the requirement of an estimate for

$$\sum_{m \le x} \chi(m),$$

where χ is a nonprincipal character modulo p. This is in fact achieved by a generalization of the early part of §7.4A using Gauss sums (see Exercise 2 of §7.4B).

The desired generalization of the Gauss sum is realized by replacing the Legendre symbol $\left(\dfrac{h}{p}\right)$ in (7.4A.1) by a nonprincipal character modulo p, $\chi(h)$. Thus for $(a, p) = 1$, we define

$$(8.4.3) \qquad \tau_\chi(a) = \sum_{h=1}^{p-1} \chi(h) e^{2\pi i a h / p}.$$

Then multiplying by $\chi(a)$ we get

$$\chi(a)\tau_\chi(a) = \sum_{h=1}^{p-1} \chi(h) e^{2\pi i h / p}$$

or

$$(8.4.4) \qquad \chi(a)\tau_\chi(a) = \tau_\chi(1).$$

Replacing χ by $\overline{\chi}$ and multiplying by $\chi(a)$, this may be rewritten as

$$(8.4.5) \qquad \tau_{\overline{\chi}}(1)\chi(a) = \tau_{\overline{\chi}}(a).$$

Also from (8.4.3),

$$\overline{\tau_\chi(a)} = \sum_{h=1}^{p-1} \overline{\chi}(h) e^{-2\pi i h a / p}$$

$$= \sum_{h=1}^{p-1} \overline{\chi}(-h) e^{2\pi i h a / p},$$

so that

$$(8.4.6) \qquad \overline{\tau_\chi(a)} = \overline{\chi}(-1)\tau_{\overline{\chi}}(a).$$

As in §7.4A, we will require the value of $|\tau_\chi(1)|$.

Lemma 8.4.1. For any integer a, $(a, p) = 1$, and any nonprinciple character χ modulo p,

$$(8.4.7) \qquad |\tau_\chi(a)| = |\tau_\chi(1)| = p^{1/2}.$$

Proof. From (8.4.4), $|\tau_\chi(a)| = |\tau_\chi(1)|$ so that for all $(a,p) = 1$, $|\tau_\chi(a)|^2 = |\tau_\chi(1)|^2$. Summing over these a yields

$$(p-1)|\tau_\chi(1)|^2 = \sum_{a=1}^{p-1} |\tau_\chi(a)|^2,$$

and since $\tau_\chi(0) = 0$, we have

$$(p-1)|\tau_\chi(1)|^2 = \sum_{a=0}^{p-1} |\tau_\chi(a)|^2 = \sum_{a=0}^{p-1} \tau_\chi(a)\overline{\tau_\chi(a)}$$

$$= \sum_{a=0}^{p-1} \sum_{h,h'=1}^{p-1} \chi(h)\overline{\chi}(h')e^{2\pi i(h-h')a/p}$$

$$= p(p-1),$$

since . . . .
From this (8.4.7) follows.

Lemma 8.4.2. For χ any nonprincipal character modulo p, x any positive integer,

(8.4.8)
$$\left| \sum_{a=1}^{x} \chi(a) \right| \le c\sqrt{p} \log p.$$

[Actually, one can establish (8.4.8) with the constant $c = 1$.]

Proof. We first note that it suffices to consider the case $0 < x < p$. Then we have

$$\tau_{\overline{\chi}}(1)\chi(a) = \tau_{\overline{\chi}}(a) = \sum_{h=1}^{p-1} \overline{\chi}(h)e^{2\pi i h a/p}$$

$$= \sum_{\substack{|h|<p/2 \\ h\ne 0}} \overline{\chi}(h)e^{2\pi i h a/p}$$

so that summing over $a = 1, \ldots, x$, we get

$$\tau_{\overline{\chi}}(1) \sum_{a=1}^{x} \chi(a) = \sum_{\substack{|h|<p/2 \\ h\ne 0}} \overline{\chi}(h) \sum_{a=1}^{x} e^{2\pi i h a/p}.$$

Taking absolute values and using (8.4.7) yields

$$p^{1/2}\left|\sum_{a=1}^{x}\chi(a)\right| \le \sum_{\substack{|h|<p/2 \\ h\ne 0}}\left|\sum_{a=1}^{x}e^{2\pi iha/p}\right| = 2\sum_{h=1}^{(p-1)/2}\left|\sum_{a=1}^{x}e^{2\pi iha/p}\right|$$

. . . .

We can now prove the following theorem.

Theorem 8.4.1. For p an odd prime and d dividing $(p - 1)$, we have

(8.4.9)
$$N_d(x) = \frac{x}{d} + O(\sqrt{p}\ \log p).$$

(As given, the O is uniform in x, d, and p.)

Proof. Use (8.4.8) in (8.4.2) and recall that there are exactly d characters χ such that $\chi^d = \chi_0$. . . . .
We can complicate the situation somewhat by considering dth power residues that are prime to a fixed integer A.

Theorem 8.4.2. For p an odd prime, d dividing $(p - 1)$, and A a given positive integer, let $N_d(x, A) =$ the number of positive dth power residues mod p that are $\le x$, and that are also prime to A. Then

(8.4.10)
$$N_d(x, A) = \frac{\phi(A)}{A}\cdot\frac{x}{d} + O\left(2^{\nu(A)}\sqrt{p}\ \log p\right).$$

(As given, the O is uniform in x, d, A, and p.)

Proof. . . . .

§8.5 PRIMITIVE ROOTS OF A PRIME

For a given odd prime p, one can ask for estimates of the number of primitive roots modulo p in any interval and, in particular, for the number of positive primitive roots $\le x$. Since there are exactly $\phi(p - 1)$ positive primitive roots $< p$, the case of special interest is where x is of smaller order than p. Naturally enough, a representation of the characteristic function of the property of being a primitive root is required. Once again, this can be built out of the characters modulo p.

In building a representation of the aforementioned characteristic function, we will first proceed directly so as to get some feeling for how one discovers such a result. Of course, once we have the desired statement a quicker proof by verification is available.

First, letting ind n [see (6.3.9)] be the index of n with respect to some fixed primitive root modulo p, we have for $(n, p) = 1$

$$\sum_{d/(\text{ind }n, p-1)}\mu(d) = \begin{cases} 1 & \text{if } n \text{ is a primitive root mod } p \\ 0 & \text{otherwise.} \end{cases}$$

That is, if $C(n)$ equals the characteristic function of the property of being a primitive root, we have

$$(8.5.1) \qquad\qquad C(n) = \sum_{d/(\text{ind } n, p-1)} \mu(d).$$

Proceeding formally, we see that if $\xi_d(n)$ is defined by

$$\xi_d(n) = \begin{cases} 1 & \text{if } d/\text{ind } n \\ 0 & \text{otherwise,} \end{cases}$$

(8.5.1) may be rewritten as

$$(8.5.2) \qquad\qquad C(n) = \sum_{d/(p-1)} \mu(d)\xi_d(n).$$

Noting that $\xi_d(n)$ is precisely the characteristic function of the property of n that it be a dth power residue modulo p, we have from (8.4.1) that

$$\xi_d(n) = \frac{1}{d} \sum_{\chi^d = \chi_0} \chi(n).$$

Inserting this in (8.5.2) yields

$$(8.5.3) \qquad\qquad C(n) = \sum_{d/(p-1)} \frac{\mu(d)}{d} \sum_{\chi^d = \chi_0} \chi(n).$$

This last equation is, in fact, a workable representation of $C(n)$. It is, however, somewhat more convenient to rewrite it so that each character mod p appears only once in the summations. To achieve this, let Γ_d denote the set of characters modulo p, that are of order d in the character group. Then

$$\sum_{\chi^d = \chi_0} \chi(n) = \sum_{\tau/d} \sum_{\chi \in \Gamma_\tau} \chi(n),$$

and introducing this in (8.5.3) we have

$$C(n) = \sum_{d/(p-1)} \frac{\mu(d)}{d} \sum_{\tau/d} \sum_{\chi \in \Gamma_\tau} \chi(n)$$

or

$$(8.5.4) \qquad\qquad C(n) = \sum_{\tau/(p-1)} \left(\sum_{\substack{d/(p-1) \\ d \equiv 0 (\text{mod } \tau)}} \frac{\mu(d)}{d} \right) \sum_{\chi \in \Gamma_\tau} \chi(n).$$

But setting $d = \tau$ we get (here q denotes a prime)

$$\sum_{\substack{d/(p-1) \\ d \equiv 0 (\bmod \tau)}} \frac{\mu(d)}{d} = \frac{\mu(\tau)}{\tau} \sum_{\substack{t/[(p-1)/\tau] \\ (t,\tau)=1}} \frac{\mu(t)}{t} = \frac{\mu(\tau)}{\tau} \prod_{\substack{q/(p-1) \\ q \nmid \tau}} \left(1 - \frac{1}{q}\right)$$

$$= \frac{\mu(\tau)}{\tau} \frac{1}{\prod_{q/\tau}(1 - (1/q))} \cdot \prod_{q/(p-1)} \left(1 - \frac{1}{q}\right)$$

$$= \frac{\mu(\tau)}{\phi(\tau)} \cdot \frac{\phi(p-1)}{p-1}$$

and inserting this in (8.5.4) yields the following lemma.

Lemma 8.5.1. For p an odd prime, letting Γ_d denote the set of characters of the character group mod p that are of order d,

$$(8.5.5) \quad \frac{\phi(p-1)}{p-1} \sum_{d/(p-1)} \frac{\mu(d)}{\phi(d)} \sum_{\chi \in \Gamma_d} \chi(n) = \begin{cases} 1 & \text{if } n \text{ is a primitive root mod } p \\ 0 & \text{otherwise.} \end{cases}$$

As remarked earlier, (8.5.5) can be easily verified directly. To see this, let $q_1, \ldots, q_r$ be the distinct primes dividing $p - 1$, and note that

$$(8.5.6) \quad \sum_{d/(p-1)} \frac{\mu(d)}{\phi(d)} \sum_{\chi \in \Gamma_d} \chi(n) = \prod_{j=1}^{r} \left(1 - \frac{1}{\phi(q_j)} \sum_{\chi \in \Gamma_{q_j}} \chi(n)\right).$$

Then for $a = \text{ind } n =$ the index of n with respect to a fixed primitive root,

$$\frac{1}{\phi(q_j)} \sum_{\chi \in \Gamma_{q_j}} \chi(n) = \frac{1}{q_j - 1} \sum_{h=1}^{q_j - 1} e^{2\pi i h a/q_j} = \begin{cases} 1 & \text{if } q_j \mid a \\ -\dfrac{1}{(q_j - 1)} & \text{if } q_j \nmid a \end{cases}$$

so that

$$1 - \frac{1}{\phi(q_j)} \sum_{\chi \in \Gamma_{q_j}} \chi(n) = \begin{cases} 0 & \text{if } q_j \mid \text{ind } n \\ \dfrac{q_j}{q_j - 1} & \text{if } q_j \nmid \text{ind } n. \end{cases}$$

Combining this last with (8.5.6) gives

$$(8.5.7) \quad \sum_{d/(p-1)} \frac{\mu(d)}{\phi(d)} \sum_{\chi \in \Gamma_d} \chi(n) = \begin{cases} 0 & \text{if } q_j \mid \text{ind } n \text{ for } some \ j \\ \dfrac{p-1}{\phi(p-1)} & \text{if } q_j \nmid \text{ind } n \text{ for } all \ j \end{cases}$$

But that $q_j \nmid$ (ind n) for all j asserts that (ind $n, p - 1) = 1$, which is precisely the condition that n be a primitive root mod p, and hence (8.5.7) yields (8.5.5).

Theorem 8.5.1. Let prim(x) = the number of positive primitive roots modulo a fixed prime p that are $\leq x$. Then

$$(8.5.8) \qquad \text{prim}(x) = \frac{\phi(p - 1)}{p - 1}(x + O(2^{\nu(p-1)} \sqrt{p} \log p)),$$

where the O is uniform in x and p.

Proof. From (8.5.5),

$$\text{prim}(x) = \sum_{1 \leq n \leq x} \frac{\phi(p - 1)}{p - 1} \sum_{d/(p-1)} \frac{\mu(d)}{\phi(d)} \sum_{\chi \in \Gamma_d} \chi(n),$$

so that

$$(8.5.9) \qquad \text{prim}(x) = \frac{\phi(p - 1)}{p - 1} \left\{ [x] + \sum_{\substack{d/(p-1) \\ d>1}} \frac{\mu(d)}{\phi(d)} \sum_{\chi \in \Gamma_d} \sum_{1 \leq n \leq x} \chi(n) \right\}$$

Then using (8.4.8), ...
From the theorem we obtain the following corollary.

Corollary. The smallest positive primitive root of an odd prime p is smaller than $c \cdot 2^{\nu(p-1)} \sqrt{p} \log p$.

Proof. . . . .

It is interesting to generalize Theorem 8.5.1 to the case where we count primitive roots modulo p that are in a given progression $kt + \ell$. If p divides k, these are $\equiv \ell(\text{mod } p)$, and the problem is completely trivial. Thus we may assume that $p \nmid k$ and can anticipate the need for a generalization of (8.4.8) as follows.

Lemma 8.5.2. For p an odd prime, $k > 0$, ℓ, integers such that $p \nmid k$, we have

$$(8.5.10) \qquad \left| \sum_{0 < kt+\ell \leq x} \chi(kt + \ell) \right| < 2\sqrt{p} \log p.$$

Proof. Since $(k, p) = 1$, we have

$$\left| \sum_{0 < kt+\ell \leq x} \chi(kt + \ell) \right| = \left| \sum_{-(\ell/k) < t \leq (x-\ell)/k} \chi(k)\chi(t + \ell k^{-1}) \right|$$

$$= \left| \sum_{\ell k^{-1} - (\ell/k) < t \leq (x-\ell)/k + \ell k^{-1}} \chi(t) \right|$$

(where k^{-1} denotes the inverse of k modulo p). Using (8.4.8), ...

Theorem 8.5.2. Let $\text{prim}(x, k, \ell) =$ the number of positive primitive roots modulo an odd prime p that are $\leq x$ and $\equiv \ell(\text{mod } k)$. For $k > 0$, ℓ, and p given, such that $(p, k) = 1$, we have

$$(8.5.11) \qquad \text{prim}(x, k, \ell) = \frac{\phi(p-1)}{p-1}\left\{\frac{x}{k} + O(2^{\nu(p-1)}\sqrt{p}\log p)\right\}.$$

Proof. From (8.5.5),

$$\text{prim}(x, k, \ell) = \frac{\phi(p-1)}{p-1}\sum_{d\mid(p-1)}\frac{\mu(d)}{\phi(d)}\sum_{\chi\in\Gamma_d}\sum_{0<kt+\ell\leq x}\chi(kt+\ell)$$

. . . .
[where we recall that the number of $\chi\in\Gamma_d$ is $\phi(d)$].

EXERCISES

1. Estimate the number of positive primitive roots $\leq x$, modulo an odd prime p that are prime to a given positive integer A. Obtain the dependence of the error estimate on A. Use this to show that for p sufficiently large, there is always a positive integer $< p$ that is both a primitive root and prime to $p - 1$.

2. Show that the number of positive squarefree primitive roots modulo p that are $\leq x$ equals

    $$\frac{\phi(p-1)}{p-1}\{cx + O(2^{\nu(p-1)}p^{1/4}(\log p)^{1/2}x^{1/2})\}$$

    where $c = \prod_p (1 - (1/p^2))$ and the O is uniform in p.

    [*Hint.* use the fact that for a nonprincipal character χ mod p,

    $$\left|\sum_{\substack{1\leq m\leq x \\ m\equiv 0(\text{mod } d^2)}}\chi(m)\right| \leq \min\left(\frac{x}{d^2}, \sqrt{p}\log p\right),$$

    to estimate $|\sum_{\substack{m\leq x \\ m\text{ squarefree}}}\chi(m)|$.]

3. Generalize Lemma 7.4B.1 and show that for χ a nonprincipal character modulo an odd prime p,

    $$\left|\sum_{u\in\mathscr{S}, v\in\mathscr{T}}\chi(u + v)\right| \leq \sqrt{pN(\mathscr{S})N(\mathscr{T})},$$

where $\mathcal{S}$ and $\mathcal{T}$ are sets of integers such that within each set no two of the integers are congruent modulo p.

4.   Use the result of Ex. 3 to show that the smallest positive primitive root of an odd prime p is smaller than $c\,2^{\nu(p-1)}\sqrt{p}$.

§8.5A   SQUAREFUL PRIMITIVE ROOTS

In this section we consider a primitive root problem that illustrates the somewhat rare phenomenon of a significant contribution coming from a character other than the principal one. The problem is that of counting the primitive roots that are squareful. The aforementioned character is the quadratic character mod p (i.e., the character of order 2). We will require the following lemma.

Lemma 8.5A.1.   Let $\mathcal{F}$ denote the set of squareful integers. Then, for χ a character modulo a prime p, such that $\chi^2 \neq \chi_0$, we have

$$(8.5A.1) \qquad\qquad \sum_{\substack{m \le x \\ m \in \mathcal{F}}} \chi(m) = O(p^{1/6}(\log p)^{1/3} x^{1/3})$$

   Proof.

$$\sum_{\substack{m \le x \\ m \in \mathcal{F}}} \chi(m) = \sum_{\substack{u^2 q^3 \le x \\ q \text{ squarefree}}} \chi(u^2 q^3) = \sum_{\substack{u^2 q^3 \le x \\ q \text{ squarefree}}} \chi^2(u)\chi^3(q)$$

$$= \sum_{q \le \sqrt[3]{x}} \chi^3(q) \sum_{u \le \sqrt{x/q^3}} \chi^2(u).$$

Taking absolute values, we have

$$\left| \sum_{\substack{m \le x \\ m \in \mathcal{F}}} \chi(m) \right| \le \sum_{q \le \sqrt[3]{x}} \left| \sum_{u \le \sqrt{x/q^3}} \chi^2(u) \right|$$

$$\le \sum_{q \le \sqrt[3]{x}} \min\left(\frac{\sqrt{x}}{q^{3/2}}, \sqrt{p}\,\log p \right)$$

$$\le \sum_{q \le \left(\frac{\sqrt{x}}{\sqrt{p}\log p} \right)^{2/3}} \sqrt{p}\,\log p + \sum_{q > \left(\frac{\sqrt{x}}{\sqrt{p}\log p} \right)^{2/3}} \frac{\sqrt{x}}{q^{3/2}}$$

$$= O((\sqrt{p}\,\log p)^{1/3} x^{1/3}).$$

Lemma 8.5A.2.   Let $\hat{\chi}$ be the nonprincipal character modulo p such that $\hat{\chi}^2 = \chi_0$. Then

$$(8.5A.2) \qquad \sum_{\substack{m \le x \\ m \in \mathcal{F}}} \hat{\chi}(m) = \sqrt{x}\left(1 - \frac{1}{p}\right) \sum_{\substack{q \\ \text{squarefree}}} \frac{\hat{\chi}(q)}{q^{3/2}} + O\left(\sqrt[3]{x}\right).$$

Proof.

$$\sum_{\substack{m \le x \\ m \in \mathcal{F}}} \hat{\chi}(m) = \sum_{u^2 q^3 \le x} \hat{\chi}(u^2 q^3) = \sum_{u^2 q^3 \le x} \hat{\chi}^2(u)\hat{\chi}^3(q)$$

$$= \sum_{\substack{u^2 q^3 \le x \\ (u,p)=1}} \hat{\chi}(q) = \sum_{q \le \sqrt[3]{x}} \hat{\chi}(q) \sum_{\substack{u \le (\sqrt{x}/q^{3/2}) \\ (u,p)=1}} 1$$

$$= \sqrt{x}\left(1 - \frac{1}{p}\right) \sum_{q \le \sqrt[3]{x}} \frac{\hat{\chi}(q)}{q^{3/2}} + O(\sqrt[3]{x}).$$

Extending the sum to infinity, ...
(8.5A.2) follows

Since in Lemma 8.5A.2, $\hat{\chi}(m) = \left(\frac{m}{p}\right)$, (8.5A.2) can be rewritten as

$$(8.5A.3) \qquad \sum_{\substack{m \le x \\ m \in \mathcal{F}}} \hat{\chi}(m) = \sqrt{x}\left(1 - \frac{1}{p}\right) \sum_{\substack{q \\ \text{squarefree}}} \frac{\left(\frac{q}{p}\right)}{q^{3/2}} + O(\sqrt[3]{x}).$$

Theorem 8.5A.1. For p an odd prime, the number of positive primitive roots $\le x$ which are squareful equals

$$(8.5A.4) \qquad \frac{\phi(p - 1)}{p - 1}\{c\sqrt{x} + O(\sqrt[3]{x}p^{1/6}(\log p)^{1/3}2^{\nu(p-1)})\}$$

where

$$(8.5A.5) \qquad c = 2\left(\sum_{\substack{q \text{ squarefree} \\ \left(\frac{q}{p}\right) = -1}} \frac{1}{q^{3/2}}\right)\left(1 - \frac{1}{p}\right).$$

Proof. Summing (8.5.5) over the squareful integers $m \in \mathcal{F}$ that are $\le x$, gives

$$\frac{\phi(p - 1)}{p - 1} \sum_{d/(p-1)} \frac{\mu(d)}{\phi(d)} \sum_{\chi \in \Gamma_d} \sum_{\substack{m \le x \\ m \in \mathcal{F}}} \chi(m)$$

$$= \frac{\phi(p - 1)}{p - 1}\left(\text{pow}_2(x,p) - \sum_{\substack{m \le x \\ m \in \mathcal{F}}} \hat{\chi}(m) + \sum_{\substack{d/(p-1) \\ d>2}} \frac{\mu(d)}{\phi(d)} \sum_{\chi \in \Gamma_d} \sum_{\substack{m \le x \\ m \in \mathcal{F}}} \chi(m)\right)$$

where $\hat{\chi}(m) = \left(\dfrac{m}{p}\right)$ is the character mod p of order 2, and $\mathrm{pow}_2(x,p) =$ the number of squareful integers $\leq x$ which are prime to p. Using Lemma 8.5A.2 to estimate the first sum above, and Lemma 8.5A.1 for the sums involving $\chi \in \Gamma_d$, $d > 2$, and generalizing (8.3.4), we arrive at $\ldots \ldots \ldots \ldots \ldots \ldots \ldots$, which yields (8.5A.4).

Note that (8.5A.4) provides that the smallest positive squareful primitive root mod p is $O(p(\log p)^2 2^{6\nu(p-1)})$. Thus, although this tells us that they exist, it does not reveal whether there is a least positive residue $< p$ with this property.

EXERCISES

1. Generalize Theorem 8.5A.1 by estimating the number of positive primitive roots $\leq x$ that are rth powerful for each fixed $r \geq 2$.

2. Prove that the constant c given in (8.5A.5) is not zero.

§8.5B   CONSECUTIVE PRIMITIVE ROOTS

We consider next the question of the existence of consecutive positive integers both of which are primitive roots modulo a given odd prime p. For $p = 3$, 2 is the only primitive root, so that no such consecutive pair exists. For $p = 5$, both 2 and 3 are primitive roots; whereas for $p = 7$, only 3 and 5 are primitive roots so that again there is no consecutive pair. However, we will see that for all sufficiently large primes p, there does exist such a pair of consecutive primitive roots. In fact, it is probably true for all $p \geq 11$.

The application of our counting principles to the problem at hand will involve one new circumstance; namely, the method works only if we deal with "complete" character sums (i.e., over a complete residue class). In particular we require the following lemma.

Lemma 8.5B.1.   For λ and ψ, two characters modulo an odd prime p, letting $\tau_\chi = \tau_\chi(1)$ for any character χ [see (8.4.3)], we have

$$(8.5B.1) \quad \sum_{n=0}^{p-1} \chi(n)\psi(n+1) = \begin{cases} p - 2 & \text{if } \chi = \psi = \chi_0 \\ -\chi(-1) & \text{if } \chi \neq \chi_0, \psi = \chi_0 \\ -1 & \text{if } \chi = \chi_0, \psi \neq \chi_0 \\ \dfrac{p\overline{\chi}(-1)\tau_{\overline{\chi\psi}}}{\tau_{\overline{\chi}}\,\tau_{\overline{\psi}}} & \text{if } \chi \neq \chi_0, \psi \neq \chi_0. \end{cases}$$

Proof.   For the case where $\chi = \psi = \chi_0$, we have $\ldots \ldots \ldots \ldots \ldots$
If $\chi \neq \chi_0$ and $\psi = \chi_0$, $\ldots \ldots \ldots \ldots \ldots \ldots \ldots$

Similarly, if $\chi = \chi_0$, $\psi \neq \chi_0$, $\cdots \cdots \cdots \cdots \cdots \cdots \cdots$
Finally, for $\chi \neq \chi_0$, $\psi \neq \chi_0$, we have, from (8.4.5),

$$(8.5B.2) \qquad \tau_{\bar\chi}\tau_{\bar\psi}\chi(n)\psi(n+1) = \sum_{\substack{1\le h\le p-1 \\ 1\le h'\le p-1}} \overline\chi(h)\overline\psi(h')e^{(2\pi i/p)(h+h')n+2\pi i h'/p}$$

Summing (8.5B.2) over n yields

$$\tau_{\bar\chi}\tau_{\bar\psi}\sum_{n=0}^{p-1}\chi(n)\psi(n+1) = \sum_{\substack{1\le h\le p-1 \\ 1\le h'\le p-1}}\overline\chi(h)\overline\psi(h')e^{2\pi i h'/p}\sum_{n=0}^{p-1}e^{(2\pi i/p)(h+h')n}$$

$$= p\sum_{h'=1}^{p-1}\overline\chi(-h')\overline\psi(h')e^{2\pi i h'/p}$$

$$= p\overline\chi(-1)\sum_{h'=1}^{p-1}\overline{\chi\psi}(h')e^{2\pi i h'/p}$$

$$= p\overline\chi(-1)\tau_{\overline{\chi\psi}}$$

which completes the proof of (8.5B.1).

Theorem 8.5B.1. For p any prime greater than 3, the number of positive primitive roots $q < p$ such that $q+1$ is also a primitive root equals

$$(8.5B.3) \qquad p\left(\frac{\phi(p-1)}{p-1}\right)^2\left(1 + \sum_{\substack{d\mid(p-1) \\ d'\mid(p-1) \\ d>1,d'>1}}\frac{\mu(d)}{\phi(d)}\frac{\mu(d')}{\phi(d')}\sum_{\substack{\chi\in\Gamma_d \\ \psi\in\Gamma_{d'}}}\frac{\chi(-1)\tau_{\overline{\chi\psi}}}{\tau_\chi\tau_\psi}\right).$$

Proof. Noting that

$$\left(\frac{\phi(p-1)}{p-1}\right)^2\left(\sum_{d\mid(p-1)}\frac{\mu(d)}{\phi(d)}\sum_{\chi\in\Gamma_d}\chi(n)\right)\left(\sum_{d'\mid(p-1)}\frac{\mu(d')}{\phi(d')}\sum_{\psi\in\Gamma_{d'}}\psi(n+1)\right)$$

is 1 or 0 according as both n and $n+1$ are primitive roots or not, the count under consideration equals

$$\left(\frac{\phi(p-1)}{p-1}\right)^2\sum_{\substack{d\mid(p-1) \\ d'\mid(p-1)}}\frac{\mu(d)}{\phi(d)}\frac{\mu(d')}{\phi(d')}\sum_{\substack{\chi\in\Gamma_d \\ \psi\in\Gamma_{d'}}}\sum_{n=0}^{p-1}\chi(n)\psi(n+1).$$

Apply Lemma 8.5B.1, $\cdots \cdots \cdots \cdots \cdots \cdots \cdots \cdots \cdots$
so that the above equals

$$(8.5\text{B}.4) \qquad \left(\frac{\phi(p-1)}{p-1}\right)^2 \times$$

$$\left(p - 2 - \sum_{\substack{d/(p-1) \\ d>1}} \frac{\mu(d)}{\phi(d)} \sum_{\chi \in \Gamma_d} \chi(-1) - \sum_{\substack{d'/(p-1) \\ d'>1}} \frac{\mu(d')}{\phi(d')} \sum_{\psi \in \Gamma_{d'}} 1 \right.$$

$$\left. + p \sum_{\substack{d/(p-1) \\ d'/(p-1) \\ d>1, d'>1}} \frac{\mu(d)}{\phi(d)} \frac{\mu(d')}{\phi(d')} \sum_{\substack{\chi \in \Gamma_d \\ \psi \in \Gamma_{d'}}} \frac{\bar{\chi}(-1)\tau_{\bar{\chi}\bar{\psi}}}{\tau_{\bar{\chi}}\tau_{\bar{\psi}}} \right).$$

Note that with respect to any primitive root $\text{ind}(-1) = (p-1)/2$,

$$\sum_{\chi \in \Gamma_d} \chi(-1) = \sum_{\substack{1 \le a \le d \\ (a,d)=1}} e^{2\pi i a/d)((p-1)/2)}$$

$$= \sum_{\substack{1 \le a \le d \\ (a,d)=1}} e^{\pi i a((p-1)/d)}$$

[for $d/(p-1)$]. Then separating the cases where $(p-1)/d$ is even or odd, we find

. . . . ,

which yields

$$(8.5\text{B}.5) \qquad \sum_{\chi \in \Gamma_d} \chi(-1) = (-1)^{(p-1)/d} \phi(d).$$

From (8.5B.5),

$$\sum_{\substack{d/(p-1) \\ d>1}} \frac{\mu(d)}{\phi(d)} \sum_{\chi \in \Gamma_d} \chi(-1) = \sum_{\substack{d/(p-1) \\ d>1}} \mu(d)(-1)^{(p-1)/d}$$

. . . . ,

so that for $p > 3$,

$$(8.5\text{B}.6) \qquad \sum_{\substack{d/(p-1) \\ d>1}} \frac{\mu(d)}{\phi(d)} \sum_{\chi \in \Gamma_d} \chi(-1) = -1.$$

It follows that

$$(8.5\text{B}.7) \qquad \sum_{\substack{d'/(p-1) \\ d'>1}} \frac{\mu(d')}{\phi(d')} \sum_{\psi \in \Gamma_{d'}} 1 = -1.$$

Combining (8.5B.4), (8.5B.6), and (8.5B.7) our count becomes

$$\left(\frac{\phi(p-1)}{p-1}\right)^2\left(p+p\sum_{\substack{d/(p-1)\\d'/(p-1)\\d>1,d'>1}}\frac{\mu(d)}{\phi(d)}\frac{\mu(d')}{\phi(d')}\sum_{\substack{\chi\in\Gamma_d\\\psi\in\Gamma_{d'}}}\frac{\overline{\chi}(-1)\,\tau_{\overline{\chi}\,\overline{\psi}}}{\tau_{\overline{\chi}}\cdot\tau_{\overline{\psi}}}\right).$$

Noting that as χ and ψ run over Γ_d and $\Gamma_{d'}$, respectively, so do $\overline{\chi}$ and $\overline{\psi}$; (8.5B.3) follows from the above.

Recalling that

$$|\tau_\chi| = \begin{cases} 1 & \text{if } \chi = \chi_0 \\ \sqrt{p} & \text{if } \chi \neq \chi_0, \end{cases}$$

it follows that in all cases where $\chi \neq \chi_0$, $\psi \neq \psi_0$,

$$\left|\frac{\chi(-1)\,\tau_{\chi\psi}}{\tau_\chi\tau_\psi}\right| \leq \frac{1}{\sqrt{p}}.$$

Thus from (8.5B.3) we have the following corollary.

Corollary. For all sufficiently large primes p, there exists at least one positive primitive root $g < p$ such that $g + 1$ is also a primitive root mod p.

Proof. . . . .

EXERCISES

1. Show that the number of positive primitive roots $g < p$ such that $g + 1$ is also a primitive root equals

$$\left(\frac{\phi(p-1)}{p-1}\right)^2 \{p + O(4^{\nu(p-1)}\sqrt{p})\}$$

2. For any fixed positive integer b, show that for all sufficiently large primes $p \geq p_0$ there exist a pair of primitive roots modulo p that differ by b. Make some estimate of how p_0 depends on b.

3. Show that the equation $xy = z$ has no solutions for which x, y, and z are all primitive roots modulo a prime p for all primes $p > 3$. (Hint: There are moments when complicated things should be ignored.)

4. Noting that for all d dividing $(p - 1)$, $\sum_{\chi^d=\chi_0}\chi(-1) = d$ or 0 according as -1 is or is not a dth power residue modulo the odd prime p, derive (8.5B.5) as a consequence of the Möbius inversion formula.

5. For any fixed integer $d > 0$, show that for all sufficiently large primes p, $x^d - y^d \equiv 1 \pmod{p}$ has a solution such that $(x,p) = (y,p) = 1$.

§8.6   COMBINATORIAL IDENTITIES

In this section we propose to briefly illustrate the use of binomial coefficients in the construction of representations of characteristic functions of properties. We consider the characteristic function $C_k(m)$ of the property of the integer m that $m = k$. That is,

$$(8.6.1) \qquad C_k(m) = \begin{cases} 1 & \text{if } m = k \\ 0 & \text{if } m \neq k. \end{cases}$$

The desired representation is given by the following lemma.

Lemma 8.6.1.   For any fixed positive integer k, for all integers $m \geq 0$,

$$(8.6.2) \qquad C_k(m) = (-1)^k \sum_{j=0}^{m} \binom{m}{j}\binom{j}{k}(-1)^j.$$

Proof.   The binomial theorem gives

$$(1 + x)^m = \sum_{j=0}^{m} \binom{m}{j} x^j.$$

Differentiating k times and dividing by $k!$ yields

$$(8.6.3) \qquad \binom{m}{k} (1 + x)^{m-k} = \sum_{j=0}^{m} \binom{m}{j}\binom{j}{k} x^{j-k}.$$

From this we obtain (8.6.2).

From (8.6.2) we have the following combinatorial counting theorem. For any set S, let $|S|$ denote the number of elements in S.

Theorem 8.6.1.   Given r finite sets $S_1, S_2, \ldots, S_r$ that are subsets of some collection of objects Ω, let Ω_k denote the elements of Ω that belong to exactly k of the sets S_i. Then for each integer $k \geq 0$,

$$(8.6.4) \qquad |\Omega_k| = (-1)^k \sum_{j=0}^{r} (-1)^j \binom{j}{k} \sum_{1 \leq i_1 < \cdots < i_j \leq r} |S_{i_1} \cap S_{i_2} \cap \cdots \cap S_{i_j}|$$

Proof.   For each element e of Ω, let $N(e)$ denote the number of S_i, $i = 1, \ldots, r$ that contain e. Then

$$|\Omega_k| = \sum_{e \in \Omega} C_k (N(e))$$

$$= (-1)^k \sum_{e \in \Omega} \sum_{j=0}^{N(e)} \binom{N(e)}{j}\binom{j}{k}(-1)^j.$$

Interchanging the order of summation on the right,

$$|\Omega_k| = (-1)^k \sum_{j=0}^{r} (-1)^j \binom{j}{k} \sum_{\substack{e \in \Omega \\ N(e) \geq j}} \binom{N(e)}{j}$$

. . . .

Q.E.D.

The case $k = 0$ of (8.6.4) is a "sieving formula," also sometimes referred to as "the inclusion–exclusion principle." The cases for $k > 0$ are generally less familiar. We illustrate the use of (8.6.4) in the following theorem.

Theorem 8.6.2. Let $1 < a_1 < a_2 < \ldots < a_r$ be given integers and define $N_k(x) =$ the number of positive integers $\leq x$ that are divisible by exactly k of the a_i. Then

$$(8.6.5) \quad \lim_{x \to \infty} \frac{N_k(x)}{x} = (-1)^k \sum_{j=0}^{r} (-1)^j \binom{j}{k} \sum_{1 \leq i_1 < \ldots < i_j \leq r} \frac{1}{\{a_{i_1}, \ldots, a_{i_j}\}},$$

(where $\{\ldots\}$ denotes the least common multiple).

Proof. We apply (8.6.4) in the case where $S_i =$ the set of positive integers $\leq x$, that are divisible by a_i.

In the special case where the $a_1, \ldots, a_r$ are relatively prime by pairs, the right side of (8.6.5) becomes

$$(-1)^k \sum_{j=0}^{r} (-1)^j \binom{j}{k} \sum_{1 \leq i_1 < \cdots < i_j \leq r} \frac{1}{a_{i_1} \cdots a_{i_j}}.$$

On the other hand, it is easily seen that, in this case, the left side of (8.6.5) equals

$$\left(\sum_{1 \leq i_1 < \cdots < i_k \leq r} \frac{1}{(a_{i_1} - 1) \cdots (a_{i_k} - 1)} \right) \prod_{j=1}^{r} \left(1 - \frac{1}{a_j} \right)$$

For $N_k(x) = \sum |\mathcal{T}_{i_1}, \ldots, i_k|$, where $\mathcal{T}_{i_1}, \ldots, i_k$ equals the set of positive integers $\leq x$ divisible by each of $a_{i_1}, \ldots, a_{i_k}$, but *not* by any of the other a_j, and

. . . .

The following general identity is suggested for any sets of real numbers $x_1, \ldots, x_r$, all $\neq 1$ [via $x_i = 1/(1 - a_i)$]:

(8.6.6)

$$\frac{\sum_{1 \leq i_1 < \cdots < i_k \leq r} x_{i_1}, \cdots x_{i_k}}{\prod_{j=1}^{r} (1 - x_j)} = \sum_{j=0}^{r} \binom{j}{k} \sum_{1 \leq i_1 < \cdots < i_j \leq r} \left(\frac{x_{i_1}}{1 - x_{i_1}} \right) \cdots \left(\frac{x_{i_j}}{1 - x_{i_j}} \right).$$

This can in fact be proved directly. . . . .

NOTES

§8.1. Though much of number theory is implicitly devoted to counting problems, very few papers appear that focus on counting problems per se. However, one such effort is that of H. Delange, 1956 [8.1]. Delange is concerned with finding "an equivalent expression for the number of integers $\leq x$ which possess one or more properties." Most of his examples involve sets of integers with $\nu(n)$ or $\Omega(n)$ fixed.

§8.2. Theorem 8.2.1 is credited to L. Gegenbauer, 1885 [8.2].

An estimate for $Q_2(x, k, l)$ is given in Landau's *Handbuch* [8.3] which is equivalent to (8.2.8). It asserts in the case $(k, l) = 1$ that

$$Q_2(x, k, l) = \frac{x}{k} \prod_{p \nmid k} \left(1 - \frac{1}{p^2}\right) + O(\sqrt{x}).$$

E. Cohen and R. Robinson [8.4] gave another equivalent form of (8.2.8). K. Prachar [8.5] gives the slightly more precise assertion

$$Q_r(x, k, l) = A_r \frac{x}{k} + O\{r^{\nu(k)}[k^{-(1/r^2)}x^{(1/r)} + k^{(1/r)}]\},$$

where $A_r = [1/\zeta(r)]\prod_{p/k}(1 - p^{-r})$ and the O is uniform in k and r.

More recent results concerning square free numbers are summarized in [8.6].

§8.2A. Theorem 8.2A.1 is attributed to K. Roth, 1951 [8.7]. Roth also proved that intervals of length $O(x^{1/4+\epsilon})$ must contain a squarefree integer. Using a result of Van der Corput, he further improved this to $O[x^{(3/13)}(\log x)^{(4/13)}]$. In 1954, H. E. Richert [8.8] replaced this by $O(x^{(2/9)} \log x)$ and in 1955, R. A. Rankin further sharpened this to $O(x^{.22198215+\epsilon})$. [8.9].

Roth and Halberstam [8.10] showed that an interval of length $O(x^{(1/2k)+\epsilon})$ must contain a kth powerfree integer.

A related problem of interest is that of determining the size of the smallest squarefree integer in an arithmetic progression $km + l$. Denoting the integer by $q(k,l)$, Prachar [8.5] obtains

$$q(k, l) \leq c_1 k^{(3/2)} 2^{\nu(k)}.$$

Erdos [8.11] improved this to

$$q(k, l) \leq c_3 k^{(3/2)}/\log k,$$

and conjectured that $q(k,l) = O(k^{1+\epsilon})$.

§8.3. Theorem 8.3.1 was proved in 1934 by P. Erdos and G. Szekeres [8.12]. In 1958 [8.13], P. Bateman and E. Grosswald showed that for $r > 2$

$$\text{pow}_r(x) = B_r x^{(1/r)} + \gamma_1 x^{(1/(r+1))} + O(x^{(1/(r+2))}).$$

For $r = 2$, they improved the error term to $O(x^{(1/6)}e^{-a\omega(x)})$ where $\omega(x) = (\log x)^{(4/7)}(\log \log x)^{-(3/7)}$. These results were also derived later by E. Cohen and J. Davis using more elementary methods [8.14].

The error term for the case $r = 2$ was improved in 1973 to

$$O[x^{(1/6)}\exp(-A(\log x)^{(3/5)}(\log \log x)^{-(1/5)}],$$

by D. Suryanarayana and R. R. Sitaramachandra [8.15].

§8.4. Theorem 8.4.1 was given by I. M. Vinogradov in 1927 [8.15]. It depends on Lemma 8.4.2, which Vinogradov proved in 1918 [8.17]. The lemma was also discovered independently by G. Polya, in the same year [8.18].

§8.5. Estimates for $g(p)$, the smallest positive primitive root modulo p, have a long history. Vinogradov originally gave the assertion of the corollary to Theorem 8.5.1, and in 1936 improved this to

$$g(p) = O(2^m p^{1/2} \log \log p),$$

where $m = \nu(p - 1)$ [8.19]. In 1942, L. K. Hua [8.20] obtained

$$g(p) = O(2^m p^{1/2}).$$

In 1945, Erdos proved that $g(p) = O[p^{1/2} (\log p)^{17}]$, and in 1957, P. Erdos and H. N. Shapiro obtained that $g(p) = O(m^c p^{1/2})$ [8.21, 8.22].

In 1959, Yuan Wang showed that $g(p) = O(p^{1/4+\varepsilon})$. Also, under the assumption of the Riemann Hypothesis, he obtained $g(p) = O[m^6 \log p)^2]$[8.23, 8.24]. In 1962, $g(p) = O(p^{1/4+\varepsilon})$ was given independently by D. A. Burgess [8.25].

More recently, in 1974, S. D. Cohen, R. W. K. Odoni, and W. W. Stothers proved that the least primitive root modulo p^2 is less than $p^{1/4+\varepsilon}$ (for all sufficiently large primes p) [8.26].

§8.5B. E. Vegh proved in 1970 that if p is prime having 2 as a primitive root, $p \equiv 1 \pmod 4$, then p must have consecutive primitive roots [8.27].

REFERENCES

8.1. H. Delange, Sur la distribution des entiers ayant certaines propriétés, *Annale Scientifique de l'Ecole Normale Supérieure*, **73**, (3), 1956, 15–74.

8.2. L. Gegenbauer, Asymptotische Gesetze der Zahlentheorie, *Denkschriften Akademie Wien*, **49**, (1), 37–80.

8.3. E. Landau, Ref. 6.30, pp. 633–636.

8.4. E. Cohen and R. Robinson, On the distribution of k-free integers in residue classes, *Acta Arithmetica*, **8**, 1962–1963, 283–293.

8.5. K. Prachar, Uber die kleinste quadratfrei Zahl einer arithmetische Reihe, *Monatshefte für Mathematik*, **62**, 1958, 173–176.

8.6. M. J. Croft, Squarefree numbers in arithmetic progressions, *Proceedings of the London Mathematical Society*, **30**, (3), 1975, 143–159.

8.7. K. F. Roth, On the gaps between squarefree numbers, *Journal of the London Mathematical Society*, **26**, 1951, 263–268.

8.8. H. E. Richert, On the difference between consecutive squarefree numbers, *Journal of the London Mathematical Society*, **29**, 1954, 16–20.

8.9. R. A. Rankin, Van der Corput's method and the theory of exponent squares, *Quarterly Journal of Mathematics*, **6**, 1955, 147–153.

8.10. H. Halberstam and K. Roth, On the gaps between consecutive k-free integers, *Journal of the London Mathematical Society*, **26**, 1951, 268–273.

8.11. P. Erdos, Uber die kleinste quadratfrei Zahl einer arithmetische Reihe, *Monatshefte für Mathematik*, **64**, 1960, 314–316.

8.12. P. Erdos and G. Szekeres, Uber die Anzahl der Abelschen Gruppen gegebener Ordnung und uber eine verwandtes zahlentheoretisches Problem, *Acta Litterarium ac Scientiarum Regiae Universitatis Hungaricae Francisco-Josephinae*, **7**, 1934, 95–102.

8.13. P. T. Bateman and E. Grosswald, On a theorem of Erdos and Szekeres, *Illinois Journal of Mathematics*, **2**, 1958, 88–98.

8.14. E. Cohen and J. Davis, Elementary estimates for certain types of integers, *Acta Litterarium ac Scientiarum Regiae Universitatis Hungaricae Francisco-Josephinae*, **31**, 1970, 363–371.

8.15. D. Suryanarayana and R. R. Sitaramachandra, The distribution of square-full integers, *Arkiv for Matematik*, **11**, 1973, 195–201.

8.16. I. M. Vinogradov, On the bound of the least non-residue of nth powers, *Transactions of the American Mathematical Society*, **29**, 1927, 218–226.

8.17. I. M. Vinogradov, On the distribution of residues and non-residues of powers, *Journal of the Physico-Mathematical Society of Perm,* **1,** 1918, 94–96.

8.18. G. Polya, Uber die Verteilung der quadratische Reste und Nichtreste, *Göttingen Nachrichten,* 1918, 21–29.

8.19. I. M. Vinogradov, On the least primitive root of a prime, *Doklady Academy of Science, U.S.S.R.,* 1930, 7–11.

8.20. L. K. Hua, On the least primitive root of a prime, *Bulletin of the American Mathematical Society,* **48,** 1942, 726–730.

8.21. P. Erdos, On the least primitive root of a prime p, *Bulletin of the American Mathematical Society,* **51,** 1945, 131–132.

8.22. P. Erdos and H. N. Shapiro, On the least primitive root of a prime, *Pacific Journal of Mathematics,* **7,** 1957, 861–865.

8.23. Y. Wang, A note on the least primitive root of a prime, *Science Record, China, N.S.,* **3,** 1959, 174–179.

8.24. Y. Wang, On the least primitive root of a prime, *Acta Mathematica Sinica,* **10,** 1961, 1–14.

8.25. D. A. Burgess, On character sums and primitive roots, *Proceedings of the London Mathematical Society,* **12,** (3), 1962, 179–192.

8.26. S. D. Cohen, R. W. K. Odoni, and W. W. Stothers, On the least primitive root modulo p^2 *Bulletin of the London Mathematical Society,* **6,** 1974, 42–46.

8.27. E. Vegh, A new condition for consecutive primitive roots of a prime, *Elements der Mathematik,* **25,** 1970, 113.

8.28. M. Hausman, Primitive Roots Satisfying a Coprime Condition, *American Mathematical Monthly,* **83,** 1976, 720–723.

9

THE ELEMENTS
OF PRIME
NUMBER THEORY

§9.1 SIMPLE BEGINNINGS

In §2.8, several very primitive results are given concerning properties of prime numbers. In this chapter we propose to pursue a somewhat more systematic development of the basic aspects of the theory of prime numbers.

To begin, we recall the counting function, $\pi(x) = $ the number of primes $\leq x$, or

$$(9.1.1) \qquad \pi(x) = \sum_{p \leq x} 1.$$

Note then that $\pi(x)$ is a step function that has a jump of 1 when x equals a prime number. Thus, this 1 measures the sensitivity of the function $\pi(x)$ to the presence of primes. If the 1 is replaced by a quantity larger than 1, the resultant function is *more* sensitive to the presence of primes, as, for example, with

$$(9.1.2) \qquad \theta(x) = \sum_{p \leq x} \log p.$$

One can also increase sensitivity to primes indirectly by adding a sensitivity to prime powers as with

$$(9.1.3) \qquad \psi(x) = \sum_{p^{\alpha} \leq x} \log p,$$

where $\log p$ is counted for every power of the prime p that is $\leq x$. Similarly, if the 1 on the right of (9.1.1) is replaced by something smaller, the resultant function is less sensitive to the presence of primes. For example,

$$(9.1.4) \qquad\qquad T(x) = \sum_{p \le x} \frac{1}{p}$$

and

$$(9.1.5) \qquad\qquad S(x) = \sum_{p \le x} \frac{\log p}{p}$$

are both less sensitive than $\pi(x)$ and $T(x)$ is less sensitive than $S(x)$ to the presence of primes.

A reasonable heuristic principle is that the "flow of information" about primes is from the more sensitive function to the less sensitive function. Thus, for example, one would expect that any sort of estimate for $S(x)$ would have implications for $T(x)$ or that information concerning $\theta(x)$ implies things about $\pi(x)$. These kinds of observations are generally valid. Moreover, in many cases, the study of two functions of the primes, with different sensitivity levels, are completely equivalent. This, in turn, usually stems from some formal relations connecting the two functions. The following theorem provides an example of this.

Theorem 9.1.1. For all $x > 0$, we have

$$(9.1.6) \qquad\qquad \psi(x) = \sum_{\alpha=1}^{\infty} \theta(x^{1/\alpha})$$

and hence

$$(9.1.7) \qquad\qquad \psi(x) = \theta(x) + 0(\sqrt{x} \log x).$$

[Note in particular that $\psi(x) \ge \theta(x)$.]

    Proof. From (9.1.3) and (9.1.2) we have

$$\psi(x) = \sum_{p^\alpha \le x} \log p = \sum_{\alpha=1}^{\infty} \sum_{p \le x^{1/\alpha}} \log p = \sum_{\alpha=1}^{\infty} \theta(x^{1/\alpha})$$

which is (9.1.6). That $\psi(x) \ge \theta(x)$ follows from this or immediately from the definitions.

    From (9.1.2) we have

$$\theta(x) = \sum_{p \le x} \log p \le (\log x) \sum_{p \le x} 1$$

or

$$(9.1.8) \qquad\qquad \theta(x) \le \pi(x) \log x.$$

Since $\pi(x) \leq x$, it follows that $\theta(x) = O(x \log x)$. Next note that on the right of (9.1.6), for $\alpha > \log x/\log 2$, $x^{1/\alpha} < 2$, so that the corresponding summands $\theta(x^{1/\alpha}) = 0$. Then

$$\sum_{\alpha=2}^{\infty} \theta(x^{1/\alpha}) = \sum_{2 \leq \alpha \leq \log x/\log 2} \theta(x^{1/\alpha})$$

$$= \sum_{2 \leq \alpha \leq \log x/\log 2} O(x^{1/\alpha} \cdot \alpha \log x)$$

$$= O(x^{1/2} \log x) + \sum_{3 \leq \alpha \leq \log x/\log 2} O(x^{1/3} \log^2 x) = O(x^{1/2} \log x)$$

and (9.1.7) follows from (9.1.6).

From (9.1.7), we see that statements concerning the order or magnitude of one of $\theta(x)$ or $\psi(x)$ yield a result for the other. In particular, we have

$$(9.1.9) \qquad \lim_{x \to \infty} \frac{\theta(x)}{x} = \lim_{x \to \infty} \frac{\psi(x)}{x}$$

and

$$(9.1.10) \qquad \overline{\lim_{x \to \infty}} \frac{\theta(x)}{x} = \overline{\lim_{x \to \infty}} \frac{\psi(x)}{x}.$$

A natural question that arises early is, "How many primes are there?" In §2.8 it was shown that there are not only infinitely many but "more than a specified order," certainly enough to insure that [see (2.8.5)] for $N > 1$

$$(9.1.11) \qquad \prod_{p \leq N} \left(1 - \frac{1}{p}\right) < \frac{1}{\log N}.$$

On the other hand, it is reasonable to consider the frequency of the primes within the set of all integers. The answer is given by the following theorem.

Theorem 9.1.2. The primes have zero "frequency" or "density" in the set of all integers. That is,

$$(9.1.12) \qquad \lim_{x \to \infty} \frac{\pi(x)}{x} = 0.$$

Proof. Fix an integer $N \geq 2$, let $A = \prod_{p \leq N} p$, and consider $x > N$. Then the primes $\leq x$ are included among the integers from 1 to N plus the integers $\leq x$ that are not divisible by any prime $p \leq N$, (i.e., prime to A). That is,

(9.1.13) $\pi(x) \leq N + \phi(x, A),$

where $\phi(x, A) =$ the number of positive integers $\leq x$ that are prime to A. From (8.1.7),

$$\phi(x, A) = \frac{\phi(A)}{A} \cdot x + 0(2^{\nu(A)}),$$

so that from (9.1.13)

$$\frac{\pi(x)}{x} \leq \frac{N}{x} + \frac{\phi(A)}{A} + 0\left(\frac{2^{\nu(A)}}{x}\right).$$

Since A and N are fixed, this gives

(9.1.14) $\displaystyle \varlimsup_{x \to \infty} \frac{\pi(x)}{x} \leq \frac{\phi(A)}{A} = \prod_{p \leq N} \left(1 - \frac{1}{p}\right).$

But (9.1.14) holds for all fixed $N \geq 2$ and from (9.1.11), $\prod_{p \leq N} (1 - (1/p))$ tends to 0, as N tends to infinity. Thus (9.1.14) implies that $\varlimsup_{x \to \infty} \pi(x)/x = 0$, which yields (9.1.12).

Though the above theorem really provides rather "weak" information, it is useful in analyzing the relationship between $\theta(x)$ and $\pi(x)$.

Theorem 9.1.3.   As $x \to \infty$ we have

(9.1.15) $\theta(x) = \pi(x) \log x + o(x).$

Proof.   We use summation by parts as follows.

$$\theta(x) = \sum_{p \leq x} \log p = \sum_{2 \leq n \leq x} (\log n)(\pi(n) - \pi(n - 1))$$

$$= (\log 2)(\pi(2) - \pi(1)) + (\log 3)(\pi(3) - \pi(2)) + \cdots$$

$$+ (\log [x])(\pi(x) - \pi(x - 1))$$

$$= \pi(2)(\log 2 - \log 3) + \pi(3)(\log 3 - \log 4) + \cdots$$

$$+ \pi(x - 1)(\log ([x] - 1) - \log [x]) + \pi(x) \log [x]$$

or

(9.1.16) $\displaystyle \theta(x) = \pi(x) \log [x] - \sum_{2 \leq n \leq x - 1} \pi(n) \log \left(1 + \frac{1}{n}\right).$

From (9.1.12), given $\epsilon > 0$, there exists an $x_0 = x_0(\epsilon)$ such that for all $n \geq x_0$, $\pi(n) \leq \epsilon n$. Using this together with the inequality $0 < \log(1 + (1/n)) < 1/n$, we obtain

$$0 < \sum_{2 \leq n \leq x} \pi(n) \log\left(1 + \frac{1}{n}\right) = \sum_{2 < n \leq x_0} \pi(n) \log\left(1 + \frac{1}{n}\right)$$

$$+ \sum_{x_0 < n \leq x} \pi(n) \log\left(1 + \frac{1}{n}\right)$$

$$< 0(1) + \epsilon \sum_{n \leq x} 1 \leq \epsilon x + 0(1).$$

Hence we conclude that

$$\sum_{n \leq x} \pi(n) \log\left(1 + \frac{1}{n}\right) = o(x)$$

so that (9.1.16) immediately yields (9.1.15).

From (9.1.15) we obtain that

$$(9.1.17) \qquad \lim_{x \to \infty} \frac{\theta(x)}{x} = \lim_{x \to \infty} \frac{\pi(x) \log x}{x}$$

and

$$(9.1.18) \qquad \varlimsup_{x \to \infty} \frac{\theta(x)}{x} = \varlimsup_{x \to \infty} \frac{\pi(x) \log x}{x}.$$

From (9.1.9), (9.1.10), (9.1.17), and (9.1.18), we see that *if any one* of the limits

$$\lim_{x \to \infty} \frac{\theta(x)}{x}, \qquad \lim_{x \to \infty} \frac{\psi(x)}{x}, \qquad \lim_{x \to \infty} \frac{\pi(x) \log x}{x},$$

exists, they *all* exist and are equal. A famous result called the *Prime Number Theorem*, which we will prove in Chapter 10, states that

$$(9.1.19) \qquad \lim_{x \to \infty} \frac{\pi(x) \log x}{x} = 1.$$

Thus we have that the Prime Number Theorem, is equivalent to each of

$$(9.1.20) \qquad \lim_{x \to \infty} \frac{\theta(x)}{x} = 1$$

and

(9.1.21)
$$\lim_{x \to \infty} \frac{\psi(x)}{x} = 1.$$

EXERCISES

1. Show that for all $x \geq 3$,

$$\pi(x) = \frac{\theta(x)}{\log [x]} + \sum_{2 \leq n \leq x-1} \theta(n)\left(\frac{1}{\log n} - \frac{1}{\log (n + 1)}\right).$$

Use this to prove that as $x \to \infty$,

$$\sum_{2 \leq n \leq x} \frac{\theta(n)}{n \log^2 n} = o\left(\frac{x}{\log x}\right).$$

2. Refine the argument used to prove that $\pi(x) = o(x)$ [i.e., (9.1.12)], so as to obtain that $\pi(x) = 0(x/\log \log x)$.

3. Prove (9.1.17) and (9.1.18) directly by using (9.1.8) and the inequality

$$\theta(x) \geq \alpha(\log x)[\pi(x) - \pi(x^\alpha)]$$

valid for $0 < \alpha < 1$.

4. Prove (9.1.17) and (9.1.18) directly using only (9.1.8) and the assertion of Ex. 1 above.

§9.1A INDIRECT COUNTING

Continuing with the discussion of the devices that lie at the very beginnings of the theory of prime numbers, there is an idea that is also useful in many other contexts. If one is focusing on a particular sequence of positive integers and one wishes to sense the presence of the integers without actually counting them, one forms and studies some other phenomenon that is sensitive to their existence. We refer to this as "indirect counting." Perhaps the genesis of this idea was present in Euclid when he realized that every integer > 1 brings a prime into existence (i.e., a prime that divides it). More generally, when divisibility questions are involved, a typical approach is to study

(9.1A.1)
$$\sum_{1 \leq n \leq x} \sum_{\substack{m \mid n \\ m \in \mathcal{S}}} f(m),$$

for some convenient choice of the function f.

We illustrate this here for $\mathcal{S} = $ the set of primes, by taking $f(m) = \log m$. Then we note that (9.1A.1) becomes

$$\sum_{1 \leq n \leq x} \sum_{p/n} \log p = \sum_{p \leq x} \log p \sum_{\substack{1 \leq n \leq x \\ n \equiv 0 (\bmod p)}} 1$$

or

(9.1A.2)
$$\sum_{1 \leq n \leq x} \sum_{p/n} \log p = \sum_{p \leq x} (\log p) \left[\frac{x}{p} \right].$$

To obtain information from (9.1A.2), we note that for $p \leq x/2$, $[x/p] \geq (x/p) - 1 \geq \frac{1}{2}(x/p)$, so that, on one hand

$$\sum_{p \leq x} \left[\frac{x}{p} \right] \log p \geq \sum_{p \leq x/2} \left[\frac{x}{p} \right] \log p$$

implies

(9.1A.3)
$$\sum_{p \leq x} \left[\frac{x}{p} \right] \log p \geq \frac{1}{2} x \sum_{p \leq x/2} \frac{\log p}{p}.$$

On the other hand,

$$\sum_{p/n} \log p \leq \log n$$

implies that

(9.1A.4)
$$\sum_{1 \leq n \leq x} \sum_{p/n} \log p \leq \sum_{1 \leq n \leq x} \log n \leq x \log x.$$

Combining (9.1A.2), (9.1A.3), and (9.1A.4) yields

$$\frac{1}{2} x \sum_{p \leq x/2} \frac{\log p}{p} \leq x \log x$$

and replacing x by $2x$ we get

(9.1A.5) $$S(x) = \sum_{p \leq x} \frac{\log p}{p} \leq 2 \log x + 2 \log 2 = 0(\log x).$$

Thus we obtain information about one of our functions $S(x)$ involving summands smaller than 1, [i.e., which is less sensitive to primes than $\pi(x)$].

Even a relatively weak result such as (9.1A.5) can be useful. We next propose to transfer the information contained in (9.1A.5) to the function $T(x)$ defined in (9.1.4). If we proceed via summation by parts, we would obtain only that

(9.1A.6) $T(x) = 0(\log \log x),$

and we leave this to the exercises. An alternative procedure that interjects a small amount of arithmetic information yields much more. Namely, we focus on a *lower bound* estimate for

$$\prod_{p \leq N} \left(1 - \frac{1}{p} \right)$$

which is the analogue of the upper bound (9.1.11). In fact, we will see that it is precisely (9.1A.5) that enables one to transform the identity underlying (9.1.11) into the desired lower bound.

Theorem 9.1A.1.   There exists a constant $c > 0$ such that for all $N \geq 2$

(9.1A.7) $$\prod_{p \leq N} \left(1 - \frac{1}{p} \right) > \frac{c}{\log N}.$$

   Proof.   We have that

(9.1A.8) $$\prod_{p \leq N} \frac{1}{1 - \dfrac{1}{p}} = \sum_{n}{}' \frac{1}{n}$$

where the summation on the right is over all positive integers n all of whose prime factors are $\leq N$. Then fixing a number $y > N$,

$$P = \prod_{p \leq N} \frac{1}{1 - \dfrac{1}{p}} = \sum_{n \leq y}{}' \frac{1}{n} + \sum_{n > y}{}' \frac{1}{n}$$

so that

(9.1A.9) $$P \leq \sum_{n \leq y} \frac{1}{n} + \sum_{n > y}{}' \frac{1}{n}.$$

For the first sum on the right of (9.1A.9), we have

$$\sum_{n \leq y} \frac{1}{n} \leq \log y + 0(1).$$

As for the second sum, since $n > y$, $(\log n)/(\log y) > 1$, and

$$\sideset{}{'}\sum_{n>y} \frac{1}{n} \le \sideset{}{'}\sum_{n>y} \frac{\log n}{\log y} \cdot \frac{1}{n} \le \frac{1}{\log y} \sideset{}{'}\sum_{n} \frac{\log n}{n}$$

$$\le \frac{1}{\log y} \sideset{}{'}\sum_{n} \frac{1}{n} \sum_{p^{\alpha}/n} \log p = \frac{1}{\log y} \sum_{\substack{p \le N \\ \alpha \ge 1}} \log p \sideset{}{'}\sum_{n \equiv 0 (p^{\alpha})} \frac{1}{n}$$

$$\le \frac{1}{\log y} \sum_{\substack{p \le N \\ \alpha \ge 1}} \frac{\log p}{p^{\alpha}} \prod_{p \le N} \frac{1}{1 - \frac{1}{p}} = \frac{P}{\log y} \sum_{p \le N} \frac{\log p}{p - 1}$$

$$\le \frac{2P}{\log y} \sum_{p \le N} \frac{\log p}{p}.$$

Thus from (9.1A.9) we obtain

$$P \le \log y + 0(1) + \frac{2P}{\log y} \sum_{p \le N} \frac{\log p}{p},$$

and using (9.1A.5) this gives

(9.1A.10) $$P \le \log y + 0(1) + 2c_1 P \frac{\log N}{\log y}.$$

Taking $y = N^{4c_1}$ ($c_1 > \frac{1}{4}$ to insure $y > N$) in (9.1A.10) yields

$$P \le 4c_1 \log N + 0(1) + \tfrac{1}{2} P$$

so that

$$P \le 8c_1 \log N + 0(1) \le c_2 \log N,$$

from which (9.1A.7) is immediate.

Having established that for all $N \ge 2$, there is a $c > 0$ such that

(9.1A.11) $$\frac{c}{\log N} \le \prod_{p \le N} \left(1 - \frac{1}{p}\right) \le \frac{1}{\log N}$$

our transfer to $T(x)$ is quite easy. Namely, taking the logarithm of (9.1A.11), we get

$$-\log \log N + 0(1) = \log \prod_{p \le N} \left(1 - \frac{1}{p}\right)$$

or

$$\sum_{p \leq N} \log \left(1 - \frac{1}{p} \right) = -\log \log N + 0(1).$$

But since $\log \left(1 - \frac{1}{p} \right) = -\frac{1}{p} + 0\left(\frac{1}{p^2} \right)$, and $\sum_p \frac{1}{p^2}$ converges, we get

$$(9.1A.12) \qquad T(N) = \sum_{p \leq N} \frac{1}{p} = \log \log N + 0(1),$$

which is clearly much sharper than (9.1A.6).

Returning to the indirect counting device (9.1A.2) which was the source of the upper bound (9.1A.5), it is natural to ask whether it can be processed so as to yield a lower bound for $S(x)$. This can be done, and depends on the fact that the left side of (9.1A.2) does not differ appreciably from

$$\sum_{1 \leq n \leq x} \log n.$$

To see this, we use the fact that every integer $n \geq 1$ has a unique representation in the form $s^2 q$, q squarefree (see Ex. 1 of §2.7). Further, if $n = s^2 q$

$$\sum_{p/n} \log p \geq \log q$$

so that

$$\sum_{1 \leq n \leq x} \sum_{p/n} \log p \geq \sum_{s^2 q \leq x} \log q.$$

But

$$\sum_{s^2 q \leq x} \log q = \sum_{s^2 q \leq x} \log (s^2 q) - \sum_{s^2 q \leq x} \log s^2$$

$$= \sum_{n \leq x} \log n - \sum_{s \leq \sqrt{x}} (\log s^2) \sum_{q \leq x/s^2} 1$$

$$= \sum_{n \leq x} \log n + 0\left(x \sum_s \frac{\log s^2}{s^2} \right)$$

$$= \sum_{n \leq x} \log n + 0(x).$$

Thus

(9.1A.13)
$$\sum_{1 \le n \le x} \sum_{p|n} \log p \ge \sum_{1 \le n \le x} \log n + 0(x).$$

Then from (9.1A.2) we obtain

$$\sum_{1 \le n \le x} \log n + 0(x) \le \sum_{p \le x} \log p \left[\frac{x}{p} \right] \le x \sum_{p \le x} \frac{\log p}{p}.$$

Finally, since

$$\sum_{1 \le n \le x} \log n \ge \sum_{x/2 \le n \le x} \log n \ge \left(\frac{x}{2} - 1 \right) \log \frac{x}{2},$$

it follows that for $x \ge 2$, and some $c > 0$

(9.1A.14)
$$\sum_{p \le x} \frac{\log p}{p} \ge c \log x.$$

EXERCISES

1. Using summation by parts, derive (9.1A.6) from (9.1A.5).

2. Refine the proof of (9.1A.5) so as to obtain that
$$S(x) \le (1 + o(1))\log x.$$

3. Refine the proof of (9.1A.14) so as to obtain that
$$S(x) \ge (1 + o(1))\log x.$$

   Hence conclude that $S(x) \sim \log x$.

4. Show that the only prime q such that
$$\sum_{2 < p \le q} \left[\frac{q}{p} \right] = q$$

   (the summation is over primes p) is $q = 41$.

§9.2 THE PROCESSING OF log $[x]$!

From the discussion of §9.1A, one notes that the sum

(9.2.1)
$$\sum_{1 \le n \le x} \sum_{p|n} \log p$$

is partially workable only because it does not differ appreciably from $\sum_{1 \leq n \leq x} \log n = \log [x]!$. This suggests that one increase the sensitivity of (9.2.1) by including the prime powers, that is, to

$$\sum_{1 \leq n \leq x} \sum_{p^{\alpha}/n} \log p.$$

Then since $\sum_{p^{\alpha}/n} \log p = \log n$, we have

(9.2.2) $$\sum_{1 \leq n \leq x} \sum_{p^{\alpha}/n} \log p = \log [x]! = K(x)$$

[i.e., we use the notation $K(x) = \log [x]!$]. Introducing the function

(9.2.3) $$\Lambda(m) = \begin{cases} \log p & \text{if } m = p^{\alpha}, \, p \text{ a prime, } \alpha \geq 1 \\ 0 & \text{otherwise} \end{cases},$$

we have

(9.2.4) $$\log n = \sum_{d/n} \Lambda(d)$$

and also note that (9.1.3) may be written as

(9.2.5) $$\psi(x) = \sum_{n \leq x} \Lambda(n).$$

Then

$$K(x) = \log [x]! = \sum_{n \leq x} \log n$$

$$= \sum_{n \leq x} \sum_{d/n} \Lambda(d)$$

or

(9.2.6) $$K(x) = \sum_{dd' \leq x} \Lambda(d).$$

The identity (9.2.6) can be rewritten in two different ways, each of which plays some role in the development of the subject. On one hand,

(9.2.7) $$K(x) = \sum_{d \le x} \Lambda(d) \left[\frac{x}{d} \right],$$

whereas reversing the order of summation of d and d'

$$K(x) = \sum_{d' \le x} \sum_{d \le x/d'} \Lambda(d)$$

or

(9.2.8) $$K(x) = \sum_{m \le x} \psi\left(\frac{x}{m} \right).$$

The usefulness of (9.2.7) and (9.2.8) in extracting information about the primes rests on the fact that the function $K(x)$ involves only integers, without any explicit reference to primes, and therefore can be easily estimated. Different techniques will, of course, yield different levels of accuracy. For example, from (3.8.8) we have

$$\sum_{d \le n} \frac{1}{d} = \log n + c + 0\left(\frac{1}{n} \right),$$

and summing this over $n \le x$ gives

$$K(x) + cx + 0(\log x) = \sum_{n \le x} \sum_{d \le n} \frac{1}{d} = \sum_{d \le x} \frac{1}{d} \sum_{d \le n \le x} 1$$

$$= \sum_{d \le x} \frac{1}{d} (x - d + 0(1)) = x \sum_{d \le x} \frac{1}{d} - x + 0(\log x)$$

$$= x\left(\log x + c + 0\left(\frac{1}{x} \right) \right) - x + 0(\log x)$$

so that

(9.2.9) $$K(x) = x \log x - x + 0(\log x).$$

Actually, the weaker result

$$K(x) = x \log x + 0(x)$$

will suffice for the arguments of this section, which focus on extracting information from (9.2.7). In §9.3, in the consideration of (9.2.8), the sharper estimate will be required.

From (9.2.7) we obtain

$$K(x) - 2K\left(\frac{x}{2}\right) = \sum_{n \leq x} \Lambda(n)\left(\left[\frac{x}{n}\right] - 2\left[\frac{x}{2n}\right]\right),$$

and since

$$\left[\frac{x}{n}\right] - 2\left[\frac{x}{2n}\right] \geq \begin{cases} 0 & \text{for all } n \leq x \\ 1 & \text{for } \frac{x}{2} < n \leq x \end{cases}$$

This gives

$$K(x) - 2K\left(\frac{x}{2}\right) \geq \sum_{x/2 < n \leq x} \Lambda(n)$$

or

(9.2.10) $$K(x) - 2K\left(\frac{x}{2}\right) \geq \psi(x) - \psi\left(\frac{x}{2}\right).$$

But from (9.2.9), $K(x) - 2K(x/2) = 0(x)$, so that from (9.2.10), there exists a constant $C > 0$ such that for all $x \geq 1$

(9.2.11) $$Cx \geq \psi(x) - \psi\left(\frac{x}{2}\right).$$

Replacing x by $x/2^i$ in (9.2.11), we get

$$C\frac{x}{2^i} \geq \psi\left(\frac{x}{2^i}\right) - \psi\left(\frac{x}{2^{i+1}}\right)$$

and summing this over all integers $i \geq 0$ [note that for sufficiently large i, $\psi(x/2^{i+1}) = 0$] yields $\psi(x) \leq 2Cx$, so that

(9.2.12) $$\psi(x) = 0(x).$$

Now, using $[x/n] = x/n + 0(1)$ in (9.2.7) yields

$$x \log x + 0(x) = \sum_{n \leq x} \Lambda(n)\left(\frac{x}{n} + 0(1)\right)$$

$$= x \sum_{n \leq x} \frac{\Lambda(n)}{n} + 0(\psi(x))$$

$$= x \sum_{n \leq x} \frac{\Lambda(n)}{n} + 0(x)$$

[where we've used (9.2.12)]. This last gives

(9.2.13)
$$\sum_{n \leq x} \frac{\Lambda(n)}{n} = \log x + 0(1),$$

which is the pivotal relationship that we've been seeking. As we shall see, it contains the most precise information about primes extracted to this point.

We first connect (9.2.13) to an estimate for $S(x)$. To achieve this, note that

$$\sum_{n \leq x} \frac{\Lambda(n)}{n} = \sum_{p^\alpha \leq x} \frac{\log p}{p^\alpha} = \sum_{p \leq x} \frac{\log p}{p} + 0\left(\sum_{p \leq x} (\log p) \sum_{\alpha=2}^{\infty} \frac{1}{p^\alpha} \right)$$

$$= \sum_{p \leq x} \frac{\log p}{p} + 0\left(\sum_{p} \frac{\log p}{p^2} \right)$$

$$= \sum_{p \leq x} \frac{\log p}{p} + 0(1)$$

(since $\sum_p \log p / p^2$ converges), so that (9.2.13) gives

(9.2.14)
$$S(x) = \sum_{p \leq x} \frac{\log p}{p} = \log x + 0(1).$$

The estimate (9.2.14) for $S(x)$ is of sufficient strength to enable a successful "transfer" to $T(x) = \sum_{p \leq x} 1/p$ simply by summation by parts. For this application, it is convenient to define $R(x)$ by

(9.2.15)
$$S(x) = \log x + R(x),$$

so that (9.2.14) asserts that $R(x) = 0(1)$. Then

$$T(x) = \sum_{p \leq x} \frac{1}{p} = \sum_{2 \leq n \leq x} \frac{1}{\log n} (S(n) - S(n-1))$$

and inserting (9.2.15) yields

(9.2.16)
$$T(x) = \sum_{2 \leq n \leq x} \frac{1}{\log n} \log \frac{n}{n-1} + \sum_{2 \leq n \leq x} \frac{1}{\log n} (R(n) - R(n-1)).$$

We next treat each of the sums in (9.2.16) separately. Summing by parts,

$$\sum_{2 \le n \le x} \frac{1}{\log n} (R(n) - R(n - 1))$$

$$= \frac{1}{\log 2} (R(2) - R(1)) + \cdots + \frac{1}{\log [x]} (R([x]) - R([x] - 1))$$

$$= \frac{R([x])}{\log [x]} - \frac{R(1)}{\log 2} + \sum_{2 \le n \le x-1} R(n) \left(\frac{1}{\log n} - \frac{1}{\log (n + 1)} \right).$$

Since $R(x) = 0(1)$,

$$R(n) \left(\frac{1}{\log n} - \frac{1}{\log (n + 1)} \right) = 0 \left(\frac{1}{\log n} - \frac{1}{\log (n + 1)} \right)$$

so that the series

$$c_1 = \sum_{n=2}^{\infty} R(n) \left(\frac{1}{\log n} - \frac{1}{\log (n + 1)} \right)$$

converges, and

$$\sum_{2 \le n \le x-1} R(n) \left(\frac{1}{\log n} - \frac{1}{\log (n + 1)} \right) = c_1 + 0 \left(\frac{1}{\log x} \right).$$

Thus we get (note that $R(1) = 0$)

$$\sum_{2 \le n \le x} \frac{1}{\log n} (R(n) - R(n - 1)) = c_1 + 0 \left(\frac{1}{\log x} \right).$$

As for the first sum in (9.2.16), setting

$$\log \frac{n}{n - 1} = \log \left(1 + \frac{1}{n - 1} \right) = \frac{1}{n} + \xi(n)$$

we have $\xi(n) = 0(1/n^2)$, so that

$$\sum_{2 \le n \le x} \frac{1}{\log n} \log \frac{n}{n - 1} = \sum_{2 \le n \le x} \frac{1}{n \log n} + \sum_{2 \le n \le x} \frac{\xi(n)}{\log n}$$

$$= \sum_{2 \le n \le x} \frac{1}{n \log n} + \sum_{n=2}^{\infty} \frac{\xi(n)}{\log n} + \sum_{n > x} 0 \left(\frac{1}{n^2} \right)$$

$$= \sum_{2 \le n \le x} \frac{1}{n \log n} + c_2 + 0 \left(\frac{1}{x} \right),$$

where

$$c_2 = \sum_{n=2}^{\infty} \frac{\xi(n)}{\log n}.$$

Inserting the above estimates in (9.2.16), we get

(9.2.17)
$$T(x) = \sum_{2 \le n \le x} \frac{1}{n \log n} + c_3 + 0\left(\frac{1}{\log x}\right)$$

where

$$c_3 = c_1 + c_2.$$

Thus there remains the task of estimating the sum that appears in (9.2.17). In this connection, we have for $x > 3$,

$$\sum_{2 \le n \le x} \frac{1}{n \log n} = \sum_{n=3}^{[x]} \int_{n-1}^{n} \frac{1}{n \log n} \, du + \frac{1}{2 \log 2}$$

$$= \int_{2}^{[x]} \frac{du}{u \log u} + \sum_{n=3}^{[x]} \int_{n-1}^{n} \left(\frac{1}{n \log n} - \frac{1}{u \log u}\right) du + \frac{1}{2 \log 2}$$

$$= \log \log [x] + c_4 + \sum_{n=3}^{[x]} \int_{n-1}^{n} \left(\frac{1}{n \log n} - \frac{1}{u \log u}\right) du$$

where $c_4 = 1/(2 \log 2) - \log \log 2$. Further, for $n \ge u \ge n - 1$ we have

$$\left|\frac{1}{n \log n} - \frac{1}{u \log u}\right| \le \left(\frac{1}{(n-1)\log(n-1)} - \frac{1}{n \log n}\right)$$

so that the series

$$c_5 = \sum_{n=3}^{\infty} \int_{n-1}^{n} \left(\frac{1}{n \log n} - \frac{1}{u \log u}\right) du$$

converges and

$$\sum_{n=3}^{[x]} \int_{n-1}^{n} \left(\frac{1}{n \log n} - \frac{1}{u \log u}\right) du = c_5 + 0\left(\frac{1}{x \log x}\right).$$

Thus setting $c_6 = c_5 + c_4$,

$$\sum_{2 \le n \le x} \frac{1}{n \log n} = \log \log x + c_6 + 0\left(\frac{1}{\log x}\right),$$

and inserting this in (9.2.17) gives

(9.2.18)
$$T(x) = \sum_{p \le x} \frac{1}{p} = \log \log x + c_7 + 0\left(\frac{1}{\log x}\right),$$

where $c_7 = c_6 + c_3$. Note the considerable improvement of this estimate of $T(x)$ as compared with (9.1A.12).

The estimate (9.2.18) can now be used to estimate

$$P(x) = \prod_{p \le x} \left(1 - \frac{1}{p}\right)$$

more sharply than previously [i.e., as in (9.1A.11)]. We have

$$\log P(x) = \sum_{p \le x} \log\left(1 - \frac{1}{p}\right),$$

and setting

$$\log\left(1 - \frac{1}{p}\right) = -\frac{1}{p} + \lambda(p),$$

where $\lambda(p) = 0(1/p^2)$, we get

$$\log P(x) = -\sum_{p \le x} \frac{1}{p} + \sum_{p \le x} \lambda(p).$$

Then, using (9.2.18),

$$\log P(x) = -\log \log x - c_7 + \sum_{p \le x} \lambda(p) + 0\left(\frac{1}{\log x}\right);$$

since $\lambda(p) = 0(1/p^2)$, $c_8 = \sum_p \lambda(p)$ converges and

$$\sum_{p \le x} \lambda(p) = c_8 + 0\left(\frac{1}{x}\right),$$

so that for $c_9 = c_8 - c_7$,

(9.2.19) $$\log P(x) = -\log \log x + c_9 + 0\left(\frac{1}{\log x}\right).$$

Exponentiating (9.2.19) we get

$$P(x) = e^{c_9} \cdot \frac{1}{\log x} \cdot \exp\left(0\left(\frac{1}{\log x}\right)\right)$$

or

(9.2.20) $$P(x) = \prod_{p \leq x}\left(1 - \frac{1}{p}\right) = \frac{c_{10}}{\log x}\left(1 + 0\left(\frac{1}{\log x}\right)\right),$$

where $c_{10} = e^{c_9}$ (and hence $c_{10} > 0$). Further, from (9.1.11) it follows that $c_{10} \leq 1$.

EXERCISES

1. Using summation by parts, show that

$$\sum_{p \leq x} \frac{\log^2 p}{p} = \frac{1}{2}(\log x)^2 + 0(\log x).$$

2. Generalize this to the case of $\sum_{p \leq x} \log^k p / p$, where k is any fixed integer $k \geq 2$.

3. Prove that

$$\sum_{pq \leq x} \frac{1}{pq} = (\log \log x)^2 + 0(\log \log x).$$

   (The sum is taken over products of two primes p and q.)

4. Prove that

$$\sum_{pq \leq x} \frac{\log p}{p} \frac{\log q}{q} + \sum_{p \leq x} \frac{\log^2 p}{p} = \log^2 x + 0(\log x).$$

5. Show that

$$\sum_p \frac{1}{p(\log p)^\alpha}$$

   converges if and only if $\alpha > 0$.

6. Letting L denote log x, Λ denote $\Lambda(n)$, K denote $K(x)$, 1 the function identically one, derive the following relationships using the convolution arithmetic notation (see Chapter 4):

$$L = 1_0 * \Lambda, \qquad \psi = 1 * \Lambda, \qquad K = 1 * L,$$
$$K = 1 * 1_0 * \Lambda, \qquad [x] = 1 * 1_0, \qquad K = [x] * \Lambda,$$
$$K = \psi * 1.$$

Compare with (9.2.4)–(9.2.8).

7. Prove that

$$\sum_{m \leq x} \frac{\Lambda(m)}{m} \log \frac{x}{m} = \frac{1}{2} \log^2 x + 0(\log x).$$

8. Improve (9.1.7) by proving that

$$\psi(x) = \theta(x) + 0(\sqrt{x}).$$

§9.2A BOUNDS AND UPPER AND LOWER LIMITS

The estimate for $S(x)$ given in (9.2.14) has many implications for the primes.

Theorem 9.2A.1. There exist constants $A > 0$, $a > 0$, such that for all $x \geq 2$

(9.2A.1) $ax < \psi(x) < Ax,$

(9.2A.2) $ax < \theta(x) < Ax,$

(9.2A.3) $a \dfrac{x}{\log x} < \pi(x) < A \dfrac{x}{\log x}.$

Proof. In (9.2.12) we derived $\psi(x) = 0(x)$, implying $\psi(x) < Ax$ for some A and all $x \geq 2$. Since $\theta(x) \leq \psi(x)$, this same upper bound holds for $\theta(x)$. Then from (9.1.15) this, in turn, implies the upper bound for $\pi(x)$ in (9.2A.3) (perhaps with a larger A).

Proceeding to the lower bounds, since $S(x) = \log x + 0(1)$, we have for $x \geq 1$, $S(x) = \log x + E(x)$, where $|E(x)| < C$. Then for $K = e^{3C}$ and $x > K$,

$$\sum_{x/K < p \leq x} \frac{\log p}{p} = S(x) - S\left(\frac{x}{K}\right) = \log K + E(x) - E\left(\frac{x}{K}\right)$$

so that

$$\sum_{x/K<p\,\le\,x} \frac{\log p}{p} > \log K - 2C = C.$$

Then since

$$\sum_{x/K<p\,\le\,x} \frac{\log p}{p} < \frac{K}{x} \sum_{x/K<p\,\le\,x} \log p \le K \cdot \frac{\theta(x)}{x},$$

it follows that for $x > K$,

$$\frac{\theta(x)}{x} > \frac{C}{K} > 0,$$

which implies the existence of an $a > 0$ such that $\theta(x) > ax$ for all $x \ge 2$. This provides the lower bound of (9.2A.2)

Since $\psi(x) \ge \theta(x)$, the lower bound of (9.2A.1) follows. Finally, from (9.1.15), $\theta(x) = \pi(x) \log x + o(x)$, so that the lower bound for $\theta(x)$ implies the lower bound of (9.2A.3) (perhaps with a smaller $a > 0$).

This same estimate for $S(x)$ provides that infinitely often (as $x \to \infty$), the bounds given above hold with constants A and a that are close to one.

Theorem 9.2A.2.

(9.2A.4) $$\varliminf_{x\to\infty} \frac{\theta(x)}{x} \le 1 \le \varlimsup_{x\to\infty} \frac{\theta(x)}{x},$$

(9.2A.5) $$\varliminf_{x\to\infty} \frac{\psi(x)}{x} \le 1 \le \varlimsup_{x\to\infty} \frac{\psi(x)}{x},$$

(9.2A.6) $$\varliminf_{x\to\infty} \frac{\pi(x) \log x}{x} \le 1 \le \varlimsup_{x\to\infty} \frac{\pi(x) \log x}{x}.$$

Proof. From (9.1.9), (9.1.10), (9.1.17), and (9.1.18), we see that it suffices to prove (9.2A.4) in order to establish all the assertions of the theorem. We write

$$S(x) = \sum_{p\le x} \frac{\log p}{p} = \sum_{2\le n\le x} \frac{1}{n} (\theta(n) - \theta(n-1)),$$

and summing by parts yields

$$S(x) = \frac{\theta(x)}{[x]} + \sum_{2\le n\le x-1} \frac{\theta(n)}{n(n+1)}.$$

Then since $S(x) = \log x + 0(1)$ and $\theta(x) = 0(x)$ [from (9.2A.2)],

(9.2A.7)
$$\sum_{2 \leq n \leq x} \frac{\theta(n)}{n(n+1)} = \log x + 0(1).$$

From (9.2A.7), it follows that for any fixed $K > 0$,

$$\sum_{K < n \leq x} \frac{\theta(n)/n}{n+1} = \log x + 0(1).$$

This, in turn, gives

$$\min_{K < n \leq x} \frac{\theta(n)}{n} \sum_{K < n \leq x} \frac{1}{n+1} \leq \log x + 0(1) \leq \max_{K < n \leq x} \frac{\theta(n)}{n} \sum_{K < n \leq x} \frac{1}{n+1}$$

and hence

$$\min_{K < n \leq x} \frac{\theta(n)}{n} \leq \frac{\log x + 0(1)}{\log x + 0(1)} \leq \max_{K < n \leq x} \frac{\theta(n)}{n}$$

or

(9.2A.8)
$$\min_{K < n \leq x} \frac{\theta(n)}{n} \leq 1 + 0\left(\frac{1}{\log x}\right) \leq \max_{K < n \leq x} \frac{\theta(n)}{n}.$$

Since

$$\varliminf_{x \to \infty} \frac{\theta(x)}{x} \leq \lim_{K \to \infty} \varliminf_{x \to \infty} \min_{K < n \leq x} \frac{\theta(n)}{n}$$

and

$$\varlimsup_{x \to \infty} \frac{\theta(x)}{x} \geq \lim_{K \to \infty} \varlimsup_{x \to \infty} \max_{K < n \leq x} \frac{\theta(n)}{n},$$

(9.2A.4) follows from (9.2A.8).

On occasion, one rephrases a result such as (9.2A.6) as implying that *if* the

$$\lim_{x \to \infty} \frac{\pi(x)\log x}{x}$$

exists, then this limit must be 1. Thus the real essence of the proof of the Prime Number Theorem is to the effect that this limit exists.

EXERCISES

1. Show that for $\pi_2(x) = $ the number of positive integers $\leq x$ that are the product of two primes, there exist constants $A > 0$, $a > 0$ such that for all $x \geq 4$

$$a \frac{x \log \log x}{\log x} < \pi_2(x) < A \frac{x \log \log x}{\log x}.$$

2. Generalize Ex. 1 to $\pi_k(x)$, $k \geq 3$, where $\pi_k(x) = $ the number of positive integers $\leq x$ that are the product of k primes.

3. Give another proof of Theorem 9.2A.2 using the fact that $T(x) = \sum_{p \leq x} 1/p = \log \log x + 0(1)$ instead of the estimate for $S(x)$.

4. Using the assertion of Ex. 3 of §9.2, prove that

$$\varliminf_{x \to \infty} \frac{\pi_2(x)\log x}{x \log \log x} \leq 1 \leq \varlimsup_{x \to \infty} \frac{\pi_2(x)\log x}{x \log \log x}.$$

5. Let $p_1 = 2$, $p_2 = 3$, and in general $p_n = $ the nth prime. Show that there exist constants $a > 0$, $A > 0$ such that for $n \leq 2$

$$a \, n \log n < p_n < A \, n \log n.$$

6. Improve (9.1.15) by showing that

$$\theta(x) = \pi(x)\log x + 0\left(\frac{x}{\log x}\right).$$

7. For $p_n = $ the nth prime, show that the Prime Number Theorem is equivalent to the assertion $p_n \sim n \log n$.

8. Prove that

$$\varliminf_{n \to \infty} \frac{p_n}{n \log n} \leq 1 \leq \varlimsup_{n \to \infty} \frac{p_n}{n \log n}.$$

§9.2B A LOWER BOUND PROPERTY OF THE EULER FUNCTION

The estimate (9.2.20) gives that as $x \to \infty$,

$$(9.2B.1) \qquad \prod_{p \leq x} \left(1 - \frac{1}{p}\right) \sim \frac{c_{10}}{\log x},$$

where $c_{10} > 0$ is a constant. A theorem of Landau transforms this into a result concerning the size of the Euler ϕ-function.

Theorem 9.2B.1.

$$(9.2\text{B}.2) \qquad \lim_{n \to \infty} \frac{\phi(n)\log \log n}{n} = c_{10}.$$

Proof. If we fix $N > 0$ and take $n = \Pi_{p \leq N}\, p$, we have from (9.2B.1) that

$$(9.2\text{B}.3) \qquad \frac{\phi(n)}{n} = \prod_{p \leq N}\left(1 - \frac{1}{p}\right) \sim \frac{c_{10}}{\log N}.$$

But

$$\log n = \sum_{p \leq N} \log p = \theta(N)$$

so that

$$(9.2\text{B}.4) \qquad \log \log n = \log \theta(N).$$

From (9.2A.2), $aN < \theta(N) < AN$, and taking logarithms we get $\log N + 0(1) = \log \theta(N)$, so that (9.2B.4) yields

$$\log N = \log \log n + 0(1).$$

Inserting this in (9.2B.2) gives

$$\frac{\phi(n)}{n} \sim \frac{c_{10}}{\log \log n + 0(1)}$$

or

$$\frac{\phi(n)\log \log n}{n} \sim c_{10}.$$

Thus there is an infinite sequence of positive integers on which $(\phi(n)\log \log n)/n$ behaves like the proposed lower limit.

In general, for any positive integer n, fix a real number ξ, $1 < \xi < n$, to be chosen later as an appropriate function of n. Then

$$\frac{\phi(n)}{n} = \prod_{p|n}\left(1 - \frac{1}{p}\right) = \prod_{\substack{p|n \\ p \leq \xi}}\left(1 - \frac{1}{p}\right) \prod_{\substack{p|n \\ p > \xi}}\left(1 - \frac{1}{p}\right)$$

$$\geq \prod_{p \leq \xi}\left(1 - \frac{1}{p}\right)\left(1 - \frac{1}{\xi}\right)^{M}$$

where M = the number of primes p/n, $p > \xi$. Since

$$n \geq \prod_{\substack{p|n \\ p>\xi}} p \geq \xi^M,$$

we have

(9.2B.5)
$$M \leq \frac{\log n}{\log \xi},$$

so that

(9.2B.6)
$$\frac{\phi(n)}{n} \geq \prod_{p\leq\xi} \left(1 - \frac{1}{p}\right)\left(1 - \frac{1}{\xi}\right)^{\log n/\log \xi}.$$

Using (9.2B.1),

$$\prod_{p\leq\xi} \left(1 - \frac{1}{p}\right)\left(1 - \frac{1}{\xi}\right)^{\log n/\log \xi} \sim \frac{c_{10}}{\log \xi} \exp\left\{-\frac{\log n}{\xi \log \xi}(1 + o(1))\right\},$$

so that for $\xi = \log n$

$$\prod_{p\leq\xi} \left(1 - \frac{1}{p}\right)\left(1 - \frac{1}{\xi}\right)^{\log n/\log \xi} \sim \frac{c_{10}}{\log \log n} \exp\left\{-\frac{1}{\log \log n}(1 + o(1))\right\}$$

$$\sim \frac{c_{10}}{\log \log n}.$$

This yields that as $n \to \infty$,

$$\frac{\phi(n)}{n} \log \log n \geq c_{10} + o(1)$$

and this completes the proof of (9.2B.1).

From the above theorem, it follows that there exists a constant $c > 0$ such that for all integers $n \geq 3$,

(9.2B.7)
$$\phi(n) > c \cdot \frac{n}{\log \log n}.$$

§9.2C THE NUMBER OF PRIME FACTORS OF AN INTEGER

Following the dictates of the idea of indirect counting (see §9.1A) suggests that we take $f(m) \equiv 1$ in (9.1A.1), where $\mathcal{S}$ = the set of primes. We are then led to consider

(9.2C.1)
$$\sum_{n\le x}\sum_{p/n} 1 = \sum_{n\le x} \nu(n),$$

where $\nu(n) =$ the number of distinct prime factors of n.

Interchanging the order of summation in (9.2C.1), we have

$$\sum_{n\le x} \nu(n) = \sum_{p\le x}\sum_{\substack{n\le x \\ n\equiv 0\,(\mathrm{mod}\,p)}} 1 = \sum_{p\le x}\left[\frac{x}{p}\right].$$

Writing $[x/p] = (x/p) + 0(1)$, this yields

$$\sum_{n\le x} \nu(n) = \sum_{p\le x}\left(\frac{x}{p} + 0(1)\right) = x\sum_{p\le x}\frac{1}{p} + 0(\pi(x)).$$

Using (9.2.18) to estimate the sum over the primes and (9.2A.3) for $\pi(x)$ gives

(9.2C.2)
$$\sum_{n\le x} \nu(n) = x\log\log x + c_7 x + 0\left(\frac{x}{\log x}\right).$$

Thus we see that the results obtained, to this point, concerning primes combine to give a fairly good evaluation of the "average order" of the number of distinct prime factors of n.

For $\Omega(n) =$ the number of prime factors of n, counting multiplicities, $\Omega(n) \ge \nu(n)$, and

$$\sum_{n\le x} \Omega(n) - \sum_{n\le x} \nu(n) = \sum_{\substack{n\le x\ p^\alpha/n \\ \alpha>1}} 1 = \sum_{\substack{p^\alpha\le x \\ \alpha>1}}\left(\frac{x}{p^\alpha} + 0(1)\right)$$

$$= x\sum_{\substack{p^\alpha\le x \\ \alpha>1}}\frac{1}{p^\alpha} + 0(\sqrt{x}\log x),$$

(since there are $0(\sqrt{x})$ primes such that $p^2 \le x$, and $0(\log x)$ possible α for each such p). But

$$\sum_{\substack{p^\alpha\le x \\ \alpha>1}}\frac{1}{p^\alpha} = \sum_{p\le x}\sum_{\alpha=2}^{\infty}\frac{1}{p^\alpha} - \sum_{p\le x}\sum_{\substack{\alpha \\ p^\alpha>x}}\frac{1}{p^\alpha} = \sum_{p\le x}\frac{1}{p(p-1)} + 0\left(\frac{1}{x}\sum_{p\le x}1\right)$$

$$= \sum_{p}\frac{1}{p(p-1)} + 0\left(\frac{1}{\log x}\right)$$

[since $\pi(x) = 0(x/\log x)$], so that the above becomes

$$\sum_{n\le x} \Omega(n) - \sum_{n\le x} \nu(n) = x \sum_{p} \frac{1}{p(p-1)} + 0\left(\frac{x}{\log x}\right).$$

Then from (9.2C.2),

(9.2C.3) $$\sum_{n\le x} \Omega(n) = x \log \log x + c_{11}x + 0\left(\frac{x}{\log x}\right),$$

where

$$c_{11} = c_7 + \sum_{p} \frac{1}{p(p-1)}.$$

Resuming the study of $\nu(n)$, one might turn to the consideration of

$$\sum_{n\le x} \nu^2(n),$$

which is more sensitive to the larger values of $\nu(n)$. For this we have

$$\sum_{n\le x} \nu^2(n) = \sum_{n\le x}\left(\sum_{p/n} 1\right)\left(\sum_{p'/n} 1\right) = \sum_{n\le x}\sum_{p/n} 1 + \sum_{\substack{pp'/n \\ p\ne p'}} 1$$

(since when $p \ne p'$, the conditions p/n and p'/n are equivalent to pp'/n). This gives

(9.2C.4) $$\sum_{n\le x} \nu^2(n) = \sum_{n\le x} \nu(n) + \sum_{\substack{pp'\le x \\ p\ne p'}} \left[\frac{x}{pp'}\right].$$

The first sum has already been evaluated in (9.2C.2) so that we need only focus on the second sum. For this we have

(9.2C.5) $$\sum_{\substack{pp'\le x \\ p\ne p}} \left[\frac{x}{pp'}\right] = x \sum_{\substack{pp'\le x \\ p\ne p'}} \frac{1}{pp'} + 0(\pi_2(x))$$

where $\pi_2(x) =$ the number of products of two primes $\le x$. A workable estimate, which is not to the full accuracy of the method, is easily deduced from (9.2C.5). To do this, note first that

(9.2C.6) $$\left(\sum_{p\le/x} \frac{1}{p}\right)^2 - \sum_{p\le/x} \frac{1}{p^2} \le \sum_{\substack{pp'\le x \\ p\ne p'}} \frac{1}{pp'} \le \left(\sum_{p\le x} \frac{1}{p}\right)^2$$

where $\left(\sum\limits_{p \le \sqrt{x}} \dfrac{1}{p^2} \right) = 0(1)$. Then from (9.2.18),

$$\left(\sum_{p \le x} \frac{1}{p} \right)^2 = (\log \log x + 0(1))^2 = (\log \log x)^2 + 0(\log \log x)$$

and

$$\left(\sum_{p \le \sqrt{x}} \frac{1}{p} \right)^2 = (\log \log \sqrt{x})^2 + 0(\log \log x)$$

$$= (\log \log x - \log 2)^2 + 0(\log \log x)$$

$$= (\log \log x)^2 + 0(\log \log x),$$

so that from (9.2C.6),

$$\sum_{\substack{pp' \le x \\ p \ne p'}} \frac{1}{pp'} = (\log \log x)^2 + 0(\log \log x).$$

Thus using $\pi_2(x) = 0(x)$, (9.2C.5) gives

$$\sum_{\substack{pp' \le x \\ p \ne p'}} \left[\frac{x}{pp'} \right] = x(\log \log x)^2 + 0(x \log \log x),$$

which combined with (9.2C.2) and inserted in (9.2C.4) yields

(9.2C.7) $$\sum_{n \le x} \nu^2(n) = x(\log \log x)^2 + 0(x \log \log x).$$

Comparing (9.2C.2) with (9.2C.7), it appears that most of the $\nu(n)$ behave almost as if they were of order $\log \log x$. This can be made precise as follows.

Theorem 9.2C.1.   For $K(x)$ any given function that tends to infinity as $x \to \infty$ (however slowly), let $\mathcal{S}_x =$ the set of positive integers $n \le x$ such that

(9.2C.8) $$|\nu(n) - \log \log x| < K(x)/\log \log x$$

fails to hold. Then as $x \to \infty$, $|\mathcal{S}_x| = o(x)$. (Recall that $|\mathcal{S}_x|$ denotes the number of elements of the set $\mathcal{S}_x$. We say that (9.2C.7) holds "for almost all integers n.")

   Proof.   Using (9.2C.2) and (9.2C.7),

$$\sum_{n \le x} (\nu(n) - \log \log x)^2 = \sum_{n \le x} \nu^2(n) - 2(\log \log x) \sum_{n \le x} \nu(n) + [x](\log \log x)^2$$

$$= x(\log \log x)^2 - 2x(\log \log x)^2 + x(\log \log x)^2 + O(x \log \log x)$$

or

(9.2C.9) $$\sum_{n \le x} |\nu(n) - \log \log x|^2 = O(x \log \log x).$$

Then for $n \in \mathcal{S}_x$ we have

$$|\nu(n) - \log \log x| \ge K(x)/\overline{\log \log x}$$

and from (9.2C.9),

$$O(x \log \log x) \ge \sum_{n \in \mathcal{S}_x} |\nu(n) - \log \log x|^2 \ge K^2(x) |\mathcal{S}_x| \log \log x,$$

which implies

$$|\mathcal{S}_x| = O\!\left(\frac{x}{K^2(x)}\right) = o(x)$$

since $K(x) \rightarrow \infty$.

As a special case of this theorem, we may take $K(x) = \epsilon/\overline{\log \log x}$, for any fixed $\epsilon > 0$. Then $\mathcal{S}_x$ is the set of $n \le x$ for which

(9.2C.10) $$(1 - \epsilon)\log \log x \le \nu(n) \le (1 + \epsilon)\log \log x$$

fails to hold; and, of course, $|\mathcal{S}_x| = o(x)$. Since $\nu(n)$ satisfies (9.2C.10) for almost all n, for every fixed $\epsilon > 0$, it is asserted that "$\nu(n)$ *has the normal order* $\log \log x$."

It would be a bit more natural, perhaps, to assert that "$\nu(n)$ has the normal order $\log \log n$" to reflect the property that for each $\epsilon > 0$,

$$|\nu(n) - \log \log n| < \epsilon \log \log n$$

holds for almost all n. This is, in fact, a trivial consequence of the above since

(9.2C.11) $$\log \log n = \log \log x + O(1)$$

for almost all $n \le x$. To see this, simply note that almost all $n \le x$ are in the interval $\sqrt{x} < n \le x$, wherein (9.2C.11) holds.

To this point we've seen that the average and normal order of $\nu(n)$ is $\log \log n$. We now consider how large it can be as a function of n.

Theorem 9.2C.2.   There exist constants $A > 0$, $a > 0$, such that

(9.2C.12)
$$a < \overline{\lim_{x \to \infty}} \frac{\nu(n) \log \log n}{\log n} < A$$

(i.e., this upper limit is positive and finite).

Proof.   Fixing an integer N, and taking $n = \Pi_{p \leq N} \, p$, we have

(9.2C.13)
$$\nu(n) = \pi(N).$$

But

$$\log n = \sum_{p \leq N} \log p = \theta(N)$$

and from (9.2A.2) there is a constant c_1 such that $\theta(N) < c_1 N$, and hence $\log n < c_1 N$ or $N > (1/c_1) \log n$. Using this in (9.2C.13), together with the monotonicity of $\pi(N)$, we get

(9.2C.14)
$$\pi\left(\frac{1}{c_1} \log n\right) \leq \nu(n).$$

Finally, since from (9.2A.3), for $x \geq 2$ there is a constant $c_2 > 0$ such that $\pi(x) > c_2(x/\log x)$, it follows from (9.2C.14) that for sufficiently large n (i.e., sufficiently large N) we have

$$\nu(n) > \frac{c_2}{c_1} \frac{\log n}{\log((1/c_1) \log n)} > c_3 \frac{\log n}{\log \log n},$$

$c_3 > 0$, which gives the lower bound in (9.2C.12).

For a general n, let ξ be a function of n (to be chosen later) such that $1 < \xi < n$. Then

$$\nu(n) = \sum_{p/n} 1 = \sum_{\substack{p/n \\ p \leq \xi}} + \sum_{\substack{p/n \\ p > \xi}} 1 \leq \sum_{p \leq \xi} 1 + \sum_{\substack{p/n \\ p > \xi}} 1 = \pi(\xi) + \sum_{\substack{p/n \\ p > \xi}} 1.$$

In (9.2B.5), we've seen that $\sum_{\substack{p/n \\ p > \xi}} 1 \leq (\log n)/(\log \xi)$, and from (9.2A.3), $\pi(x) < c_4(x/\log x)$, so that the above yields

$$\nu(n) \leq c_4 \frac{\xi}{\log \xi} + \frac{\log n}{\log \xi}.$$

Taking $\xi = \log n$, this becomes

$$\nu(n) \le (c_4 + 1)\frac{\log n}{\log \log n},$$

and the upper bound in (9.2C.12) follows.

Most of the above results are easily carried over to $\Omega(n)$ and we leave the details of this for the exercises.

EXERCISES

1. Show that Theorem 9.2C.1 holds with $\nu(n)$ replaced by $\Omega(n)$.

2. Prove that there exist constants b_1, b_2 such that

$$\sum_{pq \le x} \frac{1}{pq} = (\log \log x)^2 + b_1 \log \log x + b_2 + 0\left(\frac{\log \log x}{\log x}\right).$$

3. Prove that there exist constants b_3, b_4 such that

$$\sum_{n \le x} \nu^2(n) = x(\log \log x)^2 + b_3 x \log \log x + b_4 x + 0\left(\frac{x \log \log x}{\log x}\right).$$

4. Show that there exist constants b_5, b_6 such that

$$\sum_{n \le x} (\nu(n) - \log \log x)^2 = b_5 x \log \log x + b_6 x + 0\left(\frac{x \log \log x}{\log x}\right).$$

5. Prove that there exist constants b_7, b_8, b_9 such that

$$\prod_{pq \le x} \left(1 - \frac{1}{pq}\right) \sim \frac{b_7}{(\log x)^{b_8 \log \log x + b_9}}.$$

6. Investigate a possible lower bound for the function

$$\phi_2(n) = n \prod_{pq/n} \left(1 - \frac{1}{pq}\right).$$

7. Show that $\overline{\lim}_{n \to \infty} (\Omega(n) \log \log n)/(\log n)$ is positive and finite.

8. Assuming the Prime Number Theorem, prove that

$$\overline{\lim_{x \to \infty}} \frac{\nu(n) \log \log n}{\log n} = \overline{\lim_{x \to \infty}} \frac{\Omega(n) \log \log n}{\log n} = 1.$$

9. Assuming the Prime Number Theorem, prove that

$$\varlimsup_{x \to \infty} \frac{(\log d(n))(\log \log n)}{\log n} = \log 2,$$

where $d(n)$ = the number of divisors of n.

10.   Show that

$$\sigma(n) = O(n \log \log n),$$

for $\sigma(n)$ = the sum of the divisors of n.

§9.2D   THE SMALLEST POSITIVE QUADRATIC NONRESIDUE

Using results about primes such as (9.2.18), an upper estimate can be made for the smallest positive quadratic nonresidue of a prime p that is of much smaller order than the $(2\sqrt{p} + 1)$ given in Theorem 7.4B.1.

Theorem 9.2D.1.   Given any $\epsilon > 0$, for all sufficiently large p, the smallest positive quadratic nonresidue modulo p is less than

$$p^{(1/(2\sqrt{e}))+\epsilon}.$$

(Here $e = 2.71828. . .$ is the base of the natural logarithms.)

   Proof.   Proceeding as with the proof of Theorem 8.4.1, we note that

$$\frac{1}{2}\left(1 - \left(\frac{n}{p}\right)\right) = \begin{cases} 1 & \text{if } \left(\dfrac{n}{p}\right) = -1 \\ 0 & \text{if } \left(\dfrac{n}{p}\right) = 1. \end{cases}$$

Thus for x an integer $0 < x < p$, the number of quadratic nonresidues $\leq x$ equals

$$\sum_{n \leq x} \frac{1}{2}\left(1 - \left(\frac{n}{p}\right)\right) = \frac{x}{2} - \frac{1}{2} \sum_{n \leq x} \left(\frac{n}{p}\right)$$

and using (8.4.8), this number is greater than

(9.2D.1) $$\frac{x}{2} - \frac{1}{2}\sqrt{p} \log p,$$

Then, fixing $c > 3$ large enough so that

(9.2D.2) $$1 + \epsilon(2\sqrt{e}) > e^{1/c},$$

we take $x = [c\sqrt{p} \log p]$ and from (9.2D.1) there are at least $\frac{1}{2}(c - 2)\sqrt{p} \log p$ quadratic nonresidues mod p that are $\leq x$.

Next fix a δ, $0 < \delta < \frac{1}{2}$, to be specified later, and assume that all primes $\leq p^\delta$ are quadratic residues of p. Each of the $\frac{1}{2}(c - 2)\sqrt{p} \log p$ nonresidues $\leq x$ must have at least one prime divisor that is a nonresidue and therefore this prime would be greater than p^δ. But the total number of integers $\leq x$ divisible by a prime between p^δ and x is

$$\leq \sum_{p^\delta < q \leq x} \left(\frac{x}{q} + 1 \right) \leq x \sum_{p^\delta < q \leq x} \frac{1}{q} + \pi(x).$$

From (9.1.12) we have $\pi(x) = o(x)$, so that the upper bound equals

$$x \sum_{p^\delta < q \leq x} \frac{1}{q} + o(x).$$

The sum over primes q can be evaluated by means of (9.2.18):

$$\sum_{p^\delta < q \leq x} \frac{1}{q} = \log \log x - \log \log p^\delta + o(1) = \log \frac{\log[c\sqrt{p} \log p]}{\log p^\delta} + o(1)$$

$$\leq \log \left(\frac{1}{2\delta} + \frac{1}{\delta} \frac{\log(c \log p)}{\log p} \right) + o(1) = \log \frac{1}{2\delta} + o(1)$$

where the $o(1)$ is as $p \to \infty$. Thus we must have

$$\frac{1}{2}(c - 2)\sqrt{p} \log p \leq x \log \frac{1}{2\delta} + o(x)$$

or

$$\frac{1}{2} \frac{(c - 2)}{c} x \leq x \log \frac{1}{2\delta} + o(x)$$

which implies

$$\frac{1}{2} \left(\frac{c - 2}{c} \right) \leq \log \frac{1}{2\delta}.$$

If we choose δ so that

(9.2D.3) $$\log \frac{1}{2\delta} < \frac{1}{2} \left(1 - \frac{2}{c} \right),$$

we get a contradiction for sufficiently large p, implying the existence of a nonresidue mod p that is $\leq p^\delta$. But (9.2D.3) is equivalent to

$$\frac{1}{2\delta} < \sqrt{e}\, e^{-1/c} \qquad \text{or} \qquad \delta > \frac{1}{2\sqrt{e}} \cdot e^{1/c},$$

and from (9.2D.2), $\delta = (1/2\sqrt{e}) + \epsilon$ satisfies this and the theorem follows.

EXERCISES

1. Generalize Theorem 9.2D.1 to the case of dth power residues for any d dividing $p - 1$.

2. Given $\epsilon > 0$, for all sufficiently large primes p show that there exist at least $c \log p$ quadratic nonresidues mod p that are less than $p^{(1/(2\sqrt{e}))+\epsilon}$, where c is a positive constant (depending on ϵ).

3. Show that a solution to Exercise 1 can be given using only that for $\chi \neq \chi_0$,

$$\left| \sum_{n \leq x} \chi(n) \right| \leq b\sqrt{p} \log p$$

for some constant b (and not the particular value $b = 1$).

§9.3 BERTRAND'S POSTULATE, THE IDEAS OF CHEBYCHEV

A famous assertion of prime number theory, which is called *Bertrand's Postulate*, states that *for every integer $n \geq 2$ there is a prime between n and $2n$*. That is, there's a prime p such that

(9.3.1) $n < p < 2n.$

One easily verifies this numerically in special cases. For $n = 2$, $p = 3$ is a prime satisfying (9.3.1). For $n = 3$, $p = 5$ is the largest prime satisfying (9.3.1) and $p = 5$ serves to verify the result for $3 \leq n \leq 4$. For $n = 5$, $p = 7$ plays the same role and continuing, we get the following sequence of primes in which each is less than twice the preceding:

3, 5, 7, 13, 23, 43, 83, 163, 317, 631, 1259; 2503, 4999, 9973,

19937, 39869, 79699, 159389, 318751, 637499, 1000003,

which serves to verify Bertrand's Postulate for $n \leq 1,000,000$.

The problem of establishing Bertrand's Postulate in general was a focal point of much effort in the mid nineteenth century (i.e., after 1845 when Bertrand formulated his conjecture). The Russian mathematician Chebychev succeeded in 1852 and in this section we propose to discuss his route to the solution.

The initial idea is to exploit (9.2.8) which asserts

$$(9.3.2) \qquad K(x) = \log[x]! = \sum_{m \le x} \psi\left(\frac{x}{m}\right),$$

together with (9.2.9) which gives

$$(9.3.3) \qquad K(x) = x \log x - x + 0(\log x).$$

We note that (9.3.2) is in a form to which Möbius inversion can be applied (see Theorem 3.5.2) to yield

$$(9.3.4) \qquad \psi(x) = \sum_{m \le x} \mu(m) K\left(\frac{x}{m}\right).$$

Then, if one inserts (9.3.3) for $K(x/m)$ on the right, among other terms, one gets

$$\sum_{m \le x} 0\left(\log \frac{x}{m}\right) = 0(x),$$

so that the best that this procedure could yield would be $\psi(x) = 0(x)$ [already otherwise derived in (9.2.12)]. In fact, all lower bound information appears to have been lost. The difficulty here is that (9.3.4), as an explicit solution of the equations (9.3.2), uses too many equations. The sum on the right of (9.3.4) is over *all $m \le x$*. If, for example, it were over a finite set of m, the $0(\log x/m)$ coming from (9.3.3) would remain $0(\log x)$. However, clearly, one gives up on the objective of solving explicitly for $\psi(x)$ itself and replaces this by some linear combination of values of $\psi(x)$.

Consider then real numbers a_d, $d = 1, \ldots, r$ and form

$$(9.3.5) \qquad Z(x) = \sum_{d=1}^{r} a_d K\left(\frac{x}{d}\right).$$

From (9.3.3),

$$Z(x) = \sum_{d=1}^{r} a_d\left(\frac{x}{d} \log \frac{x}{d} - \frac{x}{d}\right) + 0(\log x)$$

$$= \left(\sum_{d=1}^{r} \frac{a_d}{d}\right)(x \log x - x) - x \sum_{d=1}^{r} \frac{a_d \log d}{d} + 0(\log x).$$

Imposing the condition

(9.3.6)
$$\sum_{d=1}^{r} \frac{a_d}{d} = 0,$$

this becomes

(9.3.7)
$$Z(x) = \lambda x + 0(\log x)$$

where

(9.3.8)
$$\lambda = -\sum_{d=1}^{r} \frac{a_d \log d}{d}.$$

On the other hand, using (9.3.2),

$$Z(x) = \sum_{d=1}^{r} a_d \sum_{m \le x/d} \psi\left(\frac{x}{md}\right) = \sum_{t \le x} \psi\left(\frac{x}{t}\right)\left(\sum_{\substack{d/t \\ d \le r}} a_d\right)$$

or

(9.3.9)
$$Z(x) = \sum_{t \le x} \psi\left(\frac{x}{t}\right) A(t)$$

where

(9.3.10)
$$A(t) = \sum_{\substack{d/t \\ d \le r}} a_d.$$

Combining (9.3.9) and (9.3.7) yields

(9.3.11)
$$\sum_{t \le x} \psi\left(\frac{x}{t}\right) A(t) = \lambda x + 0(\log x).$$

It is this relation that replaces (9.3.2). One apparent advantage is that the order of magnitude here is λx as opposed to the previously larger $x \log x$. This suggests that $A(t)$ will have achieved this by introducing changes of sign and hence that (9.3.11) might be a source of inequalities for $\psi(x)$. That these changes of sign in $A(t)$ are, in fact, forced by (9.3.6) is easily seen in that for

(9.3.12)
$$M = \text{l.c.m.}\{d\},$$
$$\substack{1 \le d \le r \\ a_d \ne 0}$$

we have

(9.3.13)
$$\sum_{t=1}^{M} A(t) = 0.$$

For

$$\sum_{t=1}^{M} A(t) = \sum_{t=1}^{M} \sum_{\substack{d/t \\ d \le r}} a_d = \sum_{\substack{d=1 \\ a_d \ne 0}}^{r} a_d \sum_{\substack{1 \le t \le M \\ t \equiv 0 (\bmod d)}} 1 = M \sum_{\substack{d=1 \\ a_d \ne 0}}^{r} \frac{a_d}{d} = 0.$$

Further, $A(t)$ is periodic with the M defined above as a period. This is immediate from

(9.3.14)
$$A(t) = \sum_{\substack{d/(t,M) \\ d \le r}} a_d.$$

Since we've replaced (9.3.4) by (9.3.5), we see that the function

$$a(m) = \begin{cases} a_m & \text{for } 1 \le m \le r \\ 0 & \text{otherwise} \end{cases}$$

is intended as an "approximation" to $\mu(m)$ and $A(t)$ is an "approximation" to

$$e(m) = \sum_{d/m} \mu(d).$$

To construct such approximations that satisfy (9.3.6), we fix an integer $s \ge 1$, and define

(9.3.15)
$$a_d = \mu(d) \qquad \text{for} \qquad 1 \le d \le s.$$

Then setting $M_s = \prod_{p \le s} p$, take $a_d = 0$, for $s < d < M_s$, and for $d = r$, $r = M_s$, choose a_r so as to satisfy

(9.3.16)
$$\sum_{d \le s} \frac{\mu(d)}{d} + \frac{a_r}{r} = 0.$$

This a_r will be an integer (except for $s = 2$) and canceling common factors with r can reduce the effective value of r. In the above, there is the possibility for small s that $M_s \le s$, in which case, take $r = s + 1$ in (9.3.16). The sets of as generated by this procedure are given below for various values of s:

(9.3.17)

s	r	NONZERO a_d								
1	2	$a_1 = 1$	$a_2 = -2$							
2	3	$a_1 = 1$	$a_2 = -1$	$a_3 = -\frac{3}{2}$						
3	6	$a_1 = 1$	$a_2 = -1$	$a_3 = -1$	$a_6 = -1$					
5	30	$a_1 = 1$	$a_2 = -1$	$a_3 = -1$	$a_5 = -1$	$a_{30} = 1$				
7	105	$a_1 = 1$	$a_2 = -1$	$a_3 = -1$	$a_5 = -1$	$a_6 = 1$	$a_7 = -1$	$a_{105} = 1$		
11	2310	$a_1 = 1$	$a_2 = -1$	$a_3 = -1$	$a_5 = -1$	$a_6 = 1$	$a_7 = -1$	$a_{10} = 1$	$a_{11} = -1$	$a_{2310} = 1$

Our objective is to use (9.3.11) to obtain bounds for $\psi(x)$ of the form

$$\underline{c}x + o(x) \leq \psi(x) \leq \bar{c}x + o(x)$$

with $\underline{c} > 0$, so that

$$\psi(x) - \psi\left(\frac{x}{2}\right) \geq \left(\underline{c} - \frac{\bar{c}}{2}\right)x + o(x)$$

for all large x. Then from (9.1.7), this implies

$$\theta(x) - \theta\left(\frac{x}{2}\right) = \sum_{x/2 < p \leq x} \log p \geq \left(\underline{c} - \frac{\bar{c}}{2}\right)x + o(x).$$

Thus we see that provided

(9.3.18) $2\underline{c} > \bar{c},$

it would follow that there is a prime between x and $x/2$ for all large x (in fact, many such primes).

We attempt first to carry this out using the sequence given under $s = 1$ in the table (9.3.17). Here, $M = 2$, and (9.3.14) gives (since $a_1 = 1$, $a_2 = -2$)

$$A(t) = \begin{cases} 1 & \text{if } t \text{ is odd} \\ -1 & \text{if } t \text{ is even,} \end{cases}$$

so that (9.3.11) becomes

(9.3.19) $\psi(x) - \psi\left(\frac{x}{2}\right) + \psi\left(\frac{x}{3}\right) - \cdots = \lambda x + 0(\log x)$

where $\lambda = \log 2$. Then, using the fact that $\psi(x)$ is nondecreasing, we note that for all integers $m \geq 1$

$$-\psi\left(\frac{x}{2m}\right) + \psi\left(\frac{x}{2m + 1}\right) \le 0,$$

so that (9.3.19) yields

(9.3.20) $$\psi(x) \ge \lambda x + 0(\log x).$$

On the other hand, since for all $m \ge 1$,

$$\psi\left(\frac{x}{2m + 1}\right) - \psi\left(\frac{x}{2m + 2}\right) \ge 0,$$

(9.3.19) gives that

$$\psi(x) - \psi\left(\frac{x}{2}\right) \le \lambda x + 0(\log x).$$

Then replacing x by $x/2^i$ we have

$$\psi\left(\frac{x}{2^i}\right) - \psi\left(\frac{x}{2^{i+1}}\right) \le \frac{\lambda}{2^i} \cdot x + 0(\log x).$$

Summing over $i \ge 0$ yields [there are $0(\log x)$ terms]

(9.3.21) $$\psi(x) \le 2\lambda x + 0(\log^2 x).$$

Thus we have estimates with $\underline{c} = \lambda$ and $\bar{c} = 2\lambda$, so (9.3.18) just fails. (Actually, we will see in §9.3B that this particular approach can be salvaged.) In any event, we note that the failure had nothing to do with the actual value of λ, but is related to constants stemming from the form of (9.3.19) and the method used to treat it. Equations (9.3.20) and (9.3.21) do yield immediately that

(9.3.22) $$\log 2 \le \underline{\lim} \frac{\psi(x)}{x} \le \overline{\lim} \frac{\psi(x)}{x} \le 2 \log 2$$

where $\log 2 = .693147 \ldots$ and $2 \log 2 = 1.386294 \ldots$.

We turn next to the sequence given under $s = 3$ in the table (9.3.17). Here $a_1 = 1$, $a_2 = -1$, $a_3 = -1$, $a_6 = -1$, $M = 6$, and (9.3.14) gives

$$A(t) = \begin{cases} 1 & \text{if } (t, 6) = 1 \\ 0 & \text{if } (t, 6) = 2, 3 \\ -2 & \text{if } (t, 6) = 6 \end{cases}$$

and (9.3.11) becomes

$$(9.3.23) \qquad \psi(x) + \psi\left(\frac{x}{5}\right) - 2\psi\left(\frac{x}{6}\right) + \psi\left(\frac{x}{7}\right) + \psi\left(\frac{x}{11}\right) - 2\psi\left(\frac{x}{12}\right) + \cdots$$

$$= \lambda x + O(\log x)$$

where $\lambda = (\log 2)/2 + (\log 3)/3 + (\log 6)/6 = 1.011404265\ldots$

Group the terms on the left side of (9.3.23) as follows:

$$\left(\psi(x) - \psi\left(\frac{x}{6}\right)\right) + \left(\psi\left(\frac{x}{5}\right) - \psi\left(\frac{x}{6}\right)\right) + \cdots$$

and note that, in this first period that is displayed, the grouped summands are all ≥ 0. Thus the sum of these first groups is $\geq \psi(x) - \psi(x/6)$. For a general period,

$$\left(\psi\left(\frac{x}{6m+1}\right) - \psi\left(\frac{x}{6m+6}\right)\right) + \left(\psi\left(\frac{x}{6m+5}\right) - \psi\left(\frac{x}{6m+6}\right)\right)$$

is always ≥ 0, so that (9.3.23) yields

$$(9.3.24) \qquad \psi(x) - \psi\left(\frac{x}{6}\right) \leq Z(x) = \lambda x + O(\log x).$$

Replacing x by $x/6^i$ and summing over i then gives

$$(9.3.25) \qquad \psi(x) \leq \frac{6}{5}\lambda x + O(\log^2 x).$$

Now, in (9.3.23), a group of terms such as

$$-2\psi\left(\frac{x}{6}\right) + \psi\left(\frac{x}{7}\right) + \psi\left(\frac{x}{11}\right) \leq 0$$

and similarly, every three consecutive terms after that, have a sum ≤ 0. Thus, from (9.2.23), we obtain

$$(9.3.26) \qquad \psi(x) + \psi\left(\frac{x}{5}\right) \geq Z(x) = \lambda x + O(\log x)$$

and from (9.3.24),

$$\psi\left(\frac{x}{5}\right) \leq \frac{6}{25}\lambda x + O(\log^2 x),$$

so that this gives

(9.3.27) $\psi(x) \geq \frac{19}{25}\lambda x + 0(\log^2 x)$.

Here we have $\underline{c} = \frac{19}{25}\lambda$, $\bar{c} = \frac{6}{5}\lambda$, and

$$2\underline{c} - \bar{c} = \frac{38}{25}\lambda - \frac{6}{5}\lambda = \frac{8}{25}\lambda > 0.$$

This, of course, serves to establish Bertrand's Postulate for all sufficiently large x. Further, from (9.3.25) and (9.3.27) it follows that

(9.3.28) $.7686672412 \ldots \leq \varliminf \frac{\psi(x)}{x} \leq \varlimsup \frac{\psi(x)}{x} \leq 1.213685118 \ldots$,

which is clearly a sharpening of (9.3.22).

Of course, there still remains the task of refining the arithmetic of the above argument so as to obtain Bertrand's Postulate for *all $x \geq 2$*. this will be carried out in §9.3A.

Also, one should note that although the heuristic ideas that led to (9.3.17) were probably part of Chebychev's motivation (and they did lead to success), they would require some modification in order to become a bona fide method. Further, as we shall see in §9.3B and §9.3C, simpler ideas provide a workable procedure.

EXERCISES

1. For the arithmetic function $a_s(m)$ defined in (9.3.15) [and (9.3.16)], show that as $s \to \infty$, $a_s(m)$ approaches $\mu(m)$ in the sense of the valuation defined in §4.4. Show also that the corresponding function $A_s(t)$ converges to the unit function $e(t)$.

2. Let β be any real number, $\beta > \frac{30}{19}$. Show that for all sufficiently large x, there is a prime between x and βx.

3. Use the a_d in (9.3.17) corresponding to $s = 5$ to show that

    $$\lambda x + 0(\log^2 x) \leq \psi(x) \leq \frac{6}{5}\lambda x + 0(\log^2 x)$$

    where $\lambda = \log 2/2 + \log 3/3 + \log 5/5 - \log 30/30 = .9212920229 \ldots$.
    Note that the constants implicit in the $0(\log x)$ errors can be explicitly estimated. From this, deduce that for any $\beta > 1.2$, for all sufficiently large x, there is a prime between x and βx.

§9.3A PROOF OF BERTRAND'S POSTULATE

To prove Bertrand's Postulate, we propose to follow through the numerical details of the method of §9.3 (using the sequence $a_1 = 1$, $a_2 = -1$, $a_3 = -1$, $a_6 = -1$) and make a determination of how large is "sufficiently large."

We first require a more precise version of (9.2.9) in which the $O(\log x)$ is replaced by numerical inequalities.

Lemma 9.3A.1.   For $K(x) = \log [x]!$, we have for all $x \geq 2$,

(9.3A.1) $K(x) \geq x \log x - x + 1 - \log x$

and

(9.3A.2) $K(x) \leq x \log x - x + (2 - 2 \log 2) + \log x.$

   Proof.   Clearly

$$K(x) = \sum_{n=2}^{[x]} \log n \geq \sum_{n=2}^{[x]} \int_{n-1}^{n} \log u \, du$$

or

$$K(x) \geq \int_{1}^{[x]} \log u \, du = \int_{1}^{x} \log u \, du - \int_{[x]}^{x} \log u \, du,$$

implying

$$K(x) \geq x \log x - x + 1 - \log x,$$

which is (9.3A.1).
   Also,

$$K(x) = \sum_{n=2}^{[x]} \log n \leq \sum_{n=2}^{[x]-1} \int_{n}^{n+1} \log u \, du + \log x$$

or

$$K(x) \leq \int_{2}^{x} \log u \, du + \log x = x \log x - x + (2 - 2 \log 2) + \log x,$$

which is (9.3A.2).
   In the case at hand, (9.3.5) becomes

(9.3A.3) $Z(x) = K(x) - K\left(\frac{x}{2}\right) - K\left(\frac{x}{3}\right) - K\left(\frac{x}{6}\right)$

and applying (9.3A.1) and (9.3A.2) yields

(9.3A.4) $Z(x) \geq \lambda x - 4 \log x + (8 \log 2 + 2 \log 3 - 5)$

$$\geq \lambda x - 4 \log x + 2.7$$

and

(9.3A.5) $Z(x) \leq \lambda x + 4 \log x + (-1-4 \log 2 - 2 \log 3) \leq \lambda x + 4 \log x$

where $\lambda = \log 2/2 + \log 3/3 + \log 6/6 = 1.01140 \ldots$.
From the first inequality of (9.3.24), (9.3A.5) implies

$$\psi(x) - \psi\left(\frac{x}{6}\right) \leq Z(x) \leq \lambda x + 4 \log x$$

and replacing x by $x/6^i$,

$$\psi\left(\frac{x}{6^i}\right) - \psi\left(\frac{x}{6^{i+1}}\right) \leq \lambda \frac{x}{6^i} + 4 \log \frac{x}{6^i}.$$

Noting that $x/6^{i+1} < 2$ is equivalent to $i + 1 > [(\log x/2)/\log 6]$, summing the above over $0 \leq i \leq \log x/\log 6$ we get

$$\psi(x) \leq \frac{6}{5}\lambda x + \frac{4 \log x}{\log 6} \cdot \log \frac{x}{6} + 4 \log x$$

implying

(9.3A.6) $\psi(x) \leq \frac{6}{5}\lambda x + (2.24)\log^2 x.$

Similarly, from the first inequality of (9.3.26) and (9.3A.4),

(9.3A.7) $\psi(x) + \psi\left(\frac{x}{5}\right) \geq Z(x) \geq \lambda x - 4 \log x + 2.7.$

Using (9.3A.6), for $x > 5$,

$$\psi\left(\frac{x}{5}\right) \leq \frac{6}{25}\lambda x + (2.24)\log^2 x,$$

so that (9.3A.7) yields

(9.3A.8) $\psi(x) \geq \frac{19}{25}\lambda x - (2.24)\log^2 x - 4 \log x + 2.7.$

From (9.3A.6) and (9.3A.8),

(9.3A.9) $\psi(2x) - \psi(x) \geq \frac{8}{25}\lambda x - (4.48)\log^2 2x - 4 \log 2x.$

Next, note that

$$\psi(2x) - \psi(x) = \sum_{x < p^\alpha \le 2x} \log p \le (\log 2x) \sum_{x < p^\alpha \le 2x} 1$$

$$\le (\log 2x)\left(\sum_{x < p \le 2x} 1 + \sqrt{2x} + \frac{\sqrt[3]{2x} \log 2x}{\log 2} \right),$$

which, when combined with (9.3A.9), gives

$$(\log 2x)(\pi(2x) - \pi(x)) \ge \frac{8}{25}\lambda x - (4.48)\log^2 2x - 4 \log 2x$$

$$- \sqrt{2x} \log 2x - \frac{\sqrt[3]{2x} \log^2 2x}{\log 2}$$

or

(9.3A.10) $$(\log 2x)(\pi(2x) - \pi(x)) \ge x\left(\frac{8}{25}\lambda - \frac{(4.48)\log^2 2x}{x} \right.$$

$$\left. - \frac{4 \log 2x}{x} - \frac{\sqrt{2} \log 2x}{\sqrt{x}} - \frac{\sqrt[3]{2} \log^2 2x}{(\log 2)x^{2/3}} \right).$$

Since for $x \ge 10^6$, the function

$$f(x) = \frac{(4.48)\log^2 2x}{x} + \frac{4 \log 2x}{x} + \frac{\sqrt{2} \log 2x}{\sqrt{x}} + \frac{\sqrt[3]{2} \log^2 2x}{(\log 2)x^{2/3}}$$

is monotone decreasing, it follows that in this range

$$f(x) \le f(10^6) = .05978 \ldots < .06.$$

Thus for $x \ge 10^6$,

$$\tfrac{8}{25}\lambda - f(x) > \tfrac{8}{25}\lambda - .06 > .26$$

and hence from (9.3A.10), for $x \ge 10^6$,

(9.3A.1i) $$\pi(2x) - \pi(x) > (.26)\frac{x}{\log 2x} > 0,$$

which verifies Bertrand's Postulate for $x \ge 10^6$. For $x < 10^6$, a numerical verification was given at the beginning of §9.3.

In the last century, when electronic calculators were not available, it was considered somewhat unaesthetic to have to verify a result numerically over such a large range (as above). Thus, Chebychev himself, in order to reduce such calculations, used the a_d in (9.3.17) corresponding to $s = 5$.

EXERCISES

1. Verify that the sequence of primes used in §9.3 to verify Bertrand's Postulate for $x \leq 10^6$ are indeed all primes.

2. Carry out a proof of Bertrand's Postulate using $a_1 = 1$, $a_2 = -1$, $a_3 = -1$, $a_5 = -1$, $a_{30} = 1$.

§9.3B   RAMANUJAN'S IDEA

We saw in §9.3 that with respect to the proposed method, the simplest sequence $a_1 = 1$, $a_2 = -2$ appeared to fail. Ramanujan salvaged this procedure by means of a simple idea. Here the basic identity (9.3.9) asserts that

$$(9.3B.1) \qquad \psi(x) - \psi\left(\frac{x}{2}\right) + \psi\left(\frac{x}{3}\right) - \psi\left(\frac{x}{4}\right) + \cdots = Z(x)$$

where

$$(9.3B.2) \qquad\qquad Z(x) = K(x) - 2K\left(\frac{x}{2}\right).$$

Our previous method consisted of introducing estimates for $Z(x)$ into (9.3B.1) and using this to get an upper and lower bound for $\psi(x)$. Ramanujan noted the appearance of the difference $\psi(x) - \psi(x/2)$ in (9.3B.1) and his idea was to obtain an upper bound for $\psi(x)$ and then aim directly at a lower bound for $\psi(x) - \psi(x/2)$ [without the intermediary step of a lower bound for $\psi(x)$]. This increases the efficiency of the argument considerably and a relatively easy proof of Bertrand's Postulate results.

From (9.3B.1), since for all $m \geq 1$,

$$\psi\left(\frac{x}{2m - 1}\right) - \psi\left(\frac{x}{2m}\right) \geq 0,$$

it follows that

$$(9.3B.3) \qquad\qquad \psi(x) - \psi\left(\frac{x}{2}\right) \leq Z(x).$$

Then from (9.3B.2), (9.3A.1), and (9.3A.2),

$$\psi(x) - \psi\!\left(\frac{x}{2}\right) \leq (\log 2)x + 3 \log x.$$

Replacing x by $x/2^i$,

$$\psi\!\left(\frac{x}{2^i}\right) - \psi\!\left(\frac{x}{2^{i+1}}\right) \leq (\log 2)\frac{x}{2^i} + 3 \log \frac{x}{2^i}$$

and summing over all i, $0 \leq i \leq (\log x)/(\log 2) - 1$, we get

(9.3B.4) $$\psi(x) \leq (2 \log 2)x + \frac{3 \log^2 x}{\log 2}.$$

From (9.3B.1), we also obtain

(9.3B.5) $$\psi(x) - \psi\!\left(\frac{x}{2}\right) + \psi\!\left(\frac{x}{3}\right) \geq Z(x),$$

so that again from (9.3B.2), (9.3A.1), and (9.3A.2),

(9.3B.6) $$\psi(x) - \psi\!\left(\frac{x}{2}\right) + \psi\!\left(\frac{x}{3}\right) \geq (\log 2)x - 3 \log x.$$

Then, from (9.3B.4) for $x > 3$,

$$\psi\!\left(\frac{x}{3}\right) \leq \left(\frac{2}{3} \log 2\right)x + \frac{3 \log^2 x}{\log 2},$$

so that introducing this in (9.3B.6) we obtain

$$\psi(x) - \psi\!\left(\frac{x}{2}\right) \geq \left(\frac{1}{3} \log 2\right)x - \frac{3 \log^2 x}{\log 2} - 3 \log x.$$

This, in turn, can be rewritten as

(9.3B.7) $$\psi(x) - \psi\!\left(\frac{x}{2}\right) \geq x\!\left(\frac{1}{3} \log 2 - \frac{3 \log^2 x}{(\log 2)x} - \frac{3 \log x}{x}\right).$$

Since for $x \geq 10^4$,

$$f(x) = \frac{3 \log^2 x}{(\log 2)x} + \frac{3 \log x}{x}$$

is monotone decreasing,

$$f(x) \le f(10^4) \le .04.$$

But since $\tfrac{1}{3} \log 2 > .2310$, (9.3B.7) implies that for $x \ge 10^4$

(9.3B.8) $$\psi(x) - \psi\!\left(\frac{x}{2}\right) \ge (.19)x.$$

Finally, we observe that

$$\psi(x) - \psi\!\left(\frac{x}{2}\right) = \sum_{\frac{x}{2} < p^\alpha \le x} \log p = \sum_{\frac{x}{2} < p \le x} \log p + \sum_{\substack{\frac{x}{2} < p^\alpha \le x \\ \alpha > 1}} \log p$$

$$\le (\log x)\!\left(\pi(x) - \pi\!\left(\frac{x}{2}\right)\right) + \sum_{2 \le i \le (\log x / \log 2)} \psi(x^{1/i})$$

and using (9.3B.4) to estimate the $\psi(x^{1/i})$, we get

$$\psi(x) - \psi\!\left(\frac{x}{2}\right) \le (\log x)\!\left(\pi(x) - \pi\!\left(\frac{x}{2}\right)\right) + (2 \log 2)\sqrt{x} + \frac{3 \log^2 x}{4 \log 2}$$

$$+ \left(\frac{\log x}{\log 2}\right)\!\left((2 \log 2)\sqrt[3]{x} + \frac{\log^2 x}{3 \log 2}\right).$$

Thus (9.3B.8) yields

$$(\log x)\!\left(\pi(x) - \pi\!\left(\frac{x}{2}\right)\right) \ge x\!\left((.19) - \frac{2 \log 2}{\sqrt{x}} - \frac{2 \log x}{x^{2/3}}\right.$$

$$\left. - \frac{3 \log^2 x}{(4 \log 2)x} - \frac{\log^3 x}{3(\log 2)^2 x}\right) \ge (.05)x$$

for $x \ge 10^4$.
    Thus for $x \ge 10^4$,

(9.3B.9) $$\pi(x) - \pi\!\left(\frac{x}{2}\right) \ge (.05)\frac{x}{\log x} > 0,$$

which establishes Bertrand's Postulate in this range.

§9.3C THE ERDOS IDEAS

If one views the use of the logarithm function as an element of transcendentality, one might wish to remove it from a central role in the proof of Bertrand's Postulate. The

logarithm makes its initial essential appearance in the identity (9.3.2) and elimi-
nating it is equivalent to interpreting (9.3.2) as expressing the unique factorization
of $N!$ (N a positive integer), in terms of primes. In fact, this is the familiar

(9.3C.1)
$$N! = \prod_{p \leq N} p^{\alpha_p(N)}$$

where

(9.3C.2)
$$\alpha_p(N) = \sum_{i=1}^{\infty} \left[\frac{N}{p^i}\right].$$

Then $(2N)!/N!$ quite naturally "traps" all the primes p such that $N < p \leq 2N$ as
divisors (the old Euclid idea). However, the same is achieved more efficiently by the
integer

(9.3C.3)
$$\binom{2N}{N} = \frac{(2N)!}{(N!)^2}.$$

The "efficiency" with which (9.3C.3) captures the primes under consideration may
be measured roughly by its size (and, in a vague way, by its form). There are many
such expressions. For example, the argument of §9.3A may be viewed as focusing
on the multinomial coefficient

(9.3C.4)
$$\frac{6N!}{(3N)! \, (2N)! \, N!} = \binom{6N}{3N}\binom{3N}{N},$$

which has all primes p, $3N < p < 6N$ as divisors.

Erdos derived Bertrand's Postulate directly from consideration of the factorization of $\binom{2N}{N}$ by means of two ideas, a small technical one plus a genuinely
arithmetic insight. The technical point is simply that one should estimate $\binom{2N}{N}$
directly instead of through upper and lower bounds [such as (9.3A.1) and (9.3A.2)],
both of which contain errors. This is easily achieved as follows.

Lemma 9.3C.1.   For every integer $N \geq 1$,

(9.3C.5)
$$\frac{1}{2}\frac{4^N}{\sqrt{N}} \leq \binom{2N}{N} < 4^N.$$

Proof.   The upper bound is immediate from

$$\sum_{k=0}^{2N} \binom{2N}{k} = 4^N.$$

As for the lower bound, for $N = 1$ it is an equality, and we proceed by induction on N, assuming the result for $N - 1$, $N > 1$. Then

$$\binom{2N}{N} = \frac{(2N)(2N-1)}{N^2}\binom{2(N-1)}{N-1} \geq \binom{2N-1}{N}\frac{4^{N-1}}{\sqrt{N-1}}$$

or

$$\binom{2N}{N} \geq \left(\frac{1}{2}\cdot 4^N\right)\left(\frac{1}{\sqrt{N-1}}\right)\left(1 - \frac{1}{2N}\right).$$

The induction will be completed if we can verify that

$$\frac{1}{\sqrt{N-1}}\left(1 - \frac{1}{2N}\right) \geq \frac{1}{\sqrt{N}}$$

or

$$\left(1 - \frac{1}{2N}\right) \geq \left(1 - \frac{1}{N}\right)^{1/2}.$$

But squaring this last yields

$$1 - \frac{1}{N} + \frac{1}{4N^2} \geq 1 - \frac{1}{N},$$

which is true and the proof of the lemma is completed.

Since each prime p, $N < p \leq 2N$ divides $\binom{2N}{N}$ to exactly the first power, we have

(9.3C.6) $$\binom{2N}{N} = \left(\prod_{N<p\leq 2N} p\right)R_N,$$

where R_N is an integer divisible only by primes $\leq N$. The aim here is to use a *lower bound* for $\binom{2N}{N}$ and an *upper bound* for R_N in (9.3C.6), thereby obtaining a lower bound for $\prod_{N<p\leq 2N} p$. Of course, any such lower bound > 1 establishes Bertrand's Postulate.

The lower bound for $\binom{2N}{N}$ is already available in Lemma 9.3C.1. As for the desired upper bound for R_N, it will result from the upper bound for $\binom{2N}{N}$, as a product of Erdos' main idea.

From (9.3C.6) and the upper bound of (9.3C.5), we obtain

$$\prod_{N<p\leq 2N} p \leq 4^N$$

for all $N > 1$. Then letting $N_0 = N$, $N_1 = (N_0 + \epsilon_1)/2$ where ϵ_1 is 0 if N_0 is even, and 1 if N_0 is odd; and, in general, $N_i = (N_{i-1} + \epsilon_{i-1})/2$ where $\epsilon_{i-1} = 0$ or 1 according as N_{i-1} is even or odd, we see that the intervals

$$N_i < p \leq 2N_i,$$

$i \geq 0$ will include all primes $\leq 2N$. Since $N_i \leq (N_{i-1}/2) + \frac{1}{2}$, it follows that $N_i \leq N_0/2^i + 1 = N/2^i + 1$, so that the number of such intervals that are required is $\leq \log N/\log 2$. For each of these we have

$$\prod_{N_i<p\leq 2N_i} p \leq 4^{N_i}$$

and multiplying them yields

$$\prod_{p\leq 2N} p \leq 4^{\Sigma N_i} \leq 4^{\Sigma((N/2^i) + 1)}$$

or

(9.3C.7)
$$\prod_{p\leq 2N} p \leq 4^{2N+(\log N)/(\log 2)} = N^2 4^{2N}.$$

Since

$$\prod_{p\leq 2N+1} p \leq \prod_{p\leq 2N+2} p \leq (N + 1)^2 4^{2N+2} = (2N + 2)^2 4^{2N+1}$$

we obtain for all integers M that

(9.3C.8)
$$\prod_{p\leq M} p \leq (M + 1)^2 4^M.$$

(Note that this is weaker than (9.3C.7) for M even.)

In analyzing the prime factorization of R_N we first note the following lemma.

Lemma 9.3C.2. For $N \geq 1$, any prime $p > \sqrt{2N}$ divides $\binom{2N}{N}$ at most to the first power. For all primes p (hence including those $p \leq \sqrt{2N}$), the power to which it divides $\binom{2N}{N}$ is at most $(\log 2N)/(\log p)$.

Proof. Since

$$(9.3C.9) \qquad \binom{2N}{N} = \frac{(2N)(2N - 1) \cdots (N + 1)}{N!},$$

any prime $p > \sqrt{2N}$ can divide a factor of the numerator at most to the first power. Now let $t^*p = $ the smallest multiple of the prime p that is $\geq N + 1$ and $\leq 2N$. (If there is no such multiple of p the result is clear.) Then for

$$2N \geq pt > N, \qquad t \neq t^*,$$

set up the 1–1 correspondence

$$pt \leftrightarrow pt - pt^* = p(t - t^*),$$

where $N > p(t - t^*) > 1$. Thus, at most, the one multiple of p, pt^*, is left unmatched in the numerator of (9.3C.9) and hence for such a prime, at most, the first power of p divides $\binom{2N}{N}$.

For an arbitrary prime p, the power to which it divides $\binom{2N}{N}$ is given by

$$\sum_{i \geq 1} \left(\left[\frac{2N}{p^i} \right] - 2 \left[\frac{N}{p^i} \right] \right).$$

Since the range of i is limited by $p^i \leq 2N$ to $i \leq (\log 2N)/(\log p)$, and

$$0 \leq \left[\frac{2N}{p^i} \right] - 2 \left[\frac{N}{p^i} \right] \leq 1,$$

it follows that the above sum is $\leq (\log 2N)/(\log p)$. This completes the proof of the lemma.

Since the primes p in the range $\sqrt{2N} < p \leq N$ that divide $\binom{2N}{N}$ divide R_N we have

$$(9.3C.10) \qquad R_N \leq \prod_{\sqrt{2N} < p \leq N} p^{\epsilon_p} \prod_{p \leq \sqrt{2N}} p^{(\log 2N)/(\log p)},$$

where each $\epsilon_p = 0$ or 1. The above inequality needs some sharpening in order to enable us to come through and it is here that Erdos' main idea enters.

Lemma 9.3C.3. For $N \geq 3$, no prime p such that $N \geq p > \frac{2}{3}N$ divides $\binom{2N}{N}$.

Proof. For such a prime p, $2N \geq 2p > \frac{4}{3}N$, and $3p > 2N$, so that p divides each of $(2N)(2N - 1) \cdots (N + 1)$ and $N!$ exactly to the first power. Thus p does not divide $\binom{2N}{N} = ((2N)(2N - 1) \cdots (N + 1))/N!$.

For $N \geq 3$, for all primes p such that $N \geq p > \frac{2}{3}N$, p does not divide R_N (i.e., $\epsilon_p = 0$). Hence (9.3C.10) yields

$$(9.3C.11) \qquad R_N \leq \left(\prod_{p \leq \frac{2}{3}N} p \right) (2N)^{\sqrt{2N}}.$$

Then applying (9.3C.8) with $M = [\frac{2}{3}N]$, this gives

$$(9.3C.12) \qquad R_N \leq (\frac{2}{3}N + 1)^2 4^{(2/3)N} (2N)^{\sqrt{2N}}.$$

Inserting (9.3C.12) in (9.3C.6) together with the lower bound for $\binom{2N}{N}$ given in (9.3C.5), we get

$$\frac{1}{2} \cdot \frac{4^N}{\sqrt{N}} \leq \left(\prod_{N < p \leq 2N} p \right) \left(\frac{2}{3}N + 1 \right)^2 4^{(2/3)N} (2N)^{\sqrt{2N}}$$

or

$$\prod_{N < p \leq 2N} p \geq \frac{(1/2)(4^{(1/3)N})}{(\sqrt{N}((2/3)N + 1)^2 (2N)^{\sqrt{2N}}}.$$

Thus, Bertrand's Postulate follows if

$$(9.3C.13) \qquad 4 > 2^{(3/N) + (3\sqrt{2}/\sqrt{N})} N^{(3/2N) + (3/2/\sqrt{N})} \left(\frac{2}{3}N + 1 \right)^{6/N}.$$

The right side of (9.3C.13) is a monotone decreasing function of N as N increases and, at $N = 516$, it equals 3.99959 . . . which verifies (9.3C.13) for all $N \geq 516$. This proves that $\pi(2N) - \pi(N) > 0$ for all $N \geq 516$ and it is easily verified for $N < 516$.

Comparing the methods of Ramanujan and Erdos, we recall that Ramanujan works with

$$(9.3C.14) \qquad Z(x) = \psi(x) - \psi\left(\frac{x}{2}\right) + \psi\left(\frac{x}{3}\right) - \cdots$$

and proceeds with the two steps:

(i) $\psi(x) - \psi(x/2) \leq Z(x)$ and an upper bound for $Z(x)$ yields an upper bound for $\psi(x)$.

(ii) $\psi(x) - \psi(x/2) + \psi(x/3) \geq Z(x)$ and a lower bound for $Z(x)$, plus an upper bound for $\psi(x/3)$ gained from (i) gives lower bound for $\psi(x) - \psi(x/2)$.

In these terms, Erdos' estimate (9.3C.8) (and the proof given) is the complete analogue of (i) above. In fact, apart from the difference of technique in estimating $Z(x)$, there is just the slight complication that arises from the special nature of the interval $(N, 2N)$.

Within the framework of Erdos' argument, the analogue of (9.3C.14) is obtained by setting $x = 2N$ in (9.3C.14):

$$(9.3C.15) \quad Z(2N) = \psi(2N) - \psi(N) + \psi\left(\frac{2N}{3}\right) - \psi\left(\frac{N}{2}\right) + \cdots .$$

Then the derivation of (9.3C.8) is quite analogous to (ii). In fact, it is easy to see from the right side of (9.3C.15) that no prime powers $p^{\alpha}, \frac{2}{3}N < p^{\alpha} \leq N$, contribute to this sum. We need only note that after the terms

$$\psi(2N) - \psi(N) = \sum_{N<p^{\alpha}\leq 2N} \log p,$$

no other terms contain such prime powers. This leaves only the possibility that for some p, $\frac{2}{3}N < p \leq N$, we have $N < p^{\alpha} \leq 2N$. But $p > \frac{2}{3}N$ implies $p^{\alpha} \geq p^2 \geq \frac{4}{9}N^2 > 2N$ for $N \geq 5$. Thus underlying Erdos' derivation is the same inequality

$$Z(2N) \leq \psi(2N) - \psi(N) + \psi(\tfrac{2}{3}N),$$

but with the prime powers greater than the first estimated away so as to give

$$Z(2N) \leq \theta(2N) - \theta(N) + \theta(\tfrac{2}{3}N) + \sqrt{2N} \log 2N.$$

EXERCISES

1. Compare the estimate (9.3C.5) with what one obtains by applying (9.3A.1) and (9.3A.2).

§9.3D   THE THEOREM OF I. SCHUR

Using the Chebychev type of argument, I. Schur obtained a theorem that represented a surprising generalization of Bertrand's Postulate. Schur's Theorem is as follows.

Theorem 9.3D.1. For every pair of integers $h, k, h \geq k \geq 1$, at least one of the integers

(9.3D.1) $h + 1, h + 2, \ldots, h + k$

is divisible by some prime $p > k$.

This theorem includes Bertrand's Postulate, since for $h = k$ it implies that there is an integer n, $h < n \leq 2h$, divisible by a prime $p > h$. But this forces $n = p$.

In general, we note that it suffices to establish Schur's Theorem for k a prime. For letting $k^* =$ the largest prime $\leq k$, if we know that for $h \geq k^*$ one of $h + 1, \ldots, h + k^*$ is divisible by a prime $p > k^*$, this prime $p > k$ and the theorem follows for k.

For $k = 1$, $h + 1 \geq 2$ and Schur's Theorem simply asserts that $h + 1$ has a prime divisor. In the case $k = 2$, at least one of $h + 1, h + 2$ is odd, and hence has a prime divisor greater than 2. For $k = 3$, assuming the theorem false, there would be an h such that $h + 1, h + 2, h + 3$ are all of the form $2^\alpha 3^\beta$. Since exactly one of these integers is divisible by 3, two of them would have to be powers of 2, which differ by, at most, 2. But $h + 1 \geq 4$, implying that consecutive powers of 2 differ by at least 4. For $k = 5$, since of five consecutive integers one is divisible by 5, and at most two are divisible by 3, negating the theorem forces at least two of these integers to be powers of 2. But the only powers of two differing by ≤ 5 are $(2, 4)$ and $(4, 8)$, both of which are precluded since $h + 1 \geq k + 1 \geq 6$. Thus we may assume that $k \geq 7$.

The proof of Schur's Theorem requires a somewhat delicate handling of numerical estimates in order to reduce the finite number of "cases left over for verification." Since this would lead us somewhat astray, we shall illustrate Schur's ideas by limiting ourselves to proving the following weaker result.

Theorem 9.3D.2. Apart from a finite number of pairs of integers (h, k) (where the size of these exceptions is effectively bounded), for $h \geq k \geq 1$, at least one of the integers

(9.3D.2) $h + 1, h + 2, \ldots, h + k$

is divisible by some prime $p > k$.

We shall require the following.

Lemma 9.3D.1. For $x \geq x_0$ (where x_0 is effectively determinable),

(9.3D.3) $$\frac{\pi(x) \log x}{x} \leq 1.22.$$

Proof. From (9.3A.6),

$$\theta(x) \leq \psi(x) \leq x\left(1.213692 + (2.24)\frac{\log^2 x}{x}\right),$$

so that for sufficiently large x, say $x > x_1$, we have

(9.3D.4) $\theta(x) \leq (1.214)x.$

From Exercise 1 of §9.1, for $x > x_1$,

$$\pi(x) = \frac{\theta([x])}{\log [x]} + \sum_{2 \leq n \leq x-1} \theta(n)\left(\frac{1}{\log n} - \frac{1}{\log (n + 1)}\right)$$

$$= \frac{\theta([x])}{\log [x]} + \sum_{2 \leq n \leq x_1} \theta(n)\left(\frac{1}{\log n} - \frac{1}{\log (n + 1)}\right)$$

$$+ \sum_{x_1 < n \leq x-1} \theta(n)\left(\frac{1}{\log n} - \frac{1}{\log (n + 1)}\right),$$

so that from (9.3D.4),

$$\pi(x) \leq (1.214)\frac{[x]}{\log [x]} + (1.214) \sum_{2 \leq n \leq x-1} n\left(\frac{1}{\log n} - \frac{1}{\log (n + 1)}\right)$$

$$+ \sum_{2 \leq n \leq x_1} \theta(n)\left(\frac{1}{\log n} - \frac{1}{\log (n + 1)}\right).$$

Since

$$0 < \frac{1}{\log n} - \frac{1}{\log (n + 1)} \leq \frac{1}{n \log^2 n},$$

this gives

$$\pi(x) \leq (1.214)\frac{x}{\log x} + (1.214) \sum_{2 \leq n \leq x} \frac{1}{\log^2 n}$$

$$+ \sum_{2 \leq n \leq x_1} \theta(n)\left(\frac{1}{\log n} - \frac{1}{\log (n + 1)}\right) \leq (1.214)\frac{x}{\log x} + 0\left(\frac{x}{\log^2 x}\right),$$

which, in turn, yields (9.3D.3) for $x \geq x_0$, where x_0 depends on x_1 (and can be computed).

Proof of Theorem 9.3D.2.   Suppose that the theorem is false for a pair (h, k), $h \geq k \geq 1$. Then the integers of (9.3D.2) have only prime divisors $p \leq k$, and this is also true for the binomial coefficient

(9.3D.5) $$\binom{h + k}{k} = \frac{(h + k)!}{h! \, k!} = \frac{(h + k) \cdots (h + 1)}{k!}.$$

Thus

$$(9.3D.6) \qquad \binom{h + k}{k} = \prod_{p \le k} p^{\alpha_p},$$

where

$$\alpha_p = \sum_{i=1}^{\frac{\log (h+k)}{\log p}} \left(\left[\frac{h + k}{p^i} \right] - \left[\frac{h}{p^i} \right] - \left[\frac{k}{p^i} \right] \right).$$

Since

$$0 \le \left[\frac{h + k}{p^i} \right] - \left[\frac{h}{p^i} \right] - \left[\frac{k}{p^i} \right] \le 1,$$

we see that

$$(9.3D.7) \qquad 0 \le \alpha_p \le \frac{\log (h + k)}{\log p}.$$

Using this in (9.3D.6) gives

$$\binom{h + k}{k} \le \prod_{p \le k} p^{\log (h+k)/\log p} = \prod_{p \le k} (h + k)$$

or

$$(9.3D.8) \qquad \binom{h + k}{k} \le (h + k)^{\pi(k)}.$$

On the other hand,

$$\binom{h + k}{k} = \frac{(h + k) \cdots (h + 1)}{k(k - 1) \cdots 1} = \prod_{i=1}^{k} \frac{h + i}{i} = \prod_{i=1}^{k} \left(1 + \frac{h}{i} \right),$$

so that

$$(9.3D.9) \qquad \binom{h + k}{k} \ge \left(1 + \frac{h}{k} \right)^k.$$

Combining (9.3D.8) and (9.3D.9) yields

$$\left(1 + \frac{h}{k}\right)^k \le (h + k)^{\pi(k)}.$$

Taking logarithms, we get

$$k \log \left(1 + \frac{h}{k}\right) \le \pi(k)\log (h + k);$$

dividing by $k \log (h + k)$, multiplying by $\log k$,

$$\frac{(\log k)(\log (1 + (h/k))}{\log (h + k)} \le \frac{\pi(k)\log k}{k}.$$

Writing $\log (h + k) = \log k + \log (1 + (h/k))$, this last may be rewritten as

(9.3D.10)
$$\frac{1}{\dfrac{1}{\log (1 + (h/k))} + \dfrac{1}{\log k}} \le \frac{\pi(k)\log k}{k}.$$

Set $c_1 = \max (x_0, 155)$, where x_0 is provided by Lemma 9.3D.1 and consider various cases.

CASE I. $h \ge 4k$ and $k \ge c_1$
   Then from (9.3D.3) and (9.3D.10),

$$\left(\frac{1}{\log 5} + \frac{1}{\log c_1}\right)^{-1} \le \frac{\pi(k)\log k}{k} \le 1.22,$$

which is impossible since the left side is at least

$$\left(\frac{1}{\log 5} + \frac{1}{\log 155}\right)^{-1} \ge 1.22008 > 1.22.$$

CASE II. $4k > h$
   In this case, $k > (1/4)h$, so that $h + k > (1.25)h$. Hence the integers $h + 1, \ldots, h + k$, include all integers n such that $h < n \le (1.25)h$. Since $1.25 > 1.2$, it follows from Exercise 3 of §9.3 that for h sufficiently large, say $h \ge h_0$ (where h_0 is effectively computable), there is a prime p between h and $(1.25)h$. Thus $p > h \ge k$ implies $p > k$, which yields the theorem for the case $4k > h \ge h_0$. For $h < h_0$, we have $k \le h < h_0$ which includes only a finite number of pairs (h, k).

CASE III. $7 \le k \le c_1$
   (Recall that we have disposed of the cases $k < 7$.) From (9.3D.3) and (9.3D.10), since $k \ge 7$,

$$\frac{1}{\dfrac{1}{\log\,(1\,+\,(h/k))}\,+\,\dfrac{1}{\log\,7}} \leq 1.22$$

or

$$\frac{1}{\log\,(1\,+\,(h/k))} \geq \frac{1}{1.22}\,-\,\frac{1}{\log\,7} > .3057737,$$

which implies

$$\frac{h}{k} \leq 25.33$$

and hence

$$1 \leq h \leq (25.33)k < (25.33)c_1.$$

thus this last case includes, at most, a finite number of pairs (h, k) and the proof of Theorem 9.3D.2 is completed.

EXERCISES

1. Verify by direct counting (i.e., similarly to the cases $k < 7$) that Schur's Theorem (Theorem 9.3D.1) is true for $k = 7$ and $k = 13$, hence true for all $k \leq 16$.

2. Assuming the Prime Number Theorem [i.e., that $\pi(x) \sim x/\log x$], prove that Theorem 9.3D.2 can be generalized to the following: for any given number $c \geq 1$, apart from a finite number of integer pairs (h, k) for $k > e^c$, $h \geq ck$, at least one of the integers $h + 1, \ldots, h + k$, is divisible by a prime $p > ck$.

§9.4 PRIMES IN ARITHMETIC PROGRESSIONS

In several different contexts we've focused on arithmetic progressions and their intersection with various classes of integers. A natural case of this is that of the primes in a given arithmetic progression $An + B$, where $A > 0$ and B are integers such that $(A, B) = 1$ and $n = 1, 2, \ldots$. For many special cases (e.g., $An + 1$, §6.1A), it is easy to prove that the progression contains infinitely many primes. The general result that all such progressions contain infinitely many primes was the object of much effort in the early part of the nineteenth century. Legendre made many attempts, and based his efforts on various other arithmetic conjectures that seemed reasonable but, in fact, later proved to be false. The result was proved by Dirichlet in 1837 and may be stated as follows.

Theorem 9.4.1. (Dirichlet's Theorem)

For $A > 0$, and B, integers such that $(A, B) = 1$, the arithmetic progression $An + B$, $n = 1, 2, \ldots$ contains infinitely many primes.

In the development of this section, we will provide a proof of Dirichlet's Theorem. Also, we will focus on extending many of the basic results obtained for all primes to the case of primes in arithmetic progressions. A quick glance back over the basic results of prime number theory reveals that (9.2.13) plays a central role. This asserts that

$$(9.4.1) \qquad \sum_{n \leq x} \frac{\Lambda(n)}{n} = \log x + 0(1).$$

For a fixed integer $A > 0$, the "admissible" progressions modulo A are precisely those of the form

$$(9.4.2) \qquad An + B, \qquad A > B > 0, \qquad (A, B) = 1$$

and hence there are precisely $\phi(A)$ such progressions. Since all the prime powers (for primes not dividing A) distribute themselves among these progressions, if we assume that there is no special preference for any one of these progressions (i.e., they equidistribute among the admissible progressions), we might guess that

$$(9.4.3) \qquad \sum_{\substack{n \leq x \\ n \equiv B(\mathrm{mod}\ A)}} \frac{\Lambda(n)}{n} = \frac{1}{\phi(A)} \log x + 0(1).$$

[for each pair (A, B) listed in (9.4.2)]. This is, in fact, true and is correspondingly the focal point of the development of the basic facts concerning primes in arithmetic progression. Note that although the $0(1)$ in (9.4.3) is bounded as $x \to \infty$, it may depend on A (which we are here holding fixed).

From the derivation of (9.2.14) from (9.2.13) we see that

$$\sum_{n \leq x} \frac{\Lambda(n)}{n} - \sum_{p \leq x} \frac{\log p}{p} = \sum_{\substack{p^\alpha \leq x \\ \alpha > 1}} \frac{\log p}{p^\alpha} = 0(1),$$

so that (9.4.3) would yield

$$(9.4.4) \qquad \sum_{\substack{p \leq x \\ p \equiv B(\mathrm{mod}\ A)}} \frac{\log p}{p} = \frac{1}{\phi(A)} \log x + 0(1).$$

Since the right side of (9.4.4) tends to infinity as $x \to \infty$, this will immediately give Dirichlet's Theorem.

In approaching the problem of establishing (9.4.3), we note first that restriction to an arithmetic progression has destroyed the basic multiplicativity that allowed us to relate all integers to the primes. This multiplicativity is restored by using the relations (6.5.16) that provide [recall $(A, B) = 1$]

$$(9.4.5) \qquad \frac{1}{\phi(A)} \sum_{\chi} \overline{\chi}(B)\chi(n) = \begin{cases} 1 & \text{if } n \equiv B \pmod{A} \\ 0 & \text{otherwise} \end{cases},$$

where the sum on the left is taken over all characters χ modulo A. It then follows that

$$(9.4.6) \qquad \sum_{\substack{n \leq x \\ n \equiv B(\bmod A)}} \frac{\Lambda(n)}{n} = \frac{1}{\phi(A)} \sum_{\chi} \overline{\chi}(B) \sum_{n \leq x} \frac{\chi(n)\Lambda(n)}{n}.$$

Thus for each character χ modulo A, we are led to consider the sum

$$\sum_{n \leq x} \frac{\chi(n)\Lambda(n)}{n}.$$

The case of $\chi = \chi_0$, the principal character, is simple since using (9.4.1),

$$\sum_{n \leq x} \frac{\chi_0(n)\Lambda(n)}{n} = \sum_{\substack{n \leq x \\ (n, A)=1}} \frac{\Lambda(n)}{n}$$

$$= \sum_{n \leq x} \frac{\Lambda(n)}{n} + 0\left(\sum_{p/A} \frac{\log p}{p - 1}\right),$$

or

$$(9.4.7) \qquad \sum_{n \leq x} \frac{\chi_0(n)\Lambda(n)}{n} = \log x + 0(1).$$

For all $\chi \neq \chi_0$, our main objective will be to show that

$$(9.4.8) \qquad \sum_{n \leq x} \frac{\chi(n)\Lambda(n)}{n} = 0(1).$$

Then (9.4.6), (9.4.7), and (9.4.8) will combine to give the key result, (9.4.3). Our efforts now focus on establishing (9.4.8) for each $\chi \neq \chi_0$. To do this, we seek to imitate our previous experience with all primes. There, however, the starting point was (9.2.6), relating the primes to $\log [x]!$. This in turn was based on (9.2.4) which gives $\log n$ in terms of $\Lambda(n)$; that is,

$$(9.4.9) \qquad\qquad \log n = \sum_{d/n} \Lambda(d).$$

We retain this objective of expressing $\sum_{n \le x} \chi(n)\Lambda(n)/n$ in terms of $\log n$, and to do this solve (9.4.9) for $\Lambda(n)$ in terms of $\log n$. This is achieved by the Möbius inversion formula (Theorem 3.5.2) and gives

$$(9.4.10) \qquad\qquad \Lambda(n) = \sum_{d/n} \mu(d) \log \frac{n}{d}.$$

Then

$$\sum_{n \le x} \frac{\chi(n)\Lambda(n)}{n} = \sum_{n \le x} \frac{\chi(n)}{n} \sum_{d/n} \mu(d) \log \frac{n}{d}$$

and setting $dd' = n$, this

$$= \sum_{dd' \le x} \frac{\chi(d)\chi(d')}{dd'} \mu(d) \log d'$$

or

$$(9.4.11) \qquad \sum_{n \le x} \frac{\chi(n)\Lambda(n)}{n} = \sum_{d \le x} \frac{\chi(d)\mu(d)}{d} \sum_{d' \le x/d} \frac{\chi(d') \log d'}{d'}.$$

Since sums such as the inner one on the right appear frequently, we provide ourselves with the following.

Lemma 9.4.1. Let $f(x) > 0$ be a function that decreases monotonically, and tends to 0 as $x \to \infty$. Then, for $\chi \ne \chi_0$, the series $\sum_n \chi(n)f(n)$ converges and

$$(9.4.12) \qquad \sum_{n \le x} \chi(n)f(n) = \sum_{n=1}^{\infty} \chi(n)f(n) + 0(f(x)).$$

Proof. Setting

$$R(x) = \sum_{n \le x} \chi(n),$$

from (6.5.15) we see that since $\chi \neq \chi_0$, $R(A) = 0$. Then, since $R(x)$ is periodic with A as a period, it follows that $R(x) = 0(1)$. Thus summing by parts,

$$\sum_{x<n\leq y} \chi(n)f(n) = \sum_{[x]<n\leq[y]} (R(n) - R(n-1))f(n)$$

$$= (R([x]+1) - R([x]))f([x]+1) + \cdots + (R([y]) - R([y]-1))f([y])$$

$$= R([y])f([y]) - R([x])f([x]+1) + \sum_{[x]<n\leq[y]-1} R(n)(f(n) - f(n+1))$$

$$= 0(f([x]+1)) + \sum_{[x]<n\leq[y]-1} 0(1)|f(n) - f(n+1)|.$$

Then since $f(n)$ is monotone decreasing, we can remove the absolute value in the last sum; which telescopes and yields

$$\sum_{x<n\leq y} \chi(n)f(n) = 0(f([x]+1)) = 0(f(x)).$$

Hence as $x \to \infty$, $\sum_{x<n\leq y} \chi(n)f(n) \to 0$, implying that the series converges and

$$\sum_{x<n} \chi(n)f(n) = 0(f(x)).$$

From this last, (9.4.12) follows.

Applying the lemma with $f(x) = (\log x)/x$, the series

(9.4.13)
$$L_1 = \sum_{n=1}^{\infty} \frac{\chi(n)\log n}{n}$$

converges and

(9.4.14)
$$\sum_{n\leq z} \frac{\chi(n)\log n}{n} = L_1 + 0\left(\frac{\log z}{z}\right).$$

Returning then to (9.4.11), we insert

$$\sum_{d'\leq x/d} \frac{\chi(d')\log d'}{d'} = L_1 + 0\left(\frac{\log x/d}{x/d}\right)$$

so that

$$\sum_{n \leq x} \frac{\chi(n)\Lambda(n)}{n} = \sum_{d \leq x} \frac{\chi(d)\mu(d)}{d} \left(L_1 + 0\left(\frac{\log x/d}{x/d}\right) \right)$$

$$= L_1 \sum_{d \leq x} \frac{\chi(d)\mu(d)}{d} + 0\left(\frac{1}{x} \sum_{d \leq x} \log \frac{x}{d}\right).$$

Since

$$\sum_{d \leq x} \log \frac{x}{d} = [x]\log x - \log [x]!$$

$$= [x]\log x - x \log x + 0(x) = 0(x),$$

the above yields that for $\chi \neq \chi_0$,

(9.4.15) $$\qquad\qquad \sum_{n \leq x} \frac{\chi(n)\Lambda(n)}{n} = L_1 \sum_{d \leq x} \frac{\chi(d)\mu(d)}{d} + 0(1).$$

With (9.4.15), the problem of establishing (9.4.8) for all $\chi \neq \chi_0$ is transferred to providing that for these χ

(9.4.16) $$\qquad\qquad \sum_{d \leq x} \frac{\chi(d)\mu(d)}{d} = 0(1).$$

For this purpose we propose to apply the general inversion formula given in Exercise 6 of §3.5 (and in the corollary to Theorem 4.3.1). It asserts, in particular, that for $P(n)$ any completely multiplicative arithmetic function such that $P(1) = 1$, the equations

(9.4.17) $$\qquad\qquad g(x) = \sum_{n \leq x} P(n)f\left(\frac{x}{n}\right),$$

for all x, imply

(9.4.18) $$\qquad\qquad f(x) = \sum_{n \leq x} P(n)\mu(n)g\left(\frac{x}{n}\right).$$

Then for $\chi \neq \chi_0$, we define

(9.4.19) $$\qquad\qquad L_0(\chi) = \sum_{n=1}^{\infty} \frac{\chi(n)}{n}$$

where the convergence of the series is given by Lemma 9.4.1 and we have the following lemma.

Lemma 9.4.2. For $\chi \neq \chi_0$, a character modulo A, we have

$$(9.4.20) \qquad L_0(\chi) \sum_{n \leq x} \frac{\chi(n)\mu(n)}{n} = 0(1).$$

Proof. Taking $P(n) = \chi(n)/n$, and $f(x) = 1$ in (9.4.17), we have, using Lemma 9.4.1,

$$(9.4.21) \qquad g(x) = \sum_{n \leq x} \frac{\chi(n)}{n} = L_0(\chi) + 0\left(\frac{1}{x}\right).$$

Inserting (9.4.21) into (9.4.18) [together with $f(x) = 1$, $P(n) = \chi(n)/n$],

$$1 = \sum_{n \leq x} \frac{\chi(n)\mu(n)}{n} \left(L_0(\chi) + 0\left(\frac{1}{x/n}\right) \right)$$

$$= L_0(\chi) \sum_{n \leq x} \frac{\chi(n)\mu(n)}{n} + 0(1)$$

and (9.4.20) follows.

Whether (9.4.20) implies (9.4.16) clearly depends on whether $L_0(\chi)$ is zero. Thus our attempt to prove (9.4.3) reduces to the problem of proving that for each $\chi \neq \chi_0$ we have $L_0(\chi) \neq 0$. The route we will follow in doing this is to first show that $L_0(\chi) = 0$ could occur for, at most, one character $\chi \neq \chi_0$. For this we require the following lemma.

Lemma 9.4.3. For $\chi \neq \chi_0$, a character modulo A, such that $L_0(\chi) = 0$, we have

$$(9.4.22) \qquad L_1(\chi) \sum_{n \leq x} \frac{\chi(n)\mu(n)}{n} = -\log x + 0(1),$$

where $L_1(\chi)$ is defined in (9.4.13).

Proof. We take $P(n) = \chi(n)/n$ and $f(x) = \log x$ in (9.4.17), which gives

$$g(x) = \sum_{n \leq x} \frac{\chi(n)}{n} \log \frac{x}{n}$$

$$= (\log x) \sum_{n \leq x} \frac{\chi(n)}{n} - \sum_{n \leq x} \frac{\chi(n)\log n}{n},$$

and using Lemma 9.4.1 on both sums,

$$g(x) = L_0(\chi)\log x - L_1(\chi) + 0\left(\frac{\log x}{x}\right).$$

Then, since $L_0(\chi) = 0$, we have

$$(9.4.23) \qquad g(x) = -L_1(\chi) + 0\left(\frac{\log x}{x}\right).$$

Inserting (9.4.23) in (9.4.18) [together with $f(x) = \log x$ and $P(n) = \chi(n)/n$] yields

$$\begin{aligned}
\log x &= \sum_{n \le x} \frac{\chi(n)\mu(n)}{n}\left(-L_1(\chi) + 0\left(\frac{\log x/n}{x/n}\right)\right) \\
&= -L_1(\chi)\sum_{n \le x}\frac{\chi(n)\mu(n)}{n} + 0\left(\frac{1}{x}\sum_{n \le x}\log\frac{x}{n}\right) \\
&= -L_1(\chi)\sum_{n \le x}\frac{\chi(n)\mu(n)}{n} + 0(1)
\end{aligned}$$

and (9.4.22) follows.

From (9.4.15), (9.4.20), and (9.4.22), we obtain for $\chi \ne \chi_0$,

$$(9.4.24) \qquad \sum_{n \le x}\frac{\chi(n)\Lambda(n)}{n} = \begin{cases} 0(1) & \text{if } L_0(\chi) \ne 0 \\ -\log x + 0(1) & \text{if } L_0(\chi) = 0. \end{cases}$$

Let $N =$ the number of characters $\chi \ne \chi_0$, modulo A, such that $L_0(\chi) = 0$. We then have, from (9.4.6) with $B = 1$,

$$\sum_{\substack{n \le x \\ n \equiv 1(\text{mod } A)}}\frac{\Lambda(n)}{n} = \frac{1}{\phi(A)}\sum_{\chi}\sum_{n \le x}\frac{\Lambda(n)\chi(n)}{n}$$

or

$$(9.4.25) \qquad \phi(A)\sum_{\substack{n \le x \\ n \equiv 1(\text{mod } A)}}\frac{\Lambda(n)}{n} = \sum_{n \le x}\frac{\chi_0(n)\Lambda(n)}{n} + \sum_{\chi \ne \chi_0}\sum_{n \le x}\frac{\chi(n)\Lambda(n)}{n}.$$

From (9.4.7), the first sum on the right $= \log x + 0(1)$; from (9.4.24), the double sum gives $-\log x + 0(1)$ for each $\chi \ne \chi_0$ (N of them) such that $L_0(\chi) = 0$, $0(1)$ for those such that $L_0(\chi) \ne 0$. Hence (9.4.25) implies

$$\phi(A) \sum_{\substack{n \le x \\ n \equiv 1 (\bmod A)}} \frac{\Lambda(n)}{n} = (1 - N) \log x + 0(1).$$

Since the sum on the left above is ≥ 0 it follows that $(1 - N) \log x + 0(1) \ge 0$, and hence that $N \le 1$. Thus there is, at most, one character $\chi \ne \chi_0$ such that $L_0(\chi) = 0$. Using this fact, we can now complete the proof of (9.4.3) with the following lemma.

Lemma 9.4.4. For all nonprincipal characters χ, modulo A, $L_0(\chi) \ne 0$.

Proof. Since the number of $\chi \ne \chi_0$ such that $L_0(\chi) = 0$ is, at most, one, that one would have to be a real character. Otherwise, $L_0(\bar{\chi}) = \overline{L_0(\chi)} = 0$ would give the two characters χ, $\bar{\chi}$, with vanishing L_0. Thus we need only show that $L_0(\chi) \ne 0$ for all real characters $\chi \ne \chi_0$.

A real character $\chi \ne \chi_0$ is such that for all n, $\chi(n) = 0$, 1, or -1. Then introducing

$$(9.4.26) \qquad\qquad F(n) = \sum_{d/n} \chi(d),$$

we note that since $\chi(d)$ is multiplicative, so is $F(n)$ (Theorem 3.4.1). Thus for $n = \prod_i p_i^{\alpha_i}$ (p_i distinct primes),

$$(9.4.27) \qquad\qquad F(n) = \prod_i F(p_i^{\alpha_i}).$$

We next note that for p a prime

$$(9.4.28) \qquad F(p^\alpha) = 1 + \sum_{i=1}^{\alpha} (\chi(p))^i = \begin{cases} \alpha + 1 & \text{if } \chi(p) = 1 \\ 1 & \text{if } \chi(p) = 0 \\ 0 & \text{if } \chi(p) = -1, \alpha \text{ odd} \\ 1 & \text{if } \chi(p) = -1, \alpha \text{ even.} \end{cases}$$

From (9.4.28) we deduce that $F(p^\alpha) \ge 0$ always and $F(p^\alpha) \ge 1$ if α is even. Thus from (9.4.27) we get that

$$(9.4.29) \qquad\qquad F(n) \ge \begin{cases} 0 & \text{for all } n \\ 1 & \text{if } n \text{ is a square.} \end{cases}$$

This implies that

$$\sum_{n \le x} \frac{F(n)}{n^{1/2}} \ge \sum_{m^2 \le x} \frac{F(m^2)}{m} \ge \sum_{m \le \sqrt{x}} \frac{1}{m} > \frac{1}{2} \log x,$$

so that as a consequence,

(9.4.30)
$$\lim_{x \to \infty} \sum_{n \le x} \frac{F(n)}{n^{1/2}} = \infty.$$

The final step in the proof of this lemma is to show that $L_0(\chi) = 0$ implies that the sum in (9.4.30) is bounded. We have

$$\sum_{n \le x} \frac{F(n)}{n^{1/2}} = \sum_{n \le x} \frac{1}{n^{1/2}} \sum_{d/n} \chi(d) = \sum_{dd' \le x} \frac{\chi(d)}{d^{1/2} d'^{1/2}}$$

and note the identity

(9.4.31)
$$\sum_{dd' \le x} \frac{\chi(d)}{d^{1/2} d'^{1/2}} = \sum_{d \le \sqrt{x}} \frac{\chi(d)}{d^{1/2}} \sum_{d' \le x/d} \frac{1}{d'^{1/2}}$$

$$+ \sum_{d' \le \sqrt{x}} \frac{1}{d'^{1/2}} \sum_{\sqrt{x} < d \le x/d'} \frac{\chi(d)}{d^{1/2}}.$$

We will show that each double sum on the right of (9.4.31) is bounded. For this we will require the estimate

(9.4.32)
$$\sum_{n \le z} \frac{1}{n^{1/2}} = 2z^{1/2} + c + 0\left(\frac{1}{z^{1/2}}\right).$$

This is seen from

$$\sum_{n \le z} \frac{1}{n^{1/2}} = 1 + \sum_{n=2}^{[z]} \int_{n-1}^{n} \frac{1}{n^{1/2}} \, du$$

$$= 1 + \int_{1}^{[z]} \frac{du}{u^{1/2}} - \sum_{n=2}^{[z]} \int_{n-1}^{n} \left(\frac{1}{u^{1/2}} - \frac{1}{n^{1/2}}\right) du,$$

and since for $n \ge u \ge n - 1$, $0 \le 1/u^{1/2} - 1/n^{1/2} \le 1/((n-1)^{1/2}) - 1/n^{1/2}$,

$$\sum_{n \le z} \frac{1}{n^{1/2}} = -1 + 2[z]^{1/2} - \sum_{n=2}^{\infty} \int_{n-1}^{n} \left(\frac{1}{u^{1/2}} - \frac{1}{n^{1/2}}\right) du + 0\left(\frac{1}{z^{1/2}}\right)$$

$$= 2z^{1/2} + c + 0\left(\frac{1}{z^{1/2}}\right),$$

which is (9.4.32).

Then using Lemma 9.4.1, for $d' \le \sqrt{x}$,

$$\sum_{\sqrt{x} < d \le x/d'} \frac{\chi(d)}{d^{1/2}} = 0\left(\frac{1}{x^{1/4}}\right)$$

so that

$$\sum_{d' \le \sqrt{x}} \frac{1}{d'^{1/2}} \sum_{\sqrt{x} < d \le x/d'} \frac{\chi(d)}{d^{1/2}} = 0\left(\frac{1}{x^{1/4}} \sum_{d' \le \sqrt{x}} \frac{1}{d'^{1/2}}\right) = 0(1).$$

Also, applying (9.4.32) we have

$$\sum_{d \le \sqrt{x}} \frac{\chi(d)}{d^{1/2}} \sum_{d' \le x/d} \frac{1}{d'^{1/2}} = \sum_{d \le \sqrt{x}} \frac{\chi(d)}{d^{1/2}} \left(2 \cdot \frac{x^{1/2}}{d^{1/2}} + c + 0\left(\frac{d^{1/2}}{x^{1/2}}\right)\right)$$

$$= 2x^{1/2} \sum_{d \le \sqrt{x}} \frac{\chi(d)}{d} + c \sum_{d \le \sqrt{x}} \frac{\chi(d)}{d^{1/2}} + 0(1).$$

But Lemma 9.4.1 gives $\sum_{d \le \sqrt{x}} \chi(d)/d^{1/2} = 0(1)$, so that the above

$$= 2x^{1/2}\left(L_0(\chi) + 0\left(\frac{1}{\sqrt{x}}\right)\right) + 0(1)$$

$$= 2x^{1/2} L_0(\chi) + 0(1) = 0(1)$$

since we've assumed $L_0(\chi) = 0$. From (9.4.31), we see that our two estimates imply that $\sum_{n \le x} F(n)/n^{1/2}$ is bounded, which gives the desired contradiction.

Thus we've completed the derivation of (9.4.3), and as a result have (9.4.4) and Dirichlet's Theorem on the infinitude of primes in an admissible arithmetic progression. Once one has established (9.4.3) and (9.4.4), the derivation of basic information concerning primes in arithmetic progressions proceeds in exactly the same fashion as (9.4.1) was processed for all primes. The analogous functions that arise are

(9.4.33) $$\psi(x, A, B) = \sum_{\substack{n \le x \\ n \equiv B \pmod A}} \Lambda(n),$$

(9.4.34) $$\theta(x, A, B) = \sum_{\substack{p \le x \\ p \equiv B \pmod A}} \log p,$$

(9.4.35) $$\pi(x, A, B) = \sum_{\substack{p \le x \\ p \equiv B \pmod A}} 1.$$

The desired generalizations of our earlier discussions to these functions are included in the following exercises.

EXERCISES

1. Show that $\psi(x, A, B) = \theta(x, A, B) + 0(\sqrt{x})$.

2. Prove that for $(A, B) = 1$,

$$\lim_{x \to \infty} \frac{\psi(x, A, B)}{x} = \lim_{x \to \infty} \frac{\theta(x, A, B)}{x} = \lim_{x \to \infty} \frac{\pi(x, A, B) \log x}{x}$$

and

$$\varliminf_{x \to \infty} \frac{\psi(x, A, B)}{x} = \varliminf_{x \to \infty} \frac{\theta(x, A, B)}{x} = \varliminf_{x \to \infty} \frac{\pi(x, A, B) \log x}{x}.$$

3. Prove that for $(A, B) = 1$,

$$\lim_{x \to \infty} \frac{\psi(x, A, B)}{x} \leq \frac{1}{\phi(A)} \leq \varlimsup_{x \to \infty} \frac{\psi(x, A, B)}{x}.$$

4. Show that there exist constants $\alpha > 0$, $\beta > 0$ [depending on A, B, $(A, B) = 1$], such that for all sufficiently large x,

$$\alpha x < \psi(x, A, B) < \beta x,$$

$$\alpha x < \theta(x, A, B) < \beta x,$$

$$\alpha \frac{x}{\log x} < \pi(x, A, B) < \beta \frac{x}{\log x}.$$

5. Show that for $(A, B) = 1$,

$$\sum_{\substack{p \leq x \\ p \equiv B \pmod{A}}} \frac{1}{p} = \frac{1}{\phi(A)} \log \log x + c_1 + 0\left(\frac{1}{\log x}\right),$$

where both the constants c_1 and the 0 depend on A and B.

6. Prove that for $(A, B) = 1$,

$$\prod_{\substack{p \leq x \\ p \equiv B \pmod{A}}} \left(1 - \frac{1}{p}\right) \sim \frac{c_2}{(\log x)^{1/\phi(A)}}$$

(where c_2 may depend on A and B).

7. For given A, B, $(A, B) = 1$, let $\nu(A, B, n) =$ the number of distinct prime divisors p of n such that $p \equiv B \pmod{A}$. Prove that

$$\sum_{n \leq x} \nu(A, B, n) = \frac{1}{\phi(A)} x \log \log x + 0(x).$$

8. For $(A, B) = 1$, show that the normal order of $\nu(A, B, n)$ is $(1/\phi(A)) (\log \log n)$.

9. Let $\epsilon_1, \epsilon_2, \ldots, \epsilon_k$, be given, where each $\epsilon_i = \pm 1$. Also given are integers $a_1, \ldots, a_k$ such that $a_1^{\alpha_1} a_2^{\alpha_2} \ldots a_k^{\alpha_k}$ is a square only if all the α_i are even. Show that there are infinitely many odd primes p such that simultaneously

$$\left(\frac{a_1}{p}\right) = \epsilon_1, \quad \left(\frac{a_2}{p}\right) = \epsilon_2, \cdots, \left(\frac{a_k}{p}\right) = \epsilon_k.$$

10. The Prime Number Theorem for Arithmetic Progressions asserts that $\pi(x, A, B) \sim (1/\phi(A)) (x/\log x)$. Show that this is equivalent to each of the following:

$$\psi(x, A, B) \sim \frac{x}{\phi(A)},$$

$$\theta(x, A, B) \sim \frac{x}{\phi(A)}.$$

11. Using (9.4.20), show that $L_0(\chi) \neq 0$ for all $\chi \neq \chi_0$ implies that for all B such that $(B, A) = 1$ we have

$$\sum_{\substack{n \leq x \\ n \equiv B(\mathrm{mod}\ A)}} \frac{\mu(n)}{n} = 0(1).$$

Conversely, show that the above assertion for all $(B, A) = 1$ implies that $L_0(\chi) \neq 0$ for all $\chi \neq \chi_0$.

§9.4A   A CHEBYCHEV APPROACH TO PRIMES IN ARITHMETIC PROGRESSIONS

The treatment of primes in arithmetic progressions given above, which is in the classical spirit, is somewhat difficult to make numerically explicit. Among other requirements are a good lower estimate for $|L_0(\chi)|$. This raises the question of whether it is possible to develop an analogue of Chebychev's methods to handle primes in arithmetic progressions and thereby prove Dirichlet's Theorem. Also, one would wish to apply such methods so as to obtain the analogue of Bertrand's Postulate for primes in an arithmetic progression. Such a program has never been carried out successfully. However, Erdos has provided a method of this kind that works for certain special progressions. More precisely, Erdos' method applies to any progression $An + B$, $(A, B) = 1$, where the modulus A satisfies

(9.4A.1)
$$\sigma = \sigma(A) = \sum_{\substack{p < A \\ p \nmid A}} \frac{1}{p} < 1$$

(the sum is over primes $p < A$ that do not divide A). Thus, for example, it applies to $A = 3, 4, 5, 6, 8$, since $\sigma(3) = \frac{1}{2}$, $\sigma(4) = \frac{1}{3}$, $\sigma(5) = \frac{5}{6}$, $\sigma(6) = \frac{1}{5}$, $\sigma(8) = \frac{71}{105}$. Though $\sigma(7) = \frac{31}{30} > 1$, the progressions modulo 7 are included since (9.4A.1) holds for $A = 14$, $[\sigma(14) = .844 \ldots < 1]$. Actually, there are only a finite number of positive integers A satisfying (9.4A.1) but there are many and, as an alternative to the classical methods, this approach merits consideration.

For given integers $A, B, A > B > 0$, $(A, B) = 1$, one begins by formulating an appropriate generalization of a binomial coefficient. One natural choice is

(9.4A.2)
$$C_N(A, B) = \frac{(A + B)(2A + B) \cdots (NA + B)}{N!}$$

However, since $(A, B) = 1$, the numerator is not divisible by any prime dividing A, $C_N(A, B)$ is, in general, not an integer. In any event, its size is easily determined since

$$N! A^N < A^N \prod_{j=1}^{N} \left(j + \frac{B}{A} \right) \leq (N + 1)! A^N$$

implies

(9.4A.3)
$$A^N < C_N(A, B) \leq (N + 1) A^N.$$

As for the primes appearing in the unique factorization of $C_N(A, B)$, let

$$C_N(A, B) = \prod_p p^{W_p(N)},$$

where $W_p(N)$ is the integer exponent (positive, negative, or zero) with which p appears. The basic information concerning $W_p(N)$ is given by the following lemma.

Lemma 9.4A.1.   For p a prime *not* dividing A,

(9.4A.4)
$$1 \leq p^{W_p(N)} \leq A(N + 1),$$

whereas for p dividing A,

(9.4A.5)
$$(p^{-1/(p-1)})^N \leq p^{W_p(N)} \leq Np^{p/(p-1)} \cdot (p^{-1/(p-1)})^N.$$

Proof. Let $U(p^r) = $ the number of factors in the numerator of (9.4A.2) that are divisible by p^r, and $V(p^r) = $ the number of integers among $1, 2, \ldots, N$ divisible by p^r. Then for p not dividing A,

$$(9.4A.6) \qquad W_p(N) = \sum_{1 < p^r \le AN+B} (U(p^r) - V(p^r)).$$

Since $Ax + B \equiv 0 \pmod{p^r}$ has a unique solution for x modulo p^r

$$(9.4A.7) \qquad \left[\frac{N}{p^r}\right] \le U(p^r) \le \left[\frac{N}{p^r}\right] + 1.$$

From (9.4A.7) and the fact that $V(p^r) = [N/p^r]$, we get that $0 \le U(p^r) - V(p^r) \le 1$; (9.4A.6) yields

$$0 \le W_p(N) \le \sum_{p^r \le AN+B} 1 \le \frac{\log (AN + B)}{\log p} \le \frac{\log A (N + 1)}{\log p};$$

and (9.4A.4) follows.

For p dividing A, p does not divide the numerator of $C_N(A, B)$, so that

$$-W_p(N) = \sum_{1 < p^r \le N} \left[\frac{N}{p^r}\right].$$

Then

$$-W_p(N) \le \sum_{r=1}^{\infty} \frac{N}{p^r} = \frac{N}{p - 1}$$

and

$$-W_p(N) \ge \sum_{1 < p^r \le N} \left(\frac{N}{p^r} - 1\right)$$

$$\ge \frac{N}{p - 1} - \frac{\log N}{\log p} - \sum_{r > \frac{\log N}{\log p}} \frac{N}{p^r}$$

$$\ge \frac{N}{p - 1} - \frac{\log N}{\log p} - \frac{p}{p - 1},$$

which establishes (9.4A.5).

For $p > \sqrt{A(N + 1)}$, $N \ge A$, it follows that $p > A$, and hence the prime p does not divide A. Then from (9.4A.4), it follows that for these primes, $W_p(N) = 0$ or 1.

Note that from (9.4A.5) we have for primes p dividing A that

$$(9.4A.8) \qquad\qquad W_p(N) \geq -\left[\frac{N}{p-1}\right].$$

Then since from (9.4A.4), for primes p not dividing A we have $W_p(N) \geq 0$, it follows that

$$(9.4A.9) \qquad\qquad \prod_{p/A} p^{[N/(p-1)]} C_N(A,B)$$

is an integer.

The main technical problem that arises is to distinguish the primes $p > \sqrt{A(N+1)}$ for which $W_p(N) = 0$, and those for which $W_p(N) = 1$. Actually, only the sufficiency of the characterizing conditions will be required, but we also establish necessity in the following.

Lemma 9.4A.2. Let $N \geq A$, consider a prime $p > \sqrt{A(N+1)}$, and let d be the integer such that $0 < d < A$ and $pd \equiv B \pmod{A}$. Then $W_p(N) = 0$ if and only if there exists an integer $k \geq 0$ such that

$$(9.4A.10) \qquad\qquad \frac{AN+B}{Ak+d} < p \leq \frac{N}{k}.$$

$\left(\text{Note that if } k = 0, \text{ (9.4A.10) is interpreted as asserting}\right.$

$$\frac{AN+B}{d} < p.\bigg)$$

Also, $W_p(N) = 1$ if and only if there exists an integer $k \geq 0$ such that

$$(9.4A.11) \qquad\qquad \frac{N}{k+1} < p < \frac{A(N+1)+B}{Ak+d}.$$

Proof.   To begin, note that if j^* is the smallest positive integer such that p divides $Aj^* + B$, then

$$(9.4A.12) \qquad\qquad Aj^* + B = pd.$$

For $j^* \leq p - 1$ implies $Aj^* + B \leq Ap - A + B < Ap$, and hence $d < A$.

We consider first the case where $p > N$. Then p does not divide the denominator of $C_N(A,B)$, so that $W_N(p) = 0$ is equivalent to p not dividing the numerator. This, in turn, is the same as $j^* > N$, or

$$pd = Aj^* + B > AN + B.$$

Since $pd \equiv B \pmod A$, $pd < A(N + 1) + B$ is equivalent to $pd \leq AN + B$. Thus here $W_p(N) = 1$ is equivalent to

$$N < p < \frac{A(N + 1) + B}{d},$$

which is the case $k = 0$ of (9.4A.11).

There remains the case of $p \leq N$. Since $p > \sqrt{A(N + 1)}$, each factor of the numerator (and denominator) of $C_N(A, B)$ is divisible at most by the first power of p. Noting that $j^* < p \leq N$, we see that $Aj^* + B$ is the smallest factor of the numerator divisible by p. Consider any other j, such that $Aj + B \equiv 0 \pmod p$, where $N \geq j > j^*$ and construct the map

$$Aj + B \leftrightarrow j - j^*.$$

This map is 1–1, and matches up the $Aj + B$, $N \geq j > j^*$ which are divisible by p, with the integer $j - j^*$, $1 \leq j - j^* \leq N - j^*$, where $j - j^*$ is also divisible by p. Thus, the only multiple of p in the numerator left unmatched is $Aj^* + B$, which is exactly divisible by the first power of p. Whether this p gets canceled is clearly equivalent to whether there is an integer m in the range

(9.4A.13) $N - j^* < m \leq N$

which is a multiple of p. In other words $W_p(N) = 0$ or 1 according, as such an m does or does not exist.

Since $j^* < p$, we have

$$N - p < N - j^* \leq N,$$

and this interval from $N - p$ to N contains some multiple of p, say kp. Then

(9.4A.14) $W_p(N) = 0$ equivalent to $N - j^* < kp \leq N$

and

(9.4A.15) $W_p(N) = 1$ equivalent to $N - p < kp \leq N - j^*$.

But $Aj^* + B = pd$, gives $j^* = (pd - B)/A$, so that $N - j^* < pk$ is the same as

$$N - \left(\frac{pd - B}{A}\right) < pk$$

which is the same as the lower inequality in (9.4A.10). Thus $W_p(N) = 0$ is equivalent to the existence of such a k such that (9.4A.10) holds. Similarly, the lower bound of (9.4A.15) asserts $p > N/(k + 1)$ and the upper bound is the same as

$$p \le \frac{AN + B}{Ak + d}.$$

But this, in turn, is equivalent to $p < (A(N + 1) + B)/(Ak + d)$ [for $p(Ak + d) \equiv d \equiv B \pmod{A}$]. Thus $W_p(N) = 1$ is equivalent to asserting (9.4A.11) for some integer k.

The next problem is that of removing from $C_N(A, B)$ the primes $p > \sqrt{A(N + 1)}$ which appear, such that $p \not\equiv B \pmod{A}$. This is achieved by the following lemma.

Lemma 9.4A.3. Let $p_1, \ldots, p_t$ be the primes less than A that do not divide A and define q_i, $0 < q_i < A$, such that $p_i q_i \equiv B \pmod{A}$. Then if $N \ge A^2$, for any prime $p > \sqrt{A(N + 1)}$ that divides the numerator of the reduced fraction form of (i.e., numerator and denominator relatively prime)

$$(9.4A.16) \qquad \pi_N(A, B) = \frac{C_N(A, B)}{C_{[N/p_1]}(A, q_1) C_{[N/p_2]}(A, q_2) \cdots C_{[N/p_t]}(A, q_t)}$$

we have $p \equiv B \pmod{A}$.

Proof. Consider a $p > \sqrt{A(N + 1)}$ that divides $C_N(A, B)$ [i.e., in fact, $W_p(N) = 1$] and is such that $p \not\equiv B \pmod{A}$. Then $pd \equiv B \pmod{A}$, where $1 < d < A$ and $(d, A) = 1$. Thus d is divisible by one of the p_i, $d = p_i c$, and

$$pp_i c \equiv B \equiv p_i q_i \pmod{A}$$

implies $pc \equiv q_i \pmod{A}$. We claim that then p divides $C_{[N/p_i]}(A, q_i)$ and will demonstrate this in the following. (As a result p cannot divide $\pi_N(A, B)$.)

Note that since $N \ge A^2$, and $p_i < A$, we have $[N/p_i] \ge A$ and also, $p > \sqrt{A(N + 1)} > \sqrt{A([N/p_i] + 1)}$. Choose the integer $k \ge 0$ such that

$$(9.4A.17) \qquad \frac{N/p_i}{k + 1} < p \le \frac{N/p_i}{k}$$

(if $k = 0$, the upper inequality is vacuous). Then from the lower inequality in (9.4A.17), it follows that

$$\frac{[N/p_i]}{k + 1} < p.$$

If we also have

$$p \ge \frac{A([N/p_i] + 1) + q_i}{Ak + c},$$

since

$$\frac{A([N/p_i] + 1) + q_i}{Ak + c} > \frac{A(N/p_i) + q_i}{Ak + c} = \frac{AN + p_i q_i}{A(kp_i) + p_i c} \geq \frac{AN + B}{A(kp_i) + d}$$

[we've used $p_i c = d$, and that since $p_i q_i \equiv B \pmod{A}$, $p_i q_i \geq B$], using the upper bound of (9.4A.17), we get

$$\frac{AN + B}{A(kp_i) + d} < p \leq \frac{N}{kp_i},$$

which is of the form (9.4A.10). By Lemma 9.4A.2, this implies that p doesn't divide $C_N(A, B)$, a contradiction. Hence (9.4A.11) is established with N and B replaced by $[N/p_i]$ and q_i respectively. Thus p divides $C_{[N/p_i]}(A, q_i)$, and the Lemma follows.

Since $W_p(N) \geq 0$ for all primes $p \nmid A$, under the hypotheses of Lemma 9.4A.3 we have

$$\pi_N(A, B) \leq \prod_{p/A} p^{W_p(N) - \sum_{i=1}^{t} W_p[\frac{N}{p_i}]} \prod_{p \leq \sqrt{A(N+1)}} A(N+1) \prod_{\substack{p < A(N+1) \\ p \equiv B(\mathrm{mod}\, A)}} p.$$

From (9.4A.5),

$$\prod_{p/A} p^{W_p(N) - \sum_{i=1}^{t} W_p([\frac{N}{p_i}])} \leq \prod_{p/A} (Np^{p/(p-1)})(p^{-1/(p-1)})^N \cdot (p^{1/(p-1)})^{\sum[\frac{N}{p_i}]}$$

$$\leq N^{\nu(A)} \left(\prod_{p/A} p^{p/(p-1)} \right) \left(\prod_{p/A} p^{-1/(p-1)} \right)^{N(1-\sigma)},$$

and inserting this in the above yields

(9.4A.18)

$$\pi_N(A, B) \leq N^{\nu(A)} \left(\prod_{p/A} p^{p/(p-1)} \right) \left(\prod_{p/A} p^{-1/(p-1)} \right)^{N(1-\sigma)} (A(N+1))^{\sqrt{A(N+1)}} \prod_{\substack{p < A(N+1) \\ p \equiv B(\mathrm{mod}\, A)}} p.$$

On the other hand, (9.4A.3) yields the lower bound

$$\pi_N(A, B) > \frac{A^N}{\prod_{i=1}^{t} ([N/p_i] + 1) A^{[N/p_i]}} \geq A^{N(1-\sigma)} \prod_{i=1}^{t} (p_i/(N + p_i))$$

and combining this with (9.4A.18), we get

(9.4A.19)

$$\prod_{\substack{p<A(N+1) \\ p\equiv B(\mathrm{mod}\ A)}} p > \left(A\prod_{p/A}p^{1/(p-1)}\right)^{N(1-\sigma)} N^{-\nu(A)}\left(\prod_{i=1}^{t}(p_i/(N+p_i))\right)$$

$$\times \left(\prod_{p/A}p^{-p/(p-1)}\right)(A(N+1))^{-\sqrt{A(N+1)}}.$$

Removing some of the explicit dependence on A, (9.4A.19) implies

(9.4A.20)
$$\prod_{\substack{p<A(N+1) \\ p\equiv B(\mathrm{mod}\ A)}} p > \left(A\prod_{p/A}p^{1/(p-1)}\right)^{N(1-\sigma)+o(N)},$$

and, if $\sigma = \sigma(A) < 1$, this gives Dirichlet's Theorem on the infinitude of primes in the progression $An + B$ $[A > B > 0, (A,B) = 1]$.

To obtain results of the "Bertrand Postulate type" for primes in an arithmetic progression, the lower bound (9.4A.20) must be supplemented by an upper bound. Taking $k = 0$ and $d = 1$ in (9.4A.11), we obtain that for $N \geq A$, all primes $p \equiv B(\mathrm{mod}\ A)$ in the interval

$$N < p < A(N+1) + B$$

appear to the first power as a divisor of the numerator in the reduced fraction form of $C_N(A,B)$. From (9.4A.9), we recall that

$$\left(\prod_{p/A}p^{[N/(p-1)]}\right)C_N(A,B)$$

is an integer, so that

$$\prod_{\substack{N<p<A(N+1)+B \\ p\equiv B(\mathrm{mod}\ A)}} p \leq \prod_{p/A}p^{[N/(p-1)]}\ C_N(A,B) \leq \left(\prod_{p/A}p^{1/(p-1)}\right)^{N}C_N(A,B).$$

Then using (9.4A.3) we get

(9.4A.21)
$$\prod_{\substack{N<p<A(N+1)+B \\ p\equiv B(\mathrm{mod}\ A)}} p \leq (N+1)\left(A\prod_{p/A}p^{1/(p-1)}\right)^{N}.$$

Next, we transform (9.4A.21) so that the limits on p in the product (on the left) have a somewhat less integral form; namely, let $\xi \geq A^2$ be any real number. Then setting $N = [\xi/A]$ we have $N \leq \xi/A$ and $A(N+1) + B > \xi + B > \xi$. Thus the interval $N < p < A(N+1) + B$ contains the interval $\xi/A < p \leq \xi$ so that from (9.4A.21) we get

$$\prod_{\substack{(\xi/A)<p\le\xi \\ p\equiv B(\mathrm{mod}\ A)}} p \le \left(A \prod_{p/A} p^{1/(p-1)}\right)^{(\xi/A)} \cdot \left(\frac{\xi}{A}+1\right)$$

or

$$(9.4A.22) \qquad \prod_{\substack{(\xi/A)<p\le\xi \\ p\equiv B(\mathrm{mod}\ A)}} p \le \left(\frac{\xi}{A}+1\right) \cdot \left(A \prod_{p/A} p^{1/(p-1)}\right)^{\xi/A}.$$

Then replacing ξ by ξ/A^{i-1}, so long as $\xi/A^i \ge A$ we have for $i \ge 1$,

$$\prod_{\substack{(\xi/A^i)<p\le\xi/A^{i-1} \\ p\equiv B(\mathrm{mod}\ A)}} p \le \left(\frac{\xi}{A^i}+1\right) \left(A \prod_{p/A} p^{1/(p-1)}\right)^{\xi/A^i}.$$

Determining i^* by $\xi/A^{i^*+1} < A \le \xi/A^{i^*}$, and multiplying the above inequalities for $i = 1, 2, \ldots, i^*$, we obtain for $\xi \ge A^3$, (note that we may assume $A \ge 2$),

$$(9.4A.23) \qquad \prod_{\substack{(\xi/A^{i^*})<p\le\xi \\ p\equiv B(\mathrm{mod}\ A)}} p \le \xi^{(\log \xi/\log A)} A^{-3} \left(A \prod_{p/A} p^{1/(p-1)}\right)^{\xi/(A-1)}.$$

But since $\xi/A^{i^*} < A^2$, we see that

$$\prod_{\substack{p\le(\xi/A^{i^*}) \\ p\equiv B(\mathrm{mod}\ A)}} p < A^3,$$

so that from (9.4A.23) we have for $\xi \ge A^3$ that

$$(9.4A.24) \qquad \prod_{\substack{p\le\xi \\ p\equiv B(\mathrm{mod}\ A)}} p \le \xi^{(\log \xi/\log A)} \left(A \prod_{p/A} p^{1/(p-1)}\right)^{\xi/(A-1)}.$$

Similarly, taking $N = ([\xi/A] - 1)$ in (9.4A.19), we get that for $\xi \ge A^3 + A$ (which insures $N \ge A^2$), $A(N + 1) \le \xi$, and hence (assuming $\sigma < 1$)

$$(9.4A.25) \prod_{\substack{p\le\xi \\ p\equiv B(\mathrm{mod}\ A)}} p > \left(A \prod_{p/A} p^{1/(p-1)}\right)^{\frac{\xi}{A}(1-\sigma)} \left(\frac{A}{\xi}\right)^{\nu(A)} \left(\prod_{i=1}^{t} \left(\frac{p_i}{((\xi/A) + p_i)}\right)\right)$$

$$\times \left(\prod_{p/A} p^{-p/(p-1)}\right)^{3-2\sigma} A^{-2(1-\sigma)} \xi^{-\sqrt{\xi}}.$$

Given a real number λ, $0 < \lambda < 1$, for $\xi \ge \max(A^3 + A, A^3/\lambda)$, (9.4A.24) and (9.4A.25) yield

$$(9.4A.26) \qquad \prod_{\substack{\lambda\xi<p\leq\xi \\ p\equiv B(\mathrm{mod}\ A)}} p > \left(A \prod_{p/A} p^{1/(p-1)}\right)^{(\xi[(1-\sigma)/A-\lambda/(A-1)])} \Delta(\xi)$$

where

$$(9.4A.27) \qquad \Delta(\xi) = \left(\prod_{i=1}^{t} \left(\frac{p_i}{((\xi/A) + p_i)}\right)\right) A^{\nu(A)-2(1-\sigma)}$$
$$\xi^{-\nu(A)-(\log\ \xi/\log\ A)-\sqrt{\xi}} \left(\prod_{p/A} p^{-p/(p-1)}\right)^{3-2\sigma}.$$

Since $\Delta = (A \prod_{p/A} p^{1/(p-1)})^{o(\xi)}$ as $\xi\to\infty$, it follows that for sufficiently large ξ, there is a prime $p \equiv B(\mathrm{mod}\ A)$ satisfying

$$\lambda\xi < p \leq \xi,$$

provided

$$\frac{1 - \sigma}{A} > \frac{\lambda}{A - 1}$$

or

$$(9.4A.28) \qquad \lambda < \left(\frac{A - 1}{A}\right)(1 - \sigma).$$

For example, for $A = 6$, $\sigma(A) = \frac{1}{5}$, and (9.4A.28) requires $\lambda < \frac{5}{6}\cdot\frac{4}{5} = \frac{2}{3}$. Thus, in this case, $\lambda = \frac{1}{2}$ is admissible. We can now sharpen this to a "Bertrand Postulate type theorem" the progressions $6n + 1$ and $6n + 5$.

Theorem 9.4A.1.   For $N \geq 6$, there is always a prime p of the form $6n + 1$, and one of the form $6n - 1$, such that $N < p < 2N$.

Proof.   For $A = 6$, max $(A^3 + A, 2A^3) = 432$, so that (9.4A.26) holds for $\xi \geq 432$. Since here

$$\frac{1 - \sigma}{A} - \frac{\lambda}{A - 1} = \frac{4}{30} - \frac{1}{10} = \frac{1}{30},$$

the right side of (9.4A.26) has a logarithm equal to

$$\frac{1}{30} \xi \log (2^2 \cdot 3^{3/2}) + \log \Delta(\xi).$$

To prove the theorem for $N \geq N_0$, we set $\xi = 2N$ and attempt to establish that for such N

$$\frac{1}{15} N \log (2^2 \cdot 3^{3/2}) + \log \Delta(2N) > 0.$$

Since this is a monotone increasing function of N certainly for $N \geq 100$, we have our desired result for $N \geq N_0 = 5000$ provided

$$\frac{1}{15} (5000) \log (2^2 \cdot 3^{3/2}) + \log \Delta(2N_0)$$

$$= \frac{1}{15} (5000) \log (2^2 \cdot 3^{3/2}) + \log 5 + \frac{2}{5} \log 6 - \frac{13}{5} \log (4\sqrt{27})$$

$$- \log \left(\frac{10^4}{6} + 5\right) - \left(2 + \frac{\log 10^4}{\log 6} + 100\right) \log 10^4 > 0.$$

But, in fact, this expression equals 11.620 . . . , and the theorem is thereby proved for $N \geq 5000$.

For $N < 5000$, we simply note the following two sequences of primes, the first of the form $6n - 1$, the second of the form $6n + 1$, where in each sequence each element is less than twice the previous one.

$$[11, 17, 29, 53, 101, 197, 389, 773, 1523, 3041, 6053]$$

$$[7, 13, 19, 37, 73, 139, 297, 547, 1093, 2179, 4357, 8713]$$

The method presented in §9.4A can be refined and extended in various ways. However, we leave it at this point as simply an illustration of one way in which the Chebychev type methods can be extended to certain arithmetic progressions.

NOTES

§9.1. Chebychev introduced the notation for the functions $\theta(x)$ and $\psi(x)$ given in (9.1.2) and (9.1.3). He did not use the notation $\pi(x)$, but denoted the prime counting function by $\phi(x)$ [9.1].

The use of $\pi(x)$ may have originated with Landau. It was *not* used by Legendre, Gauss, von Mangoldt, Riemann, Hadamard, or Dirichlet in any of their papers relating to the Prime Number Theorem (which are referenced here).

Chebychev proved Theorem 9.1.1, explicitly noting (9.1.6) [9.2].

According to Ingham [9.3], (9.1.12) was first stated in an imprecise way by Euler, who said that the primes are "infinitely fewer than the integers" [9.4]. Ingham gives a proof of (9.1.12) using a formula of Legendre [9.5, 9.6]. Though (9.1.12) is certainly a consequence of Chebychev's work, it may have been proved earlier. A proof is also given in Landau's *Handbuch* [9.7].

§9.1A. In 1737, Euler proved that $\Pi_p (1 - 1/p)^{-1}$ is divergent [9.8]. He applied this to derive that $\Sigma_p 1/p$ diverges [9.8, 9.9].

Legendre [9.6] gave the empirically determined (but incorrect) formula

$$\prod_{p \leq G} \left(1 - \frac{1}{p}\right) = \frac{1.104}{\log G - .08366}.$$

In 1851, Chebychev proved (using estimation techniques in which smaller order terms were dropped) [9.17] that

$$\prod_{p \leq x} \left(1 - \frac{1}{p}\right) = \frac{c_o}{\log x}.$$

This precedes similar work of Mertens, at least insofar as the proof of (9.1A.11) is concerned.

(9.1A.12) was proved by Chebychev in 1851 [9.2]. Mertens obtained sharper estimates, which will be discussed in the notes to §9.2.

§9.2. The name, De Polignac, is mentioned in several places in connection with the results concerning $\log [x]!$. In particular, Hardy and Wright [9.10] credit him with the assertion

$$n! = \prod_p p^{\sum_{m \geq 1} \left[\frac{n}{p^m}\right]}.$$

The identity (9.2.8) is also attributed (by some) to De Polignac [9.11].

In 1852, Chebychev gave the identity (9.2.8) [9.2]. However, Mertens seems to have been the first to apply the method used here to derive what is essentially (9.2.13) [9.12].

The history of (9.2.18) goes back to Legendre [9.6], who empirically derived

$$\sum_{2 < p \leq G} \frac{1}{p} = \log (\log G - .08366) + C.$$

Chebychev obtained [9.1]

$$\sum_{p \leq x} \frac{1}{p} = \log \log x + C + o(1).$$

The sharper form (9.2.18) was given by Mertens in 1874 [9.12].

A proof of (9.2.20) was given by Mertens, and it [as well as the weaker statement (9.2B.1)] is usually referred to as Mertens' Theorem [9.12].

§9.2A. The assertions of Theorem 9.2A.1 were first proved by Chebychev in 1852 [9.2]. He also proved Theorem 9.2A.2 [9.1], so that Landau concludes that implicitly Chebychev proved [9.7] that if $\lim_{x \to \infty} \theta(x)/x = 1$ then $\lim_{x \to \infty} \pi(x)/(x/\log x) = 1$.

§9.2B. Theorem 9.2B.1 was proved by Landau in 1903 [9.13, 9.7].

§9.2C. (9.2C.2) was given by G. H. Hardy and S. Ramanujan [9.14]. Hardy and Wright's text indicates that this may be an older result. Hardy and Ramanujan also proved (9.2C.7), and were the first to state and prove Theorem 9.2C.1. The proof given here is essentially that of P. Turan [9.15].

§9.2D. As indicated in §7.4B, C. F. Gauss proved in 1801 that if $p \equiv 1 \pmod{8}$, then there must be a prime $q < 2\sqrt{p} + 1$ such that q is a quadratic nonresidue modulo p. In the following we denote the smallest positive quadratic nonresidue modulo p by $n(p)$.

In 1919 I. M. Vinogradov proved [9.16] that

$$n(p) < p^{(1/2\sqrt{e})} (\log p)^2.$$

This was improved slightly in 1952 by H. Davenport and P. Erdos [9.17] to

$$n(p) = 0(p^{(1/2)\beta} (\log p)^\beta),$$

where $\beta = 1/\sqrt{e}$.

In the early 1950s various improvements were obtained under the *assumption* of the "extended Riemann hypothesis." Under this assumption Ju V. Linnik, 1952 [9.18], proved that $n(p) = 0(p^\epsilon)$ and N. C. Ankeny, 1957 [9.19], that $n(p) = 0[(\log p)^2]$. Without any additional hypothesis, D. A. Burgess,

1957 [9.20], obtained that $n(p) = 0(p^\alpha)$ for any fixed $\alpha > \frac{1}{4}e^{-1/2}$. According to E. Grosswald [9.21], Vinogradov conjectured that $n(p) = 0(p^\epsilon)$, for any fixed $\epsilon > 0$.

In 1966, A. I. Vinogradov and Ju V. Linnik [9.22] used Burgess' method to obtain a similar estimate for the least *prime* quadratic residue [which we will denote here by $n'(p)$]; namely, $n'(p) = 0(p^\alpha)$ for fixed $\alpha > \frac{1}{4}e^{-1/2}$. Following this, P. D. T. A. Elliot obtained improvements under certain assumptions [9.23].

These methods have been applied to the problem of finding upper bounds for the least kth power nonresidue modulo p [which we'll denote by $n_k(p)$]. In 1927 [9.24], Vinogradov proved that if p is a prime and $k > 1$ a divisor of $p - 1$, then $n_k(p) < p^{1/2l} (\log p)^2$, where $l = e^{(k-1/k)}$, and p is sufficiently large. If in addition, $k > m^m$, $m \geq 8$, then $n_k(p) < p^{1/m}$, for all sufficiently large p.

Other extensions and generalizations are given in [9.25].

§9.3. J. Bertrand [9.26] stated his conjecture (in the $n < p < 2n - 2$, $n > 3$ version) in 1845 in connection with a group theoretical investigation. (A later such application may be found in E. Artin [9.27].) Bertrand was unable to prove his "postulate," but verified it for all $n < 3,000,000$. The conjecture was first proved by Chebychev in 1852.

An expository article by R. Archibald [9.28] describes the evolution of the many proofs that followed that of Chebychev.

§9.3B. Ramanujan's proof appeared in 1919 in a very brief two-page paper [9.29].

§9.3C. Erdos' proof appeared in 1930 [9.30]. A similar proof was given in 1945 [9.31] by P. Finsler.

§9.3D. Theorem 9.3D.1 is credited to Issai Schur [9.32].

§9.4. As already noted in the notes to Chapter 6 (§6.1A and §6.5), Dirichlet proved Theorem 9.4.1 in 1837. A proof also appears in [9.33]. In 1949, Atle Selberg published an "elementary proof" [9.34]. The treatment here is taken from [9.35].

§9.4A. Erdos gave the development of this section in 1935 [9.36]. Theorem 9.4A.1 had already been proved by R. Breusch in 1932 [9.37]. Erdos was also unaware of a 1933 paper of Ricci, that established similar results by similar methods [9.38, 9.39].

This type of result has been much refined in later papers. For example, Molsen, 1941, proved that there is always a prime p such that $n < p \leq \frac{8}{7}n$ of each of the forms $3x + 1$, $3x - 1$, provided $n \geq 199$ [9.40]. Also, for $n \geq 118$, he shows that the interval $n < p \leq \frac{4}{3}n$ always contains a prime of each of the forms $12x + 1$, $12x - 1$, $12x + 5$, $12x - 5$.

REFERENCES

9.1. P. Chebychev, Sur la fonction qui detérmine la totalité des nombres premiers inférieurs à une limite donnée, *Mémoires présentés à l'Académie Impériale des Sciences de St. Pétersbourg par divers Savants*, **6**, 1851, 141–157.

9.2. P. Chebychev, Mémoire sur les nombres premiers, *Journal de Mathématique pures et appliquées*, **17**, (1), 1852, 366–390.

9.3. A. E. Ingham, *The Distribution of Prime Numbers*, Cambridge University Press, 1932.

9.4. L. Euler, Ref. 2.16.

9.5. A. M. Legendre, Ref. 7.3, pp. 412–414.

9.6. A. M. Legendre, Ref. 6.13, II, pp. 86–89.

9.7. E. Landau, Ref. 6.30.

9.8. L. Euler, Ref. 2.16, pp. 227–229.

9.9. L. Euler, Ref. 2.17.

9.10. G. H. Hardy and E. M. Wright, Ref 2.4.

9.11. J. Nagura, On the interval containing at least one prime number, *Proceedings of the Japan Academy*, **28**, 1952, 177–181.

9.12. F. Mertens, Ein Beitrag zur analytischen Zahlentheorie, *Crelle's Journal*, **78**, 1874, 46–62.

9.13. E. Landau, Uber den Verlauf der zahlentheoritischen Funktion $\phi(x)$, *Archiv der Mathematik und Physik*, **5**, (3), 1903, 86–91.

9.14. G. H. Hardy and S. Ramanujan, The normal number of prime factors of a number n, *Quarterly Journal of Mathematics*, **48**, 1917, 76–92.

9.15. P. Turan, On a theorem of Hardy and Ramanujan, *Journal of the London Mathematical Society*, **9**, 1934, 274–276; **11**, 1936, 125–133.

9.16. I. Vinogradov, Ref. 8.17.

9.17. H. Davenport and P. Erdos, The distribution of quadratic and higher residues, *Publicationes Mathematicae*, **2**, 1952, 252–265.

9.18. J. V. Linnik, A remark on the least quadratic non-residue, *Comptes Rendus de l'Academie de Sciences de l'URSS*, **36**, 1942, 119–120.

9.19. N. C. Ankeny, The least quadratic non-residue, *Annals of Mathematics*, **55**, (2), 1952, 65–72.

9.20. D. A. Burgess, The distribution of quadratic residues and non-residues, *Mathematika*, **4**, 1957, 106–112.

9.21. E. Grosswald, *Mathematical Reviews*, **37**, (6239).

9.22. A. I. Vinogradov and J. V. Linnik, Hypoelliptical curves and the least prime quadratic residue, *Akademia Nauk USSR Doklady*, **168**, 1966, 259–261.

9.23. P. D. T. A. Elliot, A note on a recent paper of J. V. Linnik and A. I. Vinogradov, *Acta Arithmetica*, **13**, 1967–1968, 103–105.

9.24. I. Vinogradov, On the bound of the least non-residue of nth powers, *Transactions of the American Mathematical Society*, **29**, 1927, 218–226.

9.25. P. D. T. A. Elliot, The least prime kth power residue, *Journal of the London Mathematical Society*, **3**, (2), 1971, 205–210.

9.26. J. Bertrand, Mémoire sur le nombre de valeurs que peut prendre une fonction quand on y permute les lettres qu'elle renferme, *Journal de l'Ecole Royale Polytechnique*, **18**, (30), 1845, 123–140.

9.27. E. Artin, Braids and permutations, *Annals of Mathematics*, **48**, 1947, 643–649.

9.28. R. Archibald, Bertrand's Postulate, *Scripta Mathematica*, **11**, 1945, 109–120.

9.29. S. Ramanujan, A proof of Bertrand's Postulate, *Journal of the Indian Mathematical Society*, **11**, 1919, 181–182.

9.30. P. Erdos, Beweis eines Satzes von Tschebyschev, *Acta Litterarum ac Scientiarum Regiae Universitatis Hungaricae Francisco-Josephinae*, **5**, 1930–1932, 194–198.

9.31. P. Finsler, Uber die Primzahlen zwischen n und $2n$, *Festschrift zum 60 Geburtstag von Prof. Dr. Andreas Speiser*, Zurich: Füssli, 1945, pp. 118–122.

9.32. I. Schur, Einige Satze uber Primzahlen mit Anwendungen auf Irreduzibilitatsfragen I, *Sitzungsberichte der Preussischen Akademie der Wissenschaften*, 1929, 125–136.

9.33. G. L. Dirichlet, Ref. 6.14.

9.34. A. Selberg, An elementary proof of Dirichlet's theorem about primes in an arithmetic progression, *Annals of Mathematics*, **50**, (2), 1947, 297–304.

9.35. H. N. Shapiro, On primes in arithmetic progression, (II), *Annals of Mathematics*, **52**, 1950, 231–243.

9.36. P. Erdos, Uber die Primzahlen gewisser arithmetischen Reihen, *Mathematische Zeitschrift*, **39**, 1935, 473–491.

9.37. R. Breusch, Zur Verallgemeinerung der Bertrandschen Postulates dass zwischen x und $2x$ stets Primzahlen liegen, *Mathematische Zeitschrift*, **34**, 1932, 505–526.

9.38. G. Ricci, Sul teorema di Dirichlet relativo alla progressione aritmetica, *Bolletino della Unione Matematica Italiana*, **12**, 1933, 304–309.

9.39. G. Ricci, Sui teoremi di Dirichlet e di Bertrand–Tchebychev relativi alla progressione aritmetica, *Bolletino della Unione Matematica Italiana*, **13**, 1934, 1–11.

9.40. K. Molsen, Zur Verallgemeinerung des Bertrandschen Postulates, *Deutsche Mathematik*, **6**, 1941, 248–256.

10

THE PRIME
NUMBER THEOREM

§10.1 STATEMENTS OF THE PRIME NUMBER THEOREM

As was noted in §9.1, the Prime Number Theorem asserts

$$(10.1.1) \qquad \pi(x) \sim \frac{x}{\log x}$$

and this is completely equivalent to either

$$(10.1.2) \qquad \psi(x) \sim x$$

or

$$(10.1.3) \qquad \theta(x) \sim x.$$

These are only a few of the many equivalent formulations of the Prime Number Theorem, some of which are fairly explicit in their relationships to primes; others are not. Among the former, we've already met (Ex. 7 of §9.2A)

$$(10.1.4) \qquad p_n \sim n \log n$$

where p_n denoted the nth prime. This is easily seen since taking $x = p_n$ in (10.1.1) implies

$$(10.1.5) \qquad n = \pi(p_n) \sim \frac{p_n}{\log p_n}.$$

Taking logarithms yields

$$\log n \sim \log p_n - \log \log p_n \sim \log p_n,$$

so that (10.1.5) gives (10.1.4). Conversely, starting with (10.1.4), letting p_n be the largest prime $\leq x$, we have $p_n < x < p_{n+1}$. From (10.1.4) we see that $p_n \sim p_{n+1}$, implying $p_n \sim x$. Then

$$\pi(x) = n \sim \frac{p_n}{\log n} \sim \frac{p_n}{\log p_n} \sim \frac{x}{\log x},$$

which is (10.1.1).

Among the less obvious equivalent formulations of the prime number theorem are those involving the Möbius μ-function. If we proceed formally (and completely without validity), we see that

(10.1.6)
$$\sum_{d=1}^{\infty} \frac{\mu(d)}{d} = \prod_{p} \left(1 - \frac{1}{p} \right) = 0,$$

where the infinite product diverges to 0, by (9.1.11). There are many questionable points, foremost of which is that we do not even know that the series on the left of (10.1.6) converges. In fact, it *does* converge (and to 0) and this is a consequence of "cancellations" produced by the distribution of $(+1)$s and (-1)s as values of $\mu(d)$. That this is intimately related to properties of primes is not a priori clear [apart from the basic definition of $\mu(d)$]. A somewhat more precise view emerges if we introduce the function

(10.1.7)
$$M(x) = \sum_{d \leq x} \frac{\mu(d)}{d}$$

and attempt to interpret some statement about primes in terms of $M(x)$.

Starting with (9.2.13), we have

(10.1.8)
$$\sum_{n \leq x} \frac{\Lambda(n)}{n} = \log x + 0(1)$$

and wish to relate $\Lambda(n)$ to $\mu(n)$. Since

$$\log n = \sum_{d \mid n} \Lambda(d),$$

the Möbius inversion formula gives

(10.1.9)
$$\Lambda(n) = \sum_{d \mid n} \mu(d) \log \frac{n}{d}.$$

Then

$$\sum_{n \le x} \frac{\Lambda(n)}{n} = \sum_{n \le x} \frac{1}{n} \sum_{d|n} \mu(d) \log \frac{n}{d} = \sum_{dd' \le x} \frac{\mu(d)}{d} \frac{\log d'}{d'}$$

or

(10.1.10)
$$\sum_{n \le x} \frac{\Lambda(n)}{n} = \sum_{d' \le x} \left(\frac{\log d'}{d'} \right) M \left(\frac{x}{d'} \right),$$

so that from (10.1.8) we get

(10.1.11)
$$\sum_{n \le x} \frac{\log n}{n} M \left(\frac{x}{n} \right) = \log x + 0(1).$$

Using (10.1.11), we can easily establish that

(10.1.12)
$$\varliminf_{x \to \infty} \sum_{d \le x} \frac{\mu(d)}{d} \le 0 \le \varlimsup_{x \to \infty} \sum_{d \le x} \frac{\mu(d)}{d}.$$

To see this, note that for any fixed $K > 0$,

$$\sum_{x/K < n \le x} \frac{\log n}{n} M \left(\frac{x}{n} \right) = 0 \left(\sum_{x/K < n \le x} \frac{\log n}{n} \right) = 0(\log x)$$

so that (10.1.11) implies

$$\sum_{1 \le n \le x/K} \frac{\log n}{n} \cdot M \left(\frac{x}{n} \right) = 0(\log x),$$

from which we obtain

$$\left(\min_{K \le u \le x} M(u) \right) \sum_{n \le x/K} \frac{\log n}{n} < 0(\log x) \le \left(\max_{K \le u \le x} M(u) \right) \sum_{n \le x/K} \frac{\log n}{n}.$$

Then since

$$\sum_{n \le x/K} \frac{\log n}{n} \ge \sum_{\sqrt{x} < n \le x/K} \frac{\log n}{n} > \frac{1}{2} (\log x) \sum_{\sqrt{x} < n \le x/K} \frac{1}{n} > c \log^2 x,$$

for a constant $c > 0$ (and x large), it follows that

(10.1.13) $\lim_{x \to \infty} \min_{K < u \leq x} M(u) \leq 0 \leq \lim_{x \to \infty} \max_{K \leq u \leq x} M(u).$

Since (10.1.13) holds for all $K > 0$, (10.1.12) follows. This, of course, hints at the connective path between the μ-function and the primes and supports the plausibility of the assertion

(10.1.14) $\lim_{x \to \infty} \sum_{d \leq x} \frac{\mu(d)}{d} = 0.$

Perhaps even more dramatic is the fact that the Prime Number Theorem is equivalent to

(10.1.15) $U(x) = \sum_{n \leq x} \mu(n) = o(x).$

Here, apart from the 0 values, the $+1$ and -1 values of $\mu(n)$ are asked to provide just enough cancellation to bring the order of magnitude under the trivial estimate $0(x)$.

In saying that two statements concerning primes are *equivalent,* we are essentially stating that we can reasonably easily deduce each from the other. In a sense then, each of the statements contains exactly the "same level of information" about primes. On occasion, it is of importance to note that two statements are not apparently so comparable. For example, (10.1.8) is not (in this vague sense) equivalent to the Prime Number Theorem. Writing

$$\sum_{n \leq x} \frac{\Lambda(n)}{n} = \sum_{n \leq x} \frac{1}{n} (\psi(n) - \psi(n - 1)),$$

setting $\psi(x) = x + R(x)$, and summing by parts,

$$\sum_{n \leq x} \frac{\Lambda(n)}{n} = \sum_{n \leq x} \frac{1}{n} + \sum_{n \leq x} \frac{1}{n} (R(n) - R(n - 1))$$

$$= \log x + 0(1) + \frac{R([x])}{[x]} + \sum_{n \leq x-1} \frac{R(n)}{n(n + 1)}.$$

Since $\psi(x) = 0(x)$ implies $R(x) = 0(x)$, the above shows that (10.1.8) is equivalent to

(10.1.16) $\sum_{n \leq x} \frac{R(n)}{n(n + 1)} = 0(1).$

This is not directly comparable to the Prime Number Theorem. Actually, an estimate such as (10.1.16) could hold without necessarily having

(10.1.17) $R(n) = o(n)$

(which is the Prime Number Theorem). Even more interestingly, (10.1.16) appears
to contain additional information to that of the Prime Number Theorem since the
Prime Number Theorem only implies that

$$\sum_{n \leq x} \frac{R(n)}{n(n + 1)} = o(\log x).$$

In fact, (10.1.16) provides that either $R(n)$ is changing sign frequently so as to
produce much cancellation or, frequently, $R(n)$ is very small with respect to n.
Since, at a change of sign, $R(n)$ is small and would tend to remain small for awhile
(i.e., precluding very rapid changes), we see that (10.1.16) implies that there are
"patches" of n where $R(n)$ is "small." In fact, it is this type of information that is
critical for all the elementary proofs of the Prime Number Theorem that are
presented in later sections. Actually, stronger assertions than the Prime Number
Theorem are known, such as

(10.1.18) $R(x) = 0 \left(\dfrac{x}{\log^\alpha x} \right)$

for *every* fixed $\alpha > 1$.

   There is a development similar to the above for primes in arithmetic progressions.
In particular, the *Prime Number Theorem for Arithmetic Progressions* states that for
all $A, B, A > B > 0$, $(A, B) = 1$ [see (9.4.33), (9.4.34), (9.4.35)],

(10.1.19) $\pi(x, A, B) = \displaystyle\sum_{\substack{p \leq x \\ p \equiv B (\mathrm{mod}\, A)}} 1 \sim \dfrac{1}{\phi(A)} \dfrac{x}{\log x}.$

Further, every element of our previous discussion has an analogue for primes in
arithmetic progressions. Since the methods of proof usually carry over, these ana-
logues will be distributed among the exercises.

EXERCISES

1. Prove that the ratio of successive primes tends to one as they tend to infinity
    (assuming the Prime Number Theorem).

2. Given $A > B > 0$, $(A, B) = 1$, assuming (10.1.9), determine the asymptotic
    behavior of the nth prime $p \equiv B (\mathrm{mod}\, A)$.

3. Defining

$$R(x, A, B) = \psi(x, A, B) - \frac{1}{\phi(A)} x,$$

prove that

$$\sum \frac{R(n, A, B)}{n(n + 1)} = 0(1).$$

4. Defining $\hat{R}(x) = \theta(x) - x$ prove that

$$\sum_{n < x} \frac{\hat{R}(n)}{n(n + 1)} = 0(1).$$

Formulate and prove an analogous result for arithmetic progressions.

5. Prove that

$$\varliminf_{x \to \infty} \frac{1}{x} \sum_{n \leq x} \mu(n) \leq 0 \leq \varlimsup_{x \to \infty} \frac{1}{x} \sum_{n \leq x} \mu(n).$$

§10.2 THE ROLE OF THE MÖBIUS INVERSION FORMULA

We saw in the previous section that statements involving the Möbius function relate to those about primes. Further, the essential source of this relationship appears to be through the Möbius inversion formula [as for example with (10.1.9)]. Clearly, this type of appearance of the μ-function arises from trying to solve appropriate systems of linear equations for functions of the primes in terms of all integers [as in (10.1.9), where $\Lambda(n)$ is expressed in terms of log n]. Then one is led to the requirement of evaluating certain expressions involving the μ-function. But these same expressions can be made to arise by applying an inversion formula to a system of linear equations *where we know the solution*. This kind of "dual" role for the inversion formula then leads to information about primes.

One example of this technique is found in the proof of Dirichlet's Theorem (given in §9.4). But the general idea described above goes back certainly to the early nineteenth century. In fact, an early example was included as Exercise 3 of §3.5 and asserts that

(10.2.1)
$$\sum_{n \leq x} \frac{\mu(n)}{n} = 0(1).$$

Reviewing a derivation of this, recall part of Theorem 3.5.1 which gives that if for two functions $f(x)$, $g(x)$ the equations

(10.2.2)
$$f(x) = \sum_{n \leq x} g\left(\frac{x}{n}\right)$$

hold for all x, then

(10.2.3)
$$g(x) = \sum_{n \le x} \mu(n) f\left(\frac{x}{n}\right)$$

for all x. Taking $g(x) \equiv 1$ in (10.2.2), we have

(10.2.4)
$$f(x) = \sum_{n \le x} 1 = [x] = x + 0(1).$$

Inserting this and $g(x) = 1$ in (10.2.3) yields

$$1 = \sum_{n \le x} \mu(n)\left(\frac{x}{n} + 0(1)\right)$$

$$= x \sum_{n \le x} \frac{\mu(n)}{n} + 0(x)$$

and (10.2.1) follows.

The temptation to continue with this game is irresistible. Taking $g(x) = x$ in (10.2.2), we have

(10.2.5)
$$f(x) = \sum_{n \le x} \frac{x}{n} = x \log x + c_1 x + 0(1).$$

Inserting this and $g(x) = x$ in (10.2.3) gives

$$x = \sum_{n \le x} \mu(n)\left(\frac{x}{n} \log \frac{x}{n} + c_1 \frac{x}{n} + 0(1)\right)$$

$$= x \sum_{n \le x} \frac{\mu(n)}{n} \log \frac{x}{n} + c_1 x \sum_{n \le x} \frac{\mu(n)}{n} + 0(x).$$

Then using (10.2.1) to estimate the second sum, we obtain

(10.2.6)
$$\sum_{n \le x} \frac{\mu(n)}{n} \log \frac{x}{n} = 0(1).$$

Though the continuation is fairly clear, in real time it was 50 years between (10.2.6) and the next step. Heuristically, we note that $g(x) = 1$ led to x in (10.2.4), and then $g(x) = x$ led to $x \log x$ in (10.2.5), so that our next try should be $g(x) = x \log x$. Then, from (10.2.2),

$$f(x) = \sum_{n \le x} \frac{x}{n} \log \frac{x}{n} = x \log x \sum_{n \le x} \frac{1}{n} - x \sum_{n \le x} \frac{\log n}{n}$$

$$= x \log x \left(\log x + c_1 + 0\left(\frac{1}{x}\right) \right) - x \left(\frac{1}{2} \log^2 x + c_2 + 0\left(\frac{\log x}{x}\right) \right),$$

or

(10.2.7) $f(x) = \frac{1}{2} x \log^2 x + c_1 x \log x - c_2 x + 0(\log x).$

Inserting (10.2.7) and $g(x) = x \log x$ in (10.2.3),

$$x \log x = \sum_{n \le x} \mu(n) \left(\frac{1}{2} \cdot \frac{x}{n} \log^2 \frac{x}{n} + c_1 \frac{x}{n} \log \frac{x}{n} - c_2 \frac{x}{n} + 0\left(\log \frac{x}{n} \right) \right)$$

$$\frac{1}{2} x \sum_{n \le x} \frac{\mu(n)}{n} \log^2 \frac{x}{n} + c_1 x \sum_{n \le x} \frac{\mu(n)}{n} \log \frac{x}{n} - c_2 x \sum_{n \le x} \frac{\mu(n)}{n} + 0(x).$$

Then using (10.2.1) and (10.2.6) to estimate the second and third sums, we get

(10.2.8) $$\sum_{n \le x} \frac{\mu(n)}{n} \log^2 \frac{x}{n} = 2 \log x + 0(1).$$

What information about primes has been extracted in deriving (10.2.1), (10.2.6), and (10.2.8)? In a basic way, this information provides (10.1.8). To see this, recall from (10.1.10) that

$$\sum_{n \le x} \frac{\Lambda(n)}{n} = \sum_{dd' \le x} \frac{\mu(d)}{d} \frac{\log d'}{d'} = \sum_{d \le x} \frac{\mu(d)}{d} \cdot \sum_{d' \le x/d} \frac{\log d'}{d'}$$

$$= \sum_{d \le x} \frac{\mu(d)}{d} \left(\frac{1}{2} \log^2 \frac{x}{d} + c_2 + 0\left(\frac{d}{x} \log \frac{x}{d} \right) \right),$$

or

(10.2.9) $$\sum_{n \le x} \frac{\Lambda(n)}{n} = \frac{1}{2} \sum_{d \le x} \frac{\mu(d)}{d} \log^2 \frac{x}{d} + c_2 \sum_{d \le x} \frac{\mu(d)}{d} + 0(1).$$

Hence, more precisely, we have that (10.1.8) is implied by (10.2.1) and (10.2.8). Moreover, assuming (10.2.1) known, it follows that (10.1.8) and (10.2.8) are equivalent. We will later see that in a sense (10.2.1), (10.2.6), and (10.2.8) already also contain the Prime Number Theorem.

EXERCISES

1. Let $A > 0$ be an integer. Show that for χ any character modulo A,

$$\sum_{n < x} \frac{\mu(n)\chi(n)}{n} = 0(1).$$

[For $\chi \neq \chi_o$ this can be deduced from (9.4.20). In any event, apply the inversion formula and conclude that (9.4.17) implies (9.4.18)].

2. Prove that for $A > 0$, any integer, $A > B > 0$, $(A, B) = 1$,

$$\sum_{\substack{n \leq x \\ n \equiv B (\text{mod} A)}} \frac{\mu(n)}{n} = 0(1).$$

3. For $A > 0$, any integer, χ any character modulo A, estimate

$$\sum_{n \leq x} \frac{\mu(n)\chi(n)}{n} \log \frac{x}{n}.$$

[*Hint:*   start with $f(x) = \log x$, $P(n) = \chi(n)/n$ in (9.4.17).]

4. For $A > B > 0$, $(A, B) = 1$, estimate

$$\sum_{\substack{n \leq x \\ n \equiv B (\text{mod} A)}} \frac{\mu(n)}{n} \log \frac{x}{n}.$$

5. For any integer $A > 0$, χ a character modulo A, estimate

$$\sum_{n \leq x} \frac{\chi(n)\mu(n)}{n} \log^2 \frac{x}{n}.$$

6. For $A > B > 0$, $(A, B) = 1$, estimate

$$\sum_{\substack{n \leq x \\ n \equiv B (\text{mod} A)}} \frac{\mu(n)}{n} \log^2 \frac{x}{n}.$$

7. Estimate

$$\sum_{n \leq x} \frac{\mu(n)}{n} \log^3 \frac{x}{n}.$$

§10.3   EQUIVALENT FORMULATIONS OF THE PRIME NUMBER THEOREM

We propose to prove in this section that the Prime Number Theorem is equivalent to each of

(10.3.1) $$M(x) = \sum_{n \leq x} \frac{\mu(n)}{n} = o(1),$$

(10.3.2)
$$U(x) = \sum_{n \le x} \mu(n) = o(x),$$

and

(10.3.3)
$$V(x) = \sum_{n \le x} \frac{\mu(n) \log n}{n} = o(\log x).$$

This will be achieved by establishing the sequence of implications [denoted by (A), (B), (C), and (D)],

$$\boxed{M(x) = o(1)} \xrightarrow{\text{(A)}} \boxed{U(x) = o(x)} \xrightarrow{\text{(B)}} \boxed{\psi(x) \sim x}$$

$$\xrightarrow{\text{(C)}} \boxed{V(x) = o(\log x)} \xrightarrow{\text{(D)}} \boxed{M(x) = o(1)}$$

(A) $M(x) = o(1)$ IMPLIES $U(x) = o(x)$

Using summation by parts, we have

$$U(x) = \sum_{n \le x} \mu(n) = \sum_{n \le x} n(M(n) - M(n-1))$$

$$= [x]M([x]) - \sum_{n \le x-1} M(n) = o(x) + \sum_{n \le x} o(1) = o(x).$$

(B) $U(x) = o(x)$ IMPLIES $\psi(x) \sim x$

For the proof of this implication, we require some identity that expresses $\psi(x)$ in terms of $\mu(n)$. From (10.1.9),

$$\Lambda(n) = \sum_{t/n} \mu(t) \log \frac{n}{t}$$

so that

(10.3.4)
$$\psi(x) = \sum_{n \le x} \Lambda(n) = \sum_{tt' \le x} \mu(t) \log t'.$$

Since

$$[x] = \sum_{n \le x} 1,$$

we wish to represent 1 in the form

$$1 = \sum_{t/n} \mu(t) f\left(\frac{n}{t}\right).$$

But the Möbius inversion formula gives that

$$f(n) = \sum_{d/n} 1 = d(n),$$

where $d(n) =$ the number of divisors of n. Thus

(10.3.5) $$[x] = \sum_{tt' \le x} \mu(t)d(t')$$

and subtracting from (10.3.4) we get

(10.3.6) $$\psi(x) - [x] = \sum_{tt' < x} \mu(t)(\log t' - d(t')).$$

Finally, since

$$1 = \sum_{n \le x} \sum_{t/n} \mu(t) = \sum_{tt' \le x} \mu(t),$$

for any constant K,

(10.3.7) $$\psi(x) - [x] - K = \sum_{tt' \le x} \mu(t)y(t')$$

where

(10.3.8) $$y(t') = \log t' - d(t') - K.$$

Then we note from

$$Y(t) = \sum_{t' \le z} y(t') = \sum_{t' \le z} (\log t' - d(t') - K)$$

$$= z \log z - z - Kz - z \log z - cz + O(z^{1/2})$$

$$= (-K - 1 - c)z + O(z^{1/2});$$

taking $K = -1 - c$, we get

(10.3.9) $$Y(z) = O(z^{1/2}).$$

We set $vw = x$ (where v and w will be chosen later as functions of x), and rewrite (10.3.7) as

(10.3.10) $$\psi(x) - [x] - K = \sum_{t \le v} \mu(t)Y\left(\frac{x}{t}\right) + \sum_{t' \le w} y(t')U\left(\frac{x}{t'}\right) - U(v)Y(w).$$

Using (10.3.9), we have

$$(10.3.11) \quad \sum_{t \le v} \mu(t) Y\left(\frac{x}{t}\right) = 0\left(\sum_{t \le v} \frac{x^{1/2}}{t^{1/2}}\right) = 0(x^{1/2} v^{1/2}) = 0\left(\frac{x}{w^{1/2}}\right).$$

Our hypothesis that $U(x) = o(x)$ can be written as

$$U(x) = x\epsilon(x),$$

where $\epsilon(x) \to 0$ as $x \to \infty$. Then using the trivial estimate

$$|y(t')| = 0(t'),$$

we have

$$\sum_{t' \le w} y(t') U\left(\frac{x}{t'}\right) = 0\left(\sum_{t' \le w} t' \left|\epsilon\left(\frac{x}{t'}\right)\right| \cdot \frac{x}{t'}\right),$$

so that

$$(10.3.12) \quad \sum_{t' \le w} y(t') U\left(\frac{x}{t'}\right) = 0\left(xw \max_{t \ge x/w} |\epsilon(t)|\right).$$

Finally, we have also

$$(10.3.13) \quad U(v)Y(w) = 0(v\epsilon(v) \cdot w^{1/2}) = 0\left(\frac{x}{w^{1/2}}\right).$$

Inserting (10.3.11), (10.3.12), and (10.3.13) into (10.3.10), we obtain

$$(10.3.14) \quad \psi(x) - [x] - K = 0\left(\frac{x}{w^{1/2}}\right) + 0\left(x \cdot w \max_{t \ge x/w} |\epsilon(t)|\right).$$

Since $\epsilon(t) \to o$ as $t \to \infty$, we can choose $w = w(x)$ going to infinity sufficiently slowly as $x \to \infty$ and such that

$$(10.3.15) \quad w \max_{t > x/w} |\epsilon(t)| \to 0$$

as $x \to \infty$. (This is an immediate consequence of the fact that (10.3.15) is true for w a constant.) For this choice of $w = w(x)$ [then $v = v(x) = x/w$], (10.3.14) yields

$$\psi(x) - [x] - K = o(x),$$

which implies $\psi(x) \sim x$.

(C) $\psi(X) \sim X$ IMPLIES $V(X) = o(\text{LOG } X)$

For the proof of this implication we need an identity that gives $V(x)$ in terms of an expression involving primes. Since

$$V(x) = \sum_{n \le x} \frac{\mu(n) \log n}{n},$$

it is natural to try to express $\mu(n) \log n$ in terms of primes. From (10.1.9)

$$\Lambda(n) = \sum_{d|n} \mu(d) \log \frac{n}{d} = -\sum_{d|n} \mu(d) \log d,$$

and applying the Möbius inversion formula to this we get

(10.3.16) $$-\mu(n) \log n = \sum_{d|n} \mu\left(\frac{n}{d}\right) \Lambda(d).$$

Using (10.3.16), we have

$$V(x) = \sum_{n \le x} \frac{\mu(n) \log n}{n} = -\sum_{n \le x} \frac{1}{n} \sum_{d|n} \mu\left(\frac{n}{d}\right) \Lambda(d)$$

$$= -\sum_{dd' \le x} \frac{\Lambda(d)}{d} \cdot \frac{\mu(d')}{d'}$$

or

(10.3.17) $$V(x) = -\sum_{d \le x} \frac{\Lambda(d)}{d} M\left(\frac{x}{d}\right).$$

This identity provides a link between $V(x)$ and the primes on which a proof of the desired implication can be based. We write $\psi(x) = x + R(x)$, so that $\psi(x) \sim x$ implies $R(x) = o(x)$. Then

$$\sum_{d \le x} \frac{\Lambda(d)}{d} M\left(\frac{x}{d}\right) = \sum_{d \le x} \frac{\psi(d) - \psi(d-1)}{d} M\left(\frac{x}{d}\right)$$

or

$$(10.3.18) \qquad -V(x) = \sum_{d \le x} \frac{1}{d} M\left(\frac{x}{d}\right) + \sum_{d \le x} \frac{R(d) - R(d-1)}{d} M\left(\frac{x}{d}\right).$$

On one hand, we have

$$\sum_{d \le x} \frac{1}{d} M\left(\frac{x}{d}\right) = \sum_{d \le x} \frac{1}{d} \sum_{d' \le x/d} \frac{\mu(d')}{d'} = \sum_{dd' \le x} \frac{\mu(d')}{dd'} = \sum_{c \le x} \frac{1}{c} \sum_{d'/c} \mu(d') = 1.$$

Whereas, summing by parts, we obtain

$$\sum_{d \le x} \frac{R(d) - R(d-1)}{d} M\left(\frac{x}{d}\right) = \frac{R([x])}{[x]} M\left(\frac{x}{[x]}\right)$$

$$+ \sum_{n \le x-1} R(n)\left(\frac{1}{n} M\left(\frac{x}{n}\right) - \frac{1}{n+1} M\left(\frac{x}{n+1}\right)\right)$$

$$= o(1) + \sum_{n \le x-1} \frac{R(n)}{n}\left(M\left(\frac{x}{n}\right) - M\left(\frac{x}{n+1}\right)\right)$$

$$+ \sum_{n \le x-1} \frac{R(n)}{n(n+1)} M\left(\frac{x}{n+1}\right)$$

From (10.2.1), $M(x) = 0(1)$, so that the second sum

$$\sum_{n \le x-1} \frac{R(n)}{n(n+1)} M\left(\frac{x}{n+1}\right) = o\left(\sum_{n \le x} \frac{1}{n}\right) = o(\log x)$$

As for the first sum, since $R(n) = o(n)$,

$$\sum_{n \le x-1} \frac{R(n)}{n}\left(M\left(\frac{x}{n}\right) - M\left(\frac{x}{n+1}\right)\right) = o\left(\sum_{n \le x-1} \left|M\left(\frac{x}{n}\right) - M\left(\frac{x}{n+1}\right)\right|\right)$$

$$= o\left(\sum_{n \le x-1} \sum_{x/(n+1) < d \le x/n} \frac{1}{d}\right)$$

$$= o\left(\sum_{d \le x} \frac{1}{d}\right) = o(\log x).$$

Thus (10.3.18) yields $V(x) = o(\log x)$, which is the desired result.

(D) $V(x) = o(\text{LOG } x)$ IMPLIES $M(x) = o(1)$

From (10.2.6) we have

$$(\log x) \sum_{n < x} \frac{\mu(n)}{n} - \sum_{n \le x} \frac{\mu(n) \log n}{n} = 0(1),$$

so that

(10.3.19) $M(x)\log x - V(x) = 0(1).$

Thus $V(x) = o(\log x)$ implies $M(x) = o(1)$ as claimed.

  This completes the proof of the asserted equivalence to the Prime Number Theorem of (10.3.1), (10.3.2), and (10.3.3). The last two implications have a special interest. They are based on the identities (10.3.17) and (10.3.19) that combine to give

(10.3.20) $M(x)\log x + \sum_{d \le x} \frac{\Lambda(d)}{d} M\left(\frac{x}{d}\right) = 0(1).$

Though this identity was known before the turn of this century, it remained totally unsuspected for over 50 years that such a relation contained a route to the Prime Number Theorem itself (when combined with (10.1.8). Actually, when such a so-called "elementary" proof of the Prime Number Theorem was discovered, it was not based on (10.3.20), but on an identity that is much more convenient to work with and which will be considered in §10.4.

EXERCISES

1. Give a detailed proof that $\lim_{x \to \infty} \epsilon(x) = 0$ implies the existence of a function $w(x)$ such that $\lim_{x \to \infty} w(x) = \infty$ and (10.3.15) holds.

2. Show that if for some $\epsilon > 0$, $\psi(x) = x + 0(x/(\log x)^{1+\epsilon})$, then

$$\sum_{n \le x} \frac{\mu(n)}{n} = 0\left(\frac{1}{\log x}\right).$$

3. Show that if for some $\epsilon > 0$, $\sum_{n \le x} \mu(n)/n = 0(1/(\log x)^{2+\epsilon})$, then

$$\sum_{n=1}^{\infty} \frac{\mu(n) \log n}{n} = -1$$

    More precisely, deduce that

$$\sum_{n \le x} \frac{\mu(n) \log n}{n} = -1 + 0\left(\frac{1}{\log^{\epsilon} x}\right).$$

4. Assuming that for some $\epsilon > 0$,

$$\sum_{n \le x} \frac{\mu(n)}{n} = 0\left(\frac{1}{(\log x)^{4+\epsilon}}\right),$$

and using the result of Exercise 7 of §10.2, show that

$$\sum_{n=1}^{\infty} \frac{\mu(n) \log^2 n}{n} = -\frac{2}{3} \gamma,$$

where γ is Euler's constant defined by

$$\gamma = \lim_{x \to \infty} \left(\sum_{n \le x} \frac{1}{n} - \log x \right).$$

5. Define

$$M(x, A, B) = \sum_{\substack{n \le x \\ n \equiv B(\mathrm{mod}\, A)}} \frac{\mu(n)}{n}$$

and

$$U(x, A, B) = \sum_{\substack{n \le x \\ n \equiv B(\mathrm{mod}\, A)}} \mu(n).$$

Show that for given $A > B > 0$, $(A, B) = 1$, the assumption that $\lim_{x \to \infty} M(x, A, B)$ exists (i.e., the series converges), implies

$$U(x, A, B) = o(x)$$

as $x \to \infty$.

6. Show that
   (a) $\psi(x, A, B) \sim x/\phi(A)$ for all $B > 0$, $(B, A) = 1$ (A fixed), implies

   (*) $$\sum_{\substack{n \le x \\ n \equiv B(\mathrm{mod}\, A)}} \frac{\mu(n) \log n}{n} = o(\log^2 x)$$

   for all B, $(B, A) = 1$.
   (b) $U(x, A, B) = o(x)$ for all $B > 0$, $(B, A) = 1$ (A fixed), also implies (*) for all B, $(B, A) = 1$.

7. For a given integer $A > 0$, assume that for all B, $A > B > 0$, $(A, B) = 1$, we have

$$\sum_{\substack{n \le x \\ n \equiv B(\mathrm{mod}\, A)}} \frac{\mu(n) \log n}{n} = o(\log^2 x).$$

Show directly that this implies that $L_o(\chi) \neq o$ for characters $\chi \neq \chi_o$, modulo A.

8. Assume that for a given integer $A > 0$, and all B, $A > B > 0$, $(A, B) = 1$, we have $U(x, A, B) = o(x)$. Show that this implies

$$\psi(x, A, B) = \sum_{\substack{n \le x \\ n \equiv B \,(\text{mod}\, A)}} \Lambda(n) \sim \frac{1}{\phi(A)} x$$

for all such A, B.

The proof begins with the following steps, modeled after the case in the text:

(a)
$$\psi(x, A, B) - \frac{1}{\phi(A)} [x] - K = \sum_{\substack{tt' \le x \\ tt' \equiv B \,(\text{mod}\, A)}} \mu(t) y(t')$$

where

$$y(t') = \log t' - \frac{A}{\phi(A)} d(t') - K.$$

(b)
$$\sum_{\substack{n \le x \\ n \equiv B \,(\text{mod}\, A)}} d(n) = \frac{1}{\phi(A)} x \log x + Cx + 0(x^{1/2})$$

where C depends on A, but *not* on B.

(c)   Show that one can choose a value of K so that for *all* B, $(B, A) = 1$,

$$Y_B(z) = \sum_{\substack{t' \le z \\ t' \equiv B \,(\text{mod}\, A)}} y(t') = 0(\sqrt{z}).$$

9. If we assume that for all B, $(B, A) = 1$, $\psi(x, A, B) \sim (1/\phi(A))\, x$, show that this implies that for these B, the series

$$\sum_{\substack{n \le x \\ n \equiv B \,(\text{mod}\, A)}} \frac{\mu(n)}{n}$$

converges.

[*Hint.* Derive the generalization of (10.3.20), which is

$$M(x, A, B)\log x = -\sum_{d \le x} \frac{\Lambda(d)}{d} M\left(\frac{x}{d}, A, Bd^{-1}\right) + c_B \log x + 0(1),$$

where d^{-1} denotes the inverse of d modulo A and c_B is a constant depending on A and B.]

§10.4   THE SELBERG SYMMETRY FORMULA

The formula derived by A. Selberg, on which the first elementary proofs of the Prime Number Theorem are based, can be viewed as a natural extension of the Chebychev theme. Chebychev's methods stem from the relation

$$(10.4.1) \qquad\qquad \log n = \sum_{d\mid n} \Lambda(d),$$

which gives $\log n$ in terms of the primes. Selberg's idea can be thought of as starting with the thought of doing the same thing for $\log^2 n$. That is, we desire a function $f(d)$, related to the primes, such that

$$(10.4.2) \qquad\qquad \log^2 n = \sum_{d\mid n} f(d).$$

Were it not for the requirement that $f(d)$ be expressed in terms of primes, the Möbius inversion formula could provide an answer:

$$(10.4.3) \qquad\qquad f(n) = \sum_{d\mid n} \mu(d) \log^2 \frac{n}{d}.$$

To find the desired representation of $f(n)$ (i.e., of $\log^2 n$), we proceed more primitively and literally square $\log n$. This gives

$$\log^2 n = \left(\sum_{p^\alpha \mid n} \log p\right)^2$$

$$= \sum_{\substack{p^\alpha q^\beta \mid n \\ p \neq q}} \log p \log q + \sum_{\substack{p^\alpha \mid n \\ p^\beta \mid n}} \log^2 p.$$

Letting $\nu_p(n) = $ the exponent of the highest power of p dividing n, the above gives

$$(10.4.4) \qquad \log^2 n = \sum_{\substack{p^\alpha q^\beta \mid n \\ p \neq q}} \log p \log q + \sum_{\substack{p \mid n \\ (\text{distinct } p)}} \nu_p^2(n)\log^2 p.$$

For a given prime p, the missing terms $(\log p)(\log p)$ in the first sum, where $p = q$, are those such that $p^{\alpha+\beta}/n$. The number of these is the number of α, β, each ≥ 1, such that $\alpha + \beta \leq \nu_p(n)$ and hence equals $\frac{1}{2}\nu_p(n)(\nu_p(n) - 1)$. Restoring these in the first sum and making the correction in the second sum yields

$$(10.4.5) \qquad \log^2 n = \sum_{p^\alpha q^\beta \mid n} \log p \log q + \sum_{\substack{p \mid n \\ (\text{distinct } p)}} \frac{1}{2}\nu_p(n)(\nu_p(n) + 1)\log^2 p.$$

But since

$$\frac{1}{2}\nu_p(n)(\nu_p(n) + 1) = \sum_{\alpha=1}^{\nu_p(n)} \alpha = \sum_{p^\alpha \mid n} \alpha,$$

the second sum equals

$$\sum_{p^{\alpha}/n} \alpha \log^2 p = \sum_{d/n} \Lambda(d)\log d,$$

and (10.4.5) becomes

$$(10.4.6) \qquad \log^2 n = \sum_{d/n} \left\{ \sum_{uv=d} \Lambda(u)\Lambda(v) + \Lambda(d)\log d \right\}.$$

That is, the desired $f(n)$ is given by

$$(10.4.7) \qquad f(n) = \Lambda(n)\log n + \sum_{dd'=n} \Lambda(d)\Lambda(d').$$

An alternate derivation of (10.4.6) may be based on the convolution arithmetic (see Chapter 4). Let L denote $\log n$, and 1_o the arithmetic function that is identically 1 on the integers. Then $L = 1_o * \Lambda$, and interpreting multiplication by L as a derivation,

$$L^2 = L(1_o * \Lambda) = L * \Lambda + 1_o * L\Lambda$$
$$= 1_o * \Lambda * \Lambda + 1_o * L\Lambda$$

or

$$L^2 = 1_o * (\Lambda * \Lambda + L\Lambda),$$

which is precisely (10.4.6).

Inserting (10.4.6) in (10.4.3) we obtain

$$(10.4.8) \qquad \Lambda(n)\log n + \sum_{dd'=n} \Lambda(d)\Lambda(d') = \sum_{d/n} \mu(d)\log^2 \frac{n}{d}.$$

Summing (10.4.8) and all $n \leq x$, the right side equals

$$\sum_{n \leq x} \sum_{d/n} \mu(d)\log^2 \frac{n}{d} = \sum_{dd' \leq x} \mu(d)\log^2 d' = \sum_{d \leq x} \mu(d) \sum_{d' \leq x/d} \log^2 d'$$

$$= \sum_{d \leq x} \mu(d)\left(\frac{x}{d} \log^2 \frac{x}{d} + b_1 \frac{x}{d} \log \frac{x}{d} + b_2 \frac{x}{d} + 0\left(\log^2 \frac{x}{d} \right) \right)$$

$$= x \sum_{d \leq x} \frac{\mu(d)}{d} \log^2 \frac{x}{d} + b_1 x \sum_{d \leq x} \frac{\mu(d)}{d} \log \frac{x}{d}$$

$$+ b_2 x \sum_{d \le x} \frac{\mu(d)}{d} + 0(x).$$

Using (10.2.1), (10.2.6), and (10.2.8), this yields

(10.4.9)
$$\sum_{n \le x} \sum_{d/n} \mu(d) \log^2 \frac{n}{d} = 2x \log x + 0(x).$$

Equating this to the summation of the left side of (10.4.8) over $n \le x$ produces

(10.4.10)
$$\sum_{n \le x} \Lambda(n) \log n + \sum_{dd' \le x} \Lambda(d)\Lambda(d') = 2x \log x + 0(x),$$

which is the *Selberg Symmetry Formula*.

This formula contains much of the information about primes derived in Chapter 9 and some of it is easily extracted. For example, replacing x by $x/2$ in (10.4.10) and subtracting,

$$\sum_{x/2 < n \le x} \Lambda(n) \log n + \sum_{x/2 < dd' \le x} \Lambda(d)\Lambda(d') = x \log x + 0(x).$$

But then

$$x \log x + 0(x) > \sum_{x/2 < n \le x} \Lambda(n) \log n \ge \left(\log \frac{x}{2} \right) \left(\psi(x) - \psi\left(\frac{x}{2} \right) \right)$$

or

$$\psi(x) - \psi\left(\frac{x}{2} \right) < x + 0\left(\frac{x}{\log x} \right).$$

Replacing x by $x/2^i$, and summing over i yields that

(10.4.11)
$$\psi(x) < 2x + o(x).$$

That is, once again we have $\psi(x) = 0(x)$.

This can be used to transform (10.4.10) into other forms. Summing by parts we have

$$\sum_{n \le x} \Lambda(n) \log n = \sum_{n \le x} (\log n)(\psi(n) - \psi(n - 1))$$

$$= \psi(x) \log[x] - \sum_{n \le x-1} \psi(n) \log\left(1 + \frac{1}{n} \right)$$

$$= \psi(x) \log x + 0\left(\frac{\psi(x)}{x} \right) + \sum_{n \le x-1} 0(1)$$

or

(10.4.12) $\displaystyle\sum_{n\le x} \Lambda(n)\log n = \psi(x)\log x + 0(x).$

Using (10.4.12) in (10.4.10), it becomes

(10.4.13) $\displaystyle\psi(x)\log x + \sum_{dd'\le x} \Lambda(d)\Lambda(d') = 2x \log x + 0(x),$

or

(10.4.14) $\displaystyle\psi(x)\log x + \sum_{d\le x} \Lambda(d)\psi\left(\frac{x}{d}\right) = 2x \log x + 0(x).$

We note in passing that (10.1.8) is easily retrieved from (10.4.14). To see this, replace x by x/d' in (10.4.14) and sum over all $d' \le x$, so that we obtain

(10.4.15) $\displaystyle\sum_{d'\le x} \psi\left(\frac{x}{d'}\right)\log\frac{x}{d'} + \sum_{dd'\le x} \Lambda(d)\psi\left(\frac{x}{dd'}\right) = x \log^2 x + 0(x \log x).$

Note then that

$$\sum_{d'\le x} \psi\left(\frac{x}{d'}\right)\log\frac{x}{d'} = \log x \sum_{d'\le x} \psi\left(\frac{x}{d'}\right) - \sum_{d'\le x} (\log d')\psi\left(\frac{x}{d'}\right)$$

$$= \log x \sum_{d'\le x} \psi\left(\frac{x}{d'}\right) - \sum_{n\le x} \psi\left(\frac{x}{n}\right) \sum_{d\mid n} \Lambda(d)$$

$$= \log x \sum_{d'\le x} \psi\left(\frac{x}{d'}\right) - \sum_{dd'\le x} \Lambda(d)\psi\left(\frac{x}{dd'}\right),$$

and writing this in (10.4.15) yields

$$\sum_{n\le x} \psi\left(\frac{x}{n}\right) = x \log x + 0(x).$$

But since

$$\sum_{n\le x} \psi\left(\frac{x}{n}\right) = \sum_{n\le x} \sum_{m\le x/n} \Lambda(m) = \sum_{mn\le x} \Lambda(m)$$

$$= \sum_{m\le x} \Lambda(m)\left[\frac{x}{m}\right] = x \sum_{m\le x} \frac{\Lambda(m)}{m} + 0(x),$$

this indeed gives back (10.1.8).

To obtain still other forms of the Symmetry Formula, recall that by (9.1.7),

$$\psi(x) = \theta(x) + O(\sqrt{x} \log x),$$

so that

(10.4.16) $\psi(x)\log x = \theta(x)\log x + O(x).$

Also,

$$\sum_{\substack{p^{\alpha}q^{\beta} \le x \\ \alpha > 1}} \log p \log q = \sum_{\substack{p^{\alpha} \le x \\ \alpha > 1}} (\log p)\psi\left(\frac{x}{p^{\alpha}}\right)$$

$$= O\left(x \sum_{\substack{p^{\alpha} \le x \\ \alpha > 1}} \frac{\log p}{p^{\alpha}}\right) = O(x),$$

so that

(10.4.17) $\displaystyle\sum_{dd' \le x} \Lambda(d)\Lambda(d') = \sum_{pq \le x} \log p \log q + O(x).$

From (10.4.16) and (10.4.17), (10.4.13) becomes

(10.4.18) $\displaystyle\theta(x) \log x + \sum_{pq \le x} \log p \log q = 2x \log x + O(x),$

or

(10.4.19) $\displaystyle\theta(x)\log x + \sum_{p \le x} (\log p)\theta\left(\frac{x}{p}\right) = 2x \log x + O(x).$

Perhaps it is the form (10.4.18) that most justifies calling these identities a "Symmetry Formula." In fact, we see clearly that the "symmetry" that is exhibited is between the distribution of primes and that of the products of two primes.

These various forms of Selberg's formula are selfproliferating in that many other useful identities can be derived from them. For example, replacing x by x/m in (10.4.10), multiplying by $\Lambda(m)$, and summing over $m \le x$ yields

$$\sum_{mn \le x} \Lambda(m)\Lambda(n)\log n + \sum_{dd'd'' \le x} \Lambda(d)\Lambda(d')\Lambda(d'')$$

$$= 2 \sum_{m \le x} \Lambda(m)\frac{x}{m} \cdot \log \frac{x}{m} + O\left(\sum_{m \le x} \Lambda(m) \cdot \frac{x}{m}\right)$$

$$= 2x \sum_{m \le x} \frac{\Lambda(m)}{m} \log \frac{x}{m} + O(x \log x),$$

so that using the assertion of Exercise 7 of §9.2,

$$(10.4.20) \quad \sum_{mn \le x} \Lambda(m)\Lambda(n)\log n + \sum_{dd'd'' \le x} \Lambda(d)\Lambda(d')\Lambda(d'') = x \log^2 x + O(x \log x).$$

This can be further simplified by using (10.4.12) as follows:

$$\sum_{mn \le x} \Lambda(m)\Lambda(n)\log n = \sum_{m \le x} \Lambda(n) \sum_{n \le x/m} \Lambda(n)\log n$$

$$= \sum_{m \le x} \Lambda(m)\left(\psi\left(\frac{x}{m}\right) \log \frac{x}{m} + O\left(\frac{x}{m}\right) \right)$$

$$= (\log x) \sum_{m \le x} \Lambda(m)\psi\left(\frac{x}{m}\right)$$

$$- \sum_{m \le x} \Lambda(m)\log m \, \psi\left(\frac{x}{m}\right) + O(x \log x)$$

$$= (\log x) \sum_{mn \le x} \Lambda(m)\Lambda(n)$$

$$- \sum_{mn \le x} \Lambda(m)\Lambda(n)\log m + O(x \log x),$$

implying

$$(10.4.21) \quad \sum_{mn \le x} \Lambda(m)\Lambda(n)\log n = \frac{1}{2} \log x \sum_{mn \le x} \Lambda(m)\Lambda(n) + O(x \log x).$$

Then (10.4.20) may be rewritten as

$$(10.4.22) \quad \frac{1}{2}(\log x) \sum_{mn \le x} \Lambda(m)\Lambda(n) + \sum_{dd'd'' \le x} \Lambda(d)\Lambda(d')\Lambda(d'')$$

$$= x \log^2 x + O(x \log x).$$

Further, multiplying (10.4.13) by $\frac{1}{2} \log x$ and using (10.4.22) gives

$$(10.4.23) \quad \psi(x)\log^2 x = 2 \sum_{dd'd'' \le x} \Lambda(d)\Lambda(d')\Lambda(d'') + O(x \log x).$$

EXERCISES

1. Derive from (10.4.19) that

$$\frac{1}{2}(\log x) \sum_{pq \leq x} \log p \, \log q + \sum_{pqr \leq x} \log q \, \log q \, \log r = x \log^2 x + 0(x \log x),$$

where p, q, r denote primes. Also obtain this as a consequence of (10.4.27).

2. Prove that

$$\theta(x)\log^2 x = \sum_{pqr \leq x} \log p \, \log q \, \log r + 0(x).$$

3. Prove that

$$\log^3 n = \sum_{d \mid n} f(d),$$

where

$$f(d) = \Lambda(d)\log^2 d + 3 \sum_{uv=d} \Lambda(u)\Lambda(v)\log v + \sum_{uvw=d} \Lambda(u)\Lambda(v)\Lambda(w).$$

4. Using the identity given in Exercise 3, evaluate

$$\sum_{d \leq x} \Lambda(d)\log^2 d + 3 \sum_{uv \leq x} \Lambda(u)\Lambda(v)\log v + \sum_{uvw \leq x} \Lambda(u)\Lambda(v)\Lambda(w),$$

with an error term of $0(x)$.

5. Among the formulas of the type discussed in this section, perhaps the simplest is

$$U(x)\log x + \sum_{d \leq x} \Lambda(d)U\left(\frac{x}{d}\right) = 0(x),$$

where

$$U(x) = \sum_{n \leq x} \mu(n).$$

Prove this. [*Hint.* Note that

$$\sum_{n \leq x} \mu(n)\log n + \sum_{dd' \leq x} \mu(d)\Lambda(d') = 0.]$$

6. Without using Dirichlet's Theorem (or characters), prove that for $A > B > 0$, $(A, B) = 1$

$$\psi(x, A, B)\log x + \sum_{\substack{mn \leq x \\ mn \equiv B \pmod{A}}} \Lambda(m)\Lambda(n) = \frac{2}{\phi(A)} x \log^2 x + 0(x),$$

and that this implies

$$\theta(x, A, B)\log x + \sum_{\substack{pq \le x \\ pq \equiv B (\mathrm{mod}\, A)}} \log p \log q = \frac{2}{\phi(A)} x \log x + 0(x),$$

7. Prove that for $A > B > 0$, $(A, B) = 1$

$$\frac{1}{2} \log x \sum_{\substack{mn \le x \\ mn \equiv B (\mathrm{mod}\, A)}} \Lambda(m)\Lambda(n) + \sum_{\substack{dd'd'' \le x \\ dd'd'' \equiv B (\mathrm{mod}\, A)}} \Lambda(d)\Lambda(d')\Lambda(d'')$$

$$= \frac{1}{\phi(A)} x \log^2 x + 0(x \log x),$$

and hence

$$\psi(x, A, B)\log^2 x = 2 \sum_{\substack{dd'd'' \le x \\ dd'd'' \equiv B (\mathrm{mod}\, A)}} \Lambda(d)\Lambda(d')\Lambda(d'') + 0(x \log x).$$

State and derive analogous results for $\theta(x, A, B)$.

8. Verify the identity

$$\frac{\zeta''(s)}{\zeta(s)} = \left(\frac{\zeta'}{\zeta}\right)' + \left(\frac{\zeta'}{\zeta}\right)^2$$

[where $\zeta(s)$ is the zeta function], and deduce (10.4.8) from it. Develop a similar identity on which to base a solution of Exercise 3.

§10.5 IMMEDIATE CONSEQUENCES OF THE SYMMETRY FORMULA

The Symmetry Formula contains information about primes that is intrinsic to its very form. Much of this information is new in that it is not produced by our previous methods. For example, we have the following theorem.

Theorem 10.5.1. For

$$A = \varlimsup_{x \to \infty} \frac{\psi(x)}{x}$$

$$a = \varliminf_{x \to \infty} \frac{\psi(x)}{x}$$

we have

(10.5.1) $A + a = 2.$

Proof. Consider a sequence $x_i \to \infty$ on which $\psi(x_i) \sim Ax_i$ (we know that A is finite). Then from (10.4.14) we get

$$(10.5.2) \qquad \sum_{d \leq x_i} \Lambda(d)\psi\left(\frac{x_i}{d}\right) \sim (2 - A)x_i \log x_i.$$

For any fixed $K > 1$, we have since $\psi(x) = 0(x)$, that

$$\sum_{x_i/K < d \leq x_i} \Lambda(d)\psi\left(\frac{x_i}{d}\right) = 0\left(\sum_{d \leq x_i} \Lambda(d)\psi(K)\right) = 0(x_i).$$

Thus (10.5.2) implies

$$(10.5.3) \qquad \sum_{d \leq x_i/K} \Lambda(d)\psi\left(\frac{x_i}{d}\right) \sim (2 - A)x_i \log x_i.$$

This, in turn, gives that

$$\left(\min_{x_i \geq u \geq K} \frac{\psi(u)}{u}\right)x_i \sum_{d \leq x_i/K} \frac{\Lambda(d)}{d} \leq (2 - A)x_i \log x_i(1 + o(1))$$

and using (10.1.8), this yields

$$(10.5.4) \qquad \lim_{x_i \to \infty} \min_{x_i \geq u \geq K} \frac{\psi(u)}{u} \leq 2 - A.$$

Since (10.5.4) holds for all $K > 1$, it follows that $a \leq 2 - A$, or $A + a \leq 2$. Similarly, one can prove $2 - a \leq A$ or $2 \leq a + A$, and (10.5.1) follows.

The proof of the above theorem shows that the Symmetry Formula, by its very form, implies that (i) if there are some primes (i.e., $a > 0$) then there aren't too many (i.e., $A < 2$); and (ii) if there aren't too many primes (i.e., $A < 2$), then there must be some primes (i.e., $a > 0$). The second observation, (ii), is a bit more startling, since an "existence" of objects results from a proof that there are not too many.

Ultimately, these points will result in a proof of the Prime Number Theorem. Here we pause only to note their intrinsic strength by observing that they can be used to "magnify" the output of Chebychev's method. Writing (10.4.14) with x replaced by y, $y \leq x$ and subtracting yields

$$(10.5.5) \qquad \psi(x)\log x - \psi(y)\log y + \sum_{d \leq x} \Lambda(d)\left(\psi\left(\frac{x}{d}\right) - \psi\left(\frac{y}{d}\right)\right)$$

$$= 2x \log x - 2y \log y + 0(x),$$

In the special case where $y = \lambda x$, $0 < \lambda < 1$, λ fixed, this becomes

$$(\psi(x) - \psi(\lambda x))\log x + \psi(\lambda x)\log\frac{1}{\lambda} + \sum_{d\leq x}\Lambda(d)\left(\psi\left(\frac{x}{d}\right) - \psi\left(\frac{\lambda x}{d}\right)\right)$$

$$= 2(1 - \lambda)x \log x + 0(x),$$

or since $\psi(\lambda x) = 0(x)$,

$$(10.5.6) \qquad (\psi(x) - \psi(\lambda x))\log x + \sum_{d\leq x}\Lambda(d)\left(\psi\left(\frac{x}{d}\right) - \psi\left(\lambda\frac{x}{d}\right)\right)$$

$$= 2(1 - \lambda)x \log x + 0(x).$$

Recall now that in §9.3 (following (9.3.20), the Chebychev method produced the inequality

$$(10.5.7) \qquad\qquad \psi(x) - \psi\left(\frac{x}{2}\right) \leq (\log 2)x + o(x)$$

from the identity (9.3.19). This didn't quite suffice to give Bertrand's Postulate (for sufficiently large x), but could be salvaged by the ideas of Ramanujan or Erdos. The identity (10.5.6) easily converts (10.5.7) into such a proof. For applying (10.5.6) with $\lambda = \frac{1}{2}$ gives

$$(10.5.8) \qquad \left(\psi(x) - \psi\left(\frac{x}{2}\right)\right)\log x + \sum_{d\leq x}\Lambda(d)\left(\psi\left(\frac{x}{d}\right) - \psi\left(\frac{1}{2}\cdot\frac{x}{d}\right)\right)$$

$$= x \log x + 0(x).$$

But from (10.5.7) and $\psi(x) = 0(x)$,

$$\sum_{d\leq x}\Lambda(d)\left(\psi\left(\frac{x}{d}\right) - \psi\left(\frac{1}{2}\frac{x}{d}\right)\right) = \sum_{d\leq x/(\log x)^{1/2}}\Lambda(d)\left(\psi\left(\frac{x}{d}\right) - \psi\left(\frac{1}{2}\frac{x}{d}\right)\right)$$

$$+ \sum_{x/(\log x)^{1/2}<d\leq x}\Lambda(d)\left(\psi\left(\frac{x}{d}\right) - \psi\left(\frac{1}{2}\frac{x}{d}\right)\right)$$

$$\leq (\log 2)x \sum_{d\leq x}\frac{\Lambda(d)}{d} + o(x \log x)$$

$$\leq (\log 2)x \log x + o(x \log x),$$

and inserting this in (10.5.8) we obtain

$$\left(\psi(x) - \psi\left(\frac{x}{2}\right) \right) \log x + (\log 2)x \log x \geq x \log x + o(x \log x)$$

or

(10.5.9) $$\psi(x) - \psi\left(\frac{x}{2}\right) \geq (1 - \log 2)x + o(x).$$

Since $1 - \log 2 > 0$, this suffices to give Bertrand's Postulate for all sufficiently large x.

Though the above illustration is purely of academic interest, it does indicate that the Symmetry Formula can be used to limit the rate of change of such prime related functions as $\psi(x)$. We formalize one such limitation as follows.

Theorem 10.5.2. For any positive integers x, y, such that $0 < c_1 < x/y < c_2$ (c_1, c_2 constants), we have

(10.5.10) $$|\psi(x) - \psi(y)| \leq 2|x - y| + 0\left(\frac{x}{\log x}\right).$$

Proof. We may assume that $x > y$. Then from (10.5.5),

$$(\psi(x) - \psi(y))\log x + \psi(y)\log \frac{x}{y} < 2(x - y)\log x + 2y \log \frac{x}{y} + 0(x)$$

or

$$(\psi(x) - \psi(y))\log x < 2(x - y)\log x + 0(x),$$

and (10.5.10) follows.

EXERCISES

1. Prove the analogue of (10.5.10) for $\theta(x)$.

2. Letting

$$a_\lambda = \lim_{x \to \infty} \frac{\psi(x) - \psi(\lambda x)}{x}$$

$$A_\lambda = \overline{\lim_{x \to \infty}} \frac{\psi(x) - \psi(\lambda x)}{x}$$

for $0 \leq \lambda \leq 1$, prove that $a_\lambda + A_\lambda = 2(1 - \lambda)$.

3. State and prove the analogue of Exercise 2 for $\theta(x)$.

4. Prove that

$$\lim_{x\to\infty} \sum_{n\leq x} \frac{\mu(n)}{n} + \overline{\lim_{x\to\infty}} \sum_{n\leq x} \frac{\mu(n)}{n} = 0.$$

[Hint: use (10.3.20).]

5. Prove that for any integers $A > B > 0$, $(A, B) = 1$,

$$\lim_{x\to\infty} \frac{\psi(x, A, B)}{x} + \overline{\lim_{x\to\infty}} \frac{\psi(x, A, B)}{x} = \frac{2}{\phi(A)}$$

and

$$\lim_{x\to\infty} \frac{\theta(x, A, B)}{x} + \overline{\lim_{x\to\infty}} \frac{\theta(x, A, B)}{x} = \frac{2}{\phi(A)}.$$

6. Derive the analogue of (10.5.10) for $\psi(x, A, B)$ and $\theta(x, A, B)$.

§10.6 SELBERG'S DERIVATION OF THE PRIME NUMBER THEOREM

Selberg's derivation of the Prime Number Theorem proceeds by treating (10.4.14) by a method similar to that of successive approximations. Setting

$$(10.6.1) \qquad\qquad \psi(x) = x + R(x)$$

in (10.4.14), we obtain

$$(10.6.2) \qquad\qquad R(x) \log x + \sum_{n\leq x} \Lambda(n) R\left(\frac{x}{n}\right) = 0(x).$$

If we convert this to an inequality

$$(10.6.3) \qquad\qquad |R(x)| \log x \leq \sum_{n\leq x} \Lambda(n) \left| R\left(\frac{x}{n}\right) \right| + 0(x),$$

we have a "balanced" inequality. Namely, if one uses an estimate $|R(u)| < cu$, valid for sufficiently large u, on the right side of (10.6.3), there results

$$(10.6.4) \qquad\qquad |R(x)| \leq cx + 0\left(\frac{x}{\log x}\right).$$

That is, essentially the same estimate is output as was input. Clearly, if this were *unbalanced* and an improved constant smaller than c occurred in (10.6.4), this method could be iterated to further improve the estimate. Then there would be the possibility that, under repeated iteration, the improved constants (valid for sufficiently large x) would tend to zero and we would have a proof of the Prime Number Theorem.

To make the above idea work requires that one forceably unbalance the inequality (10.6.3). This would be achieved if there exist disjoint intervals I_j such that for $n \in I_j$ we have a very good estimate

$$(10.6.5) \qquad \left| R\left(\frac{x}{n}\right) \right| < \epsilon \frac{x}{n},$$

and these intervals are sufficiently numerous in that

$$(10.6.6) \qquad \sum_j \sum_{n \in I_j} \frac{\Lambda(n)}{n} > \eta \log x$$

for some fixed constant $\eta > 0$. The condition (10.6.5) is not hard to satisfy on certain small intervals but that these satisfy (10.6.6) is difficult to establish. This is due to the fact that the intervals are small and the stated requirement is that they contain many primes.

Selberg overcomes this difficulty by manipulating the original Symmetry Formula so as to obtain an inequality of the type (10.6.3) in which the explicit presence of the primes is removed from the right side [in (10.6.3) it is the $\Lambda(n)$]. Then, in this new setting, the required analogue of (10.6.6) becomes a lower estimate for an expression involving sums over all integers (not just prime powers).

To achieve the above, begin by multiplying (10.4.13) by $\log x$, and use (10.4.21) so as to obtain

$$(10.6.7) \quad \psi(x)\log^2 x + 2\sum_{n \leq x} \Lambda(n)\log n\psi\left(\frac{x}{n}\right) = 2x \log^2 x + O(x \log x).$$

Inserting (10.6.1) in (10.6.7) we get

$$(10.6.8) \qquad R(x)\log^2 x + 2\sum_{n \leq x} \Lambda(n)\log n\, R\left(\frac{x}{n}\right) = O(x \log x),$$

which yields the inequality

$$(10.6.9) \qquad |R(x)| \log^2 x \leq 2\sum_{n \leq x} \Lambda(n)\log n\, \left| R\left(\frac{x}{n}\right) \right| + O(x \log x).$$

Next we write (10.4.23) as

$$(10.6.10) \qquad \psi(x)\log^2 x = 2 \sum_{dd' \le x} \Lambda(d)\Lambda(d')\psi\left(\frac{x}{dd'}\right) + 0(x \log x);$$

using (10.6.1), this gives

$$(10.6.11) \qquad R(x)\log^2 x = 2 \sum_{dd' \le x} \Lambda(d)\Lambda(d')R\left(\frac{x}{dd'}\right) + 0(x \log x),$$

which yields the inequality

$$(10.6.12) \quad |R(x)| \log^2 x \le 2 \sum_{dd' \le x} \Lambda(d)\Lambda(d') \left|R\left(\frac{x}{dd'}\right)\right| + 0(x \log x).$$

Adding (10.6.9) and (10.6.12) we have

(10.6.13)

$$|R(x)| \log^2 x \le \sum_{n \le x} \Lambda(n)\log n \left|R\left(\frac{x}{n}\right)\right| + \sum_{dd' \le x} \Lambda(d)\Lambda(d') \left|R\left(\frac{x}{dd'}\right)\right|$$

$$+ 0(x \log x).$$

But setting

$$(10.6.14) \qquad \rho(z) = \sum_{n \le z} \Lambda(n)\log n + \sum_{uv \le z} \Lambda(u)\Lambda(v),$$

the right side of (10.6.13) equals

$$\sum_{n \le x} \left|R\left(\frac{x}{n}\right)\right| (\rho(n) - \rho(n - 1))$$

$$= \rho([x])R\left(\frac{x}{[x]}\right) + \sum_{n \le x-1} \rho(n)\left(\left|R\left(\frac{x}{n}\right)\right| - \left|R\left(\frac{x}{n + 1}\right)\right|\right).$$

Since from (10.4.10), $\rho(z) = 2z \log z + 0(z)$, the above equals

$$2[x]\log[x]R\left(\frac{x}{[x]}\right) + 0(x) + \sum_{n \le x-1} 2n \log n\left(\left|R\left(\frac{x}{n}\right)\right| - \left|R\left(\frac{x}{n + 1}\right)\right|\right)$$

$$+ 0\left(\sum_{n \le x} n \left|\left|R\left(\frac{x}{n}\right)\right| - \left|R\left(\frac{x}{n + 1}\right)\right|\right|\right)$$

$$= 2 \sum_{n \le x} \left| R\left(\frac{x}{n}\right) \right| (n \log n - (n-1)\log(n-1))$$

$$+ 0\left(\sum_{n \le x} n \left| R\left(\frac{x}{n}\right) - R\left(\frac{x}{n+1}\right) \right| \right) + 0(x \log x),$$

where we've applied summation parts to the first sum and used the inequality $\|a\| - \|b\| \le |a - b|$ in the second. Further,

$$\sum_{n \le x} \left| R\left(\frac{x}{n}\right) \right| (n \log n - (n-1)\log(n-1)) = \sum_{n \le x} \left| R\left(\frac{x}{n}\right) \right| (\log n + 0(1))$$

$$= \sum_{n < x} \left| R\left(\frac{x}{n}\right) \right| \log n + 0(x \log x)$$

and

$$\sum_{n \le x} n \left| R\left(\frac{x}{n}\right) - R\left(\frac{x}{n+1}\right) \right| \le \sum_{n \le x} n \left| \psi\left(\frac{x}{n}\right) - \psi\left(\frac{x}{n+1}\right) \right| + \sum_{n \le x} n \left(\frac{x}{n} - \frac{x}{n+1}\right)$$

$$\le \sum_{n \le x} \psi\left(\frac{x}{n}\right) + 0(x \log x) = 0(x \log x).$$

Thus (10.6.13) yields

$$(10.6.15) \qquad |R(x)| \log^2 x \le 2 \sum_{n \le x} \left| R\left(\frac{x}{n}\right) \right| \log n + 0(x \log x).$$

Compare (10.6.15) with (10.6.3) and note now the objective of removing the $\Lambda(n)$ has been achieved. Note also that the inequality (10.6.15) remains balanced in the sense discussed earlier.

The existence of "patches" of integers n for which (10.6.5) is achieved is provided by the following lemmas.

Lemma 10.6.1 Given $\epsilon > 0$, there exist $K_o = K_o(\epsilon)$, $x_o = x_o(\epsilon)$, such that for any $K \ge K_o$ and $x \ge x_o$, there exists an integer n, $x < n \le Kx$, for which

$$(10.6.16) \qquad \left| \frac{R(n)}{n} \right| < \epsilon.$$

Moreover, there exists a constant c_o such that we may take

$$(10.6.17) \qquad K_o(\epsilon) = e^{c_o/\epsilon}.$$

Proof. Suppose first that $R(n)$ changes sign at some integer n, $x < n \le Kx$

(i.e., $R(n)R(n + 1) \leq 0$). Then

$$|R(n)| \leq |R(n + 1) - R(n)| = |\psi(n + 1) - \psi(n) - 1| < \log n$$

and

$$\left|\frac{R(n)}{n}\right| < \frac{\log n}{n} < \frac{\log x}{x} < \epsilon$$

for $x > e^{2/\epsilon}$ (and in this case it is independent of K).

Thus there remains only the case where $R(n)$ does not change sign in the interval $x < n < Kx$. We then turn to (10.1.16) which provides the existence of a constant C such that

$$(10.6.18) \qquad \left|\sum_{x < n \leq Kx} \frac{R(n)}{n(n + 1)}\right| < C$$

for *all* $K > 1$ and $x > 1$. Since $R(n)$ is of constant sign, this may be written as

$$\sum_{x < n \leq Kx} \frac{|R(n)|}{n(n + 1)} < C,$$

which implies

$$\left(\min_{x < n \leq Kx} \frac{|R(n)|}{n}\right)\left(\sum_{x < n \leq Kx} \frac{1}{n + 1}\right) < C$$

or

$$\left(\min_{x < n \leq Kx} \frac{|R(n)|}{n}\right)(\log K + 0(1)) < C.$$

But then

$$\min_{x < n \leq Kx} \frac{|R(n)|}{n} < \frac{C}{\log K + 0(1)} < \epsilon$$

for $K > e^{c_0/\epsilon}$, where c_0 is a sufficiently large constant depending only on C and the implied constant in the $0(1)$. This completes the proof of the lemma.

The isolated occurrence of (10.6.16) provided by Lemma 10.6.1 is next extended so as to hold in an entire subinterval.

Lemma 10.6.2.   Given ϵ, $1 > \epsilon > 0$, there exist $x_1 = x_1(\epsilon)$ and constants c_1, and λ, $0 < \lambda < 1$ (independent of both x_1 and ϵ) such that for any $K \geq e^{c_1/\epsilon}$ and $x > x_1$, there exists an x^*,

$$(10.6.19) \qquad x < x^* < (1 + \lambda\epsilon)x^* < Kx;$$

which has the property that for all real numbers u satisfying

$$(10.6.20) \qquad x^* \leq u \leq (1 + \lambda\epsilon)x^*,$$

we have

$$(10.6.21) \qquad |R(u)| < \epsilon u.$$

Proof.   We apply Lemma 10.6.1 to produce $K_o = K_o(\epsilon/3)$ and $x_o = x_o(\epsilon/3)$, so that for $x \geq x_o$, $K/2 > K_o$ we have an integer n, $x < n \leq (K/2)\,x$, for which

$$(10.6.22) \qquad \left|\frac{R(n)}{n}\right| < \frac{\epsilon}{3}.$$

Since $\psi(x) = 0(x)$, it follows that $R(x) = 0(x)$, implying that there exists a positive constant C^* for which

$$(10.6.23) \qquad \left|\frac{R(x)}{x}\right| < C^*$$

for all $x \geq 1$.

Consider δ (to be chosen later), $0 < \delta < 1$, and let u be any real number, $n \leq u \leq (1 + \delta)n$. Then

$$\left|\frac{R(u)}{u} - \frac{R(n)}{n}\right| = \left|R(u)\left(\frac{1}{u} - \frac{1}{n}\right) + \frac{R(u) - R(n)}{n}\right|$$

$$\leq \left|\frac{R(u)}{u}\right|\left(\frac{u}{n} - 1\right) + \frac{1}{n}|R(u) - R(n)|$$

$$\leq \left|\frac{R(u)}{u}\right|\left(\frac{u}{n} - 1\right) + \frac{1}{n}|\psi(u) - \psi(n) - (u - n)|$$

or

$$(10.6.24) \qquad \left|\frac{R(n)}{n} - \frac{R(n)}{n}\right| \leq \left(\frac{u}{n} - 1\right)\left(\left|\frac{R(u)}{u}\right| + 1\right) + \frac{1}{n}|\psi(u) - \psi(n)|.$$

Applying Theorem 10.5.2 (with $x = u$, $y = n$, $c_1 = 1$, $c_2 = 2 > 1 + \delta$),

$$|\psi(u) - \psi(n)| \le 2(u - n) + 0\left(\frac{u}{\log u}\right),$$

so that (10.6.24) yields

$$\left|\frac{R(u)}{u} - \frac{R(n)}{n}\right| \le \left(\frac{u}{n} - 1\right)\left(\left|\frac{R(u)}{u}\right| + 3\right) + 0\left(\frac{1}{\log x}\right)$$

$$\le \delta(C^* + 3) + 0\left(\frac{1}{\log x}\right).$$

Thus we obtain

$$(10.6.25) \qquad \left|\frac{R(u)}{u}\right| \le \left|\frac{R(n)}{n}\right| + \delta(C^* + 3) + 0\left(\frac{1}{\log x}\right).$$

From (10.6.22), $|R(n)| < \epsilon/3 \cdot n$, and for x sufficiently large the $0(1/\log x)$ term is also $< \epsilon/3$. Finally, taking $\delta = \lambda\epsilon$ with

$$(10.6.26) \qquad \lambda = \frac{1}{3(C^* + 3)},$$

we get

$$\left|\frac{R(u)}{u}\right| < \epsilon$$

for all u such that $x < n \le u \le (1 + \lambda\epsilon)n \le (1 + \lambda\epsilon)(K/2)x < Kx$. This establishes the lemma with $x^* = n$ and for $K \ge 2K_o(\epsilon/3)$. But then $K \ge 2e^{3c_o/\epsilon}$ suffices, which is implied by $K \ge e^{c_1/\epsilon}$ for a suitable constant c_1.

With Lemma 10.6.2 and the inequality (10.6.15), we have all the tools necessary to carry out a proof of the Prime Number Theorem. Suppose then that we have an estimate of the form

$$(10.6.27) \qquad |R(x)| \le \alpha x$$

valid for all sufficiently large x (say $x \ge x_2$). using this estimate in the form

$$(10.6.28) \qquad \left|R\left(\frac{x}{n}\right)\right| \le \alpha\frac{x}{n},$$

valid for $x/n \ge x_2$, or $n \le x/x_2$, on the right of (10.6.15) we have that this is

$$\le 2\alpha \sum_{n \le x/x_2} \frac{x}{n} \log n + \sum_{x/x_2 < n \le x} 0\left(\frac{x \log n}{n}\right) + 0(x \log x),$$

so that we get

$$(10.6.29) \qquad |R(x)| \log^2 x \le \alpha x \log^2 x + O(x \log x).$$

This estimate does not improve on (10.6.27) as it stands. However, we propose to improve it by subtracting a correction for any integers n where we have a better estimate than (10.6.28). More precisely, suppose that we have a collection of disjoint sets S_j in the range 1 to x, such that for $n \in S_j$

$$\left| R\!\left(\frac{x}{n}\right) \right| \le \epsilon \frac{x}{n},$$

(where $\epsilon < \alpha$). Then we can improve (10.6.29) by subtracting

$$(\alpha - \epsilon)x \sum_j \sum_{n \in S_j} \frac{\log n}{n}$$

from the right side of (10.6.29) and have

$$(10.6.30) \qquad |R(x)| \log^2 x \le \alpha x \log^2 x - (\alpha - \epsilon)x \sum_j \sum_{n \in S_j} \frac{\log n}{n} + O(x \log x).$$

Such sets S_j will be provided on the basis of Lemma 10.6.2. Letting $c_2 > 1$ be any positive constant such that $\alpha/c_2 < 1$, take $\epsilon = \alpha/c_2$ (note that this same c_2 suffices for any smaller α). Then, for $K = e^{c_1/\epsilon}$, consider the intervals

$$I_j : K^j < u \le K^{j+1}.$$

For $x_1 = x_1(\epsilon)$, $K^j \ge x_1$ where $j \ge \log x_1/\log K$, that is, for $j > j^* = [\log x_1/\log K] + 1$ and $K^{j+1} < \sqrt{x} < x$ for $j + 1 \le \frac{1}{2}\log x/\log K$ or $j \le [\frac{1}{2}\log x/\log K] - 1 = \hat{j}$. Then for $j^* \le j \le \hat{j}$, Lemma 10.6.2 provides that each of the intervals I_j contains a subinterval I_j^* given by

$$I_j^* : x_j^* \le u \le (1 + \lambda\epsilon)x_j^*,$$

in which

$$|R(u)| \le \epsilon u.$$

Thus if $x/n \in I_j^*$ [i.e., $x_j^* \le x/n < (1 + \lambda\epsilon)x_j^*$] or

$$(10.6.31) \qquad \frac{x}{K^{j+1}} < \frac{1}{1 + \lambda\epsilon}\frac{x}{x_j^*} < n < \frac{x}{x_j^*} < \frac{x}{K^j},$$

we have

(10.6.32)
$$\left| R\!\left(\frac{x}{n}\right) \right| < \epsilon \cdot \frac{x}{n}.$$

Symbolically, we can write (10.6.31) as requiring that $n \in x/I_j^* \subset x/I_j$ (where x/I denotes the set of numbers obtained by dividing x by all the numbers of I). These intervals x/I_j^*, $j^* \le j \le \hat{j}$, are clearly disjoint, and for $S_j = x/I_j^*$, we have the inequality (10.6.30). Next we note that

$$\sum_{j=j^*}^{\hat{j}} \sum_{n \in S_j} \frac{\log n}{n} = \sum_{j=j^*}^{\hat{j}} \sum_{n \in x/I_j^*} \frac{\log n}{n} > \sum_{j=j^*}^{\hat{j}} \frac{1}{2}\left(1 - \frac{1}{1+\lambda\epsilon}\right)\frac{x}{x_j^*} \cdot \frac{\log x/x_j^*}{x/x_j^*}$$

$$\ge \frac{1}{2}\frac{\lambda\epsilon}{1+\lambda\epsilon} \sum_{j=j^*}^{\hat{j}} \log \frac{x}{x_j^*} > \frac{1}{2}\frac{\lambda\epsilon}{1+\lambda\epsilon}(\hat{j} - j^*)\left(\frac{1}{2}\log x\right)$$

since $x/x_j^* \ge \sqrt{x}$. Then, since $1 + \lambda\epsilon < 2$ and $\hat{j} - j^* \ge \dfrac{1}{2}\dfrac{\log x}{\log K} - \dfrac{\log x_1}{\log K} -$

$2 > \dfrac{1}{4}\dfrac{\log x}{\log K}$ for x sufficiently large, we obtain

(10.6.33)
$$\sum_{j=j^*}^{\hat{j}} \sum_{n \in S_j} \frac{\log n}{n} \ge \frac{\lambda\epsilon \log^2 x}{32 \log K} = \frac{\lambda\epsilon^2}{32c_1} \log^2 x$$

(recall that $K = e^{c_1/\epsilon}$). Using (10.6.33), (10.6.30) yields

$$|R(x)|\log^2 x \le (x \log^2 x)\left(\alpha - (\alpha - \epsilon)\frac{\lambda\epsilon^2}{32c_1}\right) + 0(x \log x),$$

which, in turn, implies that for sufficiently large x,

(10.6.34)
$$|R(x)| \le \left((\alpha - (\alpha - \epsilon)\frac{\lambda\epsilon^2}{40c_1}\right)x.$$

Thus we've replaced the α in (10.6.27) by (recall $\epsilon = \alpha/c_2$),

$$\alpha - (\alpha - \epsilon)\frac{\lambda\epsilon^2}{40c_1} = \alpha - \left(1 - \frac{1}{c_2}\right)\left(\frac{\lambda}{40c_1 c_2^2}\right)\alpha^3 = \alpha - c_3\alpha^3$$

(where c_3 is a positive constant). Thus setting $\alpha_1 = C^*$ [given in (10.6.23)], the above argument produces a sequence α_m, much that

(10.6.35)
$$\alpha_{m+1} = \alpha_m - c_3\alpha_m^3,$$

where

$$(10.6.36) \qquad\qquad |R(x)| \leq \alpha_m x$$

for all sufficiently large x (depending on m). Since the sequence α_m is decreasing and > 0, it converges to a finite limit l. Then from (10.6.35),

$$l = l - c_3 l^3$$

so that $c_3 l^3 = 0$, implying $l = 0$. That the $\alpha_m \to 0$ provides that $R(x) = o(x)$ and hence $\psi(x) \sim x$, which is the Prime Number Theorem.

EXERCISES

1. From (10.3.20) deduce that

$$(10.6.37) \qquad M(x) \log^2 x + \sum_{n \leq x} \frac{\Lambda(n) \log n}{n} M\left(\frac{x}{n}\right)$$

$$- \sum_{mn \leq x} \frac{\Lambda(m)\Lambda(n)}{mn} M\left(\frac{x}{mn}\right) = O(\log x).$$

[Hint: replace x by x/n in (10.3.20), multiply by $\Lambda(n)/n$ and sum over all $n \leq x$.]

2. Note that (10.6.37) implies

$$(10.6.38) \qquad |M(x)| \log^2 x \leq \sum_{n \leq x} \frac{\Lambda(n) \log n}{n} \left| M\left(\frac{x}{n}\right) \right|$$

$$+ \sum_{mn \leq x} \frac{\Lambda(n)\Lambda(m)}{mn} \left| M\left(\frac{x}{mn}\right) \right| + O(\log x),$$

and using (10.4.10), derive that

$$(10.6.39) \qquad |M(x)| \log^2 x \leq 2 \sum_{n \leq x} \left| M\left(\frac{x}{n}\right) \right| \log n + O(\log x).$$

{Hint: write the right side of (10.6.39) as

$$\sum_{n \leq x} \frac{1}{n} M\left(\frac{x}{n}\right) (\rho(n) - \rho(n - 1))$$

for $\rho(n)$ defined in (10.6.14).}

3. Prove that $\sum_{n \leq x} M(n)/n = 0(1)$, and use this to provide an analogue of Lemma 10.6.2 for $M(x)$[instead of $R(x)/x$], which provides intervals in which $|M(x)| < \epsilon$.

4. Use the result of Exercise 3 together with (10.6.39), to prove that $M(x) = o(1)$, and hence provide another proof of the Prime Number Theorem. (The distinctive feature of such a proof is that when the need arises for the analogue of (10.5.10), it is obvious.)

5. Using Exercise 5 of §10.4, prove that

$$(10.6.40) \qquad U(x) \log^2 x + \sum_{n \leq x} \Lambda(n)(\log n)U\left(\frac{x}{n}\right) - \sum_{mn \leq x} \Lambda(m)\Lambda(n)U\left(\frac{x}{mn}\right)$$

$$= 0(x \log x)$$

[where $U(x) = \sum_{n \leq x} \mu(n)$]. Applying (10.4.10), deduce from (10.6.39) that

$$(10.6.41) \qquad U(x) \log^2 x + 2 \sum_{n \leq x} \Lambda(n) \log nU\left(\frac{x}{n}\right) = 0(x \log x)$$

and

$$(10.6.42) \quad U(x) \log^2 x - 2 \sum_{mn \leq x} \Lambda(m)\Lambda(n)U\left(\frac{x}{m}\right) = 0(x \log x).$$

From these deduce that

$$(10.6.43) \qquad |U(x)|\log^2 x \leq \sum_{n \leq x} \Lambda(n)\log n\left|U\left(\frac{x}{n}\right)\right|$$

$$+ \sum_{mn < x} \Lambda(m)\Lambda(n)\left|U\left(\frac{x}{mn}\right)\right| + 0(x \log x),$$

and then derive from this that

$$(10.6.44) \qquad |U(x)|\log^2 x \leq 2 \sum_{n \leq x} \left|U\left(\frac{x}{n}\right)\right|\log n + 0(x \log x).$$

6. Prove that

$$\sum_{n \leq x} \frac{U(n)}{n(n + 1)} = 0(1),$$

and use this to provide an analogue of Lemma 10.6.2 for $U(x)$ that provides intervals in which $|U(z)| \leq \epsilon z$.

7. Using the result of Exercise 6 together with (10.6.44), prove that $U(x) = o(x)$ and provide still another proof of the Prime Number Theorem. (In this context the analogue of (10.5.10) is also trivial.)

8. Setting $\psi(x, A, B) = 1/[\phi(A)]x + R(x, A, B)$ and using the assertion of Exercise 6 of §10.4, derive for $A > B > 0$, $(A, B) = 1$, that

$$|R(x, A, B)|\log^2 x \leq 2 \sum_{\substack{n \leq x \\ (n,A)=1}} (\log n)\left|R\left(\frac{x}{n}, A, Bn^{-1}\right)\right| + 0(x \log x)$$

(where n^{-1} denotes the inverse of n modulo A).

9. Prove that for $A > B > 0$, $(A, B) = 1$

$$\sum_{n \leq x} \frac{R(n, A, B)}{n(n + 1)} = 0(1),$$

and using this, derive an analogue of lemma 10.6.2 for $R(x, A, B)$ so as to provide intervals in which $|R(x, A, B)| < \epsilon x$.

10. Using the inequality of Exercise 8 and applying the lemma provided by Exercise 9 (for all B, $(B, A) = 1$), prove that $R(x, A, B) = o(x)$ and hence prove the Prime Number Theorem for arithmetic progressions:

$$\psi(x, A, B) \sim \frac{1}{\phi(A)}x.$$

§10.7 THE ERDOS DERIVATION OF THE PRIME NUMBER THEOREM

Erdos' proof of the Prime Number Theorem, though not as suggestive of generalizations as Selberg's, is perhaps the *very simplest* of all known proofs. That it is not generally recognized as such is perhaps due to its highly combinatorial flavor and the type of "density argument" that is used.

In deducing the Prime Number Theorem from the Symmetry Formula (10.4.18), Erdos followed a distinctly different route from that of Selberg. Erdos' method focuses on the relation

(10.7.1) $A + a = 2$

for

$$A = \varlimsup_{x \to \infty} \frac{\theta(x)}{x}, \qquad a = \varliminf_{x \to \infty} \frac{\theta(x)}{x},$$

established in (10.5.1) (or Exercise 3 of §10.5). Note that if $A = a$, (10.7.1) implies $A = a = 1$ [i.e., $\theta(x) \sim x$] and the Prime Number Theorem is proved. Assuming $A > a$, it is proposed to use the relation

$$(10.7.2) \qquad \theta(x) \log x + \sum_{p \leq x} \theta\left(\frac{x}{p}\right) \log p = 2\, x \log x + 0(x)$$

to deduce a contradiction (which ultimately depends on the fact that $a > 0$; see Theorem 9.2A.1). Erdos' idea is as follows:

(i) Let $x \to \infty$ on a sequence such that $\theta(x) \sim Ax$. Then from (10.7.1) and (10.7.2),

$$(10.7.3) \qquad \sum_{p \leq x} \theta\left(\frac{x}{p}\right) \log p \sim (2 - A)x \log x = a\, x \log x.$$

Using (10.7.3), one shows that there are "many" primes $p \leq x$ such that $\theta(x/p)$ is approximately $a\, x/p$. Let S_x denote the set of these primes.

(ii) Fixing a prime $p_1 \in S_x$ which is not "too large," we have that $\theta(x/p_1) \sim a\,(x/p_1)$. Repeating the same argument as above shows that there are many primes $p \leq x/p_1$ such that $\theta\,(x/pp_1)$ is approximately Ax/p_1p. Let T_x denote this set of primes.

(iii) Since for $p \in S_x, q \in T_x$, we have $\theta(x/p) \sim ax/p$ and $\theta(x/qp_1) \sim Ax/qp_1$, we cannot have x/p and x/qp_1 very close. For then $\theta\,(x/p)$ and $\theta\,(x/qp_1)$ would be close, thereby forcing ax/p and Ax/qp_1 close, which forces A and a close (contrary to $A > a$). Since for each fixed $p \in S_x$ there are no $q \in T_x$ such that x/p and x/qp_1 are close, this forces a "thinning out" of the primes $\leq x$ (in fact many "bare patches"), which conflicts with

$$(10.7.4) \qquad \sum_{p \leq x} \frac{\log p}{p} = \log x + 0(1).$$

Executing the above rough idea is relatively easy once one introduces the following simple notion.

Definition 10.7.1. Given a sequence of real numbers $x \to \infty$ that we denote by D, and a function $\epsilon(x)$ that tends to zero as $x \in D$ tends to infinity, we say that "$\alpha\, x/p$ approximates $\theta\,(x/p)$ relative to D and $\epsilon(x)$" if for $S_x =$ the set of primes $p \leq x$ such that

$$(10.7.5) \qquad \left| \theta\left(\frac{x}{p}\right) - \alpha \frac{x}{p} \right| < \epsilon(x) \frac{x}{p},$$

we have

$$(10.7.6) \qquad \sum_{p \in S_x} \frac{\log p}{p} \sim \log x$$

(as $x \in D$ tends to infinity). For this we will use the notation

$$(10.7.7) \qquad \theta\left(\frac{x}{p}\right) \sim \alpha \frac{x}{p}, \quad (D, \epsilon(x)).$$

With this notion we can now make precise the rough statements given in (i) above.

Lemma 10.7.1. Let D be a sequence of real numbers $x \to \infty$, on which we have (for $a = \underline{\lim}_{y \to \infty} \theta(y)/y$),

$$(10.7.8) \qquad \sum_{p \leq x} \theta\left(\frac{x}{p}\right) \log p \sim a x \log x.$$

Then there exists an $\epsilon(x) \mapsto 0$ as $x \to \infty$ in D such that

$$(10.7.9) \qquad \theta\left(\frac{x}{p}\right) \sim a \frac{x}{p}, \quad (D, \epsilon(x)).$$

Similarly, if on D [for $A = \overline{\lim}_{y \to \infty} \theta(y)/y$],

$$(10.7.10) \qquad \sum_{p \leq x} \theta\left(\frac{x}{p}\right) \log p \sim A x \log x,$$

there exists an $\epsilon(x) \mapsto 0$ on D such that

$$(10.7.11) \qquad \theta\left(\frac{x}{p}\right) \sim A \frac{x}{p}, \quad (D, \epsilon(x)).$$

Proof. We will carry out the details of the proof under the hypothesis (10.7.8). [The case of (10.7.10) is similar and is left as an exercise.] First note that the hypothesis (10.7.8) can be written as stating

$$(10.7.12) \qquad \sum_{p < x} \theta\left(\frac{x}{p}\right) \log p = a x \log x + \eta(x) x \log x,$$

where $\eta(x) \mapsto 0$ as $x \to \infty$ in D. For an $\epsilon(x)$, to be determined later, $x \in D$, let S_x = the set of primes $p \leq x$ such that

$$(10.7.13) \qquad \left| \theta\left(\frac{x}{p}\right) - a \frac{x}{p} \right| \leq \epsilon(x) \frac{x}{p}.$$

This requires that the $p \in S_x$ satisfy both

(10.7.14)
$$\theta\left(\frac{x}{p}\right) \geq (a - \epsilon(x))\frac{x}{p}$$

and

(10.7.15)
$$\theta\left(\frac{x}{p}\right) \leq (a + \epsilon(x))\frac{x}{p}.$$

Since $a = \underline{\lim}_{x \to \infty} \theta(x)/x$, there exists a positive function $\xi(x)$ that tends monotonically to zero as $x \to \infty$, such that

(10.7.16)
$$\frac{\theta(x)}{x} \geq a - \xi(x).$$

Thus assuming $\epsilon(x) \geq \xi(x)$, (10.7.16) implies (10.7.14) for *all* $p \leq x$. Hence S_x is really determined by the condition (10.7.15).

From (10.7.12) we have

$$a x \log x + \eta(x) x \log x \geq \sum_{p \in S_x} \left(a - \xi\left(\frac{x}{p}\right)\right) \frac{x}{p} \log p + \sum_{\substack{p \leq x \\ p \notin S_x}} (a + \epsilon(x)) \frac{x}{p} \log p$$

$$\geq ax \sum_{p \leq x} \frac{\log p}{p} - x \sum_{p \in S_x} \xi\left(\frac{x}{p}\right) \frac{\log p}{p} + x\,\epsilon(x) \sum_{\substack{p \leq x \\ p \notin S_x}} \frac{\log p}{p}$$

$$\geq ax \log x + 0(x) - x \sum_{p \in S_x} \xi\left(\frac{x}{p}\right) \frac{\log p}{p} + x\epsilon(x) \sum_{\substack{p \leq x \\ p \notin S_x}} \frac{\log p}{p}$$

or

(10.7.17) $\displaystyle \epsilon(x) \sum_{\substack{p \leq x \\ p \notin S_x}} \frac{\log p}{p} \leq 0(1) + \sum_{p \leq x} \xi\left(\frac{x}{p}\right) \frac{\log p}{p} + \eta(x) \log x.$

Since

$$\sum_{x/\log x < p \leq x} \xi\left(\frac{x}{p}\right) \frac{\log p}{p} = 0\left(\sum_{x/\log x < p \leq x} \frac{\log p}{p}\right) = 0(\log \log x),$$

we have

$$0 < \sum_{p \leq x} \xi\left(\frac{x}{p}\right) \frac{\log p}{p} = 0(\log \log x) + \sum_{p \leq x/\log x} \xi\left(\frac{x}{p}\right) \frac{\log p}{p}$$

$$\leq 0 \,(\log \log x) + \xi(\log x) \log x.$$

Inserting this in (10.7.17) yields

$$\frac{1}{\log x} \sum_{\substack{p \leq x \\ p \notin S_x}} \frac{\log p}{p} \leq \left(c \, \frac{\log \log x}{\log x} + \xi(\log x) + \eta(x)\right) \cdot \frac{1}{\epsilon(x)}.$$

Then choosing

$$\epsilon(x) = \max\left(\left(c \, \frac{\log \log x}{\log x} + \xi(\log x) + \eta(x)\right)^{1/2}, \, \xi(x)\right)$$

we have $\epsilon(x) \geq \xi(x)$, $\epsilon(x) \to 0$ as $x \to \infty$, and

(10.7.18)
$$\frac{1}{\log x} \sum_{\substack{p \leq x \\ p \notin S_x}} \frac{\log p}{p} \leq \epsilon(x),$$

and from (10.7.18) we obtain (10.7.9), as desired.

Lemma 10.7.2. Under the same hypothesis as Lemma 10.7.1, let $p_1 = p_1(x)$ denote the smallest prime in the set S_x, defined above. Then there exists a positive function $\eta(x) \to 0$ as $x \to \infty$ in D, such that

(10.7.19)
$$p_1(x) < x^{\eta(x)}.$$

Proof. Consider $\eta = \eta(x)$, $0 < \eta < 1$, and assume that for an infinite subsequence x of D we have $p_1(x) > x^\eta$. Then

$$\sum_{p \in S_x} \frac{\log p}{p} \leq \sum_{x^\eta < p \leq x} \frac{\log p}{p} \leq (1 - \eta) \log x + 0(1).$$

But (10.7.6) implies the existence of a positive function $\lambda(x) \to 0$ for which

$$\sum_{p \in S_x} \frac{\log p}{p} \geq \log x - \lambda(x) \log x.$$

Hence

$$\log x - \lambda(x) \log x \leq (1 - \eta(x)) \log x + 0(1)$$

or

$$\eta(x) \le \lambda(x) + 0 \left(\frac{1}{\log x} \right).$$

Thus $\eta(x) = c(\lambda(x) + 1/\log x)$, ($c$ a suitably large positive constant) provides a contradiction and establishes (10.7.19) for this $\eta(x)$.

From the above lemma we note that, in particular, $x/p_1(x) \to \infty$ and $\log x/p_1(x) \sim \log x$, as $x \to \infty$ on D.

We are now in a position to realize steps (i) and (ii) of Erdos' method. Letting $x \to \infty$ on a sequence D which is such that $\theta(x) \sim Ax$, we have (10.7.3) and by Lemma 10.7.1, there exist an $\epsilon_1(x) \to 0$ such that

$$(10.7.20) \qquad \qquad \theta\left(\frac{x}{p}\right) \sim a\frac{x}{p}, \qquad (D, \epsilon_1(x)).$$

Letting S_x = the set of $p \le x$ such that (10.7.13) holds [with $\epsilon(x)$ replaced by $\epsilon_1(x)$], we have (10.7.6).

Since $p_1(x) \in S_x$, (10.7.20) implies that as $x/p_1(x) \to \infty$, $x \in D$, we have

$$(10.7.21) \qquad \qquad \theta\left(\frac{x}{p_1(x)}\right) \sim a\frac{x}{p_1(x)}$$

in the usual sense. Then from (10.7.2) we have

$$(10.7.22) \qquad \sum_{p \le x/p_1(x)} \theta\left(\frac{x}{pp_1}\right) \log p \sim (2 - a)\frac{x}{p_1(x)} \log \frac{x}{p_1(x)}$$

$$= A \frac{x}{p_1(x)} \log \frac{x}{p_1(x)}.$$

Applying Lemma 10.7.1, with $D_1 = \{x/p_1(x), x \in D\}$, there exists an $\epsilon_2 \to 0$ on D_1 such that

$$(10.7.23) \qquad \qquad \theta\left(\frac{x}{pp_1}\right) \sim A\frac{x}{pp_1}, \qquad (D_1, \epsilon_2(x)).$$

This gives that for T_x = the set of primes $p < x/p_1(x)$ such that

$$(10.7.24) \qquad \left| \theta\left(\frac{x}{pp_1(x)}\right) - A\frac{x}{pp_1(x)} \right| < \epsilon_2 \left(\frac{x}{p_1(x)}\right),$$

we have

$$(10.7.25) \qquad \sum_{p \in T_x} \frac{\log p}{p} \sim \log \frac{x}{p_1(x)} \sim \log x,$$

as $x \to \infty$ on D_1.

In the following argument it is a small convenience to use $U_x = S_x \cap T_x$ and note that (10.7.6) and (10.7.25) imply

$$(10.7.26) \qquad \sum_{p \in U_x} \frac{\log p}{p} \sim \log x$$

on the sequence D_1.

We are now prepared to produce the "holes" in the distribution of primes and carry out part (iii) of the argument. For p and q two primes of $U_x = S_x \cap T_x$, we fix an $\eta > 0$ and consider two cases according to which of $p_1 p$ and q is larger.

CASE 1. $p_1 p > q$
Then $x/q \geq x/p_1 p$ so that $\theta(x/q) \geq \theta(x/p_1 p)$, which implies [from (10.7.20), (10.7.24)]

$$(a + \epsilon_1(x)) \frac{x}{q} \geq \left(A - \epsilon_2 \left(\frac{x}{p_1(x)} \right) \right) \frac{x}{p_1 p}$$

or

$$\frac{(a + \epsilon_1(x))}{(A - \epsilon_2(x/p_1(x)))} p_1 p \geq q.$$

Thus for all $x \in D$ that are sufficiently large we have

$$\left(\frac{a + \eta}{A - \eta} \right) p_1 p > q.$$

[Clearly, we require that $o < \eta < A$ and small enough so that $(a + \eta)/(A - \eta) < 1$.] From this we see that the interval $p_1 p \geq q \geq (a + \eta)/(A - \eta) p_1 p$ does not intersect $U_x = S_x \cap T_x$; that is,

$$(10.7.27) \qquad \left(p_1 p \geq n \geq \frac{a + \eta}{A - \eta} p_1 p_2 \right) \cap U_x = \emptyset.$$

CASE 2. $q > p_1 p$
Then $x/q < x/p_1 p$ and from (10.5.10), (assuming $q < (A/a) p_1 p$),

$$\theta \left(\frac{x}{p_1 p} \right) - \theta \left(\frac{x}{q} \right) \leq \psi \left(\frac{x}{p_1 p} \right) - \psi \left(\frac{x}{q} \right) \leq 2 \left(\frac{x}{p_1 p} - \frac{x}{q} \right) + o \left(\frac{x}{p_1 p} \right),$$

which implies

$$\left(A - \epsilon_2\left(\frac{x}{p_1}\right)\right)\frac{x}{p_1 p} - (a + \epsilon_1(x))\frac{x}{q} < 2\left(\frac{x}{p_1 p} - \frac{x}{q}\right) + o\left(\frac{x}{p_1 p}\right).$$

Using (10.7.1), this yields for sufficiently large $x \in D$,

$$(A - \eta)\frac{x}{q} \leq \left(a + \frac{\eta}{2}\right)\frac{x}{p_1 p} + o\left(\frac{x}{p_1 p}\right).$$

Then there exists an $x^* = x^*(\eta)$ such that if $x/p_1 p \geq x^*$, this gives

$$(A - \eta)\frac{x}{q} < (a + \eta)\frac{x}{p_1 p}$$

or

$$\frac{A - \eta}{a + \eta} p_1 p < q$$

From this we see that the interval $p_1 p \leq n \leq ((A - \eta)/(a + \eta)) p_1 p$ does not intersect $U_x = S_x \cap T_x$, provided $p \leq x/(p_1 x^*)$. That is,

$$(10.7.28) \qquad \left(p_1 p \leq n \leq \frac{A - \eta}{a + \eta} p_1 p\right) \cap U_x = \emptyset.$$

Combining (10.7.27) and (10.7.28), we have that for sufficiently large $x \in D$, and $p < x/(p_1 x^*)$, $p \in U_x = S_x \cap T_x$, for the interval $I_p = I_p(x)$ given by

$$(10.7.29) \qquad I_p = \left(\frac{A - \eta}{a + \eta} p_1 p \geq n \geq \frac{a + \eta}{A - \eta} p_1 p\right),$$

we have

$$(10.7.30) \qquad\qquad\qquad I_p \cap U_x = \emptyset.$$

Setting $U_x' = \{p \in U_x, p \leq x/(p_1 x^*)\}$, we see that

$$\sum_{p \in U_x} \frac{\log p}{p} = \sum_{p \in U_x'} \frac{\log p}{p} + \sum_{\substack{p \in U_x \\ x/(p_1 x^*) \leq p \leq x/p_1}} \frac{\log p}{p};$$

and since

$$\sum_{\substack{p \in U_x \\ x/(p_1 x^*) \leq p \leq x/p_1}} \frac{\log p}{p} \leq \sum_{x/(p_1 x^*) \leq p \leq x/p_1} \frac{\log p}{p} = 0(1),$$

(10.7.26) implies that

(10.7.31)
$$\sum_{p \in U'_x} \frac{\log p}{p} \sim \log x.$$

Note next that for an interval I_p such as given in (10.7.29), summing over primes $q \in I_p$ we have

$$\sum_{q \in I_p} \frac{\log q}{q} \geq \frac{a + \eta}{A - \eta} \cdot \frac{1}{pp_1} \sum_{q \in I_p} \log q$$

$$= \left(\frac{a + \eta}{A - \eta} \right) \frac{1}{pp_1} \left(\theta\left(\frac{A - \eta}{a + \eta} \cdot pp_1 \right) - \theta\left(\frac{a + \eta}{A - \eta} \cdot pp_1 \right) \right)$$

$$\geq \frac{a + \eta}{A - \eta} \left(a \cdot \frac{A - \eta}{a + \eta} - A \cdot \frac{a + \eta}{A - \eta} + o(1) \right).$$

where the $o(1)$ is small for p sufficiently large. Thus for p sufficiently large, say $p > p_o$, and η fixed sufficiently small,

(10.7.32)
$$\sum_{q \in I_p} \frac{\log q}{q} \geq \frac{1}{4} \frac{a}{A} (A - a).$$

Finally, there remains the task of using (10.7.31) to produce many different intervals of the type I_p. We have for a fixed $K > 2 A/a$ (K chosen sufficiently large later), that

$$\log x \sim \sum_{p \in U'_x} \frac{\log p}{p} \sim \sum_{\substack{p \in U'_x \\ p > p_o}} \frac{\log p}{p}$$

and

$$\sum_{\substack{p \in U'_x \\ p > p_o}} \frac{\log p}{p} \leq \sum_{j = j_o}^{\hat{j}} \sum_{\substack{p \in U'_x \\ K^{j-1} < p \leq K^j}} \frac{\log p}{p}$$

where $j_o = \log p_o / \log K + 0(1)$, and $\hat{j} = \log x / \log K + 0(1)$. We wish to show that many of the sets $U'_x \cap (K^{j-1} < p < K^j)$ are nonempty. Let $N =$ the number of j, $j_o \leq j \leq \hat{j}$ such that
$$V_x(j) = U'_x \cap (K^{j-1} < p \leq K^j) \neq \emptyset.$$

Since

$$\sum_{p \in V_x(j)} \frac{\log p}{p} < \sum_{K^{j-1} < p \le K^j} \frac{\log p}{p} = \log K + 0(1),$$

we have

$$(1 + o(1))\log x \le \sum_{j=j_o}^{j} \sum_{p \in V_x(j)} \frac{\log p}{p} < N (\log K + 0(1))$$

so that

(10.7.33) $$N \ge \frac{(1 + o(1)) \log x}{\log K + 0(1)} > c \log x,$$

$c > 0$, for K fixed large, and $x \in D$ sufficiently large.

For each of these N values of j there is a prime $p \in V_x(j)$. Further, at least $[N/2]$ of these values of j have no two that are consecutive. Denoting these by j_v, $v = 1, \ldots ,[N/2]$, there are primes $p_{j_v} \in U'_x$ such that

$$K^{j_v-1} < p_{j_v} < K^{j_v}.$$

Then (for K fixed large enough)

$$p_1 K^{j_v-(3/2)} < \left(\frac{a + \eta}{A - \eta} \right) p_1 K^{j_v-1} < \frac{a + \eta}{A - \eta} p_1 p_{j_v} < \frac{A - \eta}{a + \eta} p_1 p_{j_v}$$

$$< \frac{A - \eta}{a + \eta} p_1 K^{j_v} < p_1 K^{j_v+(1/2)}$$

provides that the intervals $I_{p_{j_v}}$, $v = 1, \ldots , [N/2]$ are disjoint. Using (10.7.32) and (10.7.33), we obtain

$$\sum_{v=1}^{[N/2]} \sum_{q \in I_{p_{j_v}}} \frac{\log q}{q} > \frac{1}{4} \frac{a}{A} (A - a) \left[\frac{N}{2} \right] > \frac{1}{4} \frac{a}{A} (A - a) \left(\frac{c}{4} \log x \right),$$

for x large, where since $a > 0$ this lower bound is a *positive* constant times $\log x$. But from (10.7.26) we have

$$\sum_{q \in S_x \cap T_x} \frac{\log q}{q} \sim \log x \sim \sum_{q \le x} \frac{\log q}{q},$$

so that the above estimate, together with the fact that

$$\left(\bigcup I_{p_{j_v}}\right) \cap S_x \cap T_x = \emptyset,$$

provides a contradiction. With this contradiction a proof of the Prime Number Theorem is completed.

EXERCISES

1. Carry out the details of the proof of the part of Lemma 10.7.1 that asserts that the hypothesis (10.7.10) implies (10.7.11).

2. Following the method of this section, deduce the Prime Number Theorem for Arithmetic Progressions from the Symmetry Formula for the functions $\theta(x, A, B)$ given in Exercise 7 of §10.4.

3. Consider any function $f(x)$ that for $x \geq 1$ is nonnegative, nondecreasing, has $\underline{\lim} f(x)/x > 0$, and satisfies

$$f(x) \log x + \sum_{p \leq x} f\left(\frac{x}{p}\right) \log p = 2 x \log x + 0(x).$$

Prove this implies that as $x \rightarrow \infty$, $f(x) \sim x$.

4. Using the assertion of Exercise 3 above, deduce from the formula of Exercise 5 of §10.4, that $U(x) = o(x)$.

NOTES

§10.1. The history of the Prime Number Theorem really begins with Gauss. In a letter to Encke on Christmas Eve, 1849 [10.1], Gauss mentioned that he had pondered the problem of expressing $\pi(x)$ as an elementary function of x, and had determined when he was a "boy" that $\pi(x)$ should be approximated by $\int dn/\log n$.

Meanwhile, Legendre had given the empirical formula

$$\pi(x) = \frac{x}{\log x - 1.08366}$$

[10.2]. Gauss noted in his letter that Legendre's formula was of the form

$$\pi(x) = \frac{x}{\log x - A(x)}$$

and he "could not say there is a right to expect a wholly simple limit [for $(A(x))$]; on the other hand the excess of A over 1 could be wholly reasonably a quantity of order $1/\log n$."

Analyzing the situation, E. Landau [10.3] indicates that Gauss believed that

$$\lim_{x \to \infty} \frac{A(x)}{\log x} = 0,$$

and that Gauss also presumed that $A(x)$ tended to a limit. He was simply unwilling to commit himself to a specific value for this limit. In any event, Gauss (and Legendre) did not prove any of these statements.

In 1851, Chebychev showed that Legendre's formula for $\pi(x)$ was incorrect. He proved [9.1] that *if* it existed

$$\lim_{x \to \infty} \left(\frac{x}{\pi(x)} - \log x \right) = 1.$$

This, in turn, would necessitate that $A(x)$ tended to 1, and hence

$$\pi(x) = \frac{x}{\log x - 1}$$

would be the least inaccurate formula of the Legendre type.

Chebychev was unable to prove that the above limit actually existed. In his investigations he used the Riemann Zeta function defined as the analytic continuation of

$$\zeta(z) = \sum_{n=1}^{\infty} \frac{1}{n^z},$$

defined for complex z with real part greater than one. Chebychev utilized this function only for real values of z.

Riemann, 1959 [10.4], indicated how information concerning the complex zeros of the zeta function could lead to a proof of the Prime Number Theorem.

In 1896 [10.5], J. Hadamard succeeded in proving that $\zeta(z)$ does not vanish for all complex z with real part equal 1, and this resulted in a proof of the Prime Number Theorem. Simultaneously and independently, C. J. de la Vallee Poussin also obtained a proof [10.6]. Further, he later [10.7] improved the result to

$$\pi(x) = \frac{x}{\log x} + 0\left(\frac{x}{\log^2 x} \right).$$

The Prime Number Theorem was subsequently reproved and improved by others. However, a persistent residual question was as to whether a completely "elementary" proof existed. Such a proof was discovered in 1949 by P. Erdos and A. Selberg [10.8, 10.9].

Euler, in 1737 [10.10] and again in 1748 [10.11], stated (10.1.14) for which he gave the invalid proof noted in (10.1.6). In 1884, J. P. Gram [10.12] proved the weaker statement (10.2.1). The proof of (10.1.14) was given in 1897 by von Mangoldt, who applied one of Hadamard's theorems [10.13]. Von Mangoldt also proved (10.1.15). Both of these results were derived from the Prime Number Theorem by Landau in his doctoral dissertation, 1899, by completely elementary methods [10.14]. Later, in 1908 [10.15], Landau proved that

$$\sum_{n \le x} \mu(n) = 0(xe^{-\sqrt[6]{\log x}}).$$

The Prime Number Theorem for arithmetic progressions (10.1.19), was first proved by de la Vallee Poussin in 1896. In 1903 Landau [10.16] obtained

$$\pi(x, A, B) = \frac{1}{\varphi(A)} \int_2^x \frac{du}{\log u} + 0(xe^{-\sqrt[6]{\log x}}),$$

where γ is a positive constant depending on A.

§10.3. The presentation of this section is substantially that of Landau's doctoral dissertation [10.14]. In 1899 [10.7], de la Vallee Poussin proved that

$$\sum_{n \le x} \frac{\mu(n)}{n} = 0\left(\frac{1}{\log x}\right).$$

In 1908 he proved that

$$\sum_{n \le x} \frac{\mu(n)}{n} = 0(e^{-\sqrt[\gamma]{\log x}})$$

for $\gamma > 2$ [10.17].

Landau proved in 1901 that

$$\sum_{n \le x} \frac{\mu(n) \log n}{n} = -1 + 0(e^{-c\sqrt{\log \log x}})$$

for a specific constant c [10.17]. He later improved the error term to $0(e^{-\sqrt[\gamma]{\log x}})$ for $\gamma > 2$ [10.15].

§10.4. Theorem 10.5.1 is the first indication of the "independent power" of the Symmetry Formula. In spite of its simplicity, (10.5.1) does not seem to be a consequence of any other elementary methods. Also, the Symmetry Formula is perhaps the simplest route to (10.5.10), which was given by Selberg [10.8].

§10.6. The proof here follows the method of Selberg, given in 1949 [10.8].

§10.7. Erdos' proof appeared in [10.9].

Pursuing these elementary methods, various improvements were obtained. The assertion

$$\psi(x) = x + 0\left(\frac{x}{\log^m x}\right)$$

was obtained by such methods, as follows:

P. Kuhn	$m = (1/10)$	1955 [10.18]
B. Van der Corput	$m = (1/200)$	1956 [10.19]
R. Breusch	$m = (1/6)$	1969 [10.20]
E. Wirsing	$m = (3/4)$	1962 [10.21]

In 1962, E. Wirsing announced that he had obtained this result for all positive m (by elementary methods). In that year, 1962, E. Bombieri published a proof of this [10.22]. A proof by Wirsing appeared in 1964 [10.23]. A proof was also given in 1963 by A. Dusembetov [10.24].

REFERENCES

10.1. C. F. Gauss, *Gauss an Enke, Werke*, Göttingen: Königlichen Gesellschaft der Wissenschaften, II, 1849, pp. 444–447.

10.2. A. M. Legendre, Ref. 6.13, II, p. 65.

10.3. E. Landau, Ref. 6.30.

10.4. B. Riemann,Uber die Anzahl der Primzahlen unter einer gegebenen Grosse, *Monatsberichte der königlichen Preussischen Akademie der Wissenschaften zu Berlin*, 1859, 671–680.

10.5. J. Hadamard, Sur la distribution des zéros de la fonction $\zeta(s)$ et ses conséquences arithmétiques, *Bulletin de la Société Mathématique de France*, **24**, 1896, 199–220.

10.6. C. J. de la Valle Poussin, Recherches analytiques sur la thëorie des nombres, (3 parts), *Annales de la Société Scientifique de Bruxelles*, **20**, Part II, 1896, 183–256, 281–397.

10.7. C. J. de la Vallee Poussin, Sur la fonction $\zeta(s)$ de Riemann et le nombre des nombres premiers inférieures a une limite donnée, *Memoires couronnes et outre memoires,* Academie Royal des Sciences des Lettres et des Beaux-Arts de Belgique, **59,** 1899–1900.

10.8. A. Selberg, An elementary proof of the prime number theorem, *Annals of Mathematics,* **50,** 1949, 305–313.

10.9. P. Erdos, On a new method in elementary number theory which leads to an elementary proof of the prime number theorem, *Proceedings of the National Academy of Sciences,* **35,** 1949, 374–384.

10.10. L. Euler, Ref. 2.16, pp. 241–242, Theorem 18.

10.11. L. Euler, Ref. 2.17, pp. 227.

10.12. J. P. Gram, Undersøgelser angaaende Maengden af Primtal under er given Graense, *Det Kongelige Danske Videnskabeones Selskabs Skrifter naturvidenskabelig og Matematik Afdeling, Ser. 6,* **2,** 1881–1886, 183–308.

10.13. H. von Mangoldt, Beweis der Gleichung $\sum_{k=1}^{\infty} \mu(nK)/k = 0$, *Sitzungberichte der Königlichen Preussischen Akademie der Wissenschaften zu Berlin,* 1897, 835–852.

10.14. E. Landau, Neuer Beweis der Gleichung $\sum_{k=1}^{\infty} \mu(k)/k = 0$, Inauguraldissertation, Berlin, 1899.

10.15. E. Landau, Beitrage zur analytischen Zahlentheorie, *Rendiconti del Circolo Matematico di Palermo,* **26,** 1908, 169–302.

10.16. E. Landau, Uber die Primzahlen einer arithmetischen Progression, *Sitzungsberichte der Kaiserlichen Akademie der Wissenschaften in Wien Mat.-natur. Klasse,* **112,** (2a), 1903, 493–535.

10.17. E. Landau, Uber die asymptotischen Werthe einiger Zahlentheoretischer Functionen, *Mathematische Annalen,* **54,** 1901, 570–591.

10.18. P. Kuhn, Eine Verbesserung des Restigliedes beim elementaren Beweis des Primzahlsatzes, *Mathematica Scandinavia,* **3,** 1955, 75–89.

10.19. J. G. van der Corput, *Colloques sur la Théorie des Nombres,* Liege: G. Thone, 1956.

10.20. R. Breusch, An elementary proof of the prime number theorem with remainder term, *Pacific Journal of Mathematics,* **10,** 1960, 487–497.

10.21. E. Wirsing, Elementare Beweise des Primzahlsatzes mit Restglied, I, *Journal fur die reine ind angewandte Mathematik,* **211,** 1962, 205–214.

10.22. E. Bombieri, Maggiorazioni del resto nel Primzahlsatz col metodo di Erdos–Selberg, *Instituto Lombardo di Scienze e Lettere, Rendiconti A,* **96,** 1962, 343–350.

10.23. E. Wirsing, Elementare Beweis des Primzahlsatzes mit Restglied, II, *Journal fur die reine und angewandte Mathematik,* **214–215,** 1964, 1–18.

10.24. A. Dusumbetov, An elementary proof of the asymptotic law for the distribution of prime numbers, *Izvestiia Akademiia Nauk USSR,* (2), 1963, 24–31.

INDEX